Biopigments of Microbial Origin

Innovations and Applications Across Industries

Edited by

Ayush Madan

Department of Biotechnology, School of Research & Technology, People's University, Bhanpur, Bhopal, Madhya Pradesh, India

Debasis Mitra

Department of Microbiology, Graphic Era (Deemed to be University), Dehradun, Uttarakhand, India

Snežana Anđelković

Institute for Forage Crops, Kruševac, Globoder, Serbia

CABI

CABI is a trading name of CAB International

CABI
Nosworthy Way
Wallingford
Oxfordshire OX10 8DE
UK

Tel: +44 (0)1491 832111
E-mail: info@cabi.org
Website: www.cabi.org

CABI
200 Portland Street
Boston
MA 02114
USA

T: +1 (617)682-9015
E-mail: cabi-nao@cabi.org

A catalogue record for this book is available from the British Library, London, UK.

ISBN-13: 9781836990482 (hardback)
9781836990499 (ePDF)
9781836990505 (ePub)

DOI: 10.1079/9781836990505.0000

Commissioning Editor: Ward Cooper
Editorial Assistant: Emma McCann
Production Editor: Helen Elliott

Typeset by Exeter Premedia Services Pvt Ltd, Chennai, India
Printed and bound by CPI Group (UK) Ltd, Croydon, CR0 4YY

Contents

Preface v

Editor Biographies vi

Contributors viii

1 **Introduction to Biopigments: Definition and Importance** 1
Riffat Maqsood, Muhammad Qasim, Muhammad Usman and Saqib Umer

2 **Historical Perspective on the Use of Microbial Pigments** 21
Manjoo Rani, Saty Dev and Ayush Madan

3 **Diversity of Microbial Pigments: Types and Sources** 35
Pritty Nayak, Uma Thakur, Sonam Yadav and Krishan Kumar

4 **Biosynthesis Pathways of Microbial Pigments** 49
Ramesh Koduru, Somedutta Maity, Arifullah Mohammed, Mohd Tariq and Pamuru Ramachandra Reddy

5 **Genetic and Environmental Factors Influencing Pigment Production** 60
Sony Singh, Manjoo Rani, Saurabh Ghildiyal and Ayush Madan

6 **Microbial Pigments in Nature: Roles and Functions** 76
Gundala Prasada Babu, Arifullah Mohammed, Mohd Tariq, Mukul M. Barwant and Chinthala Paramageetham

7 **Biotechnological Production of Microbial Pigments** 90
Rahul Kumar, Sourav Chattaraj, Rachan Karmakar, Chaithra M., Tanja Vasić, Beatriz Elena Guerra Sierra and Debasis Mitra

8 **Isolation and Purification Techniques for Microbial Pigments** 109
Mary Malilah, Uma Thakur, Mohammedfaizan Shaikh and Krishan Kumar

9 **Structural Characterization and Analysis of Microbial Pigments** 136
Rashid Abbas Khan, Tauqeer Ahmed Qadri, Amjid Khan and Muhammad Anas

10 **Biopigments as Natural Colorants in the Food Industry** 156
Abdullah Ahmed Butt

11 Applications of Microbial Pigments in Pharmaceuticals and Cosmetics 184
Jatin A. Tekawade, Mayuresh R. Redkar, Vinod R. Biradar, Pranita K. Pawar and Sachinkumar V. Patil

12 Microbial Pigments in Textile and Dyeing Industries 205
Mridul Umesh, Nilina James, Liya Merin Stanly, Sreehari Suresh and Adhithya Sankar Santhosh

13 Antioxidant and Antimicrobial Properties of Microbial Pigments 228
Hemanta Kumar Mondal, Mohd Tariq, Apeksha Sawant, Ramandeep Saini and Ayush Madan

14 Biopigments in Environmental and Agricultural Applications 246
Jinesh Kaneriya, Vivek Pattani, Krishna Joshi and Gaurav Sanghvi

15 Economic and Commercial Viability of Microbial Pigments 268
Rashid Abbas Khan, Tauqeer Ahmed Qadri, Amjid Khan, Muhammad Anas and Divya Jain

16 Regulatory and Safety Aspects of Using Microbial Pigments 283
Pavithra Sivakumar, Dianambika Palaniswamy, Devayani Sarmah and R.M. Vijaya Ramakrishnan

17 Case Studies and Success Stories of Microbial Pigment Applications 301
Rahul Kumar, Ayush Madan and Debasis Mitra

18 Challenges and Future Prospects in the Field of Microbial Pigments 320
Avani Joshi, Dishank Purandare, Adityaraj Bhope, Shradha Chaudhari, Arpana Patil and Satish Polshettiwar

19 Sustainability and Environmental Impact of Biopigment Production 336
Priyanka and Ayush Madan

20 Emerging Technologies and Innovations in Biopigment Research 364
Gundala Prasada Babu, Arifullah Mohammed, Ee Lee Gaik, Mukul Jain and Chinthala Paramageetham

Index 379

Preface

The vibrant world of microbial pigments has captured the imagination of scientists, industry leaders, and environmentalists alike, offering a sustainable and innovative alternative to synthetic dyes and pigments. *Biopigments of Microbial Origin: Innovations and Applications Across Industries* is a comprehensive exploration of these remarkable natural compounds produced by microorganisms such as bacteria, fungi, and algae. As global industries increasingly prioritize eco-friendly solutions to address environmental and health concerns, microbial pigments have emerged as a promising frontier, blending cutting-edge science with practical applications.

This book is designed as a definitive resource for understanding the science, production, and multifaceted applications of biopigments. It explores their biosynthesis pathways, the genetic and environmental factors shaping their production, and the advanced techniques used for their isolation, purification, and characterization. Beyond the laboratory, the book highlights the transformative potential of microbial pigments across diverse sectors, including food, pharmaceuticals, cosmetics, textiles, and agriculture. Their antioxidant and antimicrobial properties, economic viability, and regulatory considerations are thoroughly examined, providing a holistic view of their role in fostering sustainable innovation.

Through detailed case studies and success stories, this work showcases real-world examples of how microbial pigments are revolutionizing industries while reducing environmental footprints. It also addresses the challenges and future prospects of biopigment research, emphasizing the need for continued innovation to unlock their full potential. This book is intended for researchers, industry professionals, policymakers, and students who are passionate about harnessing green technologies to create a more sustainable future.

We hope this edited volume inspires further exploration and collaboration in the field of microbial pigments, driving advancements that align scientific discovery with environmental stewardship.

Editor Biographies

Ayush Madan

Ayush Madan is an Assistant Professor in the Department of Biotechnology, School of Research and Technology (SORT), People's University, Bhopal, India and a Graduate Research Assistant at the Institute of Tropical Aquaculture and Fisheries, Universiti Malaysia Terengganu, Malaysia, working on a Malaysian Government-funded project focused on integrated microalgae-shrimp cultivation for wastewater bioremediation and shrimp growth enhancement. He holds an MTech in Biotechnology from Graphic Era University, India, and has expertise in microbiology, molecular biology, recombinant DNA technology, and *in vitro* techniques such as plant tissue culture and animal cell culture. With prior experience as an Assistant Professor at Shobhit Institute of Engineering and Technology, Meerut, he has published research in reputed journals and presented at national and international conferences.

Debasis Mitra

Debasis Mitra, a distinguished Fellow of SASS and IOASD Society, India, serves as an Assistant Professor in the Department of Microbiology at Graphic Era (Deemed to be University), Uttarakhand, India. His research focuses on the practical application and formulation of plant growth-promoting microorganisms, advancing sustainable crop production, protection, and development in both laboratory and field settings. Dr Mitra has been honored with several prestigious awards from various scientific societies for his contributions to microbiology and biotechnology. He holds a PhD in Microbiology from Raiganj University, West Bengal, India, and ICAR-National Rice Research Institute (NRRI), Odisha, India. Dr Mitra further expanded his academic horizons through a virtual doctoral internship at the Agroenvironmental and Health Biotechnology Laboratory, Universidad de Santander, Colombia. He earned his MSc in Biotechnology from Graphic Era (Deemed to be University), India and his BSc in Biotechnology from the Oriental Institute of Science and Technology, Vidyasagar University, India. In his pursuit of continuous learning, Dr Mitra has acquired online course certificates in Microbiology and Forensic Science from Jordan University of Science and Technology, Jordan; Climate Change from Macquarie University, Australia; and Agriculture and the World We Live In from Massey University, New Zealand, all completed in 2018. Dr Mitra has an extensive publication record, including books, book chapters, research and review papers, popular articles, research/technical bulletins, e-publications, conference proceedings, and training materials. He has developed innovative technologies of "Cocopeat with rice husk and sawdust compost as a nursery medium for growing vegetable seedlings" at ICAR- Indian Institute of Horticultural Research, India, and was involved in developing "Microbial consortium (MC) for pest and disease management in rice-horticulture based cropping system in Sikkim" and "Bio-fertilizer consortium (BC) for nutrient management in rice-horticulture based cropping system in Sikkim" at ICAR-National Rice Research Institute, India. He serves as an editor, editorial board member, advisory board member, special issue editor, guest editor, and reviewer for esteemed national and international journals. Additionally, he actively supervises Bachelor's, Master's and PhD students, fostering cutting-edge academic research in microbiology and biotechnology.

Snežana Anđelković

Snežana Andjelković has served as a senior researcher at the Institute for Forage Crops Kruševac, Republic of Serbia since 2007. Before that, she worked in the biochemical laboratory of the Health Center in Kruševac as a graduate biologist from 2003 to 2005. She graduated from the Faculty of Science, Department of Biology, University of Pristina in 2001. She obtained her Master's degree from the Faculty of Science, University of Priština in 2007, and her PhD from the Faculty of Agriculture, University of Novi Sad, in 2013. She has several research publications in journals of international repute.

Contributors

Muhammad Anas, Department of Soil Science & Plant Nutrition, Selçuk University, Konya, Türkiye

Gundala Prasada Babu, Department of Microbiology, Sri Venkateswara University, Tirupati, Andhra Pradesh, India

Mukul M. Barwant, Department of Botany, Sanjivani Arts, Commerce and Science College Kopargoan, Ahmednagar, Maharashtra, India

Adityaraj Bhope, School of Health Sciences and Technology, Dr Vishwanath Karad MIT World Peace University, Pune, Maharashtra, India

Vinod R. Biradar, Department of Pharmaceutical Chemistry, Yashwantrao Bhonsale College of Pharmacy, Sawantwadi, Maharashtra, India

Abdullah Ahmed Butt, Department of Food Science and Technology, Government College University, Faisalabad, Pakistan

Sourav Chattaraj, Centre for Industrial Biotechnology Research, School of Pharmaceutical Science, Siksha 'O' Anusandhan University, Kalinga Nagar, Bhubaneswar, Odisha, India

Shradha Chaudhari, Department of Biosciences and Technology, School of Sciences and Environmental Studies, Faculty of Sciences and Health Sciences, Dr Vishwanath Karad MIT World Peace University, Pune, Maharashtra, India

Saty Dev, Department of Mechanical Engineering, Shri Ram Murti Smarak College of Engineering and Technology, Bareilly, Uttar Pradesh, India

Ee Lee Gaik, Institute of Tropical Biodiversity and Sustainable Development, Universiti Malaysia Terengganu, Kuala Nerus, Malaysia

Saurabh Ghildiyal, Department of Hotel Management, Shri Ram Murti Smarak College of Engineering, Technology and Research, Bareilly, Uttar Pradesh, India

Divya Jain, Department of Microbiology, School of Applied and Life Sciences, Uttaranchal University, Dehradun, Uttarakhand, India

Mukul Jain, Department of Life Sciences, Parul Institute of Applied Sciences, Parul University, Vadodara, Gujarat, India

Nilina James, Department of Botany, Teresian College, Mysore, Karnataka, India

Avani Joshi, School of Health Sciences and Technology, Dr Vishwanath Karad MIT World Peace University, Pune, Maharashtra, India

Krishna Joshi, Department of Microbiology, Atmiya University, Rajkot, Gujarat, India

Jinesh Kaneriya, Department of Microbiology, Atmiya University, Rajkot, Gujarat, India

Rachan Karmakar, Department of Environmental Science, Graphic Era Deemed to be University, Dehradun, Uttarakhand, India

Amjid Khan, UNESCO UNISA Africa Chair in Nanoscience and Nanotechnology, College of Graduate Studies, University of South Africa, Pretoria, South Africa

Rashid Abbas Khan, Department of Biology, Faculty of Science, Prince of Songkhla University, Hat Yai, Songkhla, Thailand

Ramesh Koduru, Department of Biochemistry, Yogi Vemana University, Vemanapuram, Kadapa, Andhra Pradesh, India

Krishan Kumar, Department of Life Sciences, Parul Institute of Applied Sciences, Parul University, Vadodara, Gujarat, India

Rahul Kumar, Department of Microbiology, Graphic Era Deemed to be University, Dehradun, Uttarakhand, India

Chaithra M., ICAR-Central Plantation Crop Research Institute, Research Center, Kahikuchi, Guwahati, Assam, India

Somedutta Maity, School of Engineering Sciences and Technology, University of Hyderabad, Hyderabad, Telangana, India

Mary Malilah, Department of Life Sciences, Parul Institute of Applied Sciences, Parul University, Vadodara, Gujarat, India

Riffat Maqsood, Faculty of Veterinary Sciences, University of Agriculture, Faisalabad, Pakistan

Arifullah Mohammed, Department of Agriculture, Koneru Lakshmaiah Education Foundation (KLEF), Vaddeswaram Campus, Guntur, Andhra Pradesh, India

Hemanta Kumar Mondal, Department of Life Sciences, Parul Institute of Applied Sciences, Parul University, Vadodara, Gujarat, India

Pritty Nayak, Department of Life Sciences, Parul Institute of Applied Sciences, Parul University, Vadodara, Gujarat, India

Dianambika Palaniswamy, Department of Plant Pathology, Tamil Nadu Agricultural University, Coimbatore, Tamil Nadu, India

Chinthala Paramageetham, Department of Microbiology, Sri Venkateswara University, Tirupati, Andhra Pradesh, India

Vivek Pattani, Department of Microbiology, Atmiya University, Rajkot, Gujarat, India

Arpana Patil, STES's Smt. Kashibai Navale College of Pharmacy, Kondhwa, Pune, Maharashtra, India

Sachinkumar V. Patil, Department of Pharmaceutics, Dr Ashok Gujar Institute of Pharmacy, Karad, Maharashtra, India

Pranita K. Pawar, Department of Pharmaceutics, Yashwantrao Bhonsale College of Pharmacy, Sawantwadi, Maharashtra, India

Satish Polshettiwar, School of Health Sciences and Technology, Dr Vishwanath Karad MIT World Peace University, Pune, Maharashtra, India

Dishank Purandare, School of Health Sciences and Technology, Dr Vishwanath Karad MIT World Peace University, Pune, Maharashtra, India

Tauqeer Ahmed Qadri, Department of Biosciences, COMSATS University Islamabad, Islamabad, Pakistan

Muhammad Qasim, Department of Theriogenology, University of Agriculture, Faisalabad, Pakistan

R.M. Vijaya Ramakrishnan, Department of Plant Pathology, Tamil Nadu Agricultural University, Coimbatore, Tamil Nadu, India

Manjoo Rani, Department of Pharmacy, Shri Ram Murti Smarak College of Engineering and Technology, Bareilly, Uttar Pradesh, India

Pamuru Ramachandra Reddy, Department of Biochemistry, Yogi Vemana University, Vemanapuram, Kadapa, Andhra Pradesh, India

Mayuresh R. Redkar, Department of Pharmaceutics, Yashwantrao Bhonsale College of Pharmacy, Sawantwadi, Maharashtra, India; Rajarambapu College of Pharmacy, Kasegaon, Maharashtra, India

Priyanka, Graduate Institute of Biomedical Sciences, China Medical University Hospital, Taichung City, Taiwan

Gaurav Sanghvi, Department of Microbiology, Marwadi University, Rajkot, Gujarat, India

Ramandeep Saini, Department of Biotechnology, Chandigarh College of Technology, Chandigarh Group of Colleges, Landran (140307), Greater Mohali, Punjab, India

Adhithya Sankar Santhosh, Department of Life Sciences, Christ University, Bangalore, Karnataka, India

Devayani Sarmah, Division of Plant Pathology, Indian Agricultural Research Institute, Central Rice Research Institute, Cuttack, Odisha, India

Apeksha Sawant, Department of Life Sciences, Parul Institute of Applied Sciences, Parul University, Vadodara, Gujarat, India

Mohammedfaizan Shaikh, Department of Life Sciences, Parul Institute of Applied Sciences, Parul University, Vadodara, Gujarat, India

Beatriz Elena Guerra Sierra, Facultad de Ciencias Naturales y Agropecuarias, Grupo de Investigación Biotecnología Agroambiente y Salud-Microbiota-Universidad de Santander, Bucaramanga, Colombia

Sony Singh, Department of Pharmacy, Shri Ram Murti Smarak College of Engineering and Technologies, Bareilly, Uttar Pradesh, India

Pavithra Sivakumar, Department of Plant Pathology, Tamil Nadu Agricultural University, Coimbatore, Tamil Nadu, India

Liya Merin Stanly, Department of Life Sciences, Christ University, Bangalore, Karnataka, India

Sreehari Suresh, Centre for Global Health Research, Saveetha Medical College and Hospital, Saveetha Institute of Medical and Technical Sciences (SIMATS), Chennai, Thandalam, India

Mohd Tariq, Department of Academics, Sumandeep Vidyapeeth, Deemed to be University, Vadodara, Gujarat, India; Department of Biotechnology, Graphic Era Deemed to be University, Dehradun, Uttarakhand, India

Jatin A. Tekawade, Department of Pharmaceutics, Yashwantrao Bhonsale College of Pharmacy, Sawantwadi, Maharashtra, India

Uma Thakur, Department of Life Sciences, Parul Institute of Applied Sciences, Parul University, Vadodara, Gujarat, India

Saqib Umer, Department of Theriogenology, University of Agriculture, Faisalabad, Pakistan

Mridul Umesh, Department of Life Sciences, Christ University, Bangalore, Karnataka, India

Muhammad Usman, Faculty of Veterinary Sciences, University of Agriculture, Faisalabad, Pakistan

Tanja Vasić, Faculty of Agriculture, University of Niš, Kruševac, Serbia

Sonam Yadav, Department of Life Sciences, Parul Institute of Applied Sciences, Parul University, Vadodara, Gujarat, India

1 Introduction to Biopigments: Definition and Importance

Riffat Maqsood[1], Muhammad Qasim[2], Muhammad Usman[1] and Saqib Umer[2]*
[1]*Faculty of Veterinary Sciences, University of Agriculture, Faisalabad, Pakistan;*
[2]*Department of Theriogenology, University of Agriculture, Faisalabad, Pakistan*

1.1 Introduction

Different types of biopigments will be discussed along with their sources (Fig. 1.1), uses, and synthesis. They have great industrial, economic, and cultural importance including for pharmaceuticals, environmental impact, and health. In this chapter we will also discuss the challenges faced in the biopigment industry and strategies to improve and resolve these challenges. Challenges include supply change, production and processing challenges, regulations, market and competition challenges. Improvement methods include fermentation, extraction, purification, customization, formulation, recycling, environmental impact assessment, and sustainability. The association of biopigments with animals, microorganisms, and plants is illustrated in Fig. 1.2.

1.1.1 Pigments

Pigment originates from *pigmentum*, a Latin word which means coloring substance.

Biological colors are also known as biochromes or pigments. They are created by living organisms bearing color as a consequence of absorption of a certain color. A wide range of flower and plant pigments are included in biological pigments. For instance, butterfly wings express color, as they contain cells possessing pigments (Stavenga *et al.*, 2014).

1.1.2 Classification

Biopigments are classified on varying bases, such as on the sources from which they originate. Their origins can be categorized as:

- plants
- animals
- microorganisms:
  a. bacterial
  b. fungal.

This is represented in Fig. 1.3.

1.2 Biopigments of Plant Origin

The pigments obtained from plants or those giving color to plant cells and tissues can be extracted. Plant pigments include chlorophyll and other pigments for light absorbance. These

*Corresponding author: saqib.umer@uaf.edu.pk

 Biopigments of Microbial Origin: Innovations and Applications Across Industries (eds Madan *et al.*)
DOI: 10.1079/ 9781836990505.0001

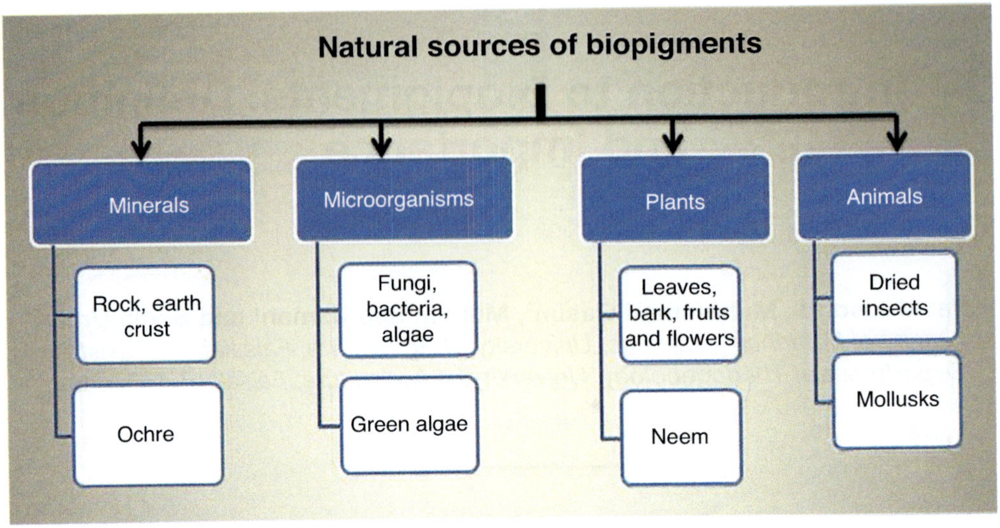

Fig. 1.1. Natural sources of biopigments.

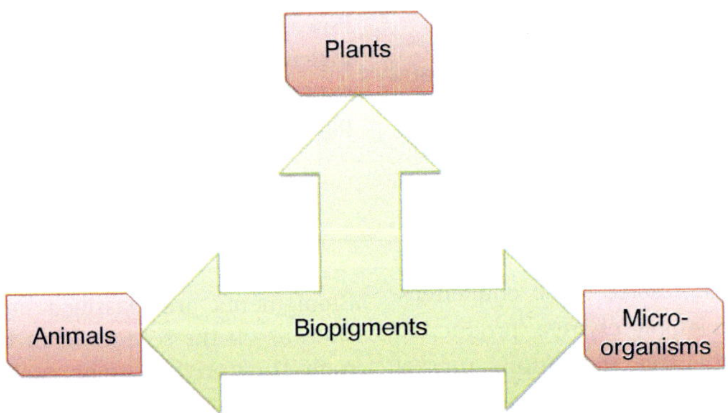

Fig. 1.2. Association of biopigments with animals, microorganisms, and plants.

are of great importance for photosynthesis in plants (Grotewold, 2006; Lee, 2019).

Additionally, pigments are involved in pollination; when pigments assemble or disappear, the color of the flower changes, alerting pollinators to which flowers are rich in pollen and nectar (Weiss, 1991).

1.2.1 Chlorophyll

The Greek words *khloros* (green) and *phyllon* (leaves) are the original sources of the English word "chlorophyll," a green pigment in plants which imparts color and helps in photosynthesis by acting as a photoreceptor. The French chemists Joseph Bienaimé Caventou and Pierre-Joseph Pelletier isolated and named chlorophyll in 1817. In 1906, scientists discovered that chlorophyll contains magnesium. German chemist Richard Willstätter provides a better approach to understand the accurate chemical composition of chlorophyll by determining the correct molecular formula for chlorophyll, $C_{55}H_{70}MgN_4O_6$ (Willstätter, 1906).

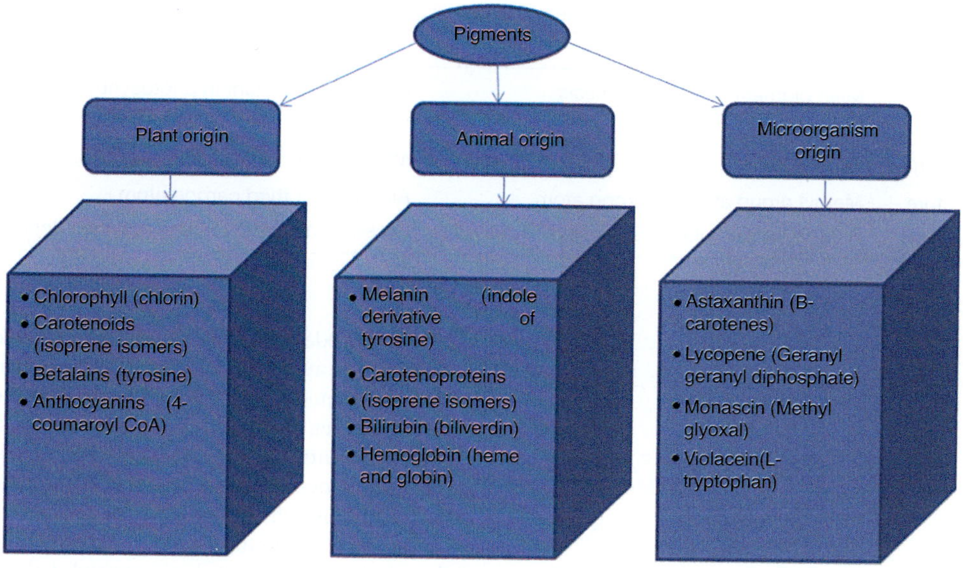

Fig. 1.3. Classification of pigments.

Basically, chlorin absorbs blue and red light wavelengths and reflects a majority of green color in plants and vegetables (Cömert *et al.*, 2020). Chlorophyll a and chlorophyll b are usually present in all terrestrial plants and green algae (Table 1.1). Chlorophyll a is present in red algae, whereas kelps and diatoms exhibit chlorophyll c substituted for chlorophyll b. Numerous artificial chlorins with altered ring structures and/or functional groups have been studied (Taniguchi and Lindsey, 2017).

The oxidation of the appropriate B (III) subbacteriochlorin or reduction of B (III) subporphyrin can result in production of contracted chlorins (Tsurumaki *et al.*, 2008).

1.2.1.1 Structure

The chemical formula of chlorophyll is $C_{55}H_{72}O_5N_4Mg$ (Fig. 1.4). Porphyrin is an offshoot of chlorophyll. A magnesium atom is an integral component of chlorophyll's fundamental chemical structure. Additionally, four nitrogen atoms surround the magnesium. There is a pyrrole ring structure with four carbons and nitrogen. A hydrocarbon end is also present. The porphyrin head is a unitary term for the four pyrrole rings containing magnesium and nitrogen. The hydrocarbon end in the hydrophobic phytol tail plays crucial role in anchoring chlorophyll within thylakoid membrane (Björkman *et al.*, 1981; Niedzwiedzki and Blankenship, 2010). Hemoglobin, the pigment that transports oxygen in mammalian red blood cells, is nearly identical to chlorophyll.

1.2.1.2 Uses

Chlorophyll has been claimed to reduce tumors and increase antioxidant activity. It is utilized as a colorant in food, medicines, and beauty products.

- It helps in photosynthesis. Green light from the electromagnetic spectrum is reflected by chlorophyll while red and blue light are absorbed. More chlorophyll allows leaves to absorb more light for photosynthesis.
- Chlorophyllin is a semisynthetic and water-soluble derivative of chlorophyll, commonly used as an additive in medicines or food colorings. It also treats wounds. Many supplements are available in this form.
- Chlorophyll reduces the inception of cancerous tumors by blocking absorbance of aflatoxins (de Vogel *et al.*, 2005).

Table 1.1. Some common differences between chlorophyll a and chlorophyll b.

	Chlorophyll a	Chlorophyll b
Role	Main pigment in photosynthesis	Accessory pigment in photosynthesis
Color	Blue-green	Yellow-green
Absorption	430–660 nm	450–650 nm
Structure	Methyl group at third carbon atom	Aldehyde at third carbon atom
Source	Photosynthetic plants, algae, cyanobacteria	Green algae, plants, some cyanobacteria

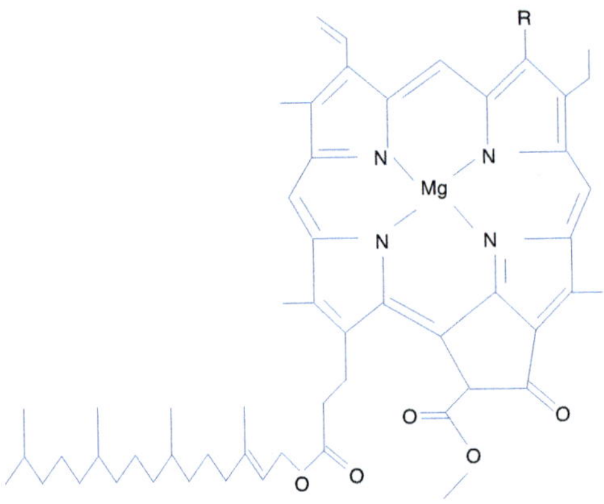

Fig. 1.4. General structure of chlorophyll a and chlorophyll b. Source: Mandal and Dutta (2020). R = CH3 (Chlorophyll a), R = CHO (Chlorophyll b)

- Some supplement makers claim that chlorophyll can help with weight loss, boost red blood cells, and neutralize toxins because of its antioxidant nature.

1.2.2 Carotenoids

Carotenoids are tetraterpenoids that are either red, yellow or orange. In photosynthesis, they act as auxiliary pigments and help in capturing light. Additionally, in order to avoid photooxidative damage, they help in oxygen scavenging. They work as precursors of the plant hormone abscisic acid. They absorb light and transmit that energy to the chlorophyll in the photosystem. Furthermore, they serve as antioxidants (McElroy and Kopsell, 2009; Falkowski and Raven, 2013).

Wackenroder discovered carotene while searching for an anthelminthic (Wackenroder, 1831). He diluted carrot juice and extracted the liquid with ether. The liquid evaporated, leaving a yellow fatty oil and carotene. In the 1930s, Karrer and Kuhn discovered the structure of β-carotene (precursor of vitamin A) and lycopene.

The two isoprene isomers dimethylallyl diphosphate (DMAPP) and isopentenyl diphosphate (IPP) are the sources of carotenoids. IPP and DMAPP from the methylerythritol phosphate (MEP) pathway are commonly used for carotenoid production in plants.

1.2.2.1 Structure

β-Carotene is a typical carotenoid, composed of two carbon rings linked by an 18-carbon chain with switching single and double bonds (Fig. 1.5). Its chemical formula is $C_{40}H_{56}$. It is an isoprenoid compound, formed via linking two C_{20} molecules

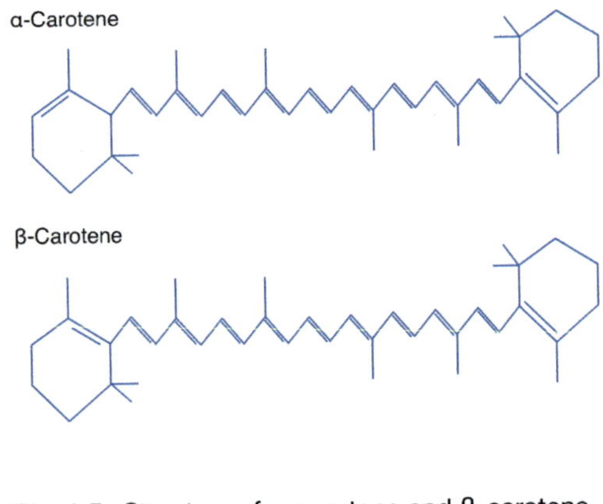

Fig. 1.5. Structure of α-carotene and β-carotene. Source: Latowski *et al.* (2014).

tail to tail to make a structure of 40-carbon isoprene molecules. Carotenoids are lipophilic because of the existence of long unsaturated aliphatic chains. The chemical formula of carotenoids is $C_{40}H_{56}O_2$. Some carotenoids contain polar hydroxyl and keto functional groups, leading to enhanced attraction for lipid/water interactions. Thus, the antioxidant activity depends on these molecules' interaction in respect to the water/lipid interface (Lehninger *et al.*, 2005).

1.2.2.2 Uses

Carotenoids are of great importance because of their photoprotective activity through nonphotochemical quenching, anticarcinogenic and antioxidative activity.

- Carotenoids transfer the energy they absorb to chlorophylls, and thus contribute to photosynthesis. They scavenge reactive oxygen species, and protect cell membranes and proteins from damage. They regulate plant development and fruit ripening.
- α-Carotene is well known for its anticarcinogenic activity (Nishino *et al.*, 2000, 2002).
- Zeaxanthin and lutein are acknowledged as antioxidants.
- β-Carotene provides a significant amount of vitamin A (Naves and Moreno, 1998).
- Carotenoids offer protection against diseases including cancer, heart disease, and cataracts (Davies, 1995).
- They can trigger the expression of genes that carry the instructions needed to produce a protein, connexin 43 (Novo *et al.*, 2013).
- Carotenoids, solely or in combination with other antioxidants, can enhance the efficiency of lipid peroxidation (McNulty *et al.*, 2007).

1.2.3 Betalains

Betalains are red or yellow, water-soluble and nitrogen-containing pigments. Betalains were first isolated in crystalline form in 1957. Their chemical composition was identified by Dr Tom Mabry at the University of Zurich in 1960. They are synthesized into two fundamental groups: yellow-orange betaxanthins and red-violet betacyanins. Chromophore is common for all betalain pigments. These pigments are found in cactus and amaranth. Betalains have a crucial role in the deep red color of beets (Lee, 2019).

The secretion of betalains in plants is stimulated by the production of L-3,4-dihydroxyphenylalanine (L-DOPA) from tyrosine in the presence of cytochrome P450 (Polturak *et al.*, 2016). Two genes, BvCYP76AD1 and BvCYP76AD5, participate in the process by which tyrosine is converted into L-DOPA. Betanin, used for synthesis of artificial coumarinic betalaine, is obtained from the red beet (Gonçalves *et al.*, 2012). Betalamic acid is produced from betanin. The combination of 7-amino-4-methylcoumarin with betalamic acid results in the production of betalain. It is used for the identification of *Plasmodium*-infected erythrocytes by live cell imaging (Gonçalves *et al.*, 2013).

1.2.3.1 Structure

Betalamic acid or 4-(2-oxoethylidene)−1, 2, 3, 4-tetrahydropyridine-2,6-dicarboxylic acid is a major constituent in betalains (Fig. 1.6). Different betacyanins are produced when betanidin undergoes modification (glycosylation and acylation) at positions 5 or 6. In the presence of CYP76AD1 (and/or tyrosinase), tyrosine is transformed into L-DOPA, and then, DOPA's activity (cleavage by DOPA 4,5-dioxygenase) and subsequent spontaneous cyclization lead to production of betalamic acid. Betacyanins are produced by the condensation between betalamic acid and cyclo-DOPA, and amines or their derivatives. Their chemical formula is $C_{24}H_{26}N_2O_{13}$ (Gandía-Herrero *et al.*, 2005).

1.2.3.2 Uses

Betalains have therapeutic uses as dietary supplements and pharmaceutical drugs by virtue of their antihepatitis, antiinflammatory, and anticancer traits. They also exhibit antimicrobial and antimalarial effects so are of great importance (Rahimi *et al.*, 2019).

- They are responsible for the dispersal of seed and have great affinity for pollinators.
- They protect plants from ultraviolet radiation, soil pathogens, and wounding.

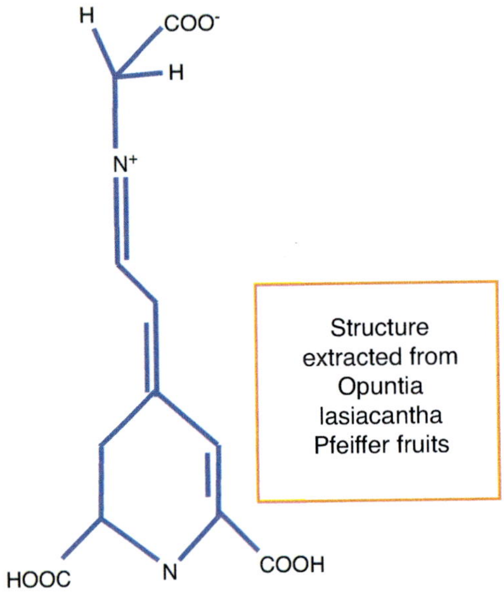

Fig. 1.6. Structure of betalains. Source: Gürses *et al.* (2024) used with permission from Elsevier.

- They help plants tolerate drought and have a role in scavenging free radicals (Borkowski *et al.*, 2005).
- They act as antioxidant molecules and protect against the injurious effects of oxidative stress. Phenolic acid and a cyclic amine that is an electron donor provide antioxidative properties (Kanner *et al.*, 2001).
- Metals, including Cu2 +and Eu2+, can be detected by betalain fluorescence following metal binding (Stintzing and Carle, 2007).
- Intravenous administration of betalains increases blood pressure and heart rate but oral administration has no effect (Krantz *et al.*, 1980).
- Betanin inhibits formation of antibodies like IgG and IgE, as its presence leads to inhibition of the allergic response (Pourrat *et al.*, 1987).

1.2.4 Anthocyanins

Anthocyanins are red to blue water-soluble flavonoid pigments, varying according to pH. Anthocyanins are derived from the Greek words *anthos* meaning flower and *kyanos* meaning dark blue. They belong to the flavonoid (flavan-like) cluster of phytochemicals, present in various products including wines, honey, vegetables, nuts, olive oil, fruits, cereals, and cocoa (Aramwit *et al.*, 2010).

The German botanist Ludwig Marquart discovered anthocyanins in 1835. Anthocyanins had been observed for centuries as 'colored cell sap' (Davies, 2004).

The main precursor of anthocyanins is 4-coumaroyl CoA, which is derived from phenylalanine. Anthocyanins are produced in the endoplasmic reticulum and stored in the vacuoles of reproductive and vegetative organs of plants (Landi *et al.*, 2015). Synthesis is controlled by structural genes in four steps. First, phenylalanine is converted to 4-coumaryl CoA via phenylalanine ammonialyase (PAL) Next, 4-coumaroyl CoA is transformed into dihydrokaempferol (DHK) in the presence of chalcone synthase (CHS). DHK can also be transformed into dihydroquercetin (DHQ). Next, dihydroflavonol 4-reductase (DFR) converts DHK and DHQ into leukocyanidin and leukopelargonidin. The formation of these leukoanthocyanidins is triggered by anthocyanidin synthase (ANS). Anthocyanidins are further modified to produce stable and water-soluble pigments (Willstätter, 1906).

1.2.4.1 Structure

The chemical formula is $C_{15}H_{11}O^{+}$. The fundamental structure consists of an aromatic ring combined with a heterocyclic ring that has an oxygen atom attached to it, as well as a third aromatic ring (Fig. 1.7). Hydroxy groups are included in these compounds to maintain the charge on the flavylium cation. When one of the phenols is replaced with glycosides, the compound is known as an anthocyanin (Balaban *et al.*, 1969).

1.2.4.2 Uses

Anthocyanins help to prevent cardiovascular diseases and oxidative stress. They play important roles as coloring agents for foods and cosmetics.

- They impart color to leaves, plant stems, roots, and petals of flowers, and fruits.

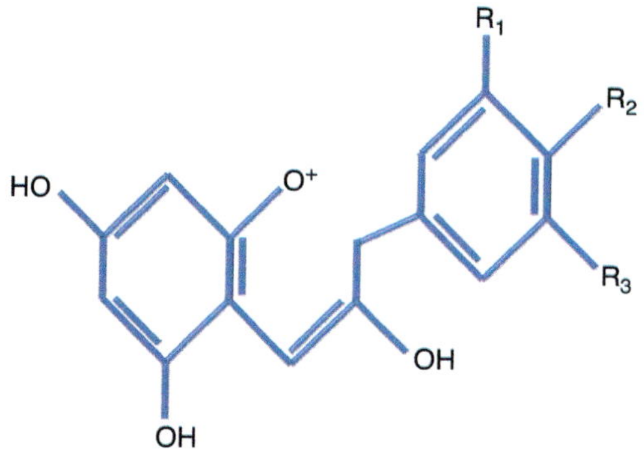

Fig. 1.7. Structure of anthocyanins. Source: Valavanidis and Vlachogianni (2013).

- They act as colorants for foods and cosmetics.
- They are odorless and astringent.
- They help in glucose metabolism, prevention against cardiovascular disease, and mediation of oxidative stress.
- The ferric reducing activity of plasma (FRAP) assay has been used to explore the antioxidant activities of synthesized anthocyanidins (Benzie and Strain, 1996).
- They are involved in maintaining brain function.
- They help to prevent cardiovascular diseases.

1.2.5 Summary

A summary of biological pigments from plants is given in Table 1.2.

1.3 Biopigments of Animal Origin

Animals use pigments for protection via mimicry or warning coloration. They also act as cues for courting and sexual behavior. As an illustration, cephalopods communicate via chromatophores. Rhodopsin intercepts light. Melanin protects the body from ultraviolet radiations. Carotenoids are also produced in animals such as spider mites, gall midges, and pea aphids. Arthropods are able to synthesize this pigment by genetic activity, via fungal horizontal gene transfer (HGT) processes (Cobbs *et al.*, 2013).

There are many abnormal conditions caused by the presence, absence or surplus synthesis of pigments. Albinism is a genetic disorder caused by partial or complete absence of melanin. Lamellar ichthyosis, named "fish scale disease," is due to the overproduction of melanin. A condition known as vitiligo occurs when melanocytes, which produce pigment, are lost.

1.3.1 Melanin

The term 'melanin' originated from an ancient Greek word *mélas* which means "black, dark." In the 1840s the Swedish chemist Berzelius extracted a dark pigment from eye membranes and named it melanin.

Melanin is produced in specific groups of cells called melanocytes found in the basal layer of the epidermis. Tyrosine oxidation followed by polymerization is the initial step in the melanogenesis process, which produces melanin. The skin becomes darker as a consequence of this process, which is initiated by ultraviolet (UV) light exposure. Melanin is present in the eyes, medulla, brainstem, and zona reticularis of the adrenal gland (Bandaranayake, 2006).

1.3.1.1 Structure

Melanin is a heteropolymer linked by non-hydrolyzable bonds. Its chemical formula is $C_{18}H_{10}N_2O_4$. Melanin is a polymer made from 5,6-dihydroxyindole, which is synthesized from tyrosine. There are five basic types derived from aerobic oxidation of phenols: eumelanin, pheomelanin, neuromelanin, allomelanin, and pyomelanin (Cao *et al.*, 2021).

1.3.1.2 Uses

- Melanin provides a defense mechanism against predators.
- It protects from stress due to UV radiations and reactive oxygen species (ROS).
- Melanin from chestnut shells is commonly used as a coloring agent and antioxidant (Yao and Qi, 2016).
- Melanin shields the DNA of mice and *Drosophila* flies from non-UV radiation damage.

Table 1.2. Summary of plant origin biological pigments.

Pigment	Discovery	Color	Origin	Chemical formula	Uses	Reference
Chlorophyll	Cavento and Pelletier	Green	Chlorin	$C_{55}H_{70}MgN_4O_6$	Photosynthesis and anticancerous	(Björkman *et al.*, 1981; Niedzwiedzki and Blankenship, 2010)
Carotenoids	Wackenroder	Red/ orange/ yellow	Isoprene Isomers	$C_{40}H_{56}O_2$	Photosynthesis, antioxidants, and expression of genes	(Novo *et al.*, 2013)
Betalains	Mabry	Red/ yellow	Tyrosine	$C_{24}H_{26}N_2O_{13}$	Antioxidants, pollination, and free radical scavenging	(Pourrat *et al.*, 1987)
Anthocyanins	Marquart	Red/blue	Coumaroyl CoA and phenylalanine	$C_{15}H_{11}O^+$	Antioxidants and food colorants	(Balaban *et al.*, 1969)

1.3.2 Carotenoproteins

Marine animals are unable to produce their own carotenoids but are dependent on plants. Carotenoproteins are the most prevalent pigments.

Dimethylallyl diphosphate and isopentenyl diphosphate are used for the synthesis of carotenoids in plants. Carotenoproteins are generated by the combination of carotenoids and proteins. They contain β-carotene, lycopene, and astaxanthin with the chemical formulae $C_{40}H_{56}$, $C_{40}H_{56}$, and $C_{40}H_{52}O_4$ respectively.

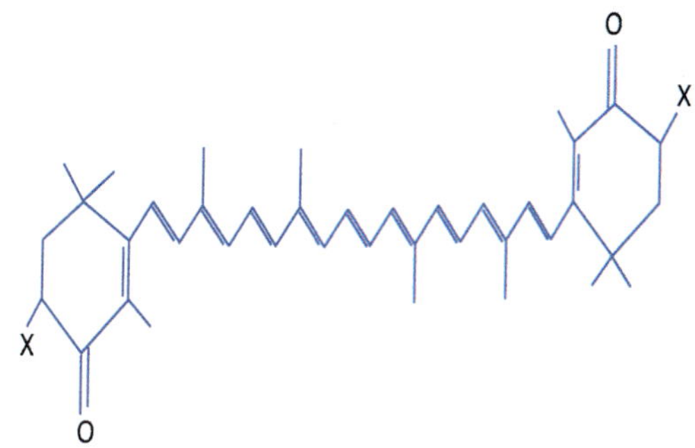

Fig. 1.8. Structure of carotenoproteins. Source: Heras *et al.* (2007).

1.3.2.1 Structure

The structure is determined by the chemical binding of protein subunits and chromogen (Fig. 1.8). This pigment is of two types: A and B. Type A is a combination of carotenoids with glycoprotein present in aquatic invertebrate skins and seashells. Type B is a combination of carotenoids with lipoproteins that is less stable. It is present in eggs, ovaries, and blood (Nadakal, 1960).

1.3.2.2 Uses

- They are precursors of vitamin A.
- They act as photoprotectors.
- They have antioxidant properties.
- They enhance immunity and are contributors to reproduction.

1.3.3 Bilirubin

The word "bilirubin" is derived from Latin meaning "red bile." It is a red-orange substance, discovered by the German scientist Rudolf Virchow in 1847, when he isolated bilirubin crystals from hematomas.

Biliverdin is the metabolic precursor of bilirubin. It is produced when a porphyrin in heme undergoes oxidative cleavage (Mujawar *et al.*, 2024). Reduction converts biliverdin into bilirubin. Bilirubin is eliminated after conjugation with glucuronic acid due to its water solubility.

1.3.3.1 Structure

Bilirubin is composed of an open-chain tetrapyrrole. Its chemical formula is $C_{33}H_{36}N_4O_6$. It is structurally similar to phycobilin and phytochrome that is utilized by plants to detect light. All have an open chain consisting of four pyrrolic rings.

1.3.3.2 Uses

- The antioxidant reaction of bilirubin is crucial in the brain because it scavenges superoxide in the neurotransmission of N-methyl-D-aspartic acid, avoiding excitotoxicity and neuronal death.
- It protects against cardiovascular disorders.
- It imparts the yellow hue to healing wounds and jaundiced skin (discoloration).
- It provides color to stool and urine.

1.3.4 Haemoglobin

In 1959, John Kendrew discovered the structure of hemoglobin and myoglobin. Hemoglobin is the red color pigment present in red blood cells (RBCs) which is responsible for transporting carbon dioxide and oxygen through blood. Heme consists of iron and a pigment called porphyrin, which also provides red color to blood. Synthesis of hemoglobin commences in proerythroblasts and continues in reticulocytes. First, succinyl-CoA, produced during the Krebs metabolic cycle, combines with glycine to generate a pyrrole molecule. Then, four pyrroles combine to make protoporphyrin IX, which subsequently joins forces with iron to make the heme molecule.

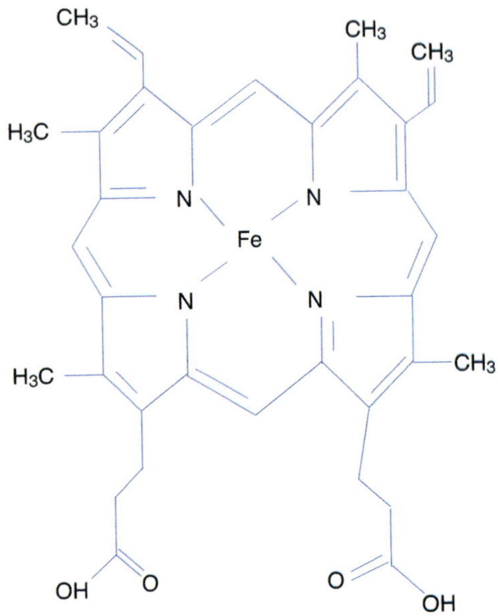

Fig. 1.9. Structure of hemoglobin.

1.3.4.1 Structure

Hemoglobin is made up of four globin proteins and four heme groups forming a tetrahedral structure (Fig. 1.9). All globin proteins are attached to a heme group composed of an iron ion and a pigment called porphyrin. It is made of two α (141 amino acids each) and twoβ (146 amino acids each) chains forming a quaternary structure. The molecular formula of hemoglobin is $C_{2952}H_{4464}N_{3248}O_{812}S_8Fe_4$ (Perutz, 1962).

1.3.4.2 Uses

- This iron-containing protein carries oxygen from tissues.
- It gives red color to red blood cells.
- Hemoglobin level testing can indicate disease such as anoxia, hypoxia, and anemia (Perutz, 1962).
- As a secondary function, hemoglobin transports carbon dioxide from the body's tissues to the lungs for exhalation.

1.3.5 Summary

A summary of biological pigments from animals is given in Table 1.3.

1.4 Biopigments of Microorganism Origin

Microorganisms, such as bacteria, fungi, algae, and actinomycetes, produce colored metabolites called microbial pigments. They include astaxanthins, lycopene, and monascin. Bacterial pigments may be soluble pigments, referred as extracellular pigments, or insoluble pigments, referred to as intracellular pigments.

Pigments including carotenoids, melanin, violacein, and zeaxanthin are produced by bacteria. For instance, cyanobacteria give rise to pigments such as chlorophyll f, a, and d. The primary coloring substance found in cyanobacteria is the blue-green pigment phycocyanin. They also produce carotenoids and cyanophycin. Bacteriochlorophyll a and b are produced by purple bacteria. Bacteriochlorophyll c and e are produced by green bacteria. Melanin is produced by *Streptomyces* (Narsing Rao *et al.*, 2017). Fusorubin is obtained from *Fusarium* (fungal origin).

1.4.1 Astaxanthin

Astaxanthin is a pigment that can be produced by both bacteria and fungi. Richard Kuhn discovered astaxanthin in 1938 in Germany. This lipid-soluble pigment can be observed in plants, yeast, and microalgae. It gives pink color to salmonids, flamingos, and crustaceans. It is also known as marine carotenoid so its precursor is β-carotene. It is a keto-carotenoid compound containing hydroxyl and ketone functional groups (Margalith, 1999). Methyl erythritol 4-phosphate (MEP) and mevalonate (MVA) pathways are responsible for the production of isopentenyl pyrophosphate (IPP). Via the action of two enzymes, ipi1 and ipi2, β-carotene is produced by isomerization, condensation, and desaturation. The production of astaxanthin is regulated by the interaction between β-carotene and cyt P450 within cytoplasm. The formation of astaxanthin is regulated by the production of cyt P450, which is controlled by crtr-b genes.

1.4.1.1 Structure

The chemical structure of astaxanthin is $C_{40}H_{52}O_4$. It contains oxygen and is considered a xanthophyll carotenoid. It is composed of a polyene

Table 1.3. Summary of animal origin biopigments.

Pigment	Discovery	Color	Origin	Chemical formula	Uses	Reference
Melanin	Berzelius	Dark, black	5,6-dihydroxy- indole → derivative of tyrosine	$C_{18}H_{10}N_2O_4$	Antioxidant	(Cao *et al*., 2021)
Caroteno-proteins	Britton	Red/ orange/ yellow	Isoprene Isomers	$C_{40}H_{56}$	Photoreceptors, vitamin precursors	(Heras *et al*., 2007)
Bilirubin	Virchow	Red-orange	Biliverdin	$C_{33}H_{36}N_4O_6$	Antioxidant, colorant for urine and feces	(Heras *et al*., 2007)
Hemoglobin	Kendrew	Red color	Heme group and globin proteins	$C_{2952}H_{4464}N_{3248}O_{812}S_8Fe_4$	Transports oxygen	(Perutz, 1962)

chain connecting two terminal β-ionone rings (Fig. 1.10). The β-ionone ring has two asymmetric carbons at the 3,3′ position, whereas the molecule's opposite end has a hydroxyl group (-OH). Both hydroxyl and keto (C = O) groups of oxygen are present. A hydroxyl group is capable of producing mono- or diesters by linking to numerous kinds of fatty acids, including palmitic, oleic, estearic, or linoleic acid (Udayan *et al.*, 2017).

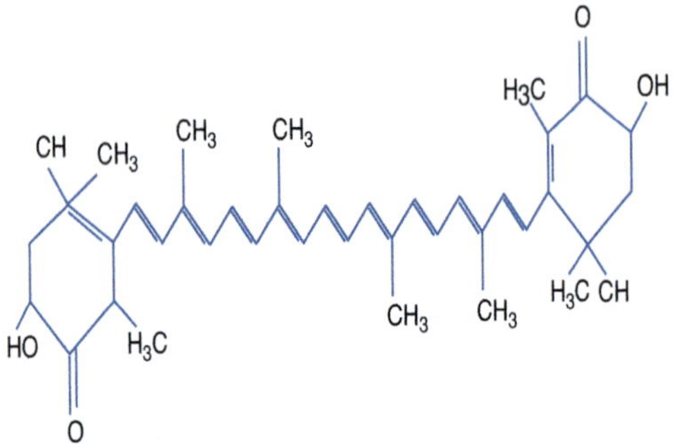

Fig. 1.10. Structure of astaxanthin. Source: Debnath *et al.* (2024).

1.4.1.2 Uses

- Astaxanthin has anticancer and antiinflammatory properties and is used as a supplement.
- It helps to prevent diabetes, cardiovascular diseases, and neurodegenerative disorders.
- It is an immunity booster acting against free radical damage.
- It is an antioxidant so helps to prevents Alzheimer and Parkinson diseases.
- It prevents peptic ulcers by reducing *H. pylori* growth.

1.4.2 Lycopene

In 1876, Millardet discovered lycopene in tomatoes but the name was given by Schunck. It is produced from fungi. Lycopene is a carotenoid, a type of hydrocarbon, responsible for the yellow, red, and orange colors in plants and algae. Geranylgeranyl diphosphate (GGPP) is a major precursor of lycopene, which is produced by the condensation of DMAPP and IPP. Lycopene is produced when mevalonic acid is transformed into dimethylallyl pyrophosphate. Condensation then forms geranylgeranyl pyrophosphate. This is condensed to phytoene which after several desaturation steps is converted to lycopene (Higdon *et al.*, 2016). It is insoluble in water and nonpolar but soluble in organic solvents.

1.4.2.1 Structure

Lycopene is an open-chain, acyclic, C40 polyisoprenoid unsaturated carotenoid with a chemical formula of $C_{40}H_{56}$ (Fig. 1.11). It comprises 13 double bonds, 11 of which are linked, arranged in a linear pattern. The formal chemical name of lycopene is 2,6,10,14,19,23,27,31 octam ethyl2,6,8,10,12,14,16,18,20,22,24,26,30-dotriacontatridecaene. Common names include Ψ,Ψcarotene, all-trans-lycopene, and (all-E)-lycopene.

1.4.2.2 Uses

- Lycopene reduces the negative effects of a high-fat diet on theplacenta, and promotes fetal development.
- It has antioxidant anticancer properties and provides sun protection.
- It is cardio-protective, neuro-protective, and hepato-protective.
- It improves fertility in males.

1.4.3 Monascin

The genus *Monascus* belongs to the family Monascaceae, and Kingdom Fungi. Monascin is a yellow-orange pigment, discovered in 1960 by Dr Yoshitake Yamano. Using 60%, 70%, and 80% ethanol, respectively, monascin has been extracted from red kojic rice. Methylglyoxal (MG) is a major precursor to monascin, a substance found in red yeast rice. The maximum rate of monascin extraction, 75.68%, was recorded at 70% ethanol concentration. Its synthesis involves various steps: malonyl-CoA and acetyl-CoA are condensed to form hexanone chromophores. These react with long-chain fatty acids to form orange pigments which are reduced to form monascin.

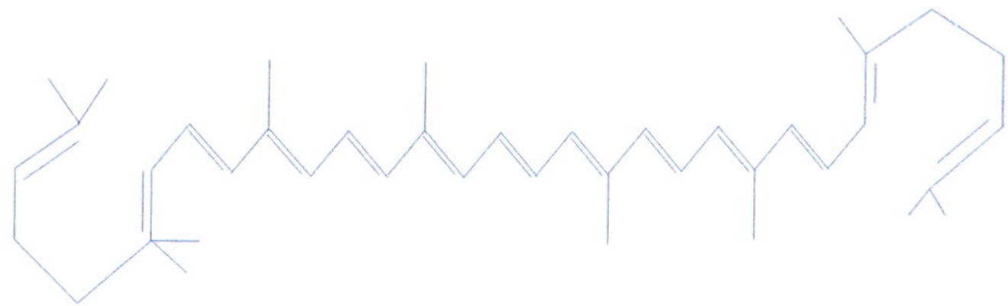

Fig. 1.11. Structure of lycopene. Source: Camara *et al.* (2013).

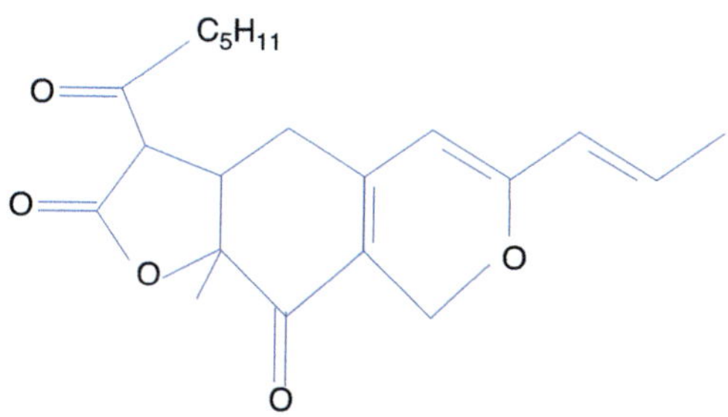

Fig. 1.12. Structure of monascin. Source: Manan *et al.* (2017).

1.4.3.1 Structure

The empirical formula is $C_{21}H_{26}O_5$ (Fig. 1.12). Monascin is an organic heterotricyclic substance, 3 a,4,8,9a-tetrahydro-2H-furo[3,2 g][2] benzopyran-2,9(3H)-dione, that is substituted at positions 3, 6, and 9 a by hexanoyl, (1E)-prop-1-en-1-yl and methyl groups, respectively (the 3S,3aR,9aR diastereoisomer).

1.4.3.2 Uses

- Monascin is used as a food additive, nitrite substitute and colorant.
- It helps to control hyperglycemia and prevent obesity.
- It has antitumor, anti-Alzheimer, and anti-inflammatory activity.
- It can inhibit activation of neuron-derived orphan receptor (NOR 1), a nitric oxide donor.
- It can lower blood lipid levels.
- It has antidepressant properties.

1.4.4 Violacein

Violacein is a water-soluble purple pigment obtained from many Gram-negative bacteria such as *Chromobacterium violaceum*. The genes required for violacein production are vioAB-CDE. It was isolated in 1882 by Boisbaudran. Its precursor is L-tryptophan that is an indole derivative. There are many factors that influence violacein production such as tryptophan supplementation, surfactant, pH, nicotinamides, and zinc. The enzymes responsible for the Vio biosynthesis pathway condense two L-tryptophan molecules to yield Vio. Deoxyviolacein is then formed having one less oxygen atom at the 6 position of the indole ring. Vio generated from wild strains consists of 10% deoxyviolacein. There is extra oxygen in oxyviolacein at the 20 position of the indole ring compared with Vio.

1.4.4.1 Structure

Three connected nitrogen-containing heterocycles make up this compound: a 2-pyrrolidone, an oxindole, and a 5-hydroxylindole (Fig. 1.13). It is formed when two tryptophan molecules condense. Its chemical formula is $C_{20}H_{13}N_3O_3$. The structure was verified by high-performance liquid chromatography-mass spectrometry (HPLC-MS) and proton nuclear magnetic resonance (1 hour NMR) analysis.

1.4.4.2 Uses

- Violacein is broad-spectrum antibacterial agent, which works well against *Staphylococcus aureus* and other Gram-negative bacteria.
- It has antiviral, antifungal, and antiprotozoal actions.
- It has immunomodulatory and antioxidant actions and thus can help to treat neurodegenerative diseases.
- As it has antimicrobial properties, it serves as a preservative for food.

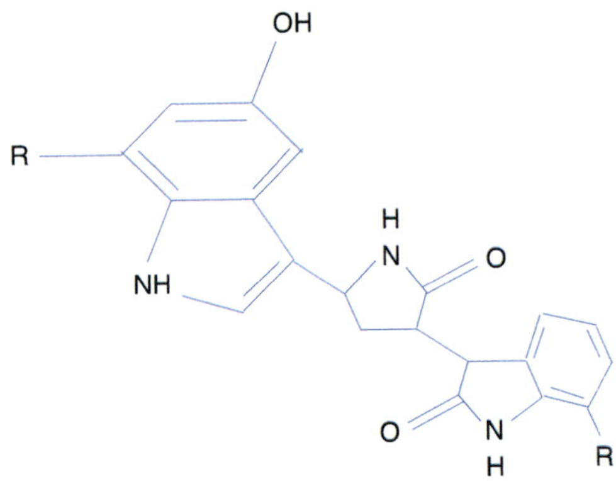

Fig. 1.13. Structure of violacein. Source: Venegas *et al.* (2019).

- It exhibits anticancer properties towards prostate, breast, lung, cervical, melanoma, ovarian, colon, and renal cancers (Durán *et al.*, 2007).

1.4.5 Summary

A summary of biological pigments from animals is given in Table 1.4, while Fig. 1.14 gives a summary of pigment production from all origins.

1.5 Importance of Biopigments

Biopigments play important in various aspects of life. They play important roles in the industrial, economical, and pharmaceutical fields and have environmental impacts, spiritual significance, and health benefits (Fig. 1.15).

In industry, these pigments have wide applications in cosmetics, agrochemicals, and food. They attract these industries because of low cost, stability, biodegradability, and high production value. The cosmetics business has shifted to more environmentally friendly production methods for both chemicals and packaging. The application of cosmetics for medical reasons gave rise to a new field known as cosmeceuticals (Pagels *et al.*, 2022). These substances have antioxidant, melanogenesis-inhibiting, and photo-protective qualities so are preferred by consumers.

They are used to add color to food and improve quality. These colors provide β-cryptoxanthin, which helps to prevent heart disease. Foods with significant amounts of indoles, sulforaphane, and isothiocyanate hinder the action of substances that cause cancer. In the textile industry, biopigments provide color and opacity by adherence to the surface of fabrics. These pigments are also used in dyes, beverages, packaging, paints, and paper making.

Economically, they are used as biofertilizers, and some microorganisms have pigments that are used for water purification. They are used in agriculture for plant protection from UV light, for plant growth, in photosynthesis, and pollination.

Biopigments have a wide role in pharmaceutical products. Bacterial pigments are used as treatments for cancer and other bacterial infections, popular because of their noncarcinogenic properties, biodegradability, and low cost (Venil *et al.*, 2013). They are used for tablet coating and capsule coloration.

Some biopigments have cytotoxic, antioxidant, antimicrobial, antimalarial, anticancer, and antitumor activities. Some show antiinflammatory, immunity booster, and antiproliferative actions. Antiaging creams are used and whitening effects can also be attained.

Biopigments also have environmental impacts. As they are naturally derived, they are less carcinogenic than synthetic pigments. Many microorganisms have pigments that allow bioremediation, biodegradability, and recycling.

Some biopigments have cultural significance. Biopigments have been used in various spiritual and ritual practices; in Native American cultures they are used to adorn the body, symbolize spiritual connection, and communicate with ancestors. In ancient egyptian culture, biopigments were imbued with symbolic meanings; for example, red iron oxide (hematite) symbolized life and fertility. They are also used in traditional medicine such as the use of turmeric (curcumin) in Ayurvedic medicine. Biopigments were used to symbolize power, wealth, and status, such as the use of Tyrian purple in ancient Roman and Greek cultures to signify royalty.

Table 1.4. Summary of microorganism origin biopigments.

Pigment	Discovery	Color	Origin	Chemical formula	Uses	Reference
Astaxanthin	Kuhn	Pink color	β-Carotene	$C_{40}H_{52}O_4$	Anticancerous and antiinflammatory	(Udayan *et al.*, 2017)
Lycopene	Millardet	Red/ orange/ yellow	Geranyl geranyl diphosphate	$C_{40}H_{56}$	Cardio-protective, neuro-protective, hepato-protective	(Camara *et al.*, 2013)
Monascin	Yamano	Yellow-orange	Methyl glyoxal	$C_{21}H_{26}O_5$	Antitumor	(Manan *et al.*, 2017)
Violacein	Boisbandran	Purple	L-tryptophan	$C_{20}H_{13}N_3O_3$	Antibacterial, antifungal, antiprotozoal	(Durán *et al.*, 2007)

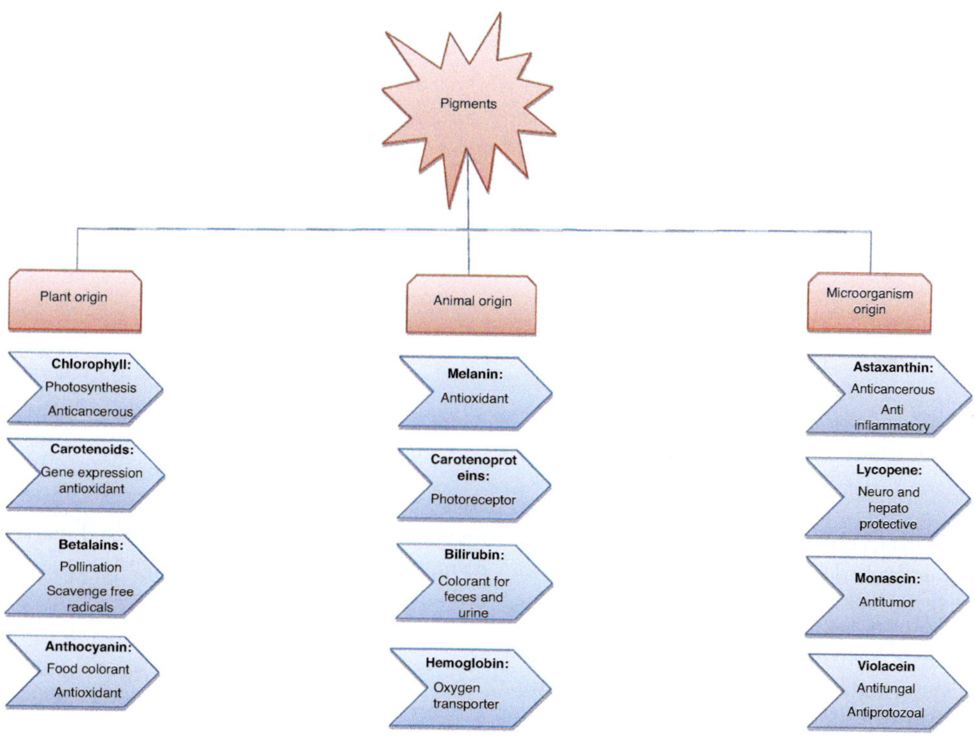

Fig. 1.14. Uses of pigments.

1.6 Challenges

There are many challenges in biopigment production (Fig. 1.16). Development of new colors for the food sector is problematic, in relation to safety, nutrient value, and minimal impact on price. The use of natural colors is five times more expensive than using synthetic pigments, and in the case of confectionery items it can be 20 times more expensive (Sigurdson *et al.*, 2017). An additional issue with natural pigment products is deodorization. Furthermore, they are susceptible to temperature,

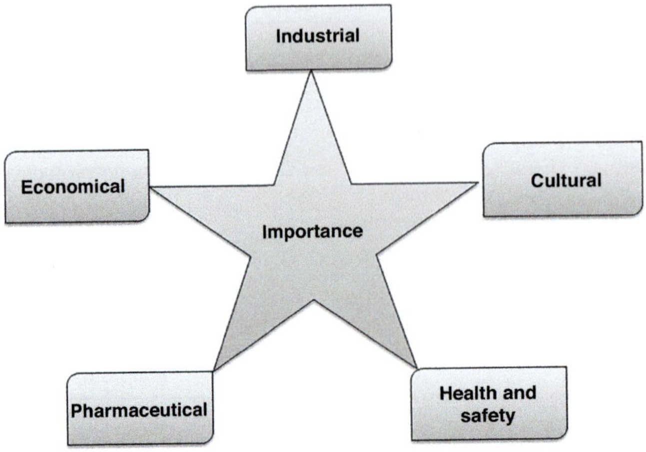

Fig. 1.15. Aspects of the importance of pigments.

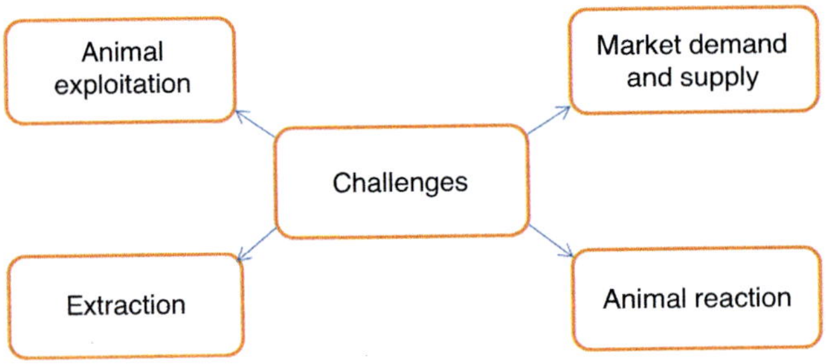

Fig. 1.16. Challenges in the pigment industry.

oxygen, heat, light, pH, and UV rays, which cause fading and color loss. Some have sensitivity to proteins, metal ions, or chemical substances. Vitamin C causes degradation of anthocyanins. Chlorophyll degrades in such conditions while isoprenoid plant compound and conjugated substances are unstable when exposed to light and oxygen (Mayne, 1996). Similarly, fungal pigments can produce mycotoxins. Extraction may also be a challenge if desired quantity or color is not obtained.

Pigments obtained from animals may lead to animal exploitation. Animal origin pigments may cause allergic reactions. Therefore, for natural pigment extraction, nontoxic and nonpathogenic strains need to be used.

There are supply challenges including raw material sources, cultivation, harvesting, storage, and transport. Processing challenges include yield and purity. Biopigments also face limited market demand and brand production competition.

1.7 Improvements

Biopigments can be improved using various methods such as genetic engineering and metabolic engineering, which is used for production of natural pigments in heterologous cell factories. Plant cell/tissue cultures can also be utilized to increase natural pigment production (Fig. 1.17).

Optimization of microbial cultivation approaches includes microbial fermentation processes to increase biopigment yields. Their stability and consistency can be enhanced by encapsulating

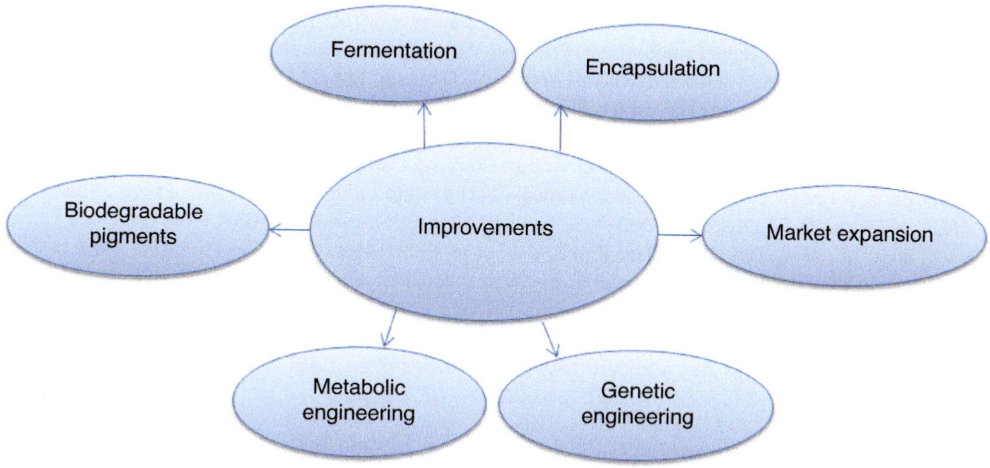

Fig. 1.17. Potential improvements in the biopigment industry.

biopigments in nanoparticles or liposomes. Production costs can be reduced to make them more competitive with synthetic pigments. Renewable feedstocks, such as agricultural waste or algae, can be used to produce biopigments. Biodegradable biopigments can reduce environmental impact.

Market expansion of biopigments should be improved by identification of new industries, applications, and geographic regions for their use. Stay up to date with market trends, forecasts, and regulatory changes. Participate in trade shows, conferences, and events to showcase biopigments and connect with customers. Educate customers about the benefits, applications, and proper use of biopigments. Training programs and workshops should be organized for sales teams, distributors, and partners to enhance their knowledge and skills.

1.8 Conclusion

Biopigments are naturally occurring pigments extracted from microbes, plants, and animals. They include anthocyanins, betalains, carotenoids, and violacein. Biopigments are used in various industries, including food, cosmetics, pharmaceuticals, and textiles. They offer potential benefits including sustainability, biodegradability, and improved safety profiles. They are used as antioxidants, anticancer, antimicrobial, antimalarial, and antiinflammatory agents. They have cytotoxic effects. There are many challenges faced by biopigments such as limited market demand, stability, and cost. These can be improved using genetic engineering, fermentation, and market expansion techniques.

References

Aramwit, P., Bang, N. and Srichana, T. (2010) The properties and stability of anthocyanins in mulberry fruits. *Food Research International* 43(4), 1093–1097. DOI: 10.1016/j.foodres.2010.01.022.

Balaban, A., Schroth, W. and Fischer, G. (1969) Pyrylium salts part I. Syntheses. *Advances in Heterocyclic Chemistry* 10, 241–326. DOI: 10.1016/S0065-2725(08)60499-7.

Bandaranayake, W.M. (2006) The nature and role of pigments of marine invertebrates. *Natural Product Reports* 23(2), 223–255. DOI: 10.1039/b307612c.

Benzie, I.F. and Strain, J.J. (1996) The Ferric Reducing Ability of Plasma (FRAP) as a measure of "antioxidant power": The FRAP assay. *Analytical Biochemistry* 239(1), 70–76. DOI: 10.1006/abio.1996.0292.

Björkman, O., Lange, O., Nobel, P., Osmond, C. and Ziegler, H. (1981) *Physiological Plant Ecology I. Responses to the Physical Environment*. Springer-Verlag, Berlin, Germany.

Borkowski, T., Szymusiak, H., Gliszczyńska-Rwigło, A., Rietjens, I.M.C.M. and Tyrakowska, B. (2005) Radical scavenging capacity of wine anthocyanins is strongly pH-dependent. *Journal of Agricultural and Food Chemistry* 53(14), 5526–5534. DOI: 10.1021/jf0478556.

Camara, M., Cortes Sánchez-Mata, M., Fernandez-Ruiz, V., Cámara, R.M., Manzoor, S. *et al*. (2013) Lycopene: A review of chemical and biological activity related to beneficial health effects. *Studies in Natural Products Chemistry* 40, 383–426. DOI: 10.1016/B978-0-444-59603-1.00011-4.

Cao, W., Zhou, X., McCallum, N.C., Hu, Z., Ni, Q.Z. *et al*. (2021) Unraveling the structure and function of melanin through synthesis. *Journal of the American Chemical Society* 143(7), 2622–2637. DOI: 10.1021/jacs.0c12322.

Cobbs, C., Heath, J., Stireman III, J.O. and Abbot, P. (2013) Carotenoids in unexpected places: Gall midges, lateral gene transfer, and carotenoid biosynthesis in animals. *Molecular Phylogenetics and Evolution* 68(2), 221–228. DOI: 10.1007/978-981-15-7360-6_19.

Cömert, E.D., Mogol, B.A. and Gökmen, V. (2020) Relationship between color and antioxidant capacity of fruits and vegetables. *Current Research in Food Science* 2, 1–10. DOI: 10.1016/j.crfs.2019.11.001.

Davies, K.J. (1995) Oxidative stress: The paradox of aerobic life. *Biochemical Society Symposia* 14. DOI: 10.1042/bss0610001.

Davies, K.M. (2004) *Plant Pigments and their Manipulation*, Vol. 14. Blackwell, Oxford, UK.

de Vogel, J., Jonker-Termont, D.S.M.L., van Lieshout, E.M.M., Katan, M.B. and van der Meer, R. (2005) Green vegetables, red meat and colon cancer: Chlorophyll prevents the cytotoxic and hyperproliferative effects of haem in rat colon. *Carcinogenesis* 26(2), 387–393. DOI: 10.1093/carcin/bgh331.

Debnath, T., Bandyopadhyay, T.K., Vanitha, K., Bobby, M.N., Nath Tiwari, O. *et al*. (2024) Astaxanthin from microalgae: A review on structure, biosynthesis, production strategies and application. *Food Research International* 176, 113841. DOI: 10.1016/j.foodres.2023.113841.

Durán, N., Justo, G.Z., Ferreira, C.V., Melo, P.S., Cordi, L. *et al*. (2007) Violacein: Properties and biological activities. *Biotechnology and Applied Biochemistry* 48(Pt 3), 127–133. DOI: 10.1042/BA20070115.

Falkowski, P.G. and Raven, J.A. (2013) *Aquatic Photosynthesis*. Princeton University Press, Princeton, NJ.

Gandía-Herrero, F., Escribano, J. and García-Carmona, F. (2005) Betaxanthins as substrates for tyrosinase. An approach to the role of tyrosinase in the biosynthetic pathway of betalains. *Plant Physiology* 138(1), 421–432. DOI: 10.1104/pp.104.057992.

Gonçalves, L.C.P., de Souza Trassi, M.A., Lopes, N.B., Dörr, F.A., dos Santos, M.T. *et al*. (2012) A comparative study of the purification of Betanin. *Food Chemistry* 131(1), 231–238. DOI: 10.1016/j.foodchem.2011.08.067.

Gonçalves, L.C.P., Tonelli, R.R., Bagnaresi, P., Mortara, R.A., Ferreira, A.G. *et al*. (2013) A nature-inspired betalainic probe for live-cell imaging of Plasmodium-infected erythrocytes. *PLoS One* 8(1), e53874. DOI: 10.1371/journal.pone.0053874.

Grotewold, E. (2006) The genetics and biochemistry of floral pigments. *Annual Review of Plant Biology* 57(1), 761–780. DOI: 10.1146/annurev.arplant.57.032905.105248.

Gürses, A., Açıkyıldız, M., Güneş, K. and Şahin, E. (2024) Natural dyes and pigments in food and beverages. In: *Renewable Dyes and Pigments*. Elsevier, New York, pp. 49–76.

Heras, H., Dreon, M.S., Ituarte, S. and Pollero, R.J. (2007) Egg carotenoproteins in neotropical Ampullariidae (Gastropoda: Arquitaenioglossa). *Comparative Biochemistry and Physiology Part C* 146(1–2), 158–167. DOI: 10.1016/j.cbpc.2006.10.013.

Higdon, J., Drake, V., Delage, B., Johnson, E. and Mayer, J. (2016) Carotenoids. α-carotene, β-carotene, β-cryptoxanthin, Lycopene, Lutein, and Zeaxanthin. Available at: https://lpi.oregonstate.edu/mic/dietary-factors/phytochemicals/carotenoids (accessed 15 August 2025).

Kanner, J., Harel, S. and Granit, R. (2001) Betalains – a new class of dietary cationized antioxidants. *Journal of Agricultural and Food Chemistry* 49(11), 5178–5185. DOI: 10.1021/jf010456f.

Krantz, C., Monier, M. and Wahlström, B. (1980) Absorption, excretion, metabolism and cardiovascular effects of beetroot extract in the rat. *Food and Cosmetics Toxicology* 18(4), 363–366. DOI: 10.1016/0015-6264(80)90190-x.

Landi, M., Tattini, M. and Gould, K.S. (2015) Multiple functional roles of anthocyanins in plant-environment interactions. *Environmental and Experimental Botany* 119, 4–17. DOI: 10.1016/j.envexpbot.2015.05.012.

Latowski, D., Szymanska, R. and Strzałka, K. (2014) Carotenoids involved in antioxidant system of chloroplasts. In: *Oxidative Damage to Plants*. Elsevier, New York, pp. 289–319.

Lee, D. (2019) *Nature's Palette: The Science of Plant Color*. University of Chicago Press, Chicago, IL.

Lehninger, A.L., Nelson, D.L. and Cox, M.M. (2005) *Lehninger Principles of Biochemistry*. Macmillan, Basingstoke, UK.

Manan, M., Mohamad, R. and Ariff, A. (2017) *Monascus spp*. a source of natural microbial color through fungal biofermentation. *Journal of Microbiology & Experimentation* 5(3), 00148. DOI: 10.15406/jmen.2017.05.00148.

Mandal, R. and Dutta, G. (2020) From photosynthesis to biosensing: Chlorophyll proves to be a versatile molecule. *Sensors International* 1, 100058. DOI: 10.1016/j.sintl.2020.100058.

Margalith, P.Z. (1999) Production of ketocarotenoids by microalgae. *Applied Microbiology and Biotechnology* 51(4), 431–438. DOI: 10.1007/s002530051413.

Mayne, S.T. (1996) Beta-carotene, carotenoids, and disease prevention in humans. *FASEB Journal* 10(7), 690–701. DOI: 10.1096/fasebj.10.7.8635686.

McElroy, J.S. and Kopsell, D.A. (2009) Physiological role of carotenoids and other antioxidants in plants and application to turfgrass stress management. *New Zealand Journal of Crop and Horticultural Science* 37(4), 327–333. DOI: 10.1080/01140671.2009.9687587.

McNulty, H.P., Byun, J., Lockwood, S.F., Jacob, R.F. and Mason, R.P. (2007) Differential effects of carotenoids on lipid peroxidation due to membrane interactions: X-ray diffraction analysis. *Biochimica et Biophysica Acta* 1768(1), 167–174. DOI: 10.1016/j.bbamem.2006.09.010.

Mujawar, T., Sevelda, P., Madea, D., Klán, P. and Švenda, J. (2024) A platform for the synthesis of oxidation products of bilirubin. *Journal of the American Chemical Society* 146(2), 1603–1611. DOI: 10.1021/jacs.3c11778.

Nadakal, A.M. (1960) Carotenoids and chlorophyllic pigments in the marine snail, Cerithidea californica Haldeman, intermediate host for several avian trematodes. *Biological Bulletin* 119(1), 98–108. DOI: 10.2307/1538938.

Narsing Rao, M.P., Xiao, M. and Li, W.-J. (2017) Fungal and bacterial pigments: Secondary metabolites with wide applications. *Frontiers in Microbiology* 8, 1113. DOI: 10.3389/fmicb.2017.01113.

Naves, M.M.V. and Moreno, F.S. (1998) β-carotene and cancer chemoprevention: From epidemiological associations to cellular mechanisms of action. *Nutrition Research* 18(10), 1807–1824. DOI: 10.1016/S0271-5317(98)00137-7.

Niedzwiedzki, D.M. and Blankenship, R.E. (2010) Singlet and triplet excited state properties of natural chlorophylls and bacteriochlorophylls. *Photosynthesis Research* 106(3), 227–238. DOI: 10.1007/s11120-010-9598-9.

Nishino, H., Tokuda, H., Murakoshi, M., Satomi, Y., Masuda, M. *et al*. (2000) Cancer prevention by natural carotenoids. *BioFactors* 13(1–4), 89–94. DOI: 10.1002/biof.5520130115.

Nishino, H., Murakosh, M., Ii, T., Takemura, M., Kuchide, M. *et al*. (2002) Carotenoids in cancer chemoprevention. *Cancer Metastasis Reviews* 21(3–4), 257–264. DOI: 10.1023/a:1021206826750.

Novo, R., Azevedo, P.S., Minicucci, M.F., Zornoff, L.A.M. and Paiva, S.A.R. (2013) Effect of beta-carotene on oxidative stress and expression of cardiac connexin 43. *Arquivos Brasileiros de Cardiologia* 101(3), 233–239. DOI: 10.5935/abc.20130160.

Pagels, F., Almeida, C., Vasconcelos, V. and Guedes, A.C. (2022) Cosmetic potential of pigments extracts from the marine cyanobacterium *Cyanobium* sp. *Marine Drugs* 20(8), 481. DOI: 10.3390/md20080481.

Perutz, M.F. (1962) *Proteins and Nucleic Acids*. Elsevier, Amsterdam, The Netherlands.

Polturak, G., Breitel, D., Grossman, N., Sarrion-Perdigones, A., Weithorn, E. *et al*. (2016) Elucidation of the first committed step in betalain biosynthesis enables the heterologous engineering of betalain pigments in plants. *New Phytologist* 210(1), 269–283. DOI: 10.1111/nph.13796.

Pourrat, A., Lejeune, B., Grand, A., Bastide, P. and Bastide, J. (1987) Propriétés du jus et des colorants de la betterave rouge. *Médecine et Nutrition* 23(3), 166–172.

Rahimi, P., Mesbah-Namin, S.A., Ostadrahimi, A., Separham, A. and Asghari Jafarabadi, M. (2019) Betalain-and betacyanin-rich supplements' impacts on the PBMC SIRT1 and LOX1 genes expression and Sirtuin-1 protein levels in coronary artery disease patients: A pilot crossover clinical trial. *Journal of Functional Foods* 60, 103401. DOI: 10.1016/j.jff.2019.06.003.

Sigurdson, G.T., Tang, P. and Giusti, M.M. (2017) Natural colorants: Food colorants from natural sources. *Annual Review of Food Science and Technology* 8(1), 261–280. DOI: 10.1146/annurev-food-030216-025923.

Stavenga, D.G., Leertouwer, H.L. and Wilts, B.D. (2014) Coloration principles of nymphaline butterflies – thin films, melanin, ommochromes and wing scale stacking. *Journal of Experimental Biology* 217(12), 2171–2180. DOI: 10.1242/jeb.098673.

Stintzing, F.C. and Carle, R. (2007) Betalains – emerging prospects for food scientists. *Trends in Food Science & Technology* 18(10), 514–525. DOI: 10.1016/j.tifs.2007.04.012.

Taniguchi, M. and Lindsey, J.S. (2017) Synthetic chlorins, possible surrogates for chlorophylls, prepared by derivatization of porphyrins. *Chemical Reviews* 117(2), 344–535. DOI: 10.1021/acs.chemrev.5b00696.

Tsurumaki, E., Saito, S., Kim, K.S., Lim, J.M., Inokuma, Y. *et al.* (2008) Synthesis and characterization of meso-aryl-substituted subchlorins. *Journal of the American Chemical Society* 130(2), 438–439. DOI: 10.1021/ja078042b.

Udayan, A., Arumugam, M. and Pandey, A. (2017) Nutraceuticals from algae and cyanobacteria. In: *Algal Green Chemistry*. Elsevier, New York, pp. 65–89.

Valavanidis, A. and Vlachogianni, T. (2013) Plant polyphenols: Recent advances in epidemiological research and other studies on cancer prevention. *Studies in Natural Products Chemistry* 39, 269–295. DOI: 10.1016/B978-0-444-62615-8.00008-4.

Venegas, F.A., Köllisch, G., Mark, K., Diederich, W.E., Kaufmann, A. *et al.* (2019) The bacterial product violacein exerts an immunostimulatory effect via TLR8. *Scientific Reports* 9(1), 13661. DOI: 10.1038/s41598-019-50038-x.

Venil, C.K., Zakaria, Z.A. and Ahmad, W.A. (2013) Bacterial pigments and their applications. *Process Biochemistry* 48(7), 1065–1079. DOI: 10.1016/j.procbio.2013.06.006.

Wackenroder, H. (1831) Ueber das Oleum radicis Dauci aetherum, das Carotin, den Carotenzucker und den officinellen succus Dauci; So wie auch über das Mannit, welches in dem Möhrensafte durch eine besondere Art der gährung gebildet wird. *Geigers Magazin Der Pharmazie* 33, 144–172.

Weiss, M.R. (1991) Floral colour changes as cues for pollinators. *Nature* 354(6350), 227–229. DOI: 10.1038/354227a0.

Willstätter, R. (1906) Zur kenntniss der zusammensetzung des chlorophylls. *Justus Liebigs Annalen Der Chemie* 350(1–2), 48–82. DOI: 10.1002/jlac.19063500103.

Yao, Z.-Y. and Qi, J.-H. (2016) Comparison of antioxidant activities of melanin fractions from chestnut shell. *Molecules* 21(4), 487. DOI: 10.3390/molecules21040487.

2 Historical Perspective on the Use of Microbial Pigments

Manjoo Rani[1], Saty Dev[2*] and Ayush Madan[3]
[1]Department of Pharmacy, Shri Ram Murti Smarak College of Engineering and Technology, Bareilly, Uttar Pradesh, India; [2]Department of Mechanical Engineering, Shri Ram Murti Smarak College of Engineering and Technology, Bareilly, Uttar Pradesh, India; [3]Department of Biotechnology, School of Research & Technology (SORT), People's University, Bhopal, Madhya Pradesh, India

2.1 Introduction

Synthetic pigments have long been a keystone of the pigment industry, notwithstanding their adverse effects on both the environment and human health. Even though these pigments continue to govern the color market, microbial pigments, which are produced as either primary or secondary metabolites, offer a sustainable and nontoxic alternative for the pigment industry. These natural pigments have vital nutritional and medicinal advantages in addition to being ecologically friendly.

The commercial success of microalgae-derived pigments in the food and dietary supplement businesses is increasingly recognized. Moreover, the pharmaceutical and textile industries are gradually adopting these pigments. Significant opportunities for growth in cosmetics, organic chemistry, synthetic biology, omics technologies, and metabolic engineering have further facilitated the large-scale production of microbial pigments, paving the way for their broader commercial utilization (Barreto *et al.*, 2023).

The historical application of pigments derived from natural sources, together with microbial pigments, spans diverse spheres such as art, medicine, and industry. In ancient times, natural pigments were integral to artistic expression, as early civilizations utilized natural pigments in paintings, textiles, and pottery. Microbial pigments, although less well documented, may have contributed to the vibrant colors in dyes and decorations used by artisans.

Natural pigments were often extracted from plants and microorganisms for their apparent therapeutic benefits, playing a vital role in wound healing and skin disorders. Industrial applications of natural pigments began with their use in dyeing fabrics, providing sustainable alternatives before synthetic dyes became prevalent (Rani, 2025).

The adaptability and bioactive potential of microbial pigments have sparked renewed interest in learning about their historical uses and how they developed into contemporary industrial applications. Studying history is essential to comprehending their development (Paillié-Jiménez *et al.*, 2020).

Figure 2.1 outlines the steps involved in microbial pigment production. Researchers start by growing microorganisms in a suitable

*Corresponding author: satyadevyadav08@gmail.com

 Biopigments of Microbial Origin: Innovations and Applications Across Industries (eds Madan *et al.*)
DOI: 10.1079/ 9781836990505.0002

nutrient solution. They then use centrifugation to separate the cells which are then purified to create the required colors.

2.2 Source of Microbial Pigments

The risks of cancer and birth defects associated with artificial dyes have lead to interest in natural alternatives. Companies have begun researching and using biochromes which are naturally occurring pigments. Pigments show their color effects because they absorb only certain light frequencies. The color you see depends on the wavelengths that these pigments pass through. Recent research shows synthetic edible colorants can lead to cancer and birth defects. Increased interest in biochromes emerged after many authorities banned synthetic dyes.

Biological pigments come from different kinds of living organisms such as minerals, plants, microalgae, fungi, yeast, and bacteria. Many biological pigments take part in chemical reactions which makes them suitable natural substitutes for artificial dyes (Aberoumand, 2011). Optimally grown microorganisms are perfect candidates for pigment production because they multiply quickly and make more pigment efficiently. Researchers study the best ways to grow microbial pigments by working with new strains and finding better growing conditions while creating advanced production methods so microbes can compete with synthetic colorants (Venil and Lakshmanaperumalsamy, 2009). Fig. 2.2 shows the sources of most microbial pigments.

2.2.1 Fungal pigments

Due to their natural metabolism, fungi generate different colored substances. Their pigment products, especially anthraquinone compounds, keep their color well and could replace synthetic dyes in textile manufacturing in an environmentally friendly way. Fungal research teams

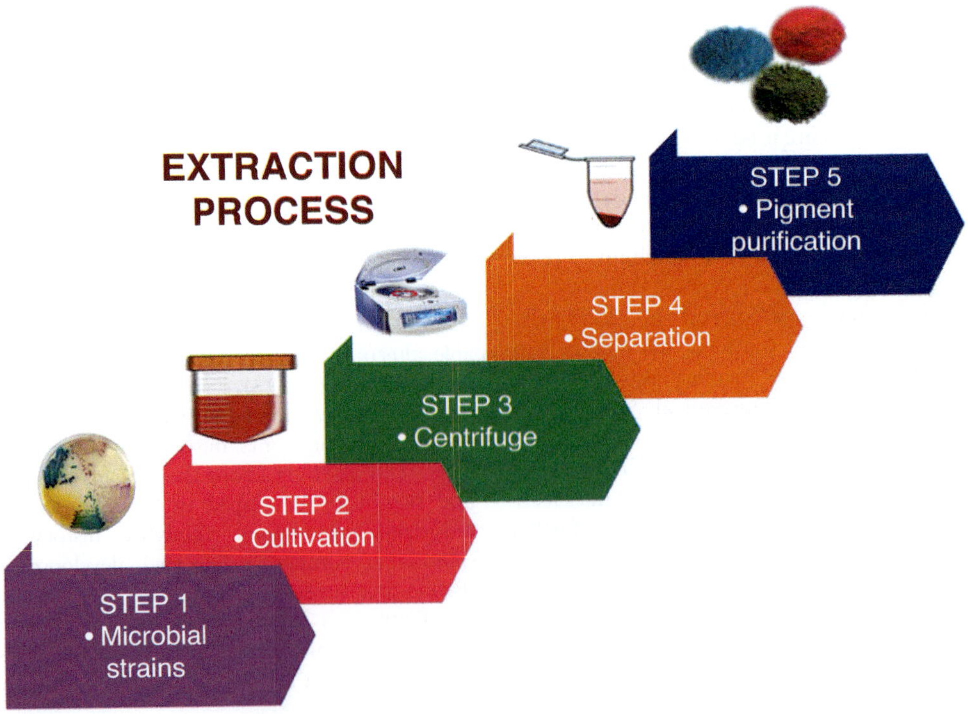

Fig. 2.1. Pigment extraction process.

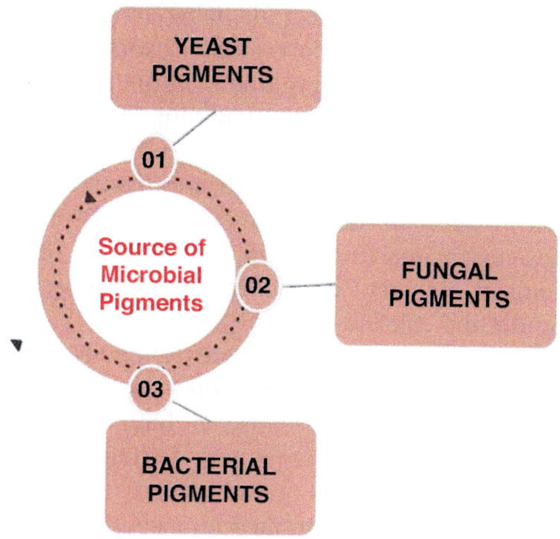

Fig. 2.2. Sources of microbial pigments.

are studying how well these pigments work for industries by using their chemical traits plus nature's plentiful supply (Joshi *et al.*, 2003).

2.2.2 Bacterial pigments

Many types of bacteria make pigments that work well for different industrial uses. *Flavobacterium* produces yellow zeaxanthin from which livestock farmers make poultry feed additives to increase the yellow hue of birds' skin and eggs. Canthaxanthin from *Bradyrhizobium* and *Halobacterium* species helps improve aquaculture feed fabrication while *Agrobacterium aurantiacum* produces astaxanthin which adds color to fish feed formulations.

2.2.3 Yeast pigments

Yeasts can produce carotenoids from low-cost materials as an effective substitute to chemical manufacturing methods. The yeast *Xanthophyllomyces dendrorhous* produces astaxanthin which serves as both a food colorant and animal feed supplement. Mata-Gómez and associates (2014) explored *Rhodotorula glutinis* yeast and related species for carotenoid productivity. Carotenoids have extensive usage as natural ingredients in cosmetics and food supplements which will maintain market growth according to Nagpal and colleagues (2011). Microorganisms show strong potential for producing many different types of bioproducts, including pigments. Scientists increasingly study microorganisms because their health benefits match well with product usage across multiple markets including agriculture, aquaculture, poultry, food, cosmetics, and pharmaceuticals.

The carotenoids market, once projected at just $1.4 billion by 2018 with modest growth rates, has now grown robustly—estimated at over $1.66 billion in 2024 and on track to reach $2.2 billion by 2032, with some projections forecasting up to $2.9 billion by 2029.

2.3 Ancient and Medieval Uses

Pigments from nature shaped all aspects of human existence, including clothing and food supplies, artistic work and social traditions. Pigments help us both make new colors and keep existing ones alive through different methods. Archaeologists found old grinding tools and pigments in a cave near Lusaka, Zambia, which experts say were produced more than 400,000 years ago. Different societies around the world made extensive use of pigments. Ancient egypt used alizarin fabric when wrapping mummies and added natural dyes to sweet treats, making these some of the oldest colored foods. The Chinese recorded textile dying techniques from plants and natural materials 5000 years ago while Indians made pigments in 2500 BC. Japanese Nara period documents from the eighth century show how people used natural pigments to color their traditional sweet treats.

Throughout history, pigments vitalized art production while supporting industrial activities and daily use, reflecting their deep cultural and practical benefits. Historically, people used natural microbes for medicine (Narsing Rao *et al.*, 2017).

Pigments made from microbial sources demonstrate antioxidant and antiinflammatory benefits (Dufossé, 2018; Pogorzelska *et al.*, 2018). These microbial pigments, including astaxanthin, prodigiosin, phycocyanin, and melanin, display high potential for use in multiple industries according to studies by Cuéllar-Bermudez *et al.* (2015) and Sigurdson *et al.* (2017). Their bio-available nature and eco-friendly properties make them great replacements for chemically created coloring materials. The development of biotechnology and microbial fermentation methods has increased production efficiency of natural pigments for industrial use according to Feng *et al.* (2016) and Kumar *et al.* (2015), as shown in Table 2.1.

2.4 Scientific Discovery and Classification

Throughout history, natural materials including isoprenoids, alkaloids, and flavonoids have supplied basic food coloring, flavors, and scents. Companies now use synthetic additives because natural products are expensive. However, many consumers now prefer natural ingredients because they realize that synthetic substances may endanger their health. Natural products have consistently held important roles in drug development throughout history. Research into new bacterial pigments has increased because the market success of these pigments has made synthetic libraries look less powerful (Barreto *et al.*, 2023).

Pigment-producing bacteria can be isolated from diverse environmental sources, utilizing various growth media to culture and purify them. Given the high expense of synthetic media, there is a strong incentive to develop cost-effective processes for pigment production. Utilizing agro-industrial residues offers a sustainable and economical solution by reducing substrate costs. Bacterial pigments are typically extracted using solvent-based methods and subsequently characterized through advanced analytical techniques, including thin layer chromatography (TLC), UV visible spectroscopy, Fourier transform infrared (FTIR) spectroscopy, electrospray ionization mass spectrometry (ESI-MS), nuclear magnetic resonance (NMR) spectroscopy, high-performance liquid chromatography (HPLC), and gel permeation chromatography.

One of the major technological challenges in bacterial pigment production is achieving pure and concentrated extracts. Bacteria produce pigments in two primary forms: those bound to bacterial mycelia and those secreted into the fermentation broth. Pigments bound to mycelia are generally extracted by disrupting the filtered mycelia with acetone, whereas secreted pigments are often recovered by extracting the aqueous broth using organic solvents such as ethyl acetate. To address environmental and health concerns related to solvent use, alternative methods such as spray drying (commonly used in the food and feed industry) and solid-phase extraction (frequently applied in the fine chemical industry) are being explored as eco-friendly and efficient alternatives.

Several years ago scientists questioned whether fermentative pigments could enter commercial markets because of high production costs and extensive safety testing needs. How society perceived products made through biological technology created an essential challenge. Today, however, several fermentative food-grade pigments are available in the market, including *Monascus* pigments, astaxanthin from *Xanthophyllomyces dendrorhous* or *Paracoccus carotinifaciens*, Arpink Red™ (Natural Red™) from *Penicillium oxalicum*, riboflavin from *Ashbya gossypii*, and β-carotene or lycopene from *Blakeslea trispora*. The demand for pigments from plants and algae shows that people will spend more on natural food ingredients that these markets value.

Research into artificial food colors has attracted attention in academic circles for many years. Modern consumers reject artificial food dyes which leads many companies to look for natural color options. More companies now use natural color agents in their products by extracting pigments from plant sources such as paprika, saffron, and leaf vegetables. Use of fermentation-derived ingredients in the food and cosmetic industries has been steadily increasing, with applications including thickening or gelling agents (e.g. xanthan, curdlan, gellan), flavor enhancers (e.g. yeast hydrolysate, monosodium glutamate), flavor compounds (e.g. γ-decalactone, diacetyl), and acidulants (e.g. lactic acid, citric acid). Research teams still work to make fermentation pigments more cost-effective than synthetic or natural pigments, search for new microorganisms, and optimize pigment production processes. The research examines current studies in food-grade pigments by both academic laboratories and private companies in the past ten years. We have found that moving scientific discoveries into market products represents a crucial roadblock in development.

Table 2.1. Microbial sources, bioactivity, and applications of natural pigments.

Pigment	Source organisms	Bioactivities known	Applications	Reference
Canthaxanthin	*Bradyrhizobium* spp., *Halobacterium* spp.	Antioxidant, anticancer	Food dye, salmon feed, poultry feed	(Hamidi *et al.*, 2017)
Astaxanthin	*Halobacterium salinarium*, *Agrobacterium aurantiacum*, *Paracoccus carotinifaciens*, yeast, microalgae	Antioxidant, photoprotection, antiinflammatory, antimicrobial	Animal feed, fish food, food colorant	(Pogorzelska *et al.*, 2018)
Prodigiosin	*Serratia marcescens*, *Pseudoalteromonas rubra*	Anticancer, antimetastatic activity, immunosuppressant, antimalarial	Yogurt, milk, carbonated drinks as colorants	(Kamble and Hiwarale, 2012)
Phycocyanin	*Aphanizomenon flos-aquae*, *Spirulina* spp., *Pseudomonas* spp.	Cytotoxicity, apoptosis induction, anti-Alzheimer activity, antioxidant	Sweets and ice creams	(Cuéllar-Bermudez *et al.*, 2015)
Violacein	*Chromobacterium violaceum*, *Janthinobacterium lividum*, *Pseudoalteromonas tunicata*	Anticancer, antioxidant, antifungal, antiviral, antituberculosis, antiparasitic, anti-HIV, antimalarial	Food, cosmetics, textile industries	(Dufossé, 2018)
Riboflavin	*Candida guilliermundii*, *Debaryomyces subglobosus*, *Eremothecium ashbyii*, *Ashbya gossypii*, *Clostridium acetobutylicum*	Anticancer, antioxidant, cardiovascular protection, role in vision	Food industry	(Dufossé, 2018)
β-Carotene	*Dunaliella salina*, *Blakeslea trispora*, *Mucor circinelloides*, *Phycomyces blakesleeanus*, *Rhodotorula* spp.	Anticancer, antioxidant, cholesterol synthesis suppression	Food coloring, vitamin A source	(Sigurdson *et al.*, 2017)
Melanin	*Saccharomyces* spp., *Neoformans* spp.	Antioxidant, antibiofilm, antimicrobial, anti-HIV	Eyeglasses, cosmetic creams, food products	(Surwase *et al.*, 2013)
Lycopene	*Lycopersicon esculentum*, *Fusarium* spp., *Blakeslea trispora*	Antioxidant, anticancer	Meat coloring	(Giovannucci *et al.*, 2002)
Arpink Red™ (now Natural Red™)	*Penicillium oxalicum*	–	Food colorant	(Kumar *et al.*, 2015)

Continued

Table 2.1. Continued

Pigment	Source organisms	Bioactivities known	Applications	Reference
Monascus pigments	*Monascus pilosus, M. purpureus, M. ruber, M. floridanus*	Antimicrobial, anticancer, antiobesity	Food colorant	(Feng *et al.*, 2016)

Producers currently produce or test methods to obtain carotenoids through yeast, bacteria, and mold cultivation. Research by Britton *et al.* in 2004 showrf that manufacturers can produce β-carotene from both microbial sources and plant extracts alongside chemical synthesis (see Table 2.1). Scientists now focus their attention on several additional carotenoids including lycopene, astaxanthin, zeaxanthin, and canthaxanthin (Paillié-Jiménez *et al.*, 2020).

2.4.1 Classification of microbial pigments

Microbial pigments, or biopigments, are secondary metabolites produced by microorganisms. These pigments hold significant potential for synthetic and commercial applications across various industries due to their diverse properties and sustainable production processes. Microbial pigments can be classified based on their structural characteristics and natural origins. Prominent examples of microbial pigments are shown in Fig. 2.3.

2.5 Industrial Applications in the 20th Century

Rhodotorula spp. intracellular carotenoids have been effectively made from a variety of agro-industrial byproducts, such as sugarcane bagasse, wheat bran, rice bran, silage, whey, raw glycerol, corn steep liquor, sugarcane molasses, waste chicken feathers, and fruit waste extract (Korumilli *et al.*, 2020). The global carotenoids market, previously expected to exceed $2 billion by 2022, is now estimated at approximately $1.7 billion in 2025 and projected to reach up to $2.6 billion by 2030. A research team tested the profit margins of manufacturing both powdered and oiled astaxanthin products by using wheat bran and olive pomace through simulated solid-state fermentation. We studied fermentation under various economic conditions to discover the best plant capacity that creates an affordable bioproduction process (Paillié-Jiménez *et al.*, 2020).

Microbial pigments from *Rhodotorula* species have potential use in food products, animal feed, medicine, and cosmetic production. Laboratories use more plant-based colorants than ever before as demand increases and industry analysts expect sector growth of 7% per year before 2020 (Andreani *et al.*, 2015; Ambati *et al.*, 2019). Compared to pigments derived from plant and animal sources, microbial pigments produced via fermentation are more dynamic, cost-effective, and environmentally sustainable (Chen *et al.*, 2019; Arashiro *et al.*, 2020). Moreover, they yield biodegradable compounds that can effectively compete with synthetic dyes, providing a viable alternative for long-term industrial applications (Baltz, 2017; Azman *et al.*, 2018).

Scientists demonstrated industrial production of astaxanthin pigments when they used agro-industrial leftovers from wheat bran and olive pomace to create solid-state fermentation systems. Technical and economic testing showed scientists how to make pigments under conditions of high yield and low manufacturing cost. Manufacturing rights and consistent color appeal under environmental pressure help make microbial pigments work in business. The use of advanced techniques, including proteomics and supercritical fluid extraction, helps both scale up production and make microbial pigments suitable for industrial use (Aslam *et al.*, 2017; Blachowicz *et al.*, 2019).

2.6 Microbial Pigments of Commercial Importance

For fermentation-based microbial pigment businesses to launch products, they must gain consumer acceptance while receiving industry approvals and getting financial backing to build their operations. The use of fermentation-produced pigments in food products was questioned due to

Fig. 2.3. Examples of microbial pigments Source: Eyal and Spencer (1992); Dufossé *et al*. (2005); Mapari *et al*. (2005); Babitha (2009); Stafsnes *et al*. (2010); Sheehan *et al*. (2012); Venil *et al*. (2013a, b).

high costs for production equipment and extensive safety tests. The public acceptance of biotech products determines how successful microbial pigments will be in their industrial applications.

Different industries have adopted microbial pigments. They serve to boost salmon's natural pink hue during aquaculture farming and act as natural antioxidants within various food items. Manufacturers offer multiple food-grade pigments to customers, including *Monascus* pigments with astaxanthin generated by *Xanthophyllomyces dendrorhous* along with Arpink Red from *Penicillium oxalicum* and pigments such as riboflavin. Scientific studies and FDA approvals confirm that these pigments meet essential safety requirements for use in food production. Consumer preference for natural ingredients has created market opportunities for microbial pigments to serve as food colorants and nutritional supplements. The rising demand for natural sustainable products means that consumers will pay more for products that meet these safety and environmental concerns.

2.7 Pigment-Producing Bacteria and Their Applications in the Textile Industry

Bacteria capable of producing pigments have gained significant attention for their potential use in the textile industry. These pigments, derived from microbial secondary metabolites, offer sustainable, eco-friendly alternatives to synthetic dyes, reducing the environmental burden associated with traditional textile dyeing processes. The use of biopigments not only aligns with environmental goals but also provides unique color profiles with added functionalities, such as antimicrobial properties. Several bacterial species are known to produce biopigments with applications in textiles as explained in Fig. 2.4.

2.8 Microbial Pigments in Medicine

Table 2.2 presents the medical functions of pigments produced by different microorganisms. The microbial pigments are sorted into carotenoids, prodiginines, phycobiliproteins, and oxyindoles with their characteristic colors and microorganisms that produce them alongside their therapeutic applications including antioxidant, anticancer, antibacterial, and photo-protective properties. Microorganisms produce β-carotene that helps to fight cancer and prodigiosin which acts as a DNA cutter with immune system suppression abilities. The listed research supports both their functionality and practical uses in scientific studies.

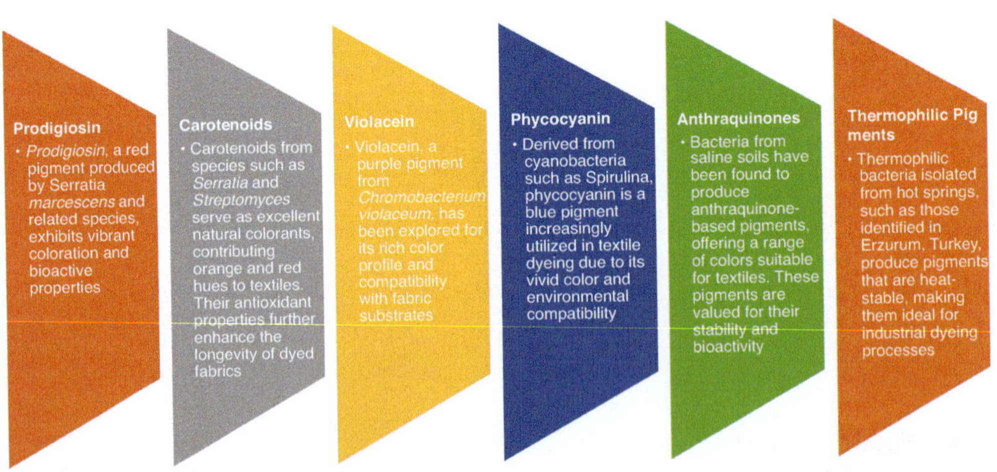

Fig. 2.4. Bacterial species of biopigments. Source: Faraag *et al.* (2017); Shetti (2017); Oztas Gulmus and Gormez (2020); Qayyum *et al.* (2020); Bashir *et al.* (2022); Kramar and Kostic (2022).

Table 2.2. Medical applications of microbial pigments.

Pigment	Color	Microorganisms	Applications	References
Carotenoids				
Canthaxanthin	Orange to deep pink	*Lactobacillus pluvialis*, *Bradyrhizobium* spp.	Antioxidant, protective against UV radiation, anticancer, and antiinflammatory properties	(Sen *et al*., 2019)
Astaxanthin	Red to orange	*Paracoccus carotinifaciens Agrobacterium aurantiacum*,	Antioxidant, UV protective, antiinflammatory, and anticancer activities	(Ambati *et al*., 2014)
Zeaxanthin	Yellow	*Staphylococcus aureus*, *Flavobacterium* spp., *Paracoccus zeaxanthinifaciens*	UV protection and antioxidant properties	(Galasso *et al*., 2017)
β-Carotene	Red	*Rhodococcus maris*, *Rhodococcus ruber*	Antioxidant activity and potential in cancer prevention	(Cappelletti *et al*., 2020)
Staphyloxanthin	Yellow	*Staphylococcus aureus*	Antioxidant activity	(Alshamaa and Issam, 2017)
Deinoxanthin	Red	*Deinococcus radiodurans*	Anticancer properties	(Tian *et al*., 2009)
Prodiginines				
Prodigiosin	Red	*Serratia marcescens*	Anticancer activity, DNA cleavage, and immune suppression	(Williamson *et al*., 2005)
Undecylprodigiosin	Red	*Streptomyces* spp.	Antimalarial, antibacterial, antioxidant, and anticancer properties	(Stankovic *et al*., 2014)
Heptyl prodigiosin	Red	β-Proteobacteria	Antiplasmodial activity	(Lazaro *et al*., 2002)
Phycobiliproteins				
Pycocyanin	Blue	*Pseudomonas* spp.	Cytotoxic effects, neutrophil apoptosis, ciliary dysfunction, and proinflammatory activity	(Baront and Rowe, 1981)
Oxyindoles				
Violacein	Purple	*Chromobacterium violaceum*, *Collimonas* spp., *Duganella* spp., *Pseudoalteromonas* spp.	Antibacterial, antifungal, antimalarial, and anticancer properties	(Durán *et al*., 2007)

2.9 Factors Affecting Pigment Production

Bacteria, fungi, and algal pigments function as food colorants and additives as reported by Nigam and Luke (2016). Microorganisms used for industrial pigment production need to adapt to different pH levels and temperatures and feed on multiple carbon sources and nitrogen materials (Kirti *et al*., 2014; Kumar *et al*., 2015). A high pigment output is necessary to make production of these pigments economically viable. Bioprospecting programs help find microorganisms that produce pigments by searching different environmental locations. Strain enhancement techniques help performance of pigment-producing microbes to produce better results while seeking novel microbes as alternatives offers additional yield potential.

The amount of microbial pigment produced during fermentation responds to both fermentation methods and the mixing ratio of its growth elements such as carbon, nitrogen, and minerals. Tuli *et al.* (2015) suggest that microbial pigments are made using two fermentation approaches: solid-state fermentation (SSF) and submerged fermentation. SSF produces higher results with better efficiency while allowing direct use of the product as a coloring agent (Babitha, 2009). The approach enables better fungal production and boosts metabolic yields while reducing water usage.

The success of pigment production by microorganisms relies on properly selected fermentation conditions and controlled process factors. The optimal temperature range for *Monascus* species to produce pigment, for instance. is between 25°C and 28°C, while *Pseudomonas* prefers 35–36°C temperatures (Kumar *et al.*, 2015). Joshi *et al.* (2003) revealed *Sarcina* sp.'s maximum growth and carotenoid production after a 72 hour period of incubation. *Micrococcus* and *Rhodotorula* have been demonstrated to produce pigment most effectively when incubated together (Attri and Joshi, 2005; Joshi and Attri, 2006). According to Attri and Joshi (2005), 48 hours was the optimal incubation period for *Chromobacter* to produce pigment. pH can affect how pigment is produced and what color it is. *Rhodotorula* produces pigments at pH 4–5.

At neutral to slightly alkaline pH, lycopene forms, whereas β-carotene forms at acidic pH (Joshi *et al.*, 2003). Carbon sources affect the color and growth of pigments from microorganisms. The best carbon source for pigment production may be monosaccharides or their polymers, depending on the species (Joshi *et al.*, 2003). A range of inorganic and organic nitrogenous compounds may be preferred by various microbes for optimal pigmentation. Factors affecting production of pigments are indicated in Fig. 2.5. Successful pigment production requires specific conditions including pH, temperature, incubation time, carbon and nitrogen sources, moisture levels, and selected fermentation methods.

2.10 Modern Perspective

2.10.1 Significance of microbial pigments as natural colorants

Microorganisms are among the most abundant and influential forms of life on Earth, playing a critical role in the ecosystem. They are integral to the food we consume, contributing to the production of certain food items through fermentation. Additionally, microorganisms serve as sources of single-cell proteins and nutritional supplements, including pigments, amino acids, vitamins, organic acids, and enzymes. As a result, pigments derived from microbial origins present a sustainable and effective alternative to synthetic colorants.

Microorganisms are known to produce a diverse array of pigments, making them a promising source of natural food colorants. Notable examples of natural pigments include carotenoids, flavonoids, tetrapyrroles, and xanthophylls such as astaxanthin. β-Carotene, one of the most widely utilized pigments in industry, is produced by certain species of microalgae and cyanobacteria. Astaxanthin, a red pigment with significant commercial value, is derived from *Phaffia rhodozyma* and *Haematococcus pluvialis*. It is extensively used in feed, pharmaceuticals, and aquaculture industries. High-yield pigment-producing microorganisms include species of *Monascus*, *Paecilomyces*, *Serratia*, *Cordyceps*, and *Streptomyces*, as well as those producing yellow, red, and blue compounds, such as *Penicillium herquei*, *Penicillium atrovenetum*, *Rhodotorula*, *Sarcina*, *Cryptococcus*, *Monascus purpureus*, *Bacillus* sp., *Achromobacter*, and *Yarrowia*.

Scientists have studied bacteria and fungi extensively because they want to develop these organisms into natural food colorants. Food pigments help with appearance and offer chemical advantages by fighting cancer and replacing vitamin A while maintaining their properties through heat and light changes. Using natural pigments offers positive results in human health and protects our environment. Our use of microbial dyes allows us to avoid environmental

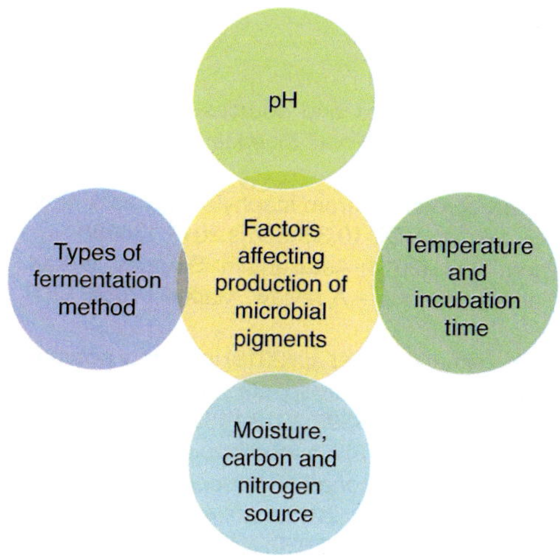

Fig. 2.5. Factors affecting production of pigments.

harm from the chemical production methods of synthetic colorants. These bio-friendly natural colorants support multiple food brands through their use in infant food products, baby cereals, condiments, pasta, processed cheese, fruit beverages, dairy items enriched with vitamins, plus energy drinks.

2.10.2 Novel practices in microbial pigment production: strain improvement

Random mutagenesis coupled with screening has helped to improve microbes for decades. In the last decade, scientists have made major strides by creating specialized tools to delete specific genes. Genetic engineering now enables exact control of gene deletion to optimize microbial system metabolism. Strain development processes benefit from complete systems and integrated methods that make production easier to execute.

Manufacturers create industrial strains for commercial success because natural microbial strains produce low levels of pigment. Our main goal is to find microbial strains that create larger pigment amounts during shorter fermentation runs. Strain enhancement is crucial in industrial fermentation when producing large amounts of important bioproducts.

The pigmentation of different microbe species responds well to treatment with common radiation mutagens including UV rays plus ethyl methane sulfonate and nitrosoguanidine. Research now studies the use of X-ray and fast neutron irradiation along with classic approaches to better enhance microbial strains. Advanced radiation methods enhanced how *Monascus purpureus* developed both color production and physical shape. The success of innovative techniques confirms the need to use traditional and advanced methods together to boost microbial pigment production.

2.11 Conclusion

The historical trajectory of microbial pigments highlights their progression from naturally occurring compounds with aesthetic and cultural significance to indispensable assets in modern industries. Initially appreciated for their natural colors, microbial pigments have steadily gained importance due to their eco-friendly nature, safety, and versatility. These attributes have positioned them as viable alternatives to synthetic dyes, which are increasingly scrutinized for their environmental and health-related drawbacks.

Advancements in biotechnology and genetic engineering have played a pivotal role in this transformation, enabling the efficient production and diversification of microbial pigments. Strain improvement, metabolic engineering, and optimized fermentation technologies have significantly enhanced pigment yields, making large-scale production economically feasible. These developments have broadened their applications across textiles, pharmaceuticals, cosmetics, and the food industry.

Looking forward, microbial pigments offer immense potential for sustainable and innovative industrial solutions. Their renewable, biodegradable, and bioactive properties align with the growing demand for environmentally responsible practices. However, further research and technological refinements are needed to address challenges such as production scalability and cost reduction. With continued scientific exploration, microbial pigments are poised to become a cornerstone of sustainable industrial development, bridging the gap between innovation and ecological stewardship.

References

Aberoumand, A. (2011) A review article on edible pigments properties and sources as natural bio-colorants in foodstuff and food industry. *World Journal of Dairy Food Science* 6, 71–78. DOI: 10.5555/20113260383.

Alshamaa, D.S. and Issam, M.M. (2017) The role of extracted carotenoid from staphylococci as antioxidant and antibacterial. *Rafidain Journal of Science* 26, 17–24. DOI: 10.33899/rjs.2017.138956.

Ambati, R.R., Phang, S.M., Ravi, S. and Aswathanarayana, R.G. (2014) Astaxanthin: Sources, extraction, stability, biological activities and its commercial applications – A review. *Marine Drugs* 12(1), 128–152. DOI: 10.3390/md12010128.

Ambati, R.R., Gogisetty, D., Aswathanarayana, R.G., Ravi, S., Bikkina, P.N. *et al*. (2019) Industrial potential of carotenoid pigments from microalgae: Current trends and future prospects. *Critical Reviews in Food Science and Nutrition* 59(12), 1880–1902. DOI: 10.1080/10408398.2018.1432561.

Andreani, N.A., Carraro, L., Martino, M.E., Fondi, M., Fasolato, L. *et al*. (2015) A genomic and transcriptomic approach to investigate the blue pigment phenotype in *Pseudomonas fluorescens*. *International Journal of Food Microbiology* 213, 88–98. DOI: 10.1016/j.ijfoodmicro.2015.05.024.

Arashiro, L.T., Boto-Ordóñez, M., Van Hulle, S.W.H., Ferrer, I., Garfí, M. *et al*. (2020) Natural pigments from microalgae grown in industrial wastewater. *Bioresource Technology* 303, 122894. DOI: 10.1016/j.biortech.2020.122894.

Aslam, B., Basit, M., Nisar, M.A., Khurshid, M., Rasool, M.H. *et al*. (2017) Proteomics: Technologies and their applications. *Journal of Chromatographic Science* 55(2), 182–196. DOI: 10.1093/chromsci/bmw167.

Attri, D. and Joshi, V.K. (2005) Optimization of apple pomace based medium and fermentation conditions for pigment production by *Micrococcus* species. *Journal of Scientific and Industrial Research* 64, 598601.

Azman, A.S., Zainal, N., Lila, M.A.M. and Kamarudin, N. (2018) Bacterial pigments: The bioactivities and as an alternative for therapeutic applications. *Natural Product Communications* 13(12), 1747–1754. DOI: 10.1177/1934578X1801301240.

Babitha, S. (2009) Microbial pigments. In: *Biotechnology for Agro-Industrial Residues Utilisation*. Springer, pp. 147–162.

Baltz, R.H. (2017) Gifted microbes for genome mining and natural product discovery. *Journal of Industrial Microbiology and Biotechnology* 44(4–5), 573–588. DOI: 10.1007/s10295-016-1815-x.

Baront, S.S. and Rowe, J.J. (1981) Antibiotic action of pyocyanin. *Antimicrobial Agents and Chemotherapy* 20, 814–820. DOI: 10.1128/AAC.20.6.814.

Barreto, J.V.D.O., Casanova, L.M., Junior, A.N., Reis-Mansur, M.C.P.P. and Vermelho, A.B. (2023) Microbial pigments: Major groups and industrial applications. *Microorganisms* 11(12), 2920. DOI: 10.3390/microorganisms11122920.

Bashir, S., Numan, M. and Shinwari, Z.K. (2022) Selective isolation of bioactive-pigmented bacteria from saline agricultural soil and assessment of their antimicrobial potential against plant pathogens. *Sustainability* 14(20), 13574. DOI: 10.3390/su142013574.

Blachowicz, A., Chiang, A.J., Romsdahl, J., Kalkum, M., Wang, C.C.C. *et al*. (2019) Proteomic characterization of *Aspergillus fumigatus* isolated from air and surfaces of the international space station. *Fungal Genetics and Biology* 124, 39–46. DOI: 10.1016/j.fgb.2019.01.001.

Cappelletti, M., Presentato, A., Piacenza, E., Firrincieli, A., Turner, R.J. *et al*. (2020) Biotechnology of *Rhodococcus* for the production of valuable compounds. *Applied Microbiology and Biotechnology* 104(20), 8567–8594. DOI: 10.1007/s00253-020-10861-z.

Chen, W., Feng, Y., Molnár, I. and Chen, F. (2019) Nature and nurture: Confluence of pathway determinism with metabolic and chemical serendipity diversifies *Monascus azaphilone* pigments. *Natural Products Report* 36, 561–572.

Cuéllar-Bermudez, S.P., Aguilar-Hernandez, I., Cardenas-Chavez, D.L., Ornelas-Soto, N., Romero-Ogawa, M.A. *et al*. (2015) Extraction and purification of high-value metabolites from microalgae: Essential lipids, astaxanthin and phycobiliproteins. *Microbial Biotechnology* 8(2), 190–209. DOI: 10.1111/1751-7915.12167.

Dufossé, L. *et al*. (2005). "Microorganisms and microalgae as sources of pigments for food use: A scientific oddity or an industrial reality?," *Trends in Food Science & Technology*, vol. 16, pp. 389-406.

Dufossé, L. (2018) Microbial production of food-grade pigments. *Food Research International* 109, 551–577. DOI: 10.1016/j.foodres.2018.04.020.

Durán, N., Justo, G.Z., Ferreira, C.V., Melo, P.S., Cordi, L. *et al.* (2007) Violacein: Properties and biological activities. *Biotechnology and Applied Biochemistry* 48(Pt 3), 127–133. DOI: 10.1042/BA20070115.

Eyal, J. and Spencer, M.G. (1992) *Morchella rotunda* sp. useful for producing natural blue pigment. Google Patents.

Faraag, A.H., El-Batal, A.I. and El-Hendawy, H.H. (2017) Characterization of prodigiosin produced by *Serratia marcescens* strain isolated from irrigation water in Egypt. *Journal of Natural Sciences* 15, 55–68. DOI: 10.7537/marsnsj150517.08.

Feng, Y., Shao, Y. and Chen, F. (2016) *Monascus* pigments. *Applied Microbiology and Biotechnology* 100(3), 859–873. DOI: 10.1007/s00253-012-4504-3.

Galasso, C., Corinaldesi, C. and Sansone, C. (2017) Carotenoids from marine organisms: Biological functions and industrial applications. *Antioxidants* 6(4), 96. DOI: 10.3390/antiox6040096.

Giovannucci, E., Clinton, S.K. and Gann, P.H. (2002) Role of antioxidant carotenoids in preventing chronic diseases: Molecular mechanisms and biochemical evidence. *American Journal of Clinical Nutrition* 76(5), 930–936. DOI: 10.1093/ajcn/76.5.930.

Hamidi, M.R., Jovanova, B. and Panovska, T.K. (2017) Toxicological evaluation of the plant products using brine shrimp (*Artemia salina* L.) model. *Macedonian Pharmaceutical Bulletin* 63(1), 37–44. DOI: 10.33320/maced.pharm.bull.2017.63.01.005.

Joshi, V.K. and Attri, D.A. (2006) Optimization of apple pomace based medium and fermentation conditions for pigment production by *Rhodotorula* species. Available at: www.cabidigitallibrary.org/doi/full/10.5555/20063219525

Joshi, V., Sharma, S. and Kumar, R. (2003) Microbial pigments. *Indian Journal of Biotechnology* 2, 362–369.

Kamble, S.P. and Hiwarale, V.D. (2012) Prodigiosin production from *Serratia marcescens* strains isolated from soil. *Journal of Environmental Research and Development* 7(1), 452–460.

Kirti, K., Amita, S., Priti, S., Mukesh Kumar, A. and Jyoti, S. (2014) Colorful world of microbes: Carotenoids and their applications. *Advances in Biology* 1, 837891. DOI: 10.1155/2014/837891.

Korumilli, T., Mishra, S. and Korukonda, J.R. (2020) Production of astaxanthin by *Xanthophyllomyces dendrorhous* on fruit waste extract and optimization of key parameters using Taguchi method. *Journal of Biochemical Technology* 11, 25–31.

Kramar, A. and Kostic, M.M. (2022) Bacterial secondary metabolites as biopigments for textile dyeing. *Textiles* 2, 252–264. DOI: 10.3390/textiles2020013.

Kumar, P., Singh, D., Kumar, A. and Kango, N. (2015) Biosynthesis of arpink red, a food-grade pigment using *Penicillium oxalicum*. *Journal of Food Science and Technology* 52(1), 568–574.

Lazaro, J.E., Nitcheu, J., Predicala, R.Z., Mangalindan, G.C., Nesslany, F. *et al.* (2002) Heptyl prodigiosin, a bacterial metabolite, is antimalarial *in vivo* and non-mutagenic *in vitro*. *Journal of Natural Toxins* 11, 367–377.

Mapari, S.A.S., Nielsen, K.F., Larsen, T.O., Frisvad, J.C., Meyer, A.S. *et al.* (2005) Exploring fungal biodiversity for the production of water-soluble pigments as potential natural food colorants. *Current Opinion in Biotechnology* 16(2), 231–238. DOI: 10.1016/j.copbio.2005.03.004.

Mata-Gómez, L.C., Montañez, J.C., Méndez-Zavala, A. and Aguilar, C.N. (2014) Biotechnological production of carotenoids by yeasts: An overview. *Microbial Cell Factories* 13, 12. DOI: 10.1016/j.copbio.2005.03.004.

Nagpal, N., Singh, A., Malhotra, P. and Kumar, M. (2011) Microbial pigments with health benefits-a mini review. *Trends in Biosciences* 4, 157–160.

Narsing Rao, M.P., Xiao, M. and Li, W.J. (2017) Fungal and bacterial pigments: Secondary metabolites with wide applications. *Frontiers in Microbiology* 8, 1–13. DOI: 10.3389/fmicb.2017.01113.

Nigam, P.S. and Luke, J.S. (2016) Food additives: Production of microbial pigments and their antioxidant properties. *Current Opinion in Food Science* 7, 93–100. DOI: 10.1016/j.cofs.2016.02.004.

Oztas Gulmus, E. and Gormez, A. (2020) Identification and characterization of novel thermophilic bacteria from hot springs, Erzurum, Turkey. *Current Microbiology* 77(6), 979–987. DOI: 10.1007/s00284-020-01880-0.

Paillié-Jiménez, M.E., Stincone, P. and Brandelli, A. (2020) Natural pigments of microbial origin. *Frontiers in Sustainable Food Systems* 4(September), 1–8. DOI: 10.3389/fsufs.2020.590439.

Pogorzelska, W., Kaczmarek-Szczepańska, B., Wieczorek, N. and Cieślak, A. (2018) Astaxanthin—A novel functional compound with a wide range of applications in human health, animal nutrition, and cosmetics. *Animals* 8(8), 128. DOI: 10.3390/ani8080128.

Qayyum, S., Basharat, S., Mian, A.H., Qayum, S., Ali, M. *et al.* (2020) Isolation, identification, and antibacterial study of pigmented bacteria. *Applied Nanoscience* 10, 4495–4503. DOI: 10.1007/s13204-020-01363-5.

Rani, M. (2025) Experimental investigation on antimicrobial properties of *Elaeocarpus ganitrus* (Rudraksha) and *Cassia occidentalis* L. Available at: https://ymerdigital.com/uploads/YMER240140.pdf (accessed 18 August 2025).

Sen, T., Barrow, C.J. and Deshmukh, S.K. (2019) Microbial pigments in the food industry-challenges and the way forward. *Frontiers in Nutrition* 6, 7–14. DOI: 10.3389/fnut.2019.00007.

Sheehan, H., Hermann, K. and Kuhlemeier, C. (2012) Color and scent: How single genes influence pollinator attraction. *Cold Spring Harbor Symposia on Quantitative Biology* 77. DOI: 10.1101/sqb.2013.77.014712.

Shetti, A.A. (2017) Antibacterial potentials of red pigment extracted from soil isolate *Serratia* sp. *International Journal of Biotechnology and Biochemistry* 3(4), 49–52. DOI: 10.9790/264X-03044952.

Sigurdson, G.T., Tang, P. and Giusti, M.M. (2017) Natural colorants: Food colorant chemistry and applications. *Annual Review of Food Science and Technology* 8, 261–280. DOI: 10.1146/annurev-food-030216-025923.

Stafsnes, M.H., Josefsen, K.D., Kildahl-Andersen, G., Valla, S., Ellingsen, T.E. *et al.* (2010) Isolation and characterization of marine pigmented bacteria from Norwegian coastal waters and screening for carotenoids with UVA-blue light absorbing properties. *Journal of Microbiology* 48(1), 16–23. DOI: 10.1007/s12275-009-0118-6.

Stankovic, N., Senerovic, L., Ilic-Tomic, T., Vasiljevic, B., Nikodinovic-Runic, J. *et al.* (2014) Properties and applications of undecylprodigiosin and other bacterial prodigiosins. *Applied Microbiology and Biotechnology* 98(9), 3841–3858. DOI: 10.1007/s00253-014-5590-1.

Surwase, S.N., Jadhav, S.B., Phugare, S.S. and Jadhav, J.P. (2013) Optimization of melanin production by *Brevundimonas* sp. SGJ using response surface methodology. *3 Biotech* 187–194. DOI: 10.1007/s13205-012-0082-4.

Tian, B., Sun, Z., Shen, S., Wang, H., Jiao, J. *et al.* (2009) Effects of carotenoids from *Deinococcus radiodurans* on protein oxidation. *Letters in Applied Microbiology* 49(6), 689–694. DOI: 10.1111/j.1472-765X.2009.02727.x.

Tuli, H.S., Chaudhary, P., Beniwal, V. and Sharma, A. (2015) Microbial pigments as natural color sources: Current trends and future perspectives. *Journal of Food Science and Technology* 52, 4669–4678. DOI: 10.1007/s13197-014-1601-6.

Venil, C.K. and Lakshmanaperumalsamy, P. (2009) An insightful overview on microbial pigment, prodigiosin. *Electronic Journal of Biology* 5, 49–61. DOI: 10.1016/j.procbio.2013.06.006.

Venil, C.K., Lakshmanaperumalsamy, P. and Kalaichelvan, P. (2013a) Bacterial pigments and their applications. *Process Biochemistry* 48, 1065–1079. DOI: 10.1016/j.procbio.2013.06.006.

Venil, C.K., Zakaria, Z.A. and Ahmad, W.A. (2013b) Bacterial pigments and their applications. *Process Biochemistry* 48, 1065–1079. DOI: 10.1016/j.procbio.2013.06.006.

Williamson, N.R., Simonsen, H.T., Ahmed, R.A.A., Goldet, G., Slater, H. *et al.* (2005) Biosynthesis of the red antibiotic, prodigiosin, in *Serratia*: Identification of a novel 2-methyl-3-n-amyl-pyrrole (MAP) assembly pathway, definition of the terminal condensing enzyme, and implications for undecylprodigiosin biosynthesis in *Streptomyces*. *Molecular Microbiology* 56(4), 971–989. DOI: 10.1111/j.1365-2958.2005.04602.x.

3 Diversity of Microbial Pigments: Types and Sources

Pritty Nayak, Uma Thakur, Sonam Yadav and Krishan Kumar*
Department of Life Sciences, Parul Institute of Applied Sciences, Parul University, Vadodara, Gujarat, India

3.1 Introduction

Microbial pigments are the most sustainable answer to the growing demand for alternatives to synthetic dyes, fueled by a general concern for the environment and a growing desire for nontoxic biodegradable colorants (Soliev *et al.*, 2011; Venil *et al.*, 2013). Unlike colorants produced by petrochemical processes, such as the vast majority of synthetic dyes, microbial pigments are a green option supplemented by bioactivity properties such as antioxidant, antibacterial, and anticancer benefits (Tuli *et al.*, 2015). It should be noted that although many researchers have investigated the microbial pigments, attempting to boost their commercialization at large scales, production costs and specialized bioprocesses often associated with microbial pigments have hampered these efforts (Mapari *et al.*, 2005). However, with improved strain quality, metabolic engineering, and low-cost industrial byproducts as a substrate, production efficiency has increased, thus making microbial pigments commercially feasible (Rodríguez-Sáiz *et al.*, 2010).

Microbial pigments analyzed in this chapter include the following: melanins from *Aspergillus niger* and *Streptomyces* spp., violacein from *Chromobacterium violaceum*, prodigiosin from *Serratia marcescens*, and carotenoids such as β-carotene and astaxanthin produced by *Blakeslea trispora* and *Phaffia rhodozyma* (Manivasagan *et al.*, 2018). It also includes phycobiliproteins of cyanobacteria and algae, as well as azaphilones (such as monascorubrin) made by *Monascus* spp. (Dufossé, 2006; Shi, 2017). It describes the individual unique properties, biosynthesis pathways, and industrial applications of each of the pigments. Growing interest in microbial pigments coincides with the sustainability goals of the world, creating more green options to replace synthetic dyes. The chemical structure of microbial pigment determines its color, solubility, and functional properties. So, on the basis of chemical structure, the microbial pigments are grouped into isoprenoids, flavin, tetrapyrrole, alkaloid pigments, polyketide, phenol-containing pigments, and melanin (Barreto *et al.*, 2023)

3.2 Isoprenoid Derivatives

These derivatives are also called the terpenoids, which are derived from the isoprene units (Delgado-Vargas *et al.*, 2000). Isoprenoids mostly constitute of carotenoids, a complex and widely applicable pigment.

*Corresponding author: krishan.kumar20140@paruluniversity.ac.in

DOI: 10.1079/ 9781836990505.0003

3.2.1 Carotenoids

Carotenoids are isoprenoid pigments that may range from yellow to orange-red in color and are considered to be omnipresent, synthesized by almost all types of microbes. They are considered to be strong antioxidants and are widely used in food dyes (Malik *et al.*, 2012). Carotenoids are produced by certain classes of microbes whose characteristics may vary by species; they tend to produce types and colors for various applications and are more reliable than synthetic products which can harm human health. Carotenoids are usually made of eight isoprenoid elements, generally inverted at the core. They are broadly classified into two groups: carotene and xanthophylls (Fig. 3.1). Carotene is a nonoxygenated molecule while xanthophylls are oxygenated molecules.

Carotenoids are produced by a wide range of sources such as plants, fungi, algaes and bacteria. Carotenoids such as β-carotene and lutein are responsible for the green color in plant tissues well as the color of leaves in autumn (Barreto *et al.*, 2023). Certain classes of fungi, including species from the genera *Rhodotorula* and *Neurospora*, are known to produce carotenes specifically.

The color found in many classes of algae such as Phaeophyta, Pyrrophyta, Chlorophyta, Chloromonadophyta, and Rhodophyta is due to α- and β-carotene (Barreto *et al.*, 2023). Certain classes of carotenoids are produced by bacteria such as Cyanobacteria, which carry saccharide-linked carotenoids such as myxoxanthophyll, and *Flavobacterium* spp. which are a source of zeaxanthin (Delgado-Vargas *et al.*, 2000). Similarly, *Arthrobacter* spp., *Serratia marcescens*, *Pseudomonas aeruginosa*, *Myxococcus*, *Agrobacterium*, and *Streptomyces* are known to produce different types of carotenoids (Delgado-Vargas *et al.*, 2000).

Carotenoids are not only known for their variety of color but also known for their wide applications. β-carotene, the most popular among the carotenoids, is also known to produce retinaldehyde which plays a crucial role in animal vision. Carotenoids in plant surfaces work like signal receptors for light, increasing the rates of photosynthesis (Britton and Britton, 1983). Carotenoids also contribute to the characteristic color of many organisms such as fishes and birds.

β-carotene also promotes the formation of spores in fungi. It has been shown to inhibit cancers in the lung and is also known to boost immunity (Kordiak *et al.*, 2022). Carotenoids such as β-carotene and lycopene generally are highly antioxidant (Kordiak *et al.*, 2022).

3.3 Flavins

Flavins are water-soluble yellow biological compounds which act as coenzymes in many enzymatic reactions. Flavin consists of a

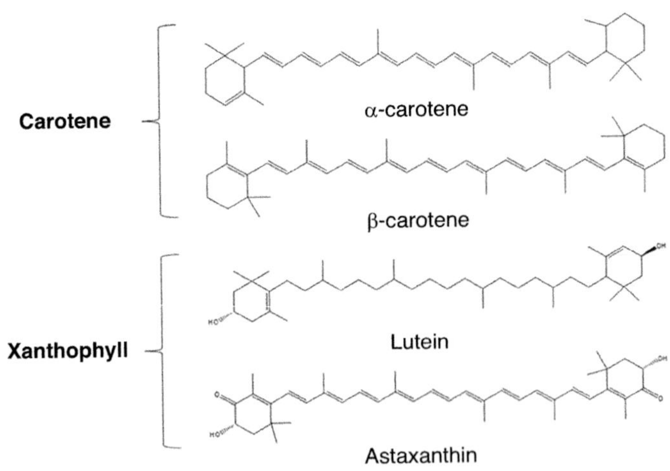

Fig. 3.1. Chemical structure of carotene and xanthophyll. Image created with ChemDraw Pro 8.0.

N-heterocylic isoalloxazine ring which is the reason for its yellow color (Fig. 3.2). The natural flavins produced by microorganisms are riboflavin, roseoflavin, and 5-deazaflavins.

3.3.1 Riboflavin

Riboflavin, commonly known as vitamin B2, is a yellow-colored, water-soluble pigment found in microbes and plants. However, higher animals cannot produce riboflavin due to the absence of specific biosynthetic pathways, making it essential to obtain from external sources (Revuelta *et al.*, 2017). It is a precursor for two important coenzymes: flavin mononucleotide (FMN) and flavin adenine dinucleotide (FAD) (Alam *et al.*, 2015). These molecules generally participate in redox reactions in most organisms, supporting the breakdown of complex macromolecules such as sugars and proteins to produce ATP (Sepúlveda Cisternas *et al.*, 2018). Riboflavin was first discovered in milk and was successfully isolated in 1879, initially named lactochrome (Beztsinna *et al.*, 2016). After its structural identification in 1933, it was renamed riboflavin. The term "flavin" was derived from the presence of an isoalloxazine ring, with *flavus* being the Latin word for yellow, while *ribo* was chosen due to the ribitol side chain (Fig. 3.2A).

Since riboflavin is water soluble, it is not stored in large amounts in the human body and must be consumed through diet. It is primarily obtained from sources such as eggs, lean meats, organ meats, muscadine grapes, mushrooms, and almonds (Unagul *et al.*, 2005). Vitamin B2 deficiency is more commonly observed in developing nations. Since riboflavin is an essential dietary component, its prolonged deficiency can lead to various health issues, including neurological problems, heart-related diseases, and even cancer (Rana *et al.*, 2021). It may also cause skin-related disorders, anaemia, renal failure, malnutrition, headaches, osteoarthritis, and several other medical conditions (Thakur *et al.*, 2017). Riboflavin is also known to enhance immunity and help fight against cancer cells, while also boosting the effectiveness of commercial anticancer medicines (Matnuri *et al.*, 2015).

Apart from chemical synthesis, the most cost-effective method of riboflavin production is through microbial fermentation, which can be achieved using flavinogenic anaerobic processes or genetically modified microbes (Revuelta *et al.*, 2017). Microbial synthesis is commonly observed in fungi such as *Ashbya gossypii*, *Eremothecium ashbyii*, *Candida albicans*, and *Candida famata*, as well as in bacteria such as *Escherichia coli*, *Bacillus subtilis*, *Lactobacillus fermentum*, and *Clostridium acetobutylicum* (Abbas and Sibirny, 2011). Another example is roseoflavin (Fig. 3.2B), produced by *Streptomyces davawensis*. It is a structural analogue of riboflavin and acts as an antibacterial compound (Barreto *et al.*, 2023). It forms an inactive complex by binding with active sites of FAD synthetase and flavokinase yielding and inhibits major cellular reactions. 7,8-Dimethyl-8-hydroxy-5 deazaflavin (Fig. 3.2C) is another example of flavin, which is found in Archaea (Barreto *et al.*, 2023). In this compound, the nitrogen in the fifth position of the isoalloxazine ring is replaced by a carbon. This compound also acts as a coenzyme for flavoproteins. It is obtained from bacteria, fungi, algae, and certain plants (Barreto *et al.*, 2023).

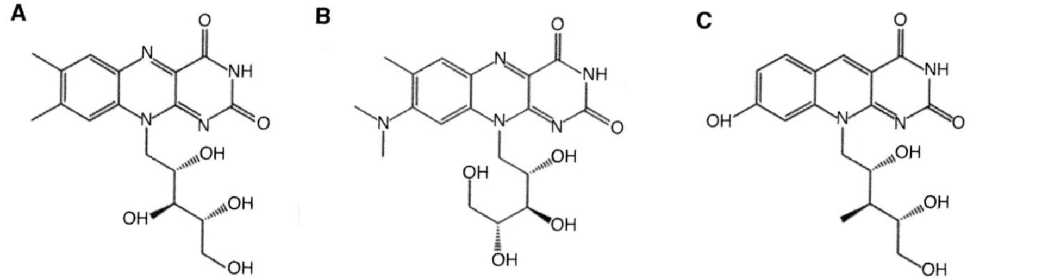

Fig. 3.2. Chemical structure of riboflavin (A), roseoflavin (B), and 7,8-dimethyl-8-hydroxy-5 deazaflavin (C).

3.4 Tetrapyrrole Derivatives

The basic structure of this pigment is made up of four pyrrole residues that are connected by a single linking carbon atom. Tetrapyrrole derivatives are grouped into linear tetrapyrroles and cyclic tetrapyrroles. Tetrapyrroles are classified into two major classes that include chlorophyll, heme groups, and vitamin B12 in the cyclic tetrapyrroles while bilins, phycobilins, and phytochromes are generally considered to be linear tetrapyrroles (Barreto *et al.*, 2023).

Chlorophyll/bacteriochlorophyll plays a major role in photosynthetic activities and consists of the pyrrole N-chelating magnesium ion (Fig. 3.3A). Chlorophyll a and b are mostly found in higher plants while classes c and d are occur in certain species of algae, and bacteriochlorophyll classes a and b usually occur in phototrophic bacteria.

Heme possesses a porphyrin outline chelating iron ions (Fig. 3.3B). Heme groups usually contribute as the coenzyme class of many hemoproteins including hemoglobin, myoglobin, cytochromes, and enzymes such as peroxidase and catalase (Amanullah *et al.*, 2022). In vitamin B12, cobalt plays the role of the chelation ion molecule. Vitamin B12 is entirely produced by microorganisms (Balabanova *et al.*, 2021).

Bilin pigments are generally produced via the breakdown of heme groups (Willows *et al.*, 2023). They are most common among the higher organisms, more precisely in mammals. Phycobilin are a subclass of bilins that are very common among various classes of algae such as Rhodophyta, Cryptophyta, and Cyanobacteria, contributing to their red and green appearance. Phycobilins contain highly pigmented aqueous proteins called phycobiliproteins, which are classified into major two groups on the basis of their characteristics; phycocyanobilin absorbs light up to 640nm (Fig. 3.3C) which phycoerythrobilin absorbs light up to 550 nm (Fig. 3.3D) (Willows *et al.*, 2023).

Tetrapyrrole derivates are a key component of chlorophyll which plays a major role in photosynthesis pathways in plants (Willows *et al.*, 2023). Heme groups are essential O_2 transporting agents; the appearance of the red feathers on the touraco species is contributed by uroporphyrin III, a type of tetrapyrrole derivative (Willows *et al.*, 2023). Bilins usually contribute as the epidermal pigments of several arthropods and invertebrates (Stavenga, 2024) and phycobilins have a major contribution to the red and blue pigments on certain classes of algae (Stavenga, 2024).

3.5 Alkaloid Derivatives

The alkaloid class of microbial pigments includes naturally occurring, low molecular weight, nitrogen-containing compounds which are produced by microorganisms such as bacteria,

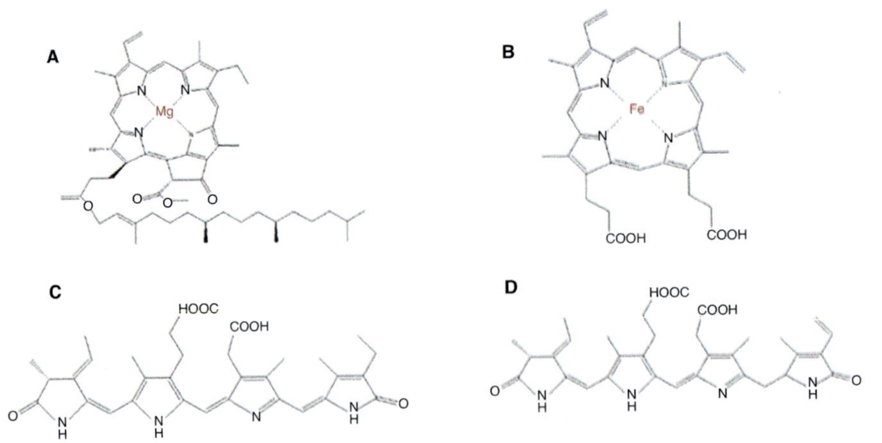

Fig. 3.3. Chemical structure of different tetrapyrrole derivatives: (A) chlorophyll a; (B) heme; (C) phycocyanin; (D) phycoerythrin. Image created with ChemDraw Pro 8.0.

fungi, and algae. They have properties of both pigments and alkaloids (Barreto *et al.*, 2023). This class includes a wide range of pigments such as betalains, phenazines, tambjamines, violaceins, and prodigiocins which are discussed below.

3.5.1 Betalains

Betalains are nitrogen-based water-soluble pigments that exist naturally in some plants and fungi, notably in plants of the Caryophyllales order and mushrooms such as *Basidiomycetes phylum*. They give many fruits and vegetables their pigmentation and are renowned for their vivid hues, which can vary from yellow to crimson.

Betanidin is an aglycone which is connected to a β-glycoside unit at the C-5 position to form the fundamental structure of betalains. For the pigment to be stable and soluble in aqueous solutions, this bond is essential. Betanidin is generated by condensation of cyclo-DOPA (cyclo-3,4-dihydroxyphenylalanine) with betalamic acid (Fig. 3.4A) (Stintzing and Carle, 2004). Production of the reddish-violet pigments called betacyanins depends on this process. The resonance structure that double bonds produce is responsible for the hue of betalains. Other betalains called betaxanthins are formed by condensation of betalamic acid with amino acids or other aminated molecules (Fig. 3.4A) (Barreto *et al.*, 2023). Betaxanthins are usually yellow whereas betacyanins are reddish-violet. The wavelengths specifically linked to these hues are around 540 nm for betacyanins and 480 nm for betaxanthins (Stintzing and Carle, 2004).

Betalains are bioactive compounds primarily characterized by their bright colors and considerable health benefits (Gandía-Herrero and García-Carmona, 2013). They have powerful antioxidant effects that help the body to protect itself from free radicals. This property is important for promoting overall health and life span by protecting cells from oxidative stress and damage (Gandía-Herrero and García-Carmona, 2013). Studies have shown that these substances reduce inflammation in the body. Regular intake of food high in betalain, such as beetroot, helps to alleviate the effects of inflammatory diseases (Wootton-Beard and Ryan, 2011). Betalains can promote healthy cardiovascular functioning with repeated use. They assist in decreasing blood pressure and oxidative stress and increasing endothelial function, which plays an important role in maintaining normal blood vessels (Wootton-Beard and Ryan, 2011). They also possess antimicrobial properties that can help prevent diseases and improve overall wellbeing (Gengatharan *et al.*, 2015).

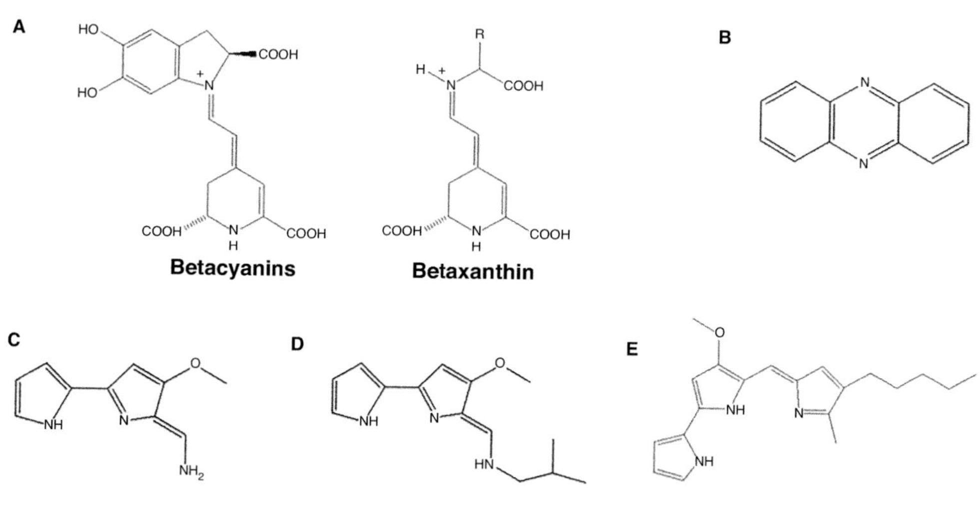

Fig. 3.4. Chemical structure of alkanoid derivatives: (A) betalains; (B) phenazines; (C) tambjamines; (D) violacein; (E) prodigiosin. Image created with ChemDraw Pro 8.0.

3.5.2 Phenazines

Certain bacteria, including fluorescent *Pseudomonas* and streptomycetes, produce phenazines, which are colored, redox-creating metabolites (Serafim *et al.*, 2023). They can combat a range of infections due to their well-known broad-spectrum antibiotic effect. Phenazines are defined by a fused ring system of which nitrogen atoms form a part (Fig. 3.4B). The distinct perservation of their pigments and functional groups is often associated with their biological activity. The various types and positioning of these functional groups can affect their properties as pigments and antibacterial agents (Serafim *et al.*, 2023).

These bacteria are found in diverse environments, including soil and water, and are associated with both plant and animal hosts. The fact that phenazines can be extracted from environmental samples therefore implies that they are indeed widespread in nature (Mavrodi *et al.*, 2010).

Phenazines work well against many diseases because of their strong antibacterial action. They inhibit the growth of fungi and bacteria, making them useful in farm pest control. This trait allows them to shield crops from microbial diseases by acting as natural pesticides (Banin *et al.*, 2008; Moreau-Marquis *et al.*, 2009). Some phenazine variants have shown anticancer effects. Studies have shown how they can cause cancer cells to die, highlighting their possible use in treating cancer. This happens because of their redox properties, which can create reactive oxygen species (ROS) that harm cancer cells (Laursen and Nielsen, 2004; Mavrodi *et al.*, 2010). Phenazines help electron movement in microbial fuel cells because they can change their charge. The textile industry uses phenazines as biocolors to dye silk. Their bright colors, which change with pH and charge state, allow their use for many color activities and varied applications (Saranya *et al.*, 2012). Some phenazines work as natural substitutes for artificial dyes in food. Their safety and efficacy make them desirable in food applications (Saranya *et al.*, 2012).

3.5.3 Tambjamines

Agents classified into a naturally formed yellow pigment class are called tambjamines, which are mostly harvested from marine sources. They are typically methoxybipyrroles bearing a flexible imine moiety and have similarities to prodiginines (Carbone *et al.*, 2010). Antibacterial, antifungal, antimalarial, and antitumor actions have been seen.

Tambjamines are characterized by two pyrrole rings carrying an enamine functional group at position C-5 and a methoxy substitute at position C-4 (Fig. 3.4C). In many of those compounds, the nitrogen atom within the enamine moiety is substituted by an alkyl chain unsuitable to be considered an amino function. Such arrangements influence their chemical properties and biological activities (Carbone *et al.*, 2010; Hernández *et al.*, 2012). The metabolic intermediate 4-methoxy-2,2'-bipyrrole-5-carbo xaldehyde (MBC) is a common feature between tambjamines and prodigiosins. This occurrence is generated by the compatible condensation of another nucleophile with several fatty acid-derived amines and 2-methyl-3-amylpyrrole (MAP) (Sakuta, 2014).

Tambjamine inhibits immunoproliferation in mice and alleviates gastritis symptoms in rabbits, demonstrating potential immunomodulatory effects. It has shown nematocidal activity on *Caenorhabditis elegans*, a model nematode (Picott *et al.*, 2019). Tambjamines exhibit diverse bioactivities, including antimicrobial, anticancer, immunomodulatory, and DNA-targeting effects. Their anion transport and metal-binding properties may contribute to their mechanisms of action. However, toxicity to normal cells and stability issues remain challenges for therapeutic development (Barreto *et al.*, 2023).

3.5.4 Violacein

The violet pigment violacein is an indole derivative, isolated mainly from bacteria of the genus *Chromobacterium*, which exhibits important antitumoral, antimicrobial, and antiparasitic properties. Violacein has a molecular mass of 343.3 g/mol. Its chemical structure is represented as 3(1,2-dihydro-5-(5-hydroxy-1H-indol -3-yl)−2-oxo-3H-pyrrol-3-ilydene)−1,3-dihydro -2H-indol-2-one, which gives it a purplish-black appearance (Fig. 3.4D) (Durán *et al.*, 2021). Violacein is insoluble in water, slightly soluble in ethanol, moderately soluble in dioxane and acetone, and soluble in DMSO, methanol, and ethyl acetate. Its melting point is 290°C (decomposition). The fluorescence emission spectrum of violacein shows an emission band at 675 nm when excited at 575 nm, indicating its potential for phototherapeutic applications. *C. violaceum* has been noted to produce violacein at a higher rate compared to other strains (Nakamura *et al.*, 2003).

Numerous important biological activities are displayed by violacein.It is a possible cancer treatment option because of its possible antitumor effects. By activating particular signaling pathways, such as TNFR1 signaling, it causes leukemia cells to undergo apoptosis (Bromberg *et al.*, 2005). Violacein offers diverse applications including antibacterial properties (Durán *et al.*, 2010), antileishmanial activity (Leon *et al.*, 2001), antimalarial activity (Lopes *et al.*, 2009), and antiviral activity (Rui *et al.*, 2012).

3.5.5 Prodigiosin

Prodigiosin is a linear tripyrrole red pigment in the prodiginine class (Fig. 3.4E). The major source of this secondary metabolite is *Serratia marcescens*, a significant human pathogen. Because of its possible anticancer, antibacterial, and even immunosuppressive properties, it is receiving considerable attention and could be used in therapeutics as well as diverse industries such as textile dyeing (Chauhan *et al.*, 2017).

Prodigiosin is biosynthesized through enzymatic condensation of 4-methoxy-2 –2-bipyrrole-5-carbaldehyde (MBC) with 2-methyl-3-n-amyl-pyrrole (MAP). The biosynthesis of these pigments is controlled by the *Pig* gene cluster (Han *et al.*, 2021). Biological systems will be affected by this molecule because it is nonpolar. Its importance lies in possible applications in the formulation as well as distribution of drugs (Chauhan *et al.*, 2017). Along with color properties, prodigiosin has demonstrated anticancer effects (Elahian *et al.*, 2013), antimicrobial properties (Yip *et al.*, 2019), and antimalarial activity (Kancharla *et al.*, 2021).

3.6 Polyketides

Polyketides are secondary metabolites secreted by microorganisms. They are formed of tetraketides and octaketides which consist of C2 units (Lin and Xu, 2020). Polyketides are synthesized through a complex enzymatic system consisting of different kind of polyketide synthases (PKSs). Genes that encode the PKSs are found in close proximity to each other (Woo *et al.*, 2003). They are abundantly produced by ascomycetous fungi, including the filamentous genera *Neurospora* and *Monascus* (Kalra *et al.*, 2020).

3.6.1 Azaphilones

Azaphilones are secondary metabolites of fungi with structural diversity and also the derivatives of polyketides. Many ascomycetous and basidiomycetous fungi, especially the former, produce azaphilone pigments, which give fruiting bodies and/or mycelia their yellow, red, or green colors (Lin and Xu, 2022). These include the genera *Aspergillus*, *Penicillium*, *Chaetomium*, *Talaromyces*, *Pestalotiopsis*, *Phomopsis*, *Emericella*, and *Epicoccum* (Lin and Xu, 2022). The azaphilonoid category includes the yellow pigments monascin and ankaflavin (Fig. 3.5A), produced from the fungal genus *Monascus*, which have been shown to strongly suppress preadipocyte development and encourage the lipolysis of mature adipocytes (Liu *et al.*, 2025). These two pigments help in the lowering of blood cholesterol levels and the production of triglycerides (Lin and Xu, 2022).

Chaetoviridins, a family of azaphilones, are produced by *Chaetomium globosum* and exhibit potent antifungal properties (Gao *et al.*, 2023). Azaphilones also exhibit a wide range of biological properties such as cytotoxic, antiinflammatory, antiviral, antimicrobial, antitumoral, and antioxidant activity (Lin and Xu, 2022).

3.6.2 Quinones

Quinones are natural substitutes for synthetic dyes. In comparison to synthetic dyes, they exhibit more effective dyeability, brightness,

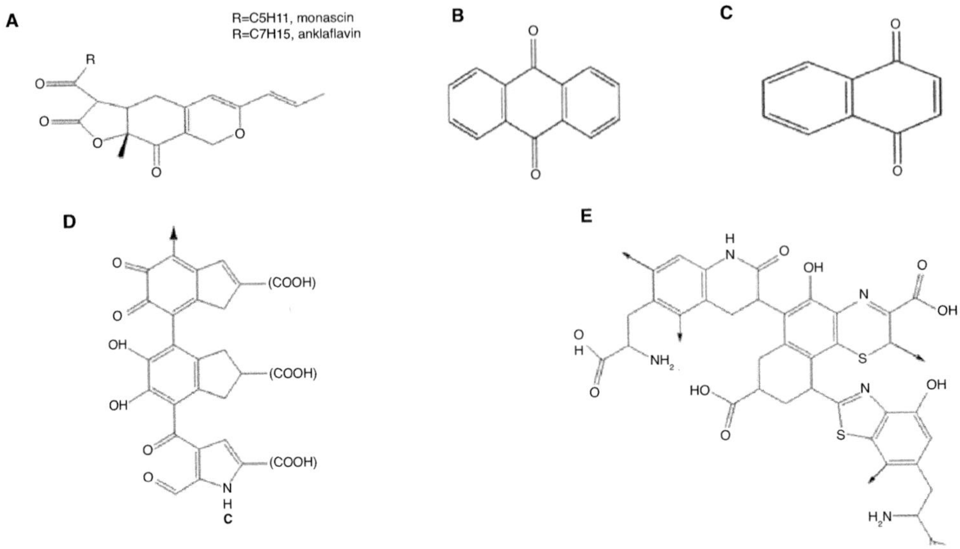

Fig. 3.5. Chemical structure of alkanoid derivetives: (A) monascin and anklaflavin; (B) anthraquionone; (C) naphthoquinone and derivatives of melanins; (D) eumelanin; (E) pheomelanin. Image created with ChemDraw Pro 8.0.

and fastness. They are ecologically suitable dyes which are simple and quick to separate, refine, and estimate. Natural quinones constitute a large class of aromatic compounds that are prevalent in nature and can be found in several families of algae, fungi (including lichens), bacteria, flowering plants, and arthropods. (Dulo *et al.*, 2021). Most of these have been claimed to have vital biological roles, such as suppressing microorganisms and limiting insect immune response. They are primarily secondary metabolites of polyketides. Due to their redox characteristics and their potential applications as antioxidants, antibacterials, and antifungals, they are of importance to humanity (Christiansen *et al.*, 2021).

For the synthesis of natural quinones in plants and other organisms, two distinct processes have been suggested (Weigl *et al.*, 2009). The polyketide or acetate-malonate route involves at least two hydroxyl groups being substituted in the R1 and R8 locations as well as systematic ring closure on both aromatic rings. This results in an octaketide chain made of acetate with monolite folding and condensing appropriately. The shikimate/chorismate/O-succinylbenzoic acid pathway is the second suggested pathway, in which pigments ultimately have at least one hydroxyl group on the R1 position and only one ring remaining unsubstituted (Caro *et al.*, 2012).

Here, quinones are classified as anthraquinone and naphthoquinone.

3.6.2.1 Anthraquinone

Anthraquinone is a quinone which has an anthracene nucleus with two carbonyl groups, typically on the B-ring. Anthracene-9,10-quinone, also known as 9,10-dioxoanthracene ($C_{14}H_8O_2$) (Fig. 3.5B), is their basis (Duval *et al.*, 2016). Numerous anthraquinone derivatives can be found in nature as a result of different substitution patterns (Duval *et al.*, 2016). The amount, kind, and location of hydroxyl and other electron-donating or accepting groups (auxochromes) affect hydrogen bonding and steric effects, determining the hue. The characteristics of dyes can be drastically changed by amino or substituted amino groups, yet most electron-donating groups have little effect on color.

Anthraquinones are found in plant families including Rubiaceae, Leguminosae, Polygonaceae, Bignoniaceae, Verbenaceae, Scrophulariaceae, and Liliaceae. These substances have been found in every component of plant tissues and organs, including heartwood, root bark, leaves, stems, pods, seed coats, fruits, nut shells, flowers, tubers, and leaf glands (Derksen, 2001). Emodin is the most commonly occurring anthraquinone because of its presence in higher fungi, molds, and lichens (Dulo *et al.*, 2021).

Anthraquinones have a crucial role in the medical field as agents with fungicidal, antibacterial, insecticidal, antiparasitic, antiviral, and anticancer properties.

3.6.2.2 Naphthoquinones

Naphthoquinones are present in higher plants, fungi, and actinomycetes in nature. They are the secondary metabolites of fungi. Detailed study of the synthesis of naphthoquinones by fungi began in 1965.They share a structural similarity with naphthalene as shown in Fig. 3.5C. Naphthoquinones exhibit important pharmacological activity against bacterial, fungal, viral, and insect diseases and pests (Mone *et al.*, 2021). They also have antipyretic and antiinflammatory properties.

A variety of naphthoquinones with a wide range of biological activity have been demonstrated to be produced by *Fusarium* spp. (Mendenstev *et al.*, 1998). As an example, bikaverin, a red heterotetracyclic pigment with the chemical structure of 6,11-dihydroxy-3,8-d imethoxy-1-methylbenzo[b]xanthene-7,10,12-trione which is produced by *Fusarium fujikuroi* and other fungi, has demonstrated antineoplastic activity, indicating its potential as a pharmaceutical agent against lymphoma, carcinoma, sarcoma, etc. (Tammam *et al.*, 2022).

3.7 Melanin

Melanins are a group of biological pigments found in almost all organisms. Melanin is produced by a particular type of cell known as a melanocyte, which is distributed unevenly (Guo *et al.*, 2023). Melanin's name originated from the Greek word *melanos*, referring to "black" or "very dark," that describes the look of melanin (Borovansky *et al.*, 2011). It is mainly dark brown to black, sometimes yellowish and red (Singh *et al.*, 2021). It is abundant in nature and can be found in many different types of bacteria and fungi, plants, cephalopod ink, and mammal skin and hair (Caldas *et al.*, 2020). Because of its diverse properties and high biocompatibility, melanin has drawn focus in the food business, medicine, and agriculture, among other disciplines. It is also involved in the protection of microorganisms from harsh environments such as high and low temperatures, low water content, increasing radioactivity, nutrient depletion, etc. (Muñoz-Torres *et al.*, 2024).

Melanin is an endogenous, naturally occurring functional molecule that comprises phenolic hydroxyl, carboxyl, and amino active groups that have the ability to absorb ultraviolet light (Diffey *et al.*, 1995). Melanin's composition and structure have a direct impact on its biological functions, including antibacterial, antitumor, chelating metal, free radical scavenging, and radiation protective properties, which have steadily emerged as the main focus of melanin activity research (Diffey *et al.*, 1995). Melanin is a naturally occurring sunscreen that absorbs broad-spectrum light and thereby shields the skin from UV light's potentially harmful effects (Diffey *et al.*, 1995).

Melanin is an amorphous, heterogeneous polymer containing phenolic or indoles with many uses. It is mainly classified into three groups: eumelanin, pheomelanin, and allomelanin (Hill, 1992).

3.7.1 Eumelanin

Eumelanin is a heterogeneous polymer that is a member of the aggregation network of oligomers and is made up of 5,6-dihydroxyindole (DHI) and 5,6-dihydroxyindole-2-carboxyl ic acid (DHICA) (Fig. 3.5D). It is synthesized from tyrosine with the involvement of enzyme tyrosinase and has 5,6- dihydroxyindole as the main precursor. Dopachromes DHI and DHICA are produced by oxidizing tyrosine or levodopa, which are then quickly oxidized to cyclodopa by cyclizing dopaquinone (Tsukamoto *et al.*, 1992).

It is also produced by some bacteria, fungi, and myxomycetes, which is a separate taxon of protist. The size of the homogeneous particles is 200 nm and they spherical in shape.

Eumelanin is found mostly in dark-skinned people. It may disperse up to 90% of the energy collected from sunlight and exhibits a very broad UV-visible absorption spectrum (Ito *et al.*, 2018).

3.7.2 Pheomelanin

Pheomelanin is more useful than eumelanin, as most of this pigment can be dissolved in alkali media. The main difference between eumelanin and pheomelanin is the presence of sulfur in pheomelanin (Singh *et al.*, 2021). The common precursor dopaquinone is the source of phenomelanin, as for eumelanin. On the other hand, dopacysteine (CD isomer), which is created when L-dopa integrates cysteine, undergoes oxidation, cycling, and rearrangement to produce 1, 4-benzothiazine. Pheomelanin is created by polymerizing the last 1,4-benzothiazine (Fig. 3.5E) (Singh *et al.*, 2021). Benzothiazine and benzothiazole, which are oligomers of sulfur elements rather than indole units, are the primary structures of pheomelanin (Napolitano *et al.*, 2013). Large amounts of benzothiazine- or benzothiazole-like pigment, which is primarily yellowish or reddish, are found in reptiles, insects, amphibians, and poultry feathers (Wolnicka-Glubisz *et al.*, 2012).

3.7.3 Allomelanin

Allomelanins are a type of nitrogen-free heterogeneous pigments present in many plants and fungi. There are numerous sources, such as catechol, dihydrofolic acid, and hyperic acid (El-Naggar, 2021). Allomelanin is a collection of diverse polymers that are produced in a range of hues by the oxidation or polymerization of 1,8-dihydroxynaphthalene (di-DHN) or 1,3,6,8-tetrahydroxynaphthalene (THN) (Plonka and Grabacka, 2006). The detoxification of phenolic chemicals that accumulate in senescing cells is accomplished by allomelanin, which is produced in the nodules of *Rhizobium* species (Cubo *et al.*, 1988).

3.8 Phenol-Containing Pigments

Phenolic pigments are mostly the styrylpyrones, usually yellow in color. In 1889, hispidin was initially isolated from *Inonotus hispidus*. They are part of the Hymenochaetaceae family, and a number of styrylpyrones can be obtained such as bisnoryangonin and hypholomin B from certain classes of fungi (Yadav *et al.*, 2023). Hispidin plays a vital role in light-emitting fungi such as the Agaricales (Palkina *et al.*, 2023). These phenolic compounds purified from lab-grown microbes are known to possess great antioxidant activities (Lee and Yun, 2011).

3.9 Conclusion

Microbial pigments are compounds derived from microbes such as bacteria, fungi, and algae. In addition to colorant properties, pigments also have bioactive properties. Microbial pigments are sustainable alternatives to synthetic colorants, with wide applications in food, textiles, pharmaceuticals, and cosmetics. Pigments produced from microorganisms belong to different groups such as isoprenoids, carotenoids, flavin, melanin, violacein, polyketides, and prodigiosin, each with unique chemical properties and specific biological functions.

Although large-scale production of natural pigments is challenging, microbial pigments offer significant economic, industrial, environmental, and commercial potential that makes them an attractive alternative to synthetic pigments.

References

Abbas, C.A. and Sibirny, A.A. (2011) Genetic control of biosynthesis and transport of riboflavin and flavin nucleotides and construction of robust biotechnological producers. *Microbiology and Molecular Biology Reviews* 75(2), 321–360. DOI: 10.1128/MMBR.00030-10.

Alam, M.M., Iqbal, S. and Naseem, I. (2015) Ameliorative effect of riboflavin on hyperglycemia, oxidative stress and DNA damage in type-2 diabetic mice: Mechanistic and therapeutic strategies. *Archives of Biochemistry and Biophysics* 584, 10–19. DOI: 10.1016/j.abb.2015.08.013.

Amanullah, S., Ghosh, C., Dey, S.G. and Dey, A. (2022) Heme proteins – structure and function. In: *Fundamentals of Porphyrin Chemistry: A 21st Century Approach*, Vol. 2. pp. 709–742.

Balabanova, L., Averianova, L., Marchenok, M., Son, O. and Tekutyeva, L. (2021) Microbial and genetic resources for cobalamin (vitamin B12) biosynthesis: From ecosystems to industrial biotechnology. *International Journal of Molecular Sciences* 22(9), 4522. DOI: 10.3390/ijms22094522.

Banin, E., Lozinski, A., Brady, K.M., Berenshtein, E., Butterfield, P.W., Moshe, M., Chevion, M., Peter Greenberg, E. and Banin, E. (2008) The potential of desferrioxamine-gallium as an anti-Pseudomonas therapeutic agent. *Proceedings of the National Academy of Sciences of the United States of America*, 105(43):16761–16766. DOI: 10.1073/pnas.0808608105.

Barreto, J.V. de O., Casanova, L.M., Junior, A.N., Reis-Mansur, M.C.P.P. and Vermelho, A.B. (2023) Microbial pigments: Major groups and industrial applications. *Microorganisms* 11(12), 2920. DOI: 10.3390/microorganisms11122920.

Beztsinna, N., Solé, M., Taib, N. and Bestel, I. (2016) Bioengineered riboflavin in nanotechnology. *Biomaterials* 80, 121–133. DOI: 10.1016/j.biomaterials.2015.11.050.

Borovanský, J. (2011) History of melanosome research. *Melanins and melanosomes: biosynthesis, biogenesis, physiological, and pathological functions*, 1–19. DOI: 10.1002/9783527636150.ch1.

Britton, G. (1983) *The Biochemistry of Natural Pigments*. Cambridge University Press.

Bromberg, N., Justo, G.Z., Haun, M., Durán, N. and Ferreira, C.V. (2005) Violacein cytotoxicity on human blood lymphocytes and effect on phosphatases. *Journal of Enzyme Inhibition and Medicinal Chemistry* 20(5), 449–454. DOI: 10.1080/14756360500273052.

Caldas, M., Santos, A.C., Veiga, F., Rebelo, R., Reis, R.L. *et al.* (2020) Melanin nanoparticles as a promising tool for biomedical applications – A review. *Acta Biomaterialia* 105, 26–43. DOI: 10.1016/j.actbio.2020.01.044.

Carbone, M., Irace, C., Costagliola, F., Castelluccio, F., Villani, G. *et al.* (2010) A new cytotoxic tambjamine alkaloid from the Azorean nudibranch *Tambja ceutae*. *Bioorganic & Medicinal Chemistry Letters* 20(8), 2668–2670. DOI: 10.1016/j.bmcl.2010.02.020.

Caro, Y., Anamale, L., Fouillaud, M., Laurent, P., Petit, T. *et al.* (2012) Natural hydroxyanthraquinoid pigments as potent food grade colorants: An overview. *Natural Products and Bioprospecting* 2(5), 174–193. DOI: 10.1007/s13659-012-0086-0.

Chauhan, R., Choudhuri, A. and Abraham, J. (2017) Evaluation of antimicrobial, cytotoxicity, and dyeing properties of prodigiosin produced by *Serratia marcescens* strain JAR8. *Asian Journal of Pharmaceutical and Clinical Research* 10(8), 279–283. DOI: 10.22159/ajpcr.2017.v10i8.18173.

Christiansen, J.V., Isbrandt, T., Petersen, C., Sondergaard, T.E., Nielsen, M.R. *et al.* (2021) Fungal quinones: Diversity, producers, and applications of quinones from Aspergillus, Penicillium, Talaromyces, Fusarium, and Arthrinium. *Applied Microbiology and Biotechnology* 105(21–22), 8157–8193. DOI: 10.1007/s00253-021-11597-0.

Cubo, M.T., Buendia-Claveria, A.M., Beringer, J.E. and Ruiz-Sainz, J.E. (1988) Melanin production by Rhizobium strains. *Applied and Environmental Microbiology* 54(7), 1812–1817. DOI: 10.1128/aem.54.7.1812-1817.1988.

Delgado-Vargas, F., Jiménez, A.R. and Paredes-López, O. (2000) Natural pigments: Carotenoids, anthocyanins, and betalains--characteristics, biosynthesis, processing, and stability. *Critical Reviews in Food Science and Nutrition* 40(3), 173–289. DOI: 10.1080/10408690091189257.

Derksen, G.C. (2001) *Red, Redder, Madder: Analysis and Isolation of Anthraquinones from Madder Roots (Rubia Tinctorum)*. Wageningen University and Research.

Diffey, B.L., Healy, E., Thody, A.J. and Rees, J.L. (1995) Melanin, melanocytes, and melanoma. *Lancet* 346(8991), 1713.

Dufossé, L. (2006) Microbial production of food grade pigments. *Food Technology and Biotechnology* 44(3), 313–321. Available at: https://hrcak.srce.hr/109913

Dulo, B., Phan, K., Githaiga, J., Raes, K. and De Meester, S. (2021) Natural quinone dyes: A review on structure, extraction techniques, analysis and application potential. *Waste and Biomass Valorization* 12(12), 6339–6374. DOI: 10.1007/s12649-021-01443-9.

Durán, M., Faljoni-Alario, A. and Durán, N. (2010) *Chromobacterium violaceum* and its important metabolites-Review. *Review Folia Microbiology* 55(6). Available at: www.biomed.cas.cz/mbu/folia/

Durán, N., Fávaro, W.J., Brocchi, M., Justo, G.Z., Castro, G.R. *et al.* (2021) Patents on violacein: A compound with great diversity of biological activities and industrial potential. *Recent Patents on Biotechnology* 15(2), 102–111. DOI: 10.2174/2213476X07666201221111655.

Duval, J., Pecher, V., Poujol, M. and Lesellier, E. (2016) Research advances for the extraction, analysis and uses of anthraquinones: A review. *Industrial Crops and Products* 94, 812–833. DOI: 10.1016/j.indcrop.2016.09.056.

El-Naggar, N.E.A. (2021) Streptomyces-based cell factories for production of biomolecules and bioactive metabolites. In: *Microbial Cell Factories Engineering for Production of Biomolecules*. Academic Press, pp. 183–234.

Elahian, F., Moghimi, B., Dinmohammadi, F., Ghamghami, M., Hamidi, M., Mirzaei, S.A. (2013) The anticancer agent prodigiosin is not a multidrug resistance protein substrate. *DNA and Cell Biology*, 32(3), 90–97. DOI: 10.1089/dna.2012.1902.

Gandía-Herrero, F. and García-Carmona, F. (2013) Biosynthesis of betalains: Yellow and violet plant pigments. *Trends in Plant Science* 18(6), 334–343. DOI: 10.1016/j.tplants.2013.01.003.

Gao, Z.J., Cao, L.L., Ren, H.P., Yu, H. and Wang, Y. (2023) Novel chlorinated and nitrogenated azaphilones with cytotoxic activities from the marine algal-derived fungus *Chaetomium globosum* 2020HZ23. *Frontiers in Microbiology* 14, 1252563. DOI: 10.3389/fmicb.2023.1252563.

Gengatharan, A., Dykes, G.A. and Choo, W.S. (2015) Betalains: Natural plant pigments with potential application in functional foods. *LWT-Food Science and Technology* 64(2), 645–649. DOI: 10.1016/j.lwt.2015.06.052.

Guo, L., Li, W., Gu, Z., Wang, L., Guo, L. *et al.* (2023) Recent advances and progress on melanin: From source to application. *International Journal of Molecular Sciences* 24(5), 4360. DOI: 10.3390/ijms24054360.

Han, R., Xiang, R., Li, J., Wang, F. and Wang, C. (2021) High-level production of microbial prodigiosin: A review. *Journal of Basic Microbiology* 61(6), 506–523. DOI: 10.1002/jobm.202100101.

Hernández, P.I., Moreno, D., Javier, A.A., Torroba, T., Pérez-Tomás, R. *et al.* (2012) Tambjamine alkaloids and related synthetic analogs: Efficient transmembrane anion transporters. *Chemical Communications* 48(10), 1556–1558. DOI: 10.1039/c1cc11300c.

Hill, H.Z. (1992) The function of melanin or six blind people examine an elephant. *BioEssays* 14(1), 49–56. DOI: 10.1002/bies.950140111.

Ito, S., Wakamatsu, K. and Sarna, T. (2018) Photodegradation of eumelanin and pheomelanin and its pathophysiological implications. *Photochemistry and Photobiology* 94(3), 409–420. DOI: 10.1111/php.12837.

Kalra, R., Conlan, X.A. and Goel, M. (2020) Fungi as a potential source of pigments: Harnessing filamentous fungi. *Frontiers in Chemistry* 8, 369. DOI: 10.3389/fchem.2020.00369.

Kancharla, P., Li, Y., Yeluguri, M., Dodean, R.A., Reynolds, K.A. *et al.* (2021) Total synthesis and antimalarial activity of 2-(*p*-hydroxybenzyl)-prodigiosins, isoheptylprodigiosin, and geometric isomers of tambjamine MYP1 isolated from marine bacteria. *Journal of Medicinal Chemistry* 64(12), 8739–8754. DOI: 10.1021/acs.jmedchem.1c00748.

Kordiak, J., Bielec, F., Jabłoński, S. and Pastuszak-Lewandoska, D. (2022) Role of beta-carotene in lung cancer primary chemoprevention: A systematic review with meta-analysis and meta-regression. *Nutrients* 14(7), 1361. DOI: 10.3390/nu14071361.

Laursen, J.B. and Nielsen, J. (2004) Phenazine natural products: Biosynthesis, synthetic analogues, and biological activity. *Chemical Reviews* 104(3), 1663–1686. DOI: 10.1021/cr020473j.

Lee, I.K. and Yun, B.S. (2011) Styrylpyrone-class compounds from medicinal fungi *Phellinus* and *Inonotus* spp., and their medicinal importance. *Journal of Antibiotics* 64(5), 349–359. DOI: 10.1038/ja.2011.2.

Leon, L.L., Miranda, C.C., Souza, A.O. and Durán, N. (2001) Antileishmanial activity of the violacein extracted from *Chromobacterium violaceum*. *Journal of Antimicrobial Chemotherapy* 48. DOI: 10.1093/jac/48.3.449.

Lin, L. and Xu, J. (2020) Fungal pigments and their roles associated with human health. *Journal of Fungi* 6(4), 280. DOI: 10.3390/jof6040280.

Lin, L. and Xu, J. (2022) Production of fungal pigments: Molecular processes and their applications. *Journal of Fungi* 9(1), 44. DOI: 10.3390/jof9010044.

Liu, C.F., Chuang, H.T., Wang, C.S., Hsu, Y.W., Pan, T.M. *et al.* (2025) *Monascus pilosus* SWM-008 red mold rice and its components, monascinol and monascin, reduce obesity in a high-fat diet-induced rat model through synergistic modulation of gut microbiota and anti-lipogenesis. *Food & Function* 16(3), 966–985. DOI: 10.1039/D4FO05435K.

Lopes, S.C.P., Blanco, Y.C., Justo, G.Z., Nogueira, P.A., Rodrigues, F.L.S. *et al.* (2009) Violacein extracted from *Chromobacterium violaceum* inhibits plasmodium growth *in vitro* and *in vivo*. *Antimicrobial Agents and Chemotherapy* 53(5), 2149–2152. DOI: 10.1128/AAC.00693-08.

Malik, K., Tokkas, J. and Goyal, S. (2012) Microbial pigments: A review. *International Journal of Microbial Resource Technology* 1(4), 361–365.

Manivasagan, P., Bharathiraja, S., Santha Moorthy, M., Mondal, S., Seo, H. *et al.* (2018) Marine natural pigments as potential sources for therapeutic applications. *Critical Reviews in Biotechnology* 38(4), 560–578. DOI: 10.1080/07388551.2017.1398713.

Mapari, S.A.S., Nielsen, K.F., Larsen, T.O., Frisvad, J.C., Meyer, A.S. *et al.* (2005) Exploring fungal biodiversity for the production of water-soluble pigments as potential natural food colorants. *Current Opinion in Biotechnology* 16(2), 231–238. DOI: 10.1016/j.copbio.2005.03.004.

Matnuri, M., Sidik, D., Bai, G., Abdukerim, M., Ahmat, K. *et al.* (2015) Correlation analysis of riboflavin, RFT2 and *Helicobater pylori* in gastric carcinoma. *International Journal of Clinical and Experimental Pathology* 8(10), 13339–13345.

Mavrodi, D.V., Peever, T.L., Mavrodi, O.V., Parejko, J.A., Raaijmakers, J.M. *et al.* (2010) Diversity and evolution of the phenazine biosynthesis pathway. *Applied and Environmental Microbiology* 76(3), 866–879. DOI: 10.1128/AEM.02009-09.

Medentsev, A.G. and Akimenko, V.K. (1998) Naphthoquinone metabolites of the fungi. *Phytochemistry*, 47(6), 935–959. DOI: 10.1016/s0031-9422(98)80053-8.

Mone, N.S., Bhagwat, S.A., Sharma, D., Chaskar, M., Patil, R.H. *et al.* (2021) Naphthoquinones and their derivatives: Emerging trends in combating microbial pathogens. *Coatings* 11(4), 434. DOI: 10.3390/coatings11040434.

Moreau-Marquis, S., O'Toole, G.A. and Stanton, B.A. (2009) Tobramycin and FDA-approved iron chelators eliminate *Pseudomonas aeruginosa* biofilms on cystic fibrosis cells. *American Journal of Respiratory Cell and Molecular Biology*, 41(3), 305–313. DOI: 10.1165/rcmb.2008-0299OC.

Muñoz-Torres, P., Cárdenas-Ninasivincha, S. and Aguilar, Y. (2024) Exploring the agricultural applications of microbial melanin. *Microorganisms* 12(7), 1352. DOI: 10.3390/microorganisms12071352.

Nakamura, Y., Asada, C. and Sawada, T. (2003) Production of antibacterial violet pigment by psychrotropic bacterium RT102 strain. In: *Biotechnology and Bioprocess Engineering*, Vol. 8.

Napolitano, A., Panzella, L., Leone, L. and d'Ischia, M. (2013) Red hair benzothiazines and benzothiazoles: Mutation-inspired chemistry in the quest for functionality. *Accounts of Chemical Research* 46(2), 519–528. DOI: 10.1021/ar300219u.

Palkina, K.A., Balakireva, A.V., Belozerova, O.A., Chepurnykh, T.V., Markina, N.M. *et al.* (2023) Domain truncation in hispidin synthase orthologs from non-bioluminescent fungi does not lead to hispidin biosynthesis. *International Journal of Molecular Sciences* 24(2), 1317. DOI: 10.3390/ijms24021317.

Picott, K.J., Deichert, J.A., deKemp, E.M., Schatte, G., Sauriol, F. *et al.* (2019) Isolation and characterization of tambjamine MYP1, a macrocyclic tambjamine analogue from marine bacterium *Pseudoalteromonas citrea*. *MedChemComm* 10(3), 478–483. DOI: 10.1039/c9md00061e.

Plonka, P. and Grabacka, M. (2006) Melanin synthesis in microorganisms – biotechnological and medical aspects. *Acta Biochimica Polonica* 53(3), 429–443. DOI: 10.18388/abp.2006_3314.

Rana, B., Bhattacharyya, M., Patni, B., Arya, M. and Joshi, G.K. (2021) The realm of microbial pigments in the food color market. *Frontiers in Sustainable Food Systems* 5, 603892. DOI: 10.3389/fsufs.2021.603892.

Revuelta, J.L., Ledesma-Amaro, R., Lozano-Martinez, P., Díaz-Fernández, D., Buey, R.M. *et al.* (2017) Bioproduction of riboflavin: A bright yellow history. *Journal of Industrial Microbiology & Biotechnology* 44(4–5), 659–665. DOI: 10.1007/s10295-016-1842-7.

Rodríguez-Sáiz, M., Barredo, J.L. and Martín, J.F. (2010) Enhanced production of prodigiosin by *Serratia marcescens* through metabolic engineering. *Biotechnology Progress* 26(6), 1340–1348. DOI: 10.1016/j.ejbt.2019.04.007.

Rui, Z., Ye, M., Wang, S., Fujikawa, K., Akerele, B. *et al.* (2012) Insights into a divergent phenazine biosynthetic pathway governed by a plasmid-born esmeraldin gene cluster. *Chemistry & Biology* 19(9), 1116–1125. DOI: 10.1016/j.chembiol.2012.07.025.

Sakuta, M. (2014) Diversity in plant red pigments: Anthocyanins and betacyanins. *In Plant Biotechnology Reports*, 8(1), 37–48). DOI: 10.1007/s11816-013-0294-z.

Saranya, R., Jayapriya, J. and Tamilselvi, A. (2012) Dyeing of silk fabric with phenazine from *Pseudomonas* species. *Coloration Technology* 128(6), 440–445. DOI: 10.1111/j.1478-4408.2012.00397.x.

Sepúlveda Cisternas, I., Salazar, J.C. and García-Angulo, V.A. (2018) Overview on the bacterial iron-riboflavin metabolic axis. *Frontiers in Microbiology* 9, 1478. DOI: 10.3389/fmicb.2018.01478.

Serafim, B., Bernardino, A.R., Freitas, F. and Torres, C.A.V. (2023) Recent developments in the biological activities, bioproduction, and applications of *Pseudomonas* spp. phenazines. *Molecules* 28(3), 1368. DOI: 10.3390/molecules28031368.

Shi, Y.C. (2017) Advances in microbial production of *Monascus* pigments. *World Journal of Microbiology and Biotechnology* 33(12), 208.

Singh, S., Nimse, S.B., Mathew, D.E., Dhimmar, A., Sahastrabudhe, H. *et al.* (2021) Microbial melanin: Recent advances in biosynthesis, extraction, characterization, and applications. *Biotechnology Advances* 53, 107773. DOI: 10.1016/j.biotechadv.2021.107773.

Soliev, A.B., Hosokawa, K. and Enomoto, K. (2011) Bioactive pigments from marine bacteria: Applications and physiological roles. *Evidence-Based Complementary and Alternative Medicine* 670349. DOI: 10.1155/2011/670349.

Stavenga, D.G. (2024) Butterfly blues and greens caused by subtractive colour mixing of carotenoids and bile pigments. *Journal of Comparative Physiology A* 210(3), 371–380. DOI: 10.1007/s00359-023-01656-4.

Stintzing, F.C. and Carle, R. (2004) Functional properties of anthocyanins and betalains in plants, food, and in human nutrition. *Trends in Food Science and Technology* 15(1), 19–38. DOI: 10.1016/j.tifs.2003.07.004.

Tammam, M.A., Sebak, M., Greco, C., Kijjoa, A. and El-Demerdash, A. (2022) Chemical diversity, biological activities and biosynthesis of fungal naphthoquinones and their derivatives: A comprehensive update. *Journal of Molecular Structure* 1268, 133711. DOI: 10.1016/j.molstruc.2022.133711.

Thakur, K., Tomar, S.K., Singh, A.K., Mandal, S. and Arora, S. (2017) Riboflavin and health: A review of recent human research. *Critical Reviews in Food Science and Nutrition* 57(17), 3650–3660. DOI: 10.1080/10408398.2016.1145104.

Tsukamoto, K., Jackson, I.J., Urabe, K., Montague, P.M. and Hearing, V.J. (1992) A second tyrosinase-related protein, TRP-2, is a melanogenic enzyme termed DOPAchrome tautomerase. *EMBO Journal* 11(2), 519–526. DOI: 10.1002/j.1460-2075.1992.tb05082.x.

Tuli, H.S., Chaudhary, P., Beniwal, V. and Sharma, A.K. (2015). Microbial pigments as natural color sources: current trends and future perspectives. *Journal of Food Science and Technology*, 52(8), 4669–4678. DOI: 10.1007/s13197-014-1601-6.

Unagul, P., Wongsa, P., Kittakoop, P., Intamas, S., Srikitikulchai, P. *et al.* (2005) Production of red pigments by the insect pathogenic fungus *Cordyceps unilateralis* BCC 1869. *Journal of Industrial Microbiology and Biotechnology* 32(4), 135–140. DOI: 10.1007/s10295-005-0213-6.

Venil, C.K., Zakaria, Z.A. and Ahmad, W.A. (2013) Bacterial pigments and their applications. *Process Biochemistry* 48(7), 1065–1079. DOI: 10.1016/B978-0-323-95251-4.00019-3.

Weigl, M., Kandelbauer, A., Hansmann, C., Pöckl, J., Müller, U. *et al.* (2009) Application of natural dyes. In: *Handbook of Natural Colorants*, Vol. 8. p. 277. DOI: 10.1002/9780470744970.

Willows, R.D., Lagarias, J.C. and Duanmu, D. (2023) Tetrapyrrole biosynthesis and signaling (chlorophyll, heme, and bilins). In: *The Chlamydomonas Sourcebook*. Academic Press, pp. 691–731.

Wolnicka-Glubisz, A., Pecio, A., Podkowa, D., Kolodziejczyk, L.M. and Plonka, P.M. (2012) Pheomelanin in the skin of *Hymenochirus boettgeri* (Amphibia: Anura: Pipidae). *Experimental Dermatology* 21(7), 537–540. DOI: 10.1111/j.1600-0625.2012.01511.x.

Woo, P.C.Y., Zhen, H., Cai, J.J., Yu, J., Lau, S.K.P. *et al.* (2003) The mitochondrial genome of the thermal dimorphic fungus *Penicillium marneffei* is more closely related to those of molds than yeasts. *FEBS Letters* 555(3), 469–477. DOI: 10.1016/s0014-5793(03)01307-3.

Wootton-Beard, P.C. and Ryan, L. (2011) A beetroot juice shot is a significant and convenient source of bioaccessible antioxidants. *Journal of Functional Foods* 3(4), 329–334. DOI: 10.1016/j.jff.2011.05.007.

Yadav, S., Tiwari, K.S., Gupta, C., Tiwari, M.K., Khan, A. *et al.* (2023) A brief review on natural dyes, pigments: Recent advances and future perspectives. *Results in Chemistry* 5, 100733. DOI: 10.1016/j.rechem.2022.100733.

Yip, C.H., Yarkoni, O., Ajioka, J., Wan, K.L. and Nathan, S. (2019) Recent advancements in high-level synthesis of the promising clinical drug, prodigiosin. *Applied Microbiology and Biotechnology* 103(4), 1667–1680. DOI: 10.1007/s00253-018-09611-z.

4 Biosynthesis Pathways of Microbial Pigments

Ramesh Koduru[1], Somedutta Maity[2], Arifullah Mohammed[3], Mohd Tariq[4,5] and Pamuru Ramachandra Reddy[1*]

[1]Department of Biochemistry, Yogi Vemana University, Vemanapuram, Andhra Pradesh, India; [2]School of Engineering Sciences and Technology, University of Hyderabad, Hyderabad, Telangana, India; [3]Department of Biotechnology, Koneru Lakshmaiah University (KLEF), Vaddeswaram Campus, Guntur, Andhra Pradesh, India; [4]Department of Academics, Sumandeep Vidyapeeth, Deemed to be University, Vadodara-391760, Gujarat, India; [5]Department of Biotechnology, Graphic Era (Deemed to be University), Dehradun, Uttarakhand-248002, India

4.1 Introduction

The area of microbial pigments represents a diverse field, illustrating a complex interaction of metabolic pathways utilized by microbes to produce vibrant hues. These pigments, recognized for their structural diversity and ecological significance, have attracted the interest of researchers for many years. This chapter analyzes the biosynthetic pathways involved in the formation of microbial pigments, clarifying the molecular framework that supports their synthesis.

Microbial pigments are bioactive chemicals produced by various microorganisms, including bacteria, fungus, and algae. These pigments exhibit a range of colors from deep reds to vivid yellows and startling blues. In addition to their visual characteristics, they serve essential roles as antioxidants, antimicrobials, and signaling molecules, thereby improving the survival and ecological success of the organisms that produce them. The biosynthesis of these pigments is regulated by complex enzymatic mechanisms that transform primary metabolites into secondary metabolites with chromogenic properties.

4.2 Key Pathways and Metabolites

The biosynthesis of microbial pigments involves primary and secondary metabolic processes that provide the necessary precursors and energy for pigment formation. The five categories of microbial pigments are (Table 4.1): carotenoids polyketides porphyrins phenazines melanin.

- carotenoids
- polyketides
- porphyrins
- phenazines
- melanin.

*Corresponding author: prcrbio@yogivemanauniversity.ac.in

 Biopigments of Microbial Origin: Innovations and Applications Across Industries (eds Madan *et al.*)
DOI: 10.1079/ 9781836990505.0004

Table 4.1. Biosynthesis pathways of microbial pigments.

Pigment class	Examples	Key precursors	Biosynthetic pathway	Enzymes involved	Microbial producers	Applications
Carotenoids	β-carotene, astaxanthin	Isopentenyl pyrophosphate (IPP), dimethylallyl pyrophosphate (DMAPP)	Mevalonate (MVA) or methylerythritol phosphate (MEP) pathway	Phytoene synthase, desaturases, cyclases	*Rhodotorula* spp., *Serratia* spp.	Food colorants, antioxidants
Polyketides	Prodigiosin, monascorubin	Acyl-CoA derivatives	Polyketide synthase pathway	Polyketide synthases (PKSs), tailoring enzymes	*Serratia marcescens*, *Monascus* spp.	Anticancer, antimicrobial agents
Porphyrins	Heme, chlorophyll	Aminolevulinic acid (ALA)	Shemin pathway or C5 pathway	ALA synthase, porphobilinogen deaminase	*Pseudomonas* spp., *Rhodobacter* spp.	Photosynthesis, oxygen transport
Phenazines	Pyocyanin, phenazine-1-carboxylic acid	Chorismate	Shikimate pathway	Phenazine synthase, chorismate mutase	*Pseudomonas aeruginosa*	Antimicrobials, redox mediators
Melanin	Dihydroxyphenylalanine (DOPA) Melanin	L-dopa	Oxidative polymerization	Tyrosinase	*Aspergillus niger*, *Cryptococcus* spp.	UV protection, heavy metal chelation

4.2.1 Carotenoid biosynthesis

There is a wide diversity of pigments recognized as carotenoids, which may be found in plants, algae, and even certain microbes. They are responsible for several critical functions in the natural world, including the defense of cells against oxidative damage, the color of fruits, vegetables, and flowers, and the production of essential molecules in animals, such as vitamin A, by acting as precursors.

Carotenoids are classified as terpenoid chemicals, which means that they are generated from the fundamental building blocks of isoprene units, more especially C5 units. These compounds are distinguished by their lengthy chains of conjugated double bonds, which are responsible for the brilliant hues (yellow, orange, and red) that are seen in them. By their structural composition, carotenoids can be divided into two distinct categories.

Carotenes are composed entirely of hydrocarbons and do not include any oxygen in their composition. Included in the list of examples are β-carotene, α-carotene, and lycopene. *Xanthophylls* are characterized by the presence of oxygen atoms in the form of hydroxyl, epoxy, or other operational groups. Lutein, zeaxanthin, and violaxanthin are some examples of such pigments.

Carotenoids are involved in the photosynthetic system via photooxidative damage: carotenoids absorb light and release the surplus as heat.

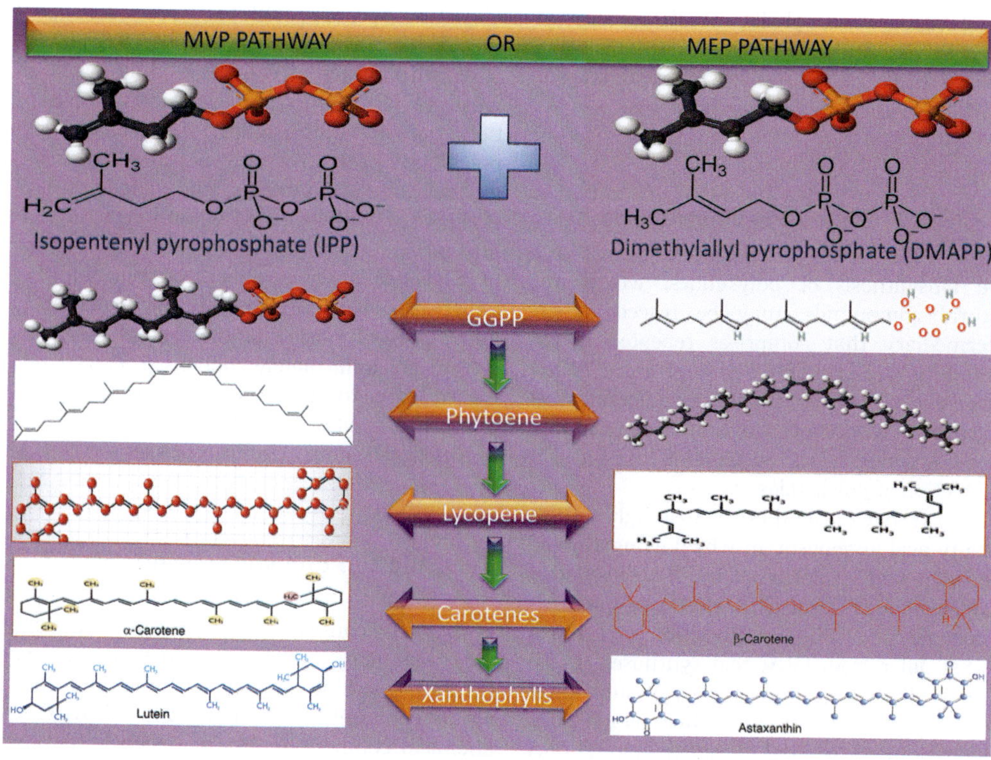

Fig. 4.1. The metabolic process for carotenoids. Phytoene synthase catalyzes the condensation of two geranyl pyrophosphate (GGPP) molecules, the first step in the biosynthetic route for carotenoids. Lycopene, a pigment with a red hue, is formed when this precursor goes through a sequence of desaturations and isomerization. Lycopene can then be cyclized by cyclases to produce molecules like α-, β-, and γ-carotenes. Carotene molecules can oxygenate enzymes like hydroxylases and ketolases to produce xanthophylls. Either the 2-C-methyl-D-erythritol-4-phosphate (MEP) route or the mevalonic acid (MVA) pathway can provide the C5 precursors (dimethylallyl diphosphate and isopentenyl diphosphate) that are necessary for GGPP synthesis. The MEP route is used by photosynthetic species, whereas the MVA pathway is used by archaea, fungi, and the majority of bacteria.

This process prevents reactive oxygen species (ROS) from being formed. The plant is protected from the damaging effects of UV radiation and other environmental stresses by the carotenoids, which function as antioxidants to neutralize free radicals and reduce oxidative stress.

The majority of carotenoids start with phytoene (C40). However, C50 and C30 carotenoids may be synthesized by some bacterial species using distinct precursors. In their most basic form, they are polyene chains that have an odd number of conjugated double bonds and two sets of terminal groups. The absorption gap of chlorophyll is 450–570 nm, which is the range of light that these chains absorb. So, carotenoids can serve as supplementary pigments during photosynthesis. One category of carotenoid structures is xanthophylls, which correspond to molecules with oxygen, and the other is carotenes, which are substances without oxygen. Fig. 4.1 shows examples from both sets.

4.2.2 Polyketide biosynthesis

The biosynthesis of polyketides, which are natural compounds, may be traced to an intermediary that comprises repeated ketide units (Fig. 4.2A). Polyketide synthases (PKSs) have been the subject of extensive genetic and enzymatic research throughout the past two decades, which has shed light on the natural production of polyketides.

Polyketide biosynthesis is strikingly similar to fatty acid anabolism, in which elongation of the fatty acid is facilitated by decarboxylative condensation of the acyl thioester unit, malonyl-coenzyme A (CoA). Polyketide synthases (PKSs) differ from fatty acid synthases (FASs) in their ability to utilize a variety of substrates, including propionyl-CoA, methylmalonyl-CoA, ethylmalonyl-CoA, and methoxymalonyl-CoA (Fig. 4.2B). Fig. 4.2C illustrates a representative chain of condensation reactions. In a first step, acyltransferase (AT) facilitates thioester exchange among malonyl-CoA and a thiol group on a pantetheinyl group linked to an acyl carrier protein (ACP), resulting in the movement of CoA. In the next step, the polyketide chain (or acetate) is transferred from the upstream ACP to an active site serine on a ketosynthase (KS) while the extender unit undergoes decarboxylation. As a result of the carbanion's nucleophilic assault on the thioester linkage on KS, the chain is finally expanded by a dual carbon atom. In FASs, the β-carbonyl group undergoes complete reduction to methylene via the sequential actions of ketoreductase (KR), enoyl reductase (ER), and dehydratase (DH) as illustrated in Fig. 4.2D. PKSs generate greater structural diversity compared to FASs because the reduction process can be paused at each stage, allowing for the production of alcohols with specific stereochemistry, olefins, or methylenes before proceeding to the subsequent condensation reaction.

These enzymes can be divided into three kinds based on gene cluster organization and PKS structure and activity. Type I PKSs are multifunctional proteins with various condensation or reduction active sites. These megasynthases often have numerous catalyst modules, each of which performs one chain extension and reduction process. Multimodular PKSs will produce polyketides with diverse structures depending on their catalytic domains. Dissociated proteins with one or two active sites make up type II PKSs. Enzymatic components repeatedly build a chain of polyketides at a given dimension. Type III PKSs, which lack ACPs, are the simplest structurally. To produce tiny aromatic metabolites, CoA thioesters are immediately condensed. Polyketides from all three routes are often converted into bioactive metabolites by oxidoreductases, cyclases, or functional component transferases.

4.2.3 Porphyrin biosynthesis

Porphyrins consist of four modified pyrrole subcomponents linked at their α carbon atoms through methine bridges (=CH−), and they are a constitute of heterocyclic, macrocyclic chemical compounds. Heme, which is a component of hemoproteins that transport oxygen throughout the body, is an important porphyrin in vertebrates. Chlorophyll, a crucial porphyrin derivative in plants, plays a role in photosynthesis by absorbing light and transferring electrons. Porphine is an extremely uncommon chemical molecule that has only ever been of theoretical importance; it is the parent of porphyrins. One name for porphyrins is substituted porphines

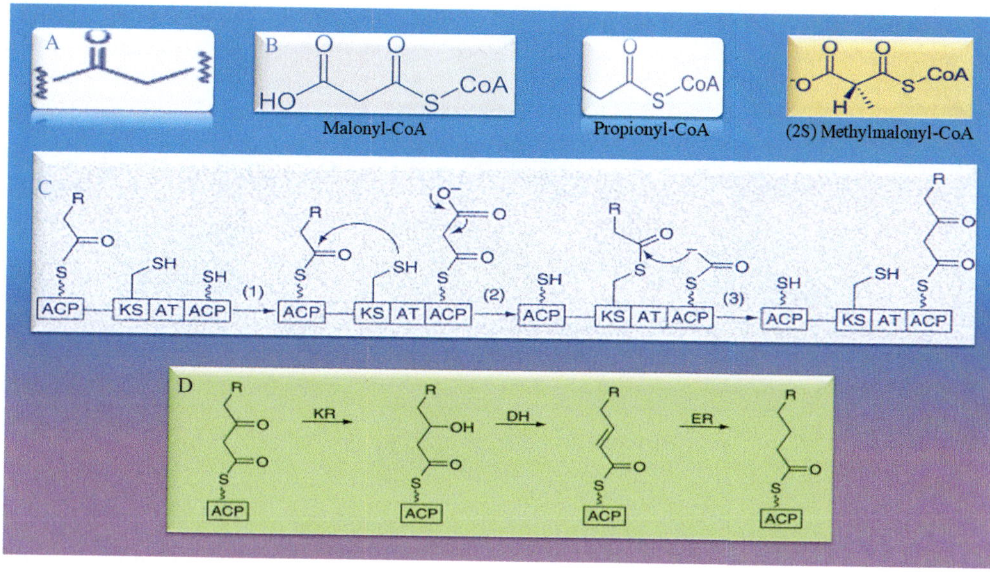

Fig. 4.2. Key concepts applicable to polyketide biosynthesis. (A) Ketide unit. (B) Substrates of polyketide. (C) FAS and PKS decarboxylative condensation. (D) FAS and PKS β-carbonyl reduction procedures.

(Rayati and Malekmohammadi, 2016). Porphyrin rings are frequently characterized as aromatic due to their structure, which comprises 26 XVI-electrons, 18 of which constitute a flat, continuous cycle (Ivanov and Boldyrev, 2014). Herein, third porphyrins are often quite colorful and absorb a lot of light in the visible part of the electromagnetic spectrum because of their extensive conjugated structure.

In eukaryotic organisms that do not undergo photosynthesis, including insects, fungi, protozoa, and bacteria belonging to the α-proteobacteria group, the first stage in biosynthesis of porphyrin is the creation of IV-aminolevulinic acid (also known as 5-ALA, IV-ALA, or dALA) through the citric acid cycle response involving the glycine amino acid. Glutamic acid is synthesized in plants, algae, bacteria (excluding the α-proteobacteria group), and archaea using glutamyl-tRNA and glutamate-1-semialdehyde. Enzymes such as glutamyl-tRNA reductase, glutamate-1-semialdehyde 2,1-aminomutase, and glutamyl-tRNA synthetase are involved in this process. The C5 or Beale route is the name given to this pathway.

Porphobilinogen synthase takes two dALA molecules and combines them to form porphobilinogen (PBG), which features a pyrrole ring. The circular tetrapyrrole uroporphyrinogen III is formed by hydrolysis of hydroxymethyl bilane (HMB), which is formed by deamination of four PBGs. Several more changes are carried out on this molecule. The major end-roduct in humans is heme, which is formed when protoporphyrin IX is coupled with iron. Other species employ intermediates to generate specific compounds. The byproducts of heme metabolism include bile pigments (Fig. 4.3).

In 1936, the Rothemund reaction was initially reported (Rothemund, 1935, 1936). It is still used today as a basis for the procedures described by Adler and Longo, which are more contemporary approaches to porphyrin synthesis (Adler *et al.*, 1967). Pyrrole and an aldehyde are the beginning points for the condensation and oxidation process that constitutes the overall scheme.

4.2.4 Phenazines

Phenazines are a big group of heterocycles containing nitrogen molecules which are made by many different types of bacteria. Man-made and natural phenazine products are both considered for how they affect biological processes

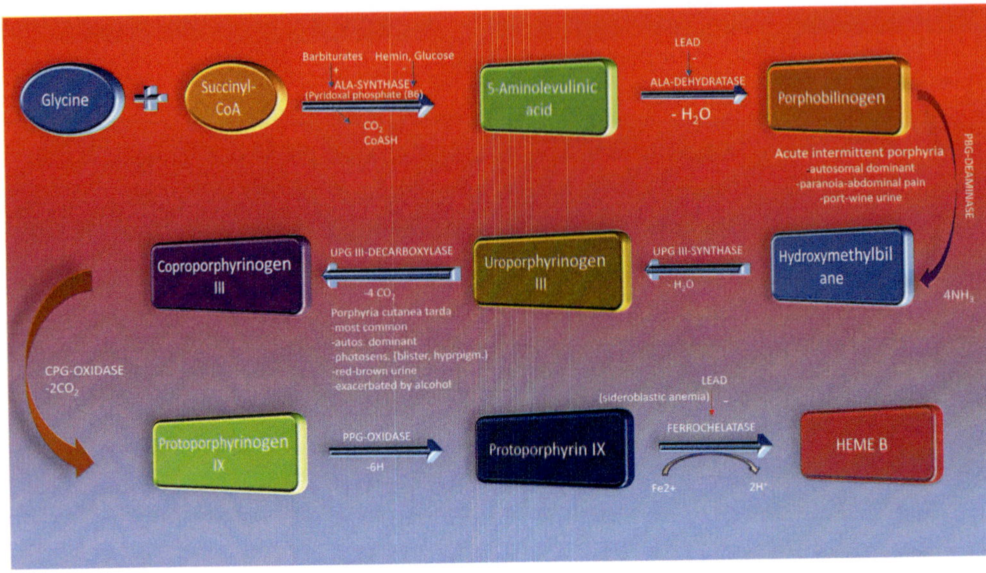

Fig. 4.3. Metabolic pathways of porphyrin biosynthesis. The enzymes involved at each step are added.

and how bacteria interact with each other. The manufacture of phenazines is a thoroughly elucidated microbial process, especially in bacteria like *Pseudomonas* spp., *Streptomyces* spp., and *Burkholderia* spp. Phenazines are secondary metabolites exhibiting a range of biological actions, such as antibacterial, anticancer, and redox characteristics (Pierson and Pierson, 2010).

Phenazines are nitrogen-containing aromatic compounds biosynthesized through the shikimate pathway, in which chorismic acid serves as a central precursor. This pathway represents a vital metabolic route in numerous microorganisms, leading to the production of aromatic amino acids and a variety of secondary metabolites, including phenazines (Fig. 4.4). The shikimic acid biosynthesis pathway is extensively well preserved and plays a role in the formation of a wide variety of metabolites that are essential for primary development. These metabolites include the three aromatic amino acids as well as paraaminobenzoic acid. A type of enzyme called 3-deoxy-d-arabinoheptuloson ate-7-phosphate (DAHP) synthase is responsible for the condensation of phosphoenolpyruvic acid and erythrose-4-phosphate in bacteria. This route is subject to tight control and is principally responsible for the initial phase of the process. There is another type of isozyme, I-DAHP synthase, present in a wide variety of bacteria. There is a possibility that one of the pathway's endproducts will give rise to feedback inhibition on each of them. Another DAHP synthase, type II-3 which is encoded by the biosynthetic operon in the third gene (phzC), shows numerous phenazine producers. This kind of DAHP synthase is more closely comparable to DAHP synthases found in solanaceous plants than it is to the DAHP synthases found in prokaryotes (Pierson *et al.*, 1990). Type II-3-DAHP synthase does not possess a loop-like region that is necessary for allosteric regulation in plants.

The expression of phzC has the potential to improve the first condensation phase, which would guarantee enough quantities of substrate for the formation of phenazine. According to Mavrodi *et al.* (2006), uncontrolled expression of phzC has the potential to deplete the cells of compounds that are necessary for primary growth, which in turn may result in a reduction in the overall fitness of the community in the rhizosphere. There is a possibility that the intricate regulation of phenazine synthesis can be explained, at least in part, by the requirement for regulated expression.

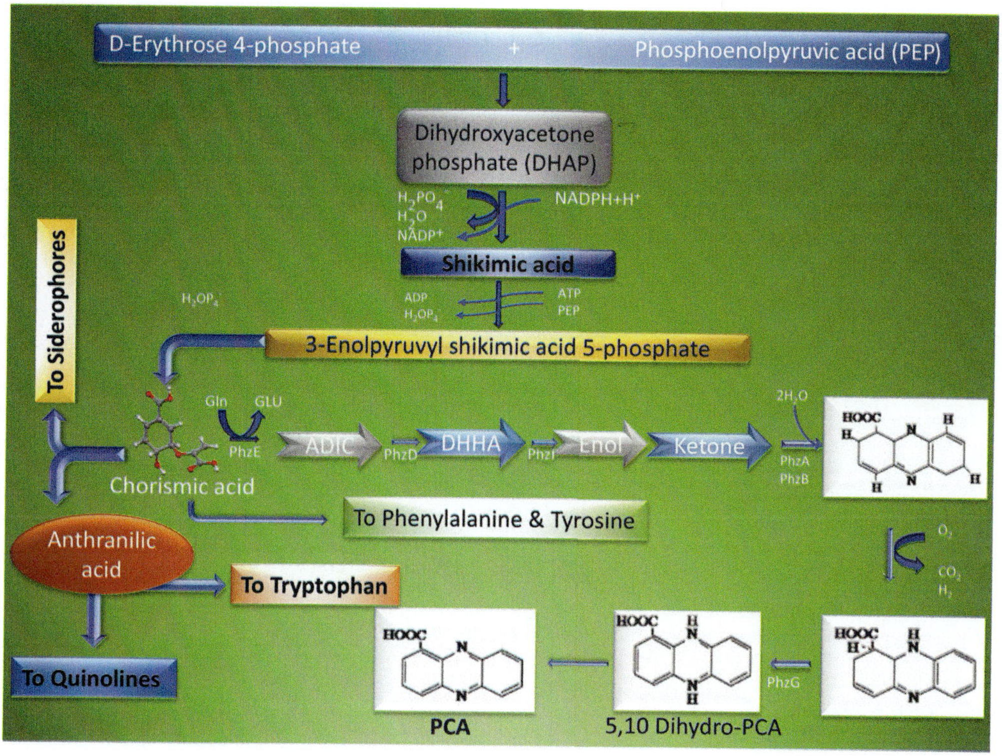

Fig. 4.4. Mentel *et al.* (2009) changed the figure to show the biosynthesis of phenazine. Most species create phenazines from the same, extremely conserved shikimic acid mechanism. After the expression of the phenazine biosynthesis genes (phzABCDEFG), which are regarded as the five core genes (phzD, phzB, phzE, phzG, and phzF), chorismic acid is used as the phenazine branch point. When it comes to the phenazine operon, the third enzyme, phzC, is likely responsible for ensuring that the shikimic acid route has enough substrate flow for phenazine production. Although the first stages of phenazine biosynthesis are starting to make more sense, many of the subsequent processes remain a mystery, and any discrepancies might be due to inherent variations in how different microbes carry out the last stages of the process. Extra terminal modifying genes generate a plethora of PCA variants (not shown). Refer to Mentel *et al.* (2009) for a comprehensive analysis of the route. Gln, glutamine; Glu, glutamic acid; DAHP, 3-deoxy-D-arabinoheptulosonate 7-phosphate; PCA, phenazine-1-carboxylic acid; ADIC, 2-amino-2-desoxyisochorismic acid; DHHA, trans-2,3-dihydro-3-hydroxyanthranilic acid.

4.2.5 Melanin

The microbial production of melanin encompasses a sequence of enzymatic and chemical processes that transform precursor molecules into melanin polymers. The processes may differ based on the organism and the specific kind of melanin synthesized. The many functions of melanin in microorganisms and fungi have been thoroughly documented in the literature (Nosanchuk and Casadevall, 2003; Plonka and Grabacka, 2006; Eisenman and Casadevall, 2012; Solano, 2014; Cordero *et al.*, 2017). The regulation of melanin production in many microbes has been explored. Given the benefits of using microbes for melanin production, including the absence of seasonal growth limitations, economic efficiency, and environmental sustainability, microbial melanin represents a significant supply of natural melanin.

Typically, the majority of microbial melanins are synthesized by the conversion of either tyrosine (DOPA pathway) or malonyl-coenzyme A (DHN pathway), mediated by distinct enzyme

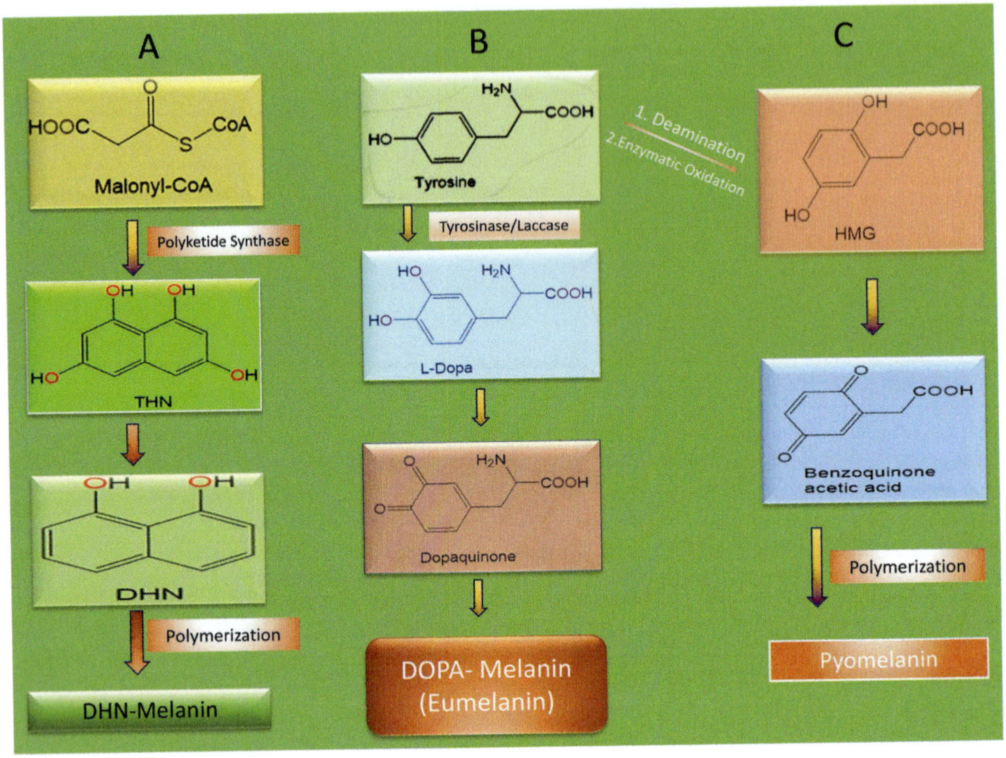

Fig. 4.5. Melanin production in microbes, schematically depicting the essential chemical reactions shared by bacteria and fungi. (A) DHN pathway. (B) DOPA pathways. (C) Different forms of melanins, such as pyomelanin, can be produced when enzymatic imbalances cause changes to the metabolic pathway.

sets (Fig. 4.5). The first step closely resembles the creation of melanin in mammals. In this method, tyrosine, a precursor of melanin, is transformed into L-dopa and subsequently into dopaquinone by the enzymes laccase and tyrosinase. However, dopaquinones have significant reactivity and undergo spontaneous oxidation and autopolymerization to produce melanin. The process of melanin synthesis through the DOPA route is known as DOPA-melanin or eumelanin. During the catabolism of aromatic compounds, enzymatic disruptions or alterations may result in the accumulation of hydroxylated aromatic intermediates, which can exhibit effects analogous to those of homogentisic acid, potentially leading to the formation of various forms of melanins. In the next process, the related precursor, which is malonyl-coenzyme A, is synthesized endogenously. The successive decarboxylative condensation of 5 malonyl-coenzyme A molecules, facilitated by polyketide synthases, produces 1,3,6,8-tetrahydroxynaphthalene (THN). THN subsequently undertakes a sequence of reduction and dehydration events to yield 1,8-dihydroxy naphthalene (DHN). The polymerization of DHN yields DHN-melanin as the endproduct (Plonka and Grabacka, 2006; Eisenman and Casadevall, 2012; Pavan *et al.*, 2020).

Both mechanisms are present in bacteria and fungus. However, many bacteria and specific classes of basidiomycetous fungi synthesize melanin using the DOPA route, but ascomycetous and some defective fungi, including nonmicroscopic fungi such as *Tuber* spp., utilize the DHN pathway for melanin production.

Regarding high-yield melanin creation, microorganisms utilizing the DHN route are not favored; meanwhile, this pathway results in endogenous pigment synthesis that is firmly affixed to the inner side of the cell wall. This renders melanin extraction exceedingly challenging and may produce artefacts resulting from aggressive extracting agents. Otherwise, melanogenesis through the DOPA route is a process employed by microbes to neutralize hazardous phenolic chemicals from the environment, including those generated by bacteria during host defense (Schmaler-Ripcke *et al.*, 2009; Almeida-Paes *et al.*, 2012). As a result, several bacteria rely on external tyrosine or its derivatives as substrates for melanin production. This is highly relevant for researchers investigating microbial melanization, as melanin is formed extracellularly, thus allowing for the avoidance of harsh extraction methods. Other catecholamines including dopamine and norepinephrine can be used as melanin substrates, although tyrosine is the most common. Keep in mind that different substrates might cause distinct catabolic processes with different enzymes, which can lead to melanins with diverse structures.

To optimize the synthesis of microbial melanins, it is possible to tune the physicochemical parameters. Multiple dietary variables and physicochemical circumstances control the regulation of melanin production enzymes, which is crucial for melanin formation. Carbon and nitrogen sources that are commonly utilized include peptone, glucose, and yeast extract. In order to reduce production costs while maintaining a high yield, research has explored the use of agricultural leftovers, including fruit leftover extract, corn steep liquor, and wheat fiber extract (Hamano and Kilikian, 2006; Silveira *et al.*, 2008; Zou and Tian, 2017).

Due to its function as a cofactor for tyrosinases and laccases, copper is a crucial element in the formation of melanin (Sendovski *et al.*, 2011; Reiss *et al.*, 2013; Yang *et al.*, 2017). Different fungal and bacterial species exhibit uneven pigmentation when exposed to varying amounts of additional copper (Held and Kutzner, 1990; Griffith *et al.*, 2007). While copper is one metal that may stimulate melanin production, other metals can do the same. A study conducted by Wang *et al.* (2019) demonstrated that the addition of iron and nickel significantly enhanced tyrosinase activity and melanin formation. Gowri and Srivastava (1996) found that bacteria may produce melanin in response to stress caused by metals. In contrast, other types of stress, such as hyperosmotic pressure, high temperatures, growth medium deficient in nutrients, etc., can actually enhance melanin formation (Coyne and al-Harthi, 1992; Fogarty and Tobin, 1996;Cordero and Casadevall, 2017).

There is no one-size-fits-all solution for cultivating melanogenic microbes as there are so many different variables that influence melanin formation. Rather, it is necessary to determine the microbe-specific composition and ratio of each component. Similarly, while cultivating, it is important to take into account environmental conditions such as temperature, pH, oxygen levels, light, pressure, and irradiation, all of which can have a significant impact on cell development and pigment production. Multifactorial experiments are commonly designed using statistical tools like the Taguchi method, the Plackett–Burman design, or response surface methodology to assess the impact of each production process factor (Surwase *et al.*, 2013; Saini and Melo, 2015; Sun *et al.*, 2016; El-Naggar and El-Ewasy, 2017).

Research on melanin that used other microorganisms, such as *Aspergillus carbonarius* or *Streptomyces glaucescens* (El-Naggar and El-Ewasy, 2017), failed to produce enough melanin for industrial use (<1g L^{-1} medium). On the other hand, research aimed at improving melanin synthesis has made use of bacteria and fungi that can convert tyrosine into melanin. It is possible to create melanin pigments in large quantities using the optimized techniques, which often include an external source of tyrosine, copper, and/or other metal ions. An example is the fact that *Armillaria cepistipes* can now be cultured in a lab setting to create melanin at gram scale (28g L^{-1} medium), which opens the door to industrial scaling up and potential future uses of melanin (Ribera *et al.*, 2019).

4.3 Regulation of Biosynthetic Pathways

The production of pigments is tightly regulated by environmental factors and genetic control. Key regulatory mechanisms include the following.

- *Quorum sensing*: in *Serratia marcescens*, prodigiosin synthesis is controlled by quorum-sensing molecules that signal high cell density.
- *Stress responses*: many pigments are overproduced under stress conditions, such as UV exposure, nutrient limitation, or oxidative stress.
- *Genetic regulation*: operons encoding biosynthetic enzymes are often regulated by transcription factors responsive to environmental cues.

4.4 Conclusion

The biosynthesis routes of microbial pigments exemplify nature's inventiveness. Each metabolic route, from carotenoids to polyketides, exemplifies a distinct adaptation that sustains microbial existence. Deciphering these routes reveals ecological enigmas and facilitates advancements in medicine, nutrition, and environmental science. In the subsequent chapter, we will examine the industrial-scale production and extraction methodologies for these microbial phenomena.

References

Adler, A.D., Longo, F.R., Finarelli, J.D., Goldmacher, J., Assour, J. *et al.* (1967) A simplified synthesis for meso-tetraphenyl porphine. *Journal of Organic Chemistry* 32(2), 476–476. DOI: 10.1021/jo01288a053.

Almeida-Paes, R., Frases, S., Araújo, G. de S., de Oliveira, M.M.E., Gerfen, G.J. *et al.* (2012) Biosynthesis and functions of a melanoid pigment produced by species of the *Sporothrix* complex in the presence of *L*-tyrosine. *Applied and Environmental Microbiology* 78(24), 8623–8630. DOI: 10.1128/AEM.02414-12.

Cordero, R.J.B. and Casadevall, A. (2017) Functions of fungal melanin beyond virulence. *Fungal Biology Reviews* 31, 99–112. DOI: 10.1016/j.fbr.2016.12.003.

Cordero, R.J.B., Vij, R. and Casadevall, A. (2017) Microbial melanins for radioprotection and bioremediation. *Microbial Biotechnology* 10(5), 1186–1190. DOI: 10.1111/1751-7915.12807.

Coyne, V.E. and al-Harthi, L. (1992) Induction of melanin biosynthesis in *Vibrio cholerae*. *Applied and Environmental Microbiology* 58(9), 2861–2865. DOI: 10.1128/aem.58.9.2861-2865.1992.

Eisenman, H.C. and Casadevall, A. (2012) Synthesis and assembly of fungal melanin. *Applied Microbiology and Biotechnology* 93(3), 931–940. DOI: 10.1007/s00253-011-3777-2.

El-Naggar, N.E.-A. and El-Ewasy, S.M. (2017) Bioproduction, characterization, anticancer and antioxidant activities of extracellular melanin pigment produced by newly isolated microbial cell factories streptomyces glaucescens NEAE-h. *Scientific Reports* 7, 42129. DOI: 10.1038/srep42129.

Fogarty, R.V. and Tobin, J.M. (1996) Fungal melanins and their interactions with metals. *Enzyme and Microbial Technology* 19(4), 311–317. DOI: 10.1016/0141-0229(96)00002-6.

Gowri, P.M. and Srivastava, S. (1996) Encapsulation as a response of *Azospirillum brasilense* sp7 to zinc stress. *World Journal of Microbiology & Biotechnology* 12, 319–322. DOI: 10.1007/BF00340207.

Griffith, G.W., Easton, G.L., Detheridge, A.P., Roderick, K., Edwards, A. *et al.* (2007) Copper deficiency in potato dextrose agar causes reduced pigmentation in cultures of various fungi. *FEMS Microbiology Letters* 276, 165–171. DOI: 10.1111/j.1574-6968.2007.00923.x.

Hamano, P.S. and Kilikian, B.V. (2006) Production of red pigments by *Monascus ruber* in culture media containing corn steep liquor. *Brazilian Journal of Chemical Engineering* 23, 443–449. DOI: 10.1590/S0104-66322006000400002.

Held, T. and Kutzner, H.J. (1990) Transcription of the tyrosinase gene in *Streptomyces michiganensis* DSM 40015 is induced by copper and repressed by ammonium. *Journal of General Microbiology* 136, 2413–2419. DOI: 10.1099/00221287-136-12-2413.

Ivanov, A.S. and Boldyrev, A.I. (2014) Deciphering aromaticity in porphyrinoids via adaptive natural density partitioning. *Organic & Biomolecular Chemistry* 12, 6145–6150. DOI: 10.1039/C4OB01018C.

Mavrodi, D.V., Blankenfeldt, W. and Thomashow, L.S. (2006) Phenazine compounds in Fluorescent *Pseudomonas* spp. biosynthesis and regulation. *Annual Review of Phytopathology* 44, 417–445. DOI: 10.1146/annurev.phyto.44.013106.145710.

Mentel, M., Ahuja, E.G., Mavrodi, D.V., Breinbauer, R., Thomashow, L.S. *et al.* (2009) Of two make one: the biosynthesis of phenazines. *ChemBioChem* 10(14), 2295–2304. DOI: 10.1002/cbic.200900323.

Nosanchuk, J.D. and Casadevall, A. (2003) The contribution of melanin to microbial pathogenesis. *Cellular Microbiology* 5(4), 203–223. DOI: 10.1046/j.1462-5814.2003.00268.x.

Pavan, M.E., López, N.I. and Pettinari, M.J. (2020) Melanin biosynthesis in bacteria, regulation and production perspectives. *Applied Microbiology and Biotechnology* 104(4), 1357–1370. DOI: 10.1007/s00253-019-10245-y.

Pierson, L.S. and Pierson, E.A. (2010) Metabolism and function of phenazines in bacteria: Impacts on the behavior of bacteria in the environment and biotechnological processes. *Applied Microbiology and Biotechnology* 86(6), 1659–1670. DOI: 10.1007/s00253-010-2509-3.

Pierson, B.K., Sands, V.M. and Frederick, J. (1990) Spectral irradiance and distribution of pigments in a highly layered marine microbial mat. *Applied and Environmental Microbiology* 56(8), 2327–2340. DOI: 10.1128/aem.56.8.2327-2340.1990.

Plonka, P.M. and Grabacka, M. (2006) Melanin synthesis in microorganisms – biotechnological and medical aspects. *Acta Biochimica Polonica* 53(3), 429–443. DOI: 10.18388/abp.2006_3314.

Rayati, S. and Malekmohammadi, S. (2016) Catalytic activity of multi-wall carbon nanotube supported manganese (III) porphyrin: An efficient, selective and reusable catalyst for oxidation of alkenes and alkanes with urea–hydrogen peroxide. *Journal of Experimental Nanoscience* 11, 872–883. DOI: 10.1080/17458080.2016.1179802.

Reiss, R., Ihssen, J., Richter, M., Eichhorn, E., Schilling, B. *et al.* (2013) Laccase versus laccase-like multi-copper oxidase: A comparative study of similar enzymes with diverse substrate spectra. *PLoS ONE* 8, e65633. DOI: 10.1371/journal.pone.0065633.

Ribera, J., Panzarasa, G., Stobbe, A., Osypova, A., Rupper, P. *et al.* (2019) Scalable biosynthesis of melanin by the basidiomycete *Armillaria cepistipes*. *Journal of Agricultural and Food Chemistry* 67(1), 132–139. DOI: 10.1021/acs.jafc.8b05071.

Rothemund, P. (1935) Formation of porphyrins from pyrrole and aldehydes. *Journal of the American Chemical Society* 57, 2010–2011. DOI: 10.1021/ja01313a510.

Rothemund, P. (1936) A new porphyrin synthesis. The synthesis of porphin 1. *Journal of the American Chemical Society* 58, 625–627. DOI: 10.1021/ja01295a027.

Saini, A.S. and Melo, J.S. (2015) One-pot green synthesis of eumelanin: Process optimization and its characterization. *RSC Advances* 5, 47671–47680. DOI: 10.1039/C5RA01962A.

Schmaler-Ripcke, J., Sugareva, V., Gebhardt, P., Winkler, R., Kniemeyer, O. *et al.* (2009) Production of pyomelanin, a second type of melanin, via the tyrosine degradation pathway in aspergillus fumigatus in *Aspergillus fumigatus*. *Applied and Environmental Microbiology* 75, 493–503. DOI: 10.1128/AEM.02077-08.

Sendovski, M., Kanteev, M., Ben-Yosef, V.S., Adir, N. and Fishman, A. (2011) First structures of an active bacterial tyrosinase reveal copper plasticity. *Journal of Molecular Biology* 405, 227–237. DOI: 10.1016/j.jmb.2010.10.048.

Silveira, J.E.P.S., Pereda, M. del C.V., Eberlin, S., Dieamant, G.C. and Di Stasi, L.C. (2008) Effects of *Coccoloba uvifera* L. on UV-stimulated melanocytes. *Photodermatology, Photoimmunology & Photomedicine* 24(6), 308–313. DOI: 10.1111/j.1600-0781.2008.00382.x.

Solano, F. (2014) Melanins: Skin pigments and much more—types, structural models, biological functions, and formation routes. *New Journal of Science* 1–28. DOI: 10.1155/2014/498276.

Sun, S., Zhang, X., Chen, W., Zhang, L. and Zhu, H. (2016) Production of natural edible melanin by *Auricularia auricula* and its physicochemical properties. *Food Chemistry* 196, 486–492. DOI: 10.1016/j.foodchem.2015.09.069.

Surwase, S.N., Jadhav, S.B., Phugare, S.S. and Jadhav, J.P. (2013) Optimization of melanin production by *Brevundimonas* sp. SGJ using response surface methodology. *SGJ Using Response Surface Methodology* 187–194. DOI: 10.1007/s13205-012-0082-4.

Wang, L., Li, Y. and Li, Y. (2019) Metal ions driven production, characterization and bioactivity of extracellular melanin from *Streptomyces* sp. zl-24. *International Journal of Biological Macromolecules* 123, 521–530. DOI: 10.1016/j.ijbiomac.2018.11.061.

Yang, J., Li, W., Ng, T.B., Deng, X., Lin, J. *et al.* (2017) Laccases: Production, expression regulation, and applications in pharmaceutical biodegradation. *Frontiers in Microbiology* 8. DOI: 10.3389/fmicb.2017.00832.

Zou, Y. and Tian, M. (2017) Fermentative production of melanin by *Auricularia auricula*. *Journal of Food Processing and Preservation* 41, e12909. DOI: 10.1111/jfpp.12909.

5 Genetic and Environmental Factors Influencing Pigment Production

Sony Singh[1], Manjoo Rani[1*], Saurabh Ghildiyal[2] and Ayush Madan[3]
[1]Department of Pharmacy, Shri Ram Murti Smarak College of Engineering and Technologies, Bareilly, Uttar Pradesh, India; [2]Department of Hotel Management, Shri Ram Murti Smarak College of Engineering, Technology and Research, Bareilly, Uttar Pradesh, India; [3]Department of Biotechnology, School of Research & Technology (SORT), People's University, Bhanpur, Bhopal, Madhya Pradesh, India

5.1 Introduction

Pigments are naturally occurring or synthetic materials that give color to living materials and objects by selectively taking in and reflecting wavelengths of light (Paillié-Jiménez *et al.*, 2020). Pigments are biomolecules (usually secondary metabolites) in a biological system that convey color in plants, animals, and microorganisms. They are found in many forms which include chlorophylls, carotenoids, flavonoids and melanin with different optical properties and specific biological roles (Vidyalakshmi *et al.*, 1999; Lee *et al.*, 2006). Pigments are important well beyond their coloration (Fimbres-Olivarría *et al.*, 2016). During photosynthesis, pigments such as chlorophyll in plants and some microorganisms act as sinks of solar energy to be used in conversion of light energy to chemical energy vital in growth and maintenance of life (Młodzińska, 2009; Joshi and Pandey, 2018).

Pigments safeguard organisms against the detrimental effects of ultraviolet (UV) radiation and oxidative stress by adsorbing excess light energy and scavenging reactive oxygen species. Ecologically, key pigment interactions include attracting pollinators and seed dispersers, counteracting herbivorous threats, and creating camouflage or warning signs. When organisms face stress, such as intense light, drought or pathogen infection, pigment formation increases (Navarro *et al.*, 2001).

In addition to their biological significance, pigments have global industrial and economic worth. They are used on a large scale in the food sector, textiles, cosmetics, and pharmaceutical industry as well as in biotechnological processes. Sustainable production of natural pigments can be considered as biodegradable, low-toxic, and even potentially beneficial to human health (Lyu *et al.*, 2022). Pigments are therefore a fundamental synthesis between biology, ecology, and technology (Klass, 1977; Chaplin, 2004).

5.1.1 Pigments: nature's color architects

Pigments are substances that impart color to materials, playing crucial roles not only in the

*Corresponding author: manjoo.yadav@gmail.com

 Biopigments of Microbial Origin: Innovations and Applications Across Industries (eds Madan *et al.*)
DOI: 10.1079/ 9781836990505.0005

natural world but also in human life and culture (Meyer *et al.*, 1992). They influence how we perceive our surroundings, protect living organisms, and inspire creativity in art and design (Casagrande *et al.*, 2009; Sturm and Duffy, 2012; Saini and Keum, 2018; Jablonski, 2021).

The key functions of pigments include the following.

- *Colour production*: pigments give materials their distinctive colors by selectively absorbing certain wavelengths of light and reflecting others. For example, a red pigment absorbs most colors except red, which it reflects to our eyes.
- *Light absorption*: in biological systems, pigments capture light energy. For instance, in plants, chlorophyll absorbs light during photosynthesis, enabling the plant to produce food.
- *Protection*: pigments can safeguard organisms from UV radiation. Melanin in human skin, for example, absorbs UV light, reducing DNA damage and lowering the risk of sunburn.
- *Camouflage and attraction*: in nature, pigments help species survive and reproduce. Some animals use pigments for camouflage, blending into their surroundings, while flowers display vivid colors to attract pollinators.
- *Art and aesthetic expression*: in human culture, pigments are central to painting, textiles, and design. They help create visual effects, evoke emotions, and produce artworks that are both beautiful and meaningful.

5.1.2 Differences between pigments and dyes

Dyes and pigments impart color to a wide variety of substances. While both serve the fundamental purpose of adding color, their selection depends on the intended application and the required properties (Table 5.1). Pigments are often preferred where long-lasting color, lightfastness, and stability are essential (Lu and Zhang, 2000).

Table 5.1. Differentiation between dyes and pigments.

Properties	Pigments	Dyes	References
Physical State	Insoluble material, converted into fine powder and used as a colorant	Small soluble colorant	(Iqbal, 2008; Burkinshaw, 2016)
Size	Comparatively large	Very small	(Gordon and Gregory, 1987)
Transparency	Less transparent	More transparent	(Chakraborty, 2010)
Solubility	Not soluble in water and many solvents	Soluble in water	(Carmen and Daniela, 2012; Chequer *et al.*, 2013)
Type of compound	Can be either organic or inorganic	Mostly organic	(Gordon and Gregory, 1987; Burkinshaw, 2016)
Auxochrome group	No auxochrome groups present	Auxochrome groups may be present	(Griffiths, 1990)
Availability	Less available	Available in large numbers	(Iqbal, 2008)
Affinity with material	No direct affinity to the material	Direct affinity to the material	(Chequer *et al.*, 2013)
Binding agents	Require binding agents	Do not require binding agents	(Carmen and Daniela, 2012)
Diffusion	Diffuses on the fabric	Diffuses in the fabric	(Chequer *et al.*, 2013)

The primary distinction between dyes and pigments lies in their solubility (Kim *et al.*, 1999). Dyes are generally soluble in water and many other solvents, allowing them to penetrate and chemically bond with the substrate. In contrast, pigments are insoluble in water due to their larger particle size; they remain suspended rather than dissolved. To apply pigments in liquid form, a binder is required to hold the particles in place, whereas dyes can dissolve directly in a suitable solvent (Parekh *et al.*, 2000).

5.2 Classification of Pigments

Pigments may be classified into natural and synthetics as shown in Fig. 5.1. Natural pigments are made by plants or animals, with advantages of eco-friendly sources of color, whereas synthetic pigments are made chemically through thermal and chemical reactions, often resulting in more intense colours, better stability, and greater availability.

Natural or artificial pigments may be either inorganic or organic. White, colored, and black pigments, fillers and special pigments are inorganic pigments and tend to be of value by virtue of their opacity, resistance to the environment and service life (Stange *et al.*, 2019). Organic pigments on the other hand have high color and tint strength, and are also available as colored pigments, black pigments, special pigments, and fillers. This typology is useful to determine the right pigment according to the aesthetic effect required, the stability and use.

5.3 Differentiation Between Organic and Inorganic Pigments

Pigments can be broadly classified into organic and inorganic types, each with distinct sources, properties, and applications. The choice between them depends on the desired color intensity, durability, stability, and cost. While organic pigments are carbon based and often derived from natural or synthetic sources, inorganic pigments are noncarbon based, typically metallic or mineral in origin. Table 5.2 shows the examples of organic and inorganic pigments.

5.3.1 Organic pigments

These pigments are primarily carbon-based compounds. Although early pigments were obtained from natural sources such as plants and animals, most modern organic pigments are synthetically produced. They are valued for their bright, intense colors and high tinting strength, but they generally have lower heat and light stability compared to inorganic pigments.

5.3.2 Inorganic pigments

These pigments do not contain carbon in their molecular structure and are often metallic in nature. They are known for their excellent heat stability, larger particle size, and superior resistance to environmental degradation. Inorganic pigments are commonly derived from minerals, metals, or other nonliving materials, and tend to have more subdued colors compared to organic pigments.

Pigment performance varies according to their chemical composition and particle structure. Natural and synthetic inorganic pigments tend to be more stable and durable, whereas organic colors, in particular high-performance grades, give brighter colors. The main differences between inorganic pigments and two types of organic pigments, classical and high-performance, are summarized in Table 5.3.

This comparison illustrates that inorganic pigments excel in stability and durability, while high-performance organic pigments balance brilliance and resistance, making them suitable for demanding applications.

5.4 Properties of an Ideal Pigment

An ideal pigment should possess a combination of physical, chemical, and safety-related characteristics to ensure optimum performance in its intended application (Yamaguchi and Hearing, 2009). The following properties are generally considered essential for an ideal pigment.

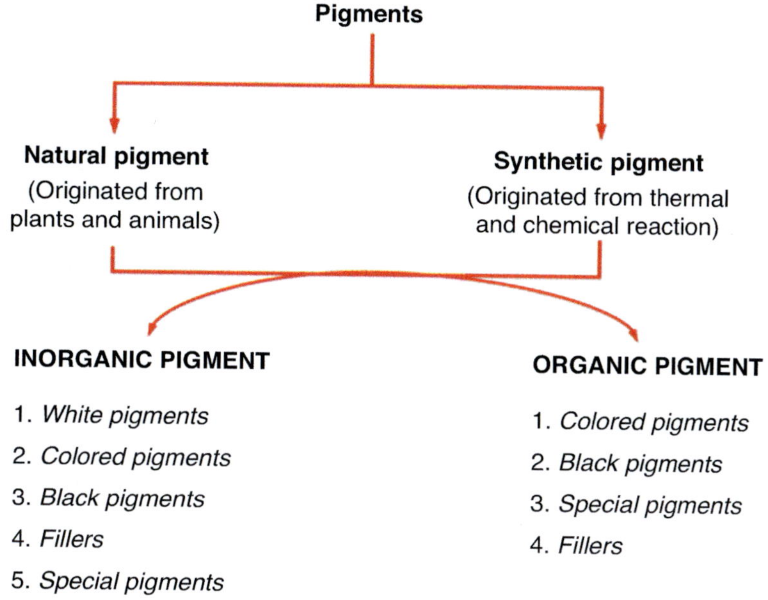

Fig. 5.1. Classification of pigments.

Table 5.2. Examples of organic and inorganic pigments.

Organic pigments	Inorganic pigments	References
Azo pigments	Mixed metal oxide	(Herbst *et al*., 2004)
Diazo	Bi-vanadate	(Herbst *et al*., 2004)
Napthol	Manganese oxide	(Bilal *et al*., 2020)
Azo lakes	Phthalo chrome green	(Herbst *et al*., 2004)
Benzimidazolone	Iron oxide	(Hauri and Hohl, 2015)
Metal complex	Cadmium	(Herbst *et al*., 2004)
Phthalocyanine	Lead chromate	(Herbst *et al*., 2004)
Thioindigo	Iron blue	(Bilal *et al*., 2020)
Dioxazine	Titanium dioxide white	(Hauri and Hohl, 2015)

- *Maximum covering power*: the pigmentss-hould provide excellent opacity, ensuring that a minimal quantity can effectively cover the substrate.
- *Optimal particle size*: particle size should ideally range between 0.2 and 0.4 μm for balanced color strength, dispersion, and stability.
- *Chemical stability*: the pigment should resist reacting with environmental factors such as moisture, oxygen, acids, or alkalis.
- *High chemical resistance*: it should withstand exposure to chemicals without degradation or discoloration.
- *Nontoxic and safe*: the pigment must be nontoxic, biologically inert, and safe for human health, without causing allergic reactions.
- *Lightfastness*: it should resist fading or color changes when exposed to light over time.

Table 5.3. Properties of pigments.

Property	Inorganic pigments	Organic pigments – classical	Organic pigments – high performance	Reference
Color, purity	Dull	Brighter	Brighter	(Boonmahome and Mongkolthanaruk, 2023)
Opacity	Higher	Transparency either more or less	Transparency either more or less	(Brinkmann *et al.*, 2022)
Color strength	Medium–low	Higher	Higher	(Caminade *et al.*, 2019)
Light fastness (blue scale)	Good to high	Low to medium	Good to high	(Celedón and Díaz, 2021)
Heat resistance	≥500°C (rarely ≤200°C)	150–200°C	200–300°C	(Chander, 2022)
Weather resistance	Varies (depending on chemistry)	Insufficient	Middle to high	(Chatragadda and Dufossé, 2021)
Resistance to chemicals	Varies (depending on chemistry)	High	High	(Chen *et al.*, 2021)
Fastness to solvents/bleed resistance	High	Medium to good	Good to high	(Choi *et al.*, 2021)
Price	Low to medium	Medium	High	(Corral *et al.*, 2020)

- *Chemical inertness*: the pigment should remain unaffected by the chemical composition of the medium or substrate.
- *Excellent dispersion*: t must disperse uniformly in the medium, avoiding clumping or settling.

5.5 Microbial Pigments and Their Color Shades

Pigments can be produced by a great number of various microorganisms, such as bacteria and fungi (molds, yeasts, and algae). The color tone of these microbial pigments varies and they find applications in food, pharmaceuticals, cosmetics, or textiles because of their biocompatibility and naturality. Table 5.4 shows some examples of microorganisms, the pigments, and colors they synthesize.

5.6 Industrial Aspects of Microbial Pigments

Microbial pigments are produced by microorganisms such as bacteria, fungi, and algae (Silva-Castro *et al.*, 2019). These pigments offer several unique characteristics and advantages which can be applied in various industries (Richmond, 2004). Fig. 5.2 illustrates the pigments used in various industrial areas. Microbial color is use in the fish industry to enhanced the pink color of farmed salmon (Kyong-Hwa *et al.*, 2011). Some natural colorants are used as antioxidants, anticancer, antibiotic, and enzymatic agents, and in the food, textiles, and cosmetics industries.

Table 5.4. Microorganisms producing different color shades.

Microorganism group	Species	Pigment color(s)	Reference
Bacteria	*Achromobacter*	Creamy	(Venil *et al*., 2013)
	Bacillus sp.	Brown	(Venil *et al*., 2013)
	Brevibacterium sp.	Orange, yellow	(Delgado-Vargas *et al*., 2000)
	Pseudomonas sp.	Yellow, blue-green (pyocyanin)	(El Feghali and Nawas, 2018; Darwesh *et al*., 2019; Govindraj *et al*., 2020)
	Streptomyces sp.	Yellow, red, blue	(Venil *et al*., 2013)
	Janthinobacterium lividum	Violet (violacein)	(Cortés-Osorio *et al*., 2017)
Molds	*Aspergillus* sp.	Orange, red	(Delgado-Vargas *et al*., 2000)
	Helminthosporium catenarium	Red	(Delgado-Vargas *et al*., 2000)
	H. gramineum	Red	(Delgado-Vargas *et al*., 2000)
	H. cynodontis	Bronze	(Delgado-Vargas *et al*., 2000)
	Monascus purpureus	Yellow, orange, red	(Venil *et al*., 2013)
	Penicillium cyclopium	Orange	(Delgado-Vargas *et al*., 2000)
Yeasts	*Cryptococcus* sp.	Red	(Venil *et al*., 2013)
	Phaffia rhodozyma	Red	(Delgado-Vargas *et al*., 2000)
	Rhodotorula lipolytica	Red	(Venil *et al*., 2013)
Algae	*Dunaliella salina*	Red	(Delgado-Vargas *et al*., 2000)

5.7 Mechanism of Pigment Color Production

The mechanism by which pigments produce color is rooted in their ability to interact with light, particularly through absorption and reflection. Fig. 5.3 illustrates that different light rays are absorbed and reflected by pigments.The exact mechanism varies depending on the type of pigment (organic and inorganic) but the basic principles are the same (Dufossé, 2009).

Figure 5.4 show how pigments interact with light to produce color. Pigments absorb certain wavelengths of light and reflect or transmit others (Venil *et al.*, 2013). Some wavelengths of light are absorbed by pigment particles (represented here as red and green arrows) and others are reflected (light blue arrow). The colors seen are dependent on the wavelengths reflected. For example, when the blue light is bounced off and the other colors are absorbed, the pigment appears blue to the human eye.

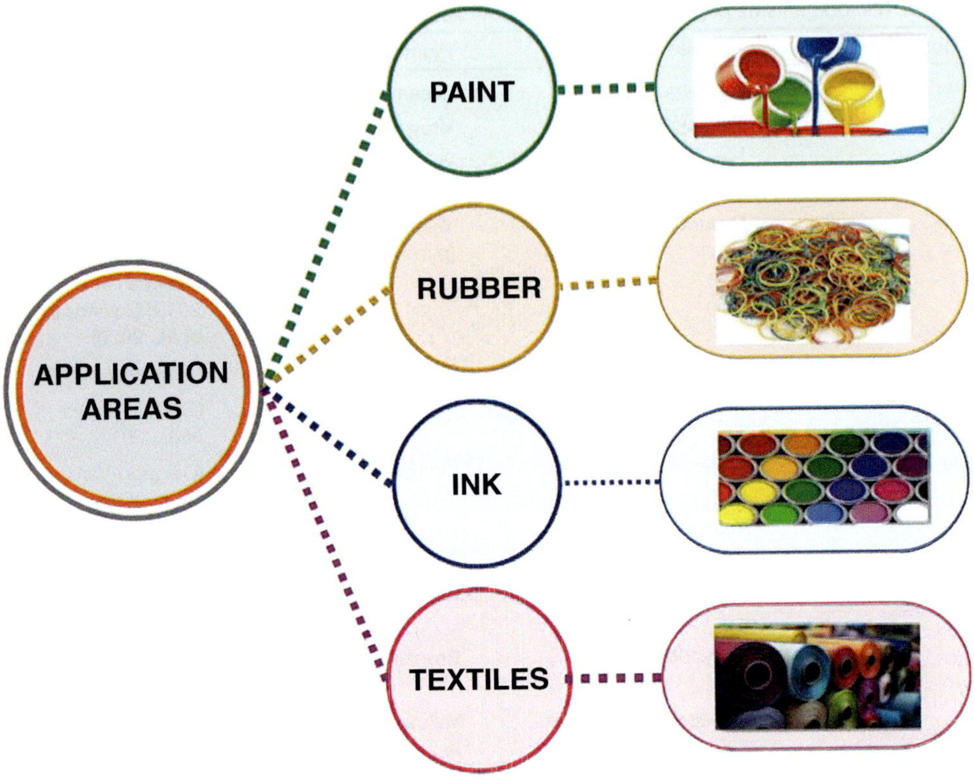

Fig. 5.2. Pigments applied in different industrial areas.

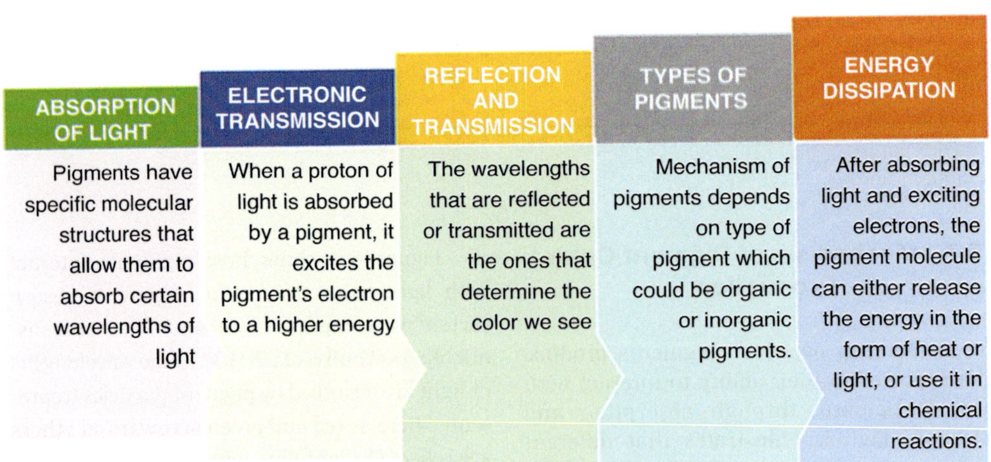

Fig. 5.3. Different methods of color production.

Fig. 5.4. Mechanism of pigment color perception.

5.8 Factors Influencing Pigment Production

The genetic make-up of an organism is one of the fundamental aspects that dictate pigment production (Gerrish and Cáceres, 2003; Lin *et al.*, 2023); Marin-Recinos and Pucker, 2024). The genes encoding enzymes and structural proteins that perform pigment synthesis specify which classes of pigments, e.g. chlorophylls, carotenoids, anthocyanins or melanins, are made. Any differences in these genes, such as mutational variation, allelic variation, etc., may result in pigmentation patterns causing a change in color saturation, hue or pattern. Pigment diversity may or may not increase through duplication of genes or decrease through their loss in a given species but also affects its ecological adaptation and survival (Johler *et al.*, 2010; Matrková and Remeš, 2012).

The biosynthesis of pigments is strongly regulated at the transcriptional, translational, and posttranslational levels. Biosynthesis pathway genes are regulated by regulatory genes and transcription factors to respond to the environment and developmental stimuli. Light-dependent transcription factors are an example of this whereby, under certain lighting conditions, chlorophyll or anthocyanin synthesis may be activated by certain transcription factors. Feedback inhibition, compartmentalization in cellular organelles, and cofactor contact also regulate enzyme activity in pigment pathways to fine-tune pigment production to physiological needs without unnecessarily expending resources.

Genetic variability, natural selection, and genetic drift in the course of evolution have resulted in the tremendous diversity of pigments in nature. Changes in genes related to pigmentation have allowed organisms to survive in new environments, e.g. UV-absorbing pigments in plants on high mountains or camouflage pigments in prey species. There is also horizontal gene transfer between organisms, especially microorganisms, which has also assisted the propagation of pigment production ability between unrelated organisms. Ecological relationships, species recognition, and reproductive success rather than functional adaptations are the causes of evolutionary diversification of pigment biosynthesis.

The production of pigments can be influenced through intrinsic and extrinsic factors which are crucial for optimizing their industrial production (Carlile, 1956; Ahmad *et al.*, 2012). Fig. 5.5 illustrates numerous factors which affect the production of pigments.

5.8.1 Nutrient availability

5.8.1.1 Carbon source

Various carbon sources influence pigment production, such as glucose, sucrose, and organic acids. Some microbes produce many pigments when they metabolize certain carbon sources (Giwa *et al.*, 2017).

Organic carbon (glucose, sucrose, fructose)

- *Energy source*: organic carbon sources, such as sugars, serve as a direct energy source for biochemical processes, including pigment synthesis.
- *Precursors for pigment biosynthesis*: many pigment biosynthesis pathways require specific organic compounds, and carbon from sugars or other organic molecules is incorporated into pigment molecules during synthesis (Richmond, 2004; Lefsih *et al.*, 2017).
- *Enhanced growth and metabolism*: the presence of organic carbon often boosts metabolic activity, leading to increased enzyme production and enhanced pigment synthesis (Yong *et al.*, 2019).

Inorganic carbon (carbon dioxide)

- *Photosynthesis*: in plants, algae, and certain bacteria, CO_2 is used in photosynthesis to produce glucose, which is then utilized for pigment production, particularly chlorophyll (Gong and Bassi, 2016). Chlorophyll is essential for capturing light energy, and without sufficient CO_2, the process

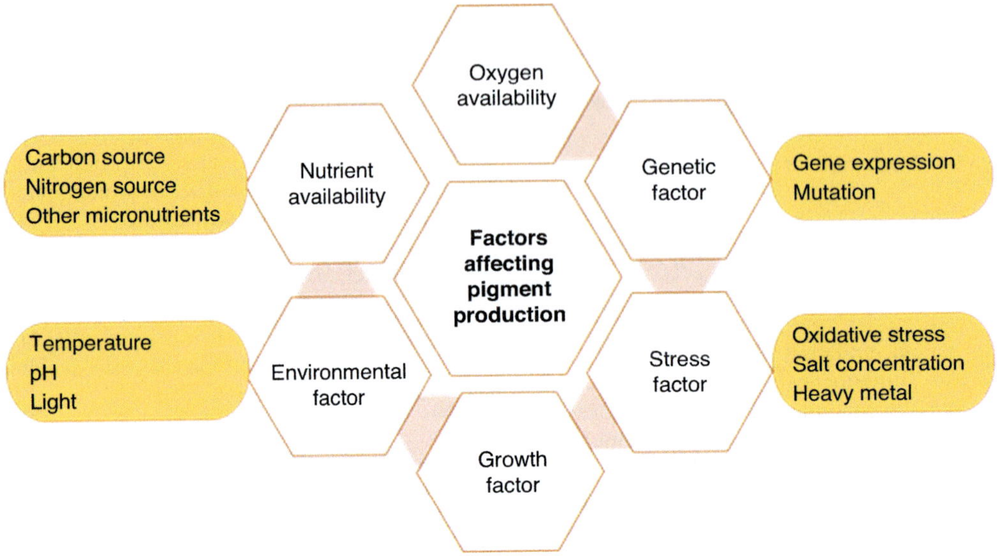

Fig. 5.5. Factors affecting the production of pigments.

of photosynthesis slows down, leading to reduced chlorophyll and other pigments.

- *Fermentation or anaerobic conditions*: in some cases, organisms rely on fermentation in the absence of oxygen, using an organic carbon source to produce energy. Under anaerobic conditions, certain pigment production pathways may be altered to adapt to the lack of oxygen or light (Mustafa *et al.*, 2019).

5.8.1.2 Nitrogen source

Source nitrogen can be available from ammonium salts and nitrates and plays major role in affecting pigment production (Yong *et al.*, 2019). A loss of nitrogen can sometimes lead to increased pigment production in certain microorganisms (Araujo and Garcia, 2005).

Forms of nitrogen

- *Ammonium*: ammonium is a common nitrogen source for many plants, algae, and microorganisms. It can be directly absorbed and used in metabolic pathways, including those responsible for pigment biosynthesis.
- *Nitrate*: nitrate is another common nitrogen source, especially for plants. Once absorbed, it is converted to ammonium by plant enzymes before being incorporated into organic molecules.

Nitrogen's role in pigment biosynthesis

- *Chlorophyll production*: nitrogen is a crucial element in the production of chlorophyll. Chlorophyll contains a porphyrin ring structure with a central magnesium atom, and nitrogen is integral to this structure (Muller-Feuga *et al.*, 2007).
- *Amino acids and proteins*: nitrogen is a key component of amino acids, the building blocks of proteins. Some pigments are involved in the synthesis of pigments, and a lack of nitrogen can impair the production of pigment-related enzymes.
- *Carotenoids and anthocyanins*: nitrogen availability can also influence the production of other pigments, such as carotenoids (which produce yellow, orange, and red colors) and anthocyanins (which produce purple, blue, and red colors).

Nitrogen deficiencies and pigment production

- *Chlorosis*: nitrogen deficiencies can lead to chlorosis, where plants turn yellow because of the breakdown of chlorophyll and other pigments. This results in a decrease in photosynthesis and overall growth.

- *Reduced pigment synthesis*: in nitrogen-deficient conditions, plants may prioritize essential functions over pigment production, leading to reduced synthesis of chlorophyll and other pigments (Subhasree *et al.*, 2011). However, certain stress conditions (such as nitrogen limitation) might induce the production of secondary metabolites like anthocyanins as a protective mechanism.

Nitrogen excess and pigment production

- *Promotes growth*: excess nitrogen typically promotes vegetative growth and can increase chlorophyll content, resulting darker green foliage. However, too much nitrogen may also lead to excessive leaf growth at the expense of flowering and fruiting (Gonzali and Perata, 2021).
- *Altered pigment profile*: in some cases, excessive nitrogen can affect the synthesis of other pigments like carotenoids. The balance between nitrogen and other nutrients (like phosphorus and potassium) is crucial for optimal pigment production.

Microbial pigment production

- *Bacteria and fungi*: many bacteria and fungi produce pigments, such as carotenoids and melanin, that rely on nitrogen for synthesis. The availability of nitrogen in microbial cultures influences the type and amount of pigment produced.
- *Nitrogen limitation*: nitrogen limitation in some microorganisms can lead to increased production of certain pigments as part of a stress response.

5.8.1.3 *Trace metals*

Elements like iron, copper, and magnesium are major cofactors for enzymes involved in production of pigments and their availability can affect pigment synthesis pathways.

5.8.2 Environmental conditions

5.8.2.1 *Temperature*

Ambient temperature can affect the enzymatic activity involved in synthesis of pigments (Arikan *et al.*, 2020). Varying temperature can either enhance or inhibit production depending upon the microorganism. Production of pigments from *Monascus* sp. requires 25–28°C and *Pseudomonas* sp. require 35–36°C (Silbir and Goksungur, 2019).

5.8.2.2 *pH*

The role of pH in pigment synthesis is central because microbes require an optimum pH range. If pH range deviates it can hinder the synthesis of pigments. pH requirement differs from one microorganism to another microorganism because some microbes need an acidic medium and some an alkaline medium for the synthesis of pigments. A minute change can affect the shade of color or appearance of pigments (Bhosale and Gadre, 2002).

5.8.2.3 *Light*

Intensity and quality of light can significantly influence the synthesis of pigments such as chlorophylls and carotenoids. This factor is important for some photosynthetic microbes like cyanobacteria and algae (Pelczar and Chan, 1977).

5.8.3 Oxygen availability

Some microbial pigments are synthesized in the presence or absence of oxygen, particularly in anaerobic or microaerophilic environments that may be influenced by availability of oxygen. Some pigments like bacteriochlorophyll are synthesized under lower oxygen demand (Oser, 1965).

5.8.4 Growth phase

Microbial growth phase can influence the production of pigments. Some microorganisms may produce more pigments during the stationary phase (when nutrients are scarce) than during the exponential growth phase.

5.8.5 Stress condition

5.8.5.1 Oxidative stress

The production of pigments can be influenced by reactive oxygen species which in some microbes are sensitive to the presence of oxygen as a protective response.

5.8.5.2 Salt concentration

High salt concentrations may induce the synthesis of some pigments, as some microbes adapt to osmotic stress by synthesizing protective pigments.

5.8.5.3 Heavy metals

Heavy metals exposure can promote the production of certain pigments, which act as protective microorganisms.

5.8.6 Genetic factors

Pigment production is stimulated by various genetic factors, which can determine both type and amount of pigment produced by the microorganism. Genetic factors play a major role in enhancing or controlling microbial pigment production (Wang *et al.*, 2012). Key genetic factors involved in pigment production are discussed below.

5.8.6.1 Biosynthetic pathways

The genes encoding enzymes involved in the biosynthesis of pigments are crucial.

- Polyketide synthase (PKS) and nonribosomal peptide synthetase (NRPS) genes are essential for the production of complex pigments like actinorhodin (a blue pigment in *Streptomyces* species).
- Carotenoid biosynthesis genes are involved in the synthesis of carotenoids, such as lycopene found in various bacteria and fungi.
- Chlorophyll synthesis genes are critical in cyanobacteria and other photosynthetic microbes for pigment production related to photosynthesis.

5.8.6.2 Regulatory genes

Genes that control the expression of pigment biosynthesis pathways play a crucial role in pigment production.

- *Transcription factor*: this factor cannot control the expression of genes involved in pigment synthesis. For example, act genes in *Streptomyces* are controlled by global regulators that control the production of pigments and antibiotics.
- *Two-component regulatory system*: various microbes use two-component systems to sense environmental cues (like oxygen, light or growth factor) which can stimulate the expression of pigment-producing genes.
- *Environment sensing genes*: microorganisms often adjust pigment production in response to changes in their environment and genes that help microbes sense and respond to environmental conditions play an important role in this process.
- *Light-responsive genes*: promote production ofpigments like carotenoids in response to UV or visible light.
- *Quorum sensing*: some microbes regulate pigment production based on population density, using signaling molecules that influence the expression of genes.

5.8.6.3 Secondary metabolite clusters

Various pigment-producing microbes possess specialized secondary metabolite gene clusters that are not involved in primary metabolism. These clusters often include genes for biosynthetic enzymes, transporters, and regulators. For example, the gene cluster for prodigiosin production in *Serratia marcescens* contains multiple genes responsible for the biosynthesis and controlling of this red pigment.

5.8.6.4 Horizontal gene transfer

In some cases, production of microbial pigment genes may be acquired through horizontal gene transfer (Fontana *et al.*, 1996). For example, pigments like anthraquinone and indigoid dyes are sometimes encoded by genes that have been horizontally transferred among different microbial species.

5.8.6.5 Genetic engineering

In laboratory settings, researchers may use genetic manipulation techniques to increase pigment production in microbes. This might involve overexpressing certain biosynthetic genes, knocking out regulatory genes, or introducing genes from other species to diversify the range of pigment produced.

5.9 Ecological and Evolutionary Implications

The manufacture of pigment has several adaptive benefits in various environments in which the organisms live. Pigment compositions and concentrations also vary so that plants, microorganisms, and animals may fine-tune their ability to absorb light, balance heat absorption, or produce camouflage (Hodgson *et al.*, 2004). To take a few examples, alpine plants with large levels of anthocyanins are resistant to strong UV radiation, and camouflage pigments in some animals decrease their chances of becoming prey. Pigment adaptation also enables organisms in aquatic environments to achieve optimal light absorption to ensure photosynthetic energy or communication with others depending on the depth of the organism (Mulders *et al.*, 2014).

A number of pigments including carotenoids, flavonoids, and melanin act as natural antioxidants in their neutralization of reactive oxygen species produced in situations of stress. Pigments such as chlorophyll-binding carotenoids trap excess light energy in photosynthetic organisms, avoiding photoinhibition and the formation of reactive oxygen species that damage cells. This photoprotective property is more significant in the environment of high light, drought, or heat stress, during which oxidative stress may limit growth and metabolic activity. Pigments also help in long-term survival and reproduction by causing a decrease in oxidative damage (Hanna, 1953).

The difference in color between species and within species is the result of evolutionary mechanisms set in motion by natural selection, sexual selection, and genetic drift. The diversity in pigmentation develops under ecological conditions, including predator–prey relationships, pollinator preferences, and light conditions. As an example, bright pigmentation in flowers can evolve to attract a particular pollinator; camouflage color in prey can evolve to prevent being detected by a predator. Genetic mutation, duplicating genes, and horizontal gene transfer help in the formation of new pigment types and patterns over time. Not only is biodiversity governed by these evolutionary dynamics, ecosystem structure and stability are also affected.

5.10 Uses and Prospects

5.10.1 Industrial, agricultural, and biotechnological applications

Owing to their color, bioactivity, and functions, pigments have wide applications in various industries. There is extensive use of natural pigments in food coloring, cosmetics, dyestuffs, and drugs, among others in the industrial sector, and they provide safer and more sustainable pigments compared to synthetic ones. In agriculture, pigments have been used to improve crops, stress tolerance, and pest resistance, with specific pigments serving as natural protective pigments against UV radiation and pathogens. In the biotechnology sector, microbial pigments are attracting interest for their antimicrobial, antioxidant, and anticancer activities and also as bioreceptors and biofunctional materials.

5.10.2 Trends and issues in research

New studies are being directed towards improving the yield and stability of pigments and concentrating on the bioactivity of pigments by modern modes of cultivation, metabolic engineering, and environmentally sound methods of extraction. It is possible that novel pigments could be discovered among extremophiles and rare microbial species. However, there are still difficulties associated with large-scale manufacturing and affordability and retaining pigment stability during processing and storage. Commercialization of new pigment sources is also affected by regulatory approval processes and consumer acceptance.

5.10.3 Potential of genetic engineering in pigment optimization

Genetic engineering has the potential to affect the development of pigment production based on adjustment of biosynthetic pathways, introduction of new genes, or the control of transcription factors engaged in pigment production. CRISPR-Cas gene editing can be used to increase pigment levels, change the coloring characteristics or enhance crop and microorganism stress resistance. The techniques of synthetic biology also enable the creation of customized pigments with desired functionality and aesthetics, opening up new avenues of food technology, medicine, and sustainable materials innovation.

5.11 Conclusion

As secondary metabolites, pigments are vital to biological systems and are best known as coloring materials but also as modulators of physiological, ecological, and defense processes. Genetic and environmental factors interact in microorganisms to promote pigment biosynthesis. The genetic factors determine the possible range and amount of pigment production that an organism is capable of, whereas the actual pigment production is determined by environmental factors such as the amount and intensity of light, temperature, and water supply as well as composition of nutrients.

Stress conditions (UV radiation, oxidative stress, pollution, and microbial competition) can either stimulate or inhibit pigment production in accordance with the adaptive requirements of an organism. Metabolic pathways depend on factors such as the pH, oxygen levels and the balance between carbon and nitrogen sources, whereas biosynthetic control is refined by hormones and growth regulators. Ecological pressures imposed by a habitat of microbial interactions in relation to pigment diversity and functions are also added.

Understanding this chain of effects is critical in maximizing pigment synthesis in an industrial situation. Microbial pigments have demonstrated good bioactivity including antimicrobial, antiinflammatory, and even anticancer properties. For example, prodigiosin from *Pseudoalteromonas* sp. and pyocyanin show cytotoxicity towards leukemia and inhibition on *Escherichia coli* biofilms respectively. These results stress both the functional ability of pigments as adaptive materials in nature and their importance as bioresources for applied science and technology.

References

Ahmad, W.A., Wan Ahmad, W.Y., Zakaria, Z.A. and Yusof, N.Z. (2012) Application of bacterial pigments as colorant. In: *Application of Bacterial Pigments as Colorant: The Malaysian Perspective*. Springer, Berlin, Germany, pp. 57–74. DOI: 10.1007/978-3-642-24520-6_4.

Araújo, S. and Garcia, V.M.T. (2005) Growth and biochemical composition of the diatom *Chaetoceros cf. wighamii* brightwell under different temperature, salinity and carbon dioxide levels. I. Protein, carbohydrates and lipids. *Aquaculture* 246(1–4), 405–412. DOI: 10.1016/j.aquaculture.2005.02.051.

Arikan, E.B., Canli, O., Caro, Y., Dufossé, L. and Dizge, N. (2020) Production of bio-based pigments from food processing industry by-products (apple, pomegranate, black carrot, red beet pulps) using *Aspergillus carbonarius*. *Journal of Fungi* 6(4), 240. DOI: 10.3390/jof6040240.

Bhosale, P.B. and Gadre, R.V. (2002) Manipulation of temperature and illumination conditions for enhanced beta-carotene production by mutant 32 of *Rhodotorula glutinis*. *Letters in Applied Microbiology* 34(5), 349–353. DOI: 10.1046/j.1472-765x.2002.01095.x.

Bilal, M., Mehmood, S. and Iqbal, H.M.N. (2020) The beast of beauty: Environmental and health concerns of toxic components in cosmetics. *Cosmetics* 7(1), 13. DOI: 10.3390/cosmetics7010013.

Boonmahome, P. and Mongkolthanaruk, W. (2023) Characterization of indole-3-acetic acid biosynthesis and stability from *Micrococcus luteus*. *Journal of Applied Biology & Biotechnology* 11(3), 187–192. DOI: 10.7324/JABB.2023.117202.

Brinkmann, S., Spohn, M.S. and Schäberle, T.F. (2022) Bioactive natural products from Bacteroidetes. *Natural Product Reports* 39(5), 1045–1065. DOI: 10.1039/d1np00072a.

Burkinshaw, S.M. (2016) *Physico-Chemical Aspects of Textile Coloration*. Wiley, Hoboken, NJ.

Caminade, C., McIntyre, K.M. and Jones, A.E. (2019) Impact of recent and future climate change on vector-borne diseases. *Annals of the New York Academy of Sciences* 1436(1), 157–173. DOI: 10.1111/nyas.13950.

Carlile, M.J. (1956) A study of the factors influencing non-genetic variation in a strain of *Fusarium oxysporum*. *Journal of General Microbiology* 14(3), 643–654. DOI: 10.1099/00221287-14-3-643.

Carmen, Z. and Daniela, S. (2012) Textile organic dyes—characteristics, polluting effects and separation/elimination procedures from industrial effluents—a critical overview. In: Puzyn, T. (ed.) *Organic Pollutants Ten Years after the Stockholm Convention—Environmental and Analytical Update*. InTech Press, Croatia, pp. 55–86. DOI: 10.5772/32373.

Casagrande, S., Costantini, D., Tagliavini, J. and Dell'Omo, G. (2009) Phenotypic, genetic, and environmental causes of variation in yellow skin pigmentation and serum carotenoids in Eurasian kestrel nestlings. *Ecological Research* 24(2), 273–279. DOI: 10.1007/s11284-008-0503-3.

Celedón, R.S. and Díaz, L.B. (2021) Natural pigments of bacterial origin and their possible biomedical applications. *Microorganisms* 9(4), 739. DOI: 10.3390/microorganisms9040739.

Chakraborty, J.N. (2010) Colouring materials. In: Chakraborty, J.N. (ed.) *Fundamentals and Practices in Colouration of Textiles*. Woodhead Publishing, India. DOI: 10.1533/9780857092823.

Chander, M. (2022) Microbial production and applications of pigments. *International Research Journal of Engineering and Technology (IRJET)* 9(9), 366–375.

Chaplin, G. (2004) Geographic distribution of environmental factors influencing human skin coloration. *American Journal of Physical Anthropology* 125(3), 292–302. DOI: 10.1002/ajpa.10263.

Chatragadda, R. and Dufossé, L. (2021) Ecological and biotechnological aspects of pigmented microbes: A way forward in development of food and pharmaceutical grade pigments. *Microorganisms* 9(3), 637. DOI: 10.3390/microorganisms9030637.

Chen, P.-H., Chiang, P.-C., Lo, W.-C., Su, C.-W., Wu, C.-Y. *et al.* (2021) A novel fucoidan complex-based functional beverage attenuates oral cancer through inducing apoptosis, G2/m cell cycle arrest and retarding cell migration/invasion. *Journal of Functional Foods* 85, 104665. DOI: 10.1016/j.jff.2021.104665.

Chequer, F.M.D., Oliveira, G.A.R., Ferraz, E.R.A., Cardoso, J.C., Zanoni, M.V.B. *et al.* (2013) Textile dyes: Dyeing process and environmental impact. In: Gunay, M. (ed.) *Eco-Friendly Textile Dyeing and Finishing*. InTech Press, Croatia.

Choi, S.Y., Lim, S., Yoon, K.H., Lee, J.I. and Mitchell, R.J. (2021) Biotechnological activities and applications of bacterial pigments violacein and prodigiosin. *Journal of Biological Engineering* 15(1), 10. DOI: 10.1186/s13036-021-00262-9.

Corral, P., Amoozegar, M.A. and Ventosa, A. (2020) Halophiles and their biomolecules: Recent advances and future applications in biomedicine. *Marine Drugs* 18(1), 33. DOI: 10.3390/md18010033.

Cortés-Osorio, N., Cardoso-Cardoso, M.A., Chavarro-Anzola, V., Prada-Salcedo, L.D., Robles, J. *et al.* (2017) Influence of environmental factors on the production of violacein synthesized by *Janthinobacterium lividum*. *International Journal of Engineering and Science* 6, 76–83. DOI: 10.9790/1813-0601037683.

Darwesh, O.M., Barakat, K., Mattar, M.Z., Sabae, S.Z. and Hassan, S.H. (2019) Production of antimicrobial blue-green pigment Pyocyanin by marine pseudomonas aeruginosa. *Biointerface Research in Applied Chemistry* 9(5), 4334–4339. DOI: 10.33263/BRIAC95.334339.

Delgado-Vargas, F., Jiménez, A.R. and Paredes-López, O. (2000) Natural pigments: Carotenoids, anthocyanins, and betalains – characteristics, biosynthesis, processing, and stability. *Critical Reviews in Food Science and Nutrition* 40(3), 173–289. DOI: 10.1080/10408690091189257.

Dufossé, L. (2009) Pigments, microbial. In: *Encyclopedia of Microbiology*, 4th edn. Academic Press, New York, pp. 457–471. DOI: 10.1016/B978-012373944-5.00155-3.

El Feghali, P.A.R. and Nawas, T. (2018) Extraction and purification of pyocyanin: A simpler and more reliable method. *MOJ Toxicology* 4(6), 417–422. DOI: 10.15406/mojt.2018.04.00139.

Fimbres-Olivarría, D., López-Elías, J.A., Carvajal-Millán, E., Márquez-Escalante, J.A., Martínez-Córdova, L.R. *et al.* (2016) *Navicula* sp. sulfated polysaccharide gels induced by Fe (III): Rheology and microstructure. *International Journal of Molecular Sciences* 17(8), 1238. DOI: 10.3390/ijms17081238.

Fontana, J.D., Guimarães, M.F., Martins, N.T., Fontana, C.A. and Baron, M. (1996) Culture of the astaxanthinogenic yeast *Phaffia rhodozyma* in low-cost media. *Applied Biochemistry and Biotechnology* 57–58, 413–422. DOI: 10.1007/BF02941721.

Gerrish, G.A. and Cáceres, C.E. (2003) Genetic versus environmental influence on pigment variation in the ephippia of *Daphnia pulicaria*. *Freshwater Biology* 48(11), 1971–1982. DOI: 10.1046/j.1365-2427.2003.01136.x.

Giwa, A., Dufour, V., Al Marzooqi, F., Al Kaabi, M. and Hasan, S.W. (2017) Brine management methods: Recent innovations and current status. *Desalination* 407, 1–23. DOI: 10.1016/j.desal.2016.12.008.

Gong, M. and Bassi, A. (2016) Carotenoids from microalgae: A review of recent developments. *Biotechnology Advances* 34, 1396–1412. DOI: 10.1016/j.biotechadv.2016.10.005.

Gonzali, S. and Perata, P. (2021) Fruit colour and novel mechanisms of genetic regulation of pigment production in tomato fruits. *Horticulturae* 7(8), 259. DOI: 10.3390/horticulturae7080259.

Gordon, P.F. and Gregory, P. (1987) *Organic Chemistry in Colour*. Springer, Berlin.

Govindraj, C., Rugamoorthi, U. and Ramarethinam, S. (2020) Isolation of *Pseudomonas aeruginosa* for bacterial pigment production and its application on synthetic knitted fabric. *Indian Journal of Fibre & Textile Research* 46(2), 168–173. DOI: 10.56042/ijftr.v46i2.31716.

Griffiths, J. (1990) Introduction: The evolution of present-day dye technology. In: Waring, D.R. and Hallas, G. (eds) *The Chemistry and Application of Dyes*. Plenum Press, New York, pp. 1–16.

Hanna, B.L. (1953) On the relative importance of genetic and environmental factors in the determination of human hair pigment concentration. *American Journal of Human Genetics* 5(4), 293–321.

Hauri, U. and Hohl, C. (2015) Photostability and breakdown products of pigments currently used in tattoo inks. *Current Problems in Dermatology* 48, 164–169. DOI: 10.1159/000369225.

Herbst, W., Hunger, K., Wilker, G., Ohleier, H. and Winter, R. (2004) Review of chemical structures and chemical reactions. In: Herbst, W., Hunger, K., Wilker, G., Ohleier, H. and Winter, R. (eds) *Industrial Organic Pigments: Production, Properties, Applications*. Wiley-VCH, Weinheim, Germany, pp. 601–636.

Hodgson, D.A., Vyverman, W., Verleyen, E., Sabbe, K., Leavitt, P.R. *et al.* (2004) Environmental factors influencing the pigment composition of *in situ* benthic microbial communities in East Antarctic lakes. *Aquatic Microbial Ecology* 37(3), 247–263. DOI: 10.3354/ame037247.

Iqbal, M. (2008) *Textile dyes*. Rahber Publishers, Pakistan.

Jablonski, N.G. (2021) The evolution of human skin pigmentation involved the interactions of genetic, environmental, and cultural variables. *Pigment Cell & Melanoma Research* 34(4), 707–729. DOI: 10.1111/pcmr.12976.

Johler, S., Stephan, R., Hartmann, I., Kuehner, K.A. and Lehner, A. (2010) Genes involved in yellow pigmentation of *Cronobacter sakazakii* ES5 and influence of pigmentation on persistence and growth under environmental stress. *Applied and Environmental Microbiology* 76(4), 1053–1061. DOI: 10.1128/AEM.01420-09.

Joshi, A.K. and Pandey, S.P. (2018) Genetic factors affecting photosynthesis. In: Pessarakli, M. (ed.) *Handbook of Photosynthesis*. CRC Press, Boca Raton, FL, pp. 539–568.

Kim, C.H., Kim, S.W. and Hong, S.I. (1999) An integrated fermentation–separation process for the production of red pigment by Serratia sp. KH-95. *Process Biochemistry* 35(5), 485–490. DOI: 10.1016/S0032-9592(99)00091-6.

Klass, M.R. (1977) Aging in the nematode caenorhabditis elegans: Major biological and environmental factors influencing life span. *Mechanisms of Ageing and Development* 6(6), 413–429. DOI: 10.1016/0047-6374(77)90043-4.

Kyong-Hwa, K., Zhong-Ji, Q., BoMi, R. and Se-Kwon, K. (2011) Characterization of growth and protein contents from microalgae *Navicula incerta* with the investigation of antioxidant activity of enzymatic hydrolysates. *Food Science and Biotechnology* 20(1), 183–191. DOI: 10.1007/s10068-011-0025-6.

Lee, B., Hayashi, K., Hirata, M., Kuroda, E., Suzuki, E. *et al.* (2006) Antiviral sulfated polysaccharide from *Navicula directa*, a diatom collected from deep-sea water in Toyama Bay. *Biological and Pharmaceutical Bulletin* 29, 2135–2139. DOI: 10.1248/bpb.29.2135.

Lefsih, K., Giacomazza, D., Dahmoune, F., Rosalia Mangione, M., Bulone, D. *et al.* (2017) Pectin from *Opuntia ficus indica*: Optimization of microwave-assisted extraction and preliminary characterization. *Food Chemistry* 221, 91–99. DOI: 10.1016/j.foodchem.2016.10.073.

Lin, L., Zhang, T. and Xu, J. (2023) Genetic and environmental factors influencing the production of select fungal colorants: Challenges and opportunities in industrial applications. *Journal of Fungi* 9(5), 585. DOI: 10.3390/jof9050585.

Lu, C. and Zhang, J. (2000) Role of light in the response of PSII photochemistry to salt stress in the cyanobacterium *Spirulina platensis*. *Journal of Experimental Botany* 51, 911–917. DOI: 10.1093/jxb/51.346.911.

Lyu, X., Lyu, Y., Yu, H., Chen, W., Ye, L. *et al.* (2022) Biotechnological advances for improving natural pigment production: A state-of-the-art review. *Bioresources and Bioprocessing* 9(1), 8. DOI: 10.1186/s40643-022-00497-4.

Marin-Recinos, M.F. and Pucker, B. (2024) Genetic factors explaining anthocyanin pigmentation differences. *BMC Plant Biology* 24(1), 627. DOI: 10.1186/s12870-024-05316-w.

Matrková, J. and Remeš, V. (2012) Environmental and genetic effects on pigment-based vs. structural component of yellow feather colouration. *PLoS ONE* 7(5), e36640. DOI: 10.1371/journal.pone.0036640.

Meyer, P., Linn, F., Heidmann, I., Meyer, H., Niedenhof, I. *et al.* (1992) Endogenous and environmental factors influence 35S promoter methylation of a maize A1 gene construct in transgenic petunia and its colour phenotype. *Molecular & General Genetics* 231(3), 345–352. DOI: 10.1007/BF00292701.

Młodzińska, E. (2009) Survey of plant pigments: Molecular and environmental determinants of plant colors. *Acta Biologica Cracovienca Series Botanica* 51(1), 7–16.

Mulders, K.J.M., Lamers, P.P., Martens, D.E. and Wijffels, R.H. (2014) Phototrophic pigment production with microalgae: Biological constraints and opportunities. *Journal of Phycology* 50(2), 229–242. DOI: 10.1111/jpy.12173.

Muller-Feuga, A., Moal, J. and Kaas, R. (2007) The microalgae of aquaculture. In: Støttrup, J.G. and McEvoy, L.A. (eds) *Live Feeds in Marine Aquaculture*. Blackwell Science, Oxford, pp. 207–252.

Mustafa, T., Khazi, M.I., Demirel, Z. and Dalay, M.C. (2019) Variation in growth, fucoxanthin, fatty acids profile and lipid content of marine *Diatoms nitzschia* sp. and *Nanofrustulum shiloi* in response to nitrogen and iron. *Biocatalysis and Agricultural Biotechnology* 17, 390–398. DOI: 10.1016/j.bcab.2018.12.023.

Navarro, E., Lorca-Pascual, J.M., Quiles-Rosillo, M.D., Nicolás, F.E., Garre, V. *et al.* (2001) A negative regulator of light-inducible carotenogenesis in *Mucor circinelloides*. *Molecular Genetics and Genomics* 266(3), 463–470. DOI: 10.1007/s004380100558.

Oser, B.L. (1965) *Hawk's Physiological Chemistry*, Vol. 1. Academic Press, New York, pp. 529–544.

Paillié-Jiménez, M.E., Stincone, P. and Brandelli, A. (2020) Natural pigments of microbial origin. *Frontiers in Sustainable Food Systems* 4, 590439. DOI: 10.3389/fsufs.2020.590439.

Parekh, S., Vinci, V.A. and Strobel, R.J. (2000) Improvement of microbial strains and fermentation processes. *Applied Microbiology and Biotechnology* 54(3), 287–301. DOI: 10.1007/s002530000403.

Pelczar, M.J. and Chan, E.C. (1977) *Laboratory Exercises in Microbiology*, 14th edn. McGraw-Hill, New York, pp. 59–61.

Richmond, A. (2004) Biological principles of mass cultivation. In: Richmond, A. (ed.) *Handbook of Microalgal Mass Culture: Biotechnology and Applied Phycology*. Blackwell Science, Oxford, pp. 147–239. DOI: 10.1002/9780470995280.

Saini, R.K. and Keum, Y.S. (2018) Significance of genetic, environmental, and pre- and postharvest factors affecting carotenoid contents in crops: A review. *Journal of Agricultural and Food Chemistry* 66(21), 5310–5324. DOI: 10.1021/acs.jafc.8b01613.

Silbir, S. and Goksungur, Y. (2019) Natural red pigment production by *Monascus purpureus* in submerged fermentation systems using a food industry waste: Brewer's spent grain. *Foods* 8(5), 161. DOI: 10.3390/foods8050161.

Silva-Castro, G.A., Moyo, A.C., Khumalo, L., van Zyl, L.J., Petrik, L.F. *et al.* (2019) Factors influencing pigment production by halophilic bacteria and its effect on brine evaporation rates. *Microbial Biotechnology* 12(2), 334–345. DOI: 10.1111/1751-7915.13319.

Stange, S., Steudler, S., Delenk, H., Werner, A., Walther, T. *et al.* (2019) Influence of environmental growth factors on the biomass and pigment production of *Chlorociboria aeruginascens*. *Journal of Fungi* 5(2), 46. DOI: 10.3390/jof5020046.

Sturm, R.A. and Duffy, D.L. (2012) Human pigmentation genes under environmental selection. *Genome Biology* 13(9), 248. DOI: 10.1186/gb-2012-13-9-248.

Subhasree, R.S., Babu, P.D., Vidyalakshmi, R. and Mohan, V.C. (2011) Effect of carbon and nitrogen sources on stimulation of pigment production by *Monascus purpureus* on jackfruit seeds. *International Journal of Microbiology Research* 2, 184–187.

Venil, C.K., Zakaria, Z.A. and Ahmad, W.A. (2013) Bacterial pigments and their applications. *Process Biochemistry* 48(7), 1065–1079. DOI: 10.1016/j.procbio.2013.06.006.

Vidyalakshmi, R., Paranthaman, R., Murugesh, S., Singaravadivel, K. and O'Carroll, P. (1999) *Microbial bioconversion of rice broken to food grade pigments—naturally exciting colours*. World Ingredients, pp. 39–42.

Wang, Y., Nakajima, A., Hosokawa, K., Soliev, A.B., Osaka, I. *et al.* (2012) Cytotoxic prodigiosin family pigments from *Pseudoalteromonas* sp. 1020R isolated from the Pacific Coast of Japan. *Bioscience, Biotechnology, and Biochemistry* 76(6), 1229–1232. DOI: 10.1271/bbb.110984.

Yamaguchi, Y. and Hearing, V.J. (2009) Physiological factors that regulate skin pigmentation. *BioFactors* 35(2), 193–199. DOI: 10.1002/biof.29.

Yong, W.-K., Lim, P.-E., Vello, V., Sim, K.-S., Abdul Majid, N. *et al.* (2019) Metabolic and physiological regulation of Chlorella sp. (*Trebouxiophyceae, Chlorophyta*) under nitrogen deprivation. *Journal of Oceanology and Limnology* 37(1), 186–198. DOI: 10.1007/s00343-019-7263-5.

6 Microbial Pigments in Nature: Roles and Functions

Gundala Prasada Babu[1], Arifullah Mohammed[2], Mohd Tariq[3,4], Mukul M. Barwant[5] and Chinthala Paramageetham[1]*

[1]Department of Microbiology, Sri Venkateswara University, Tirupati, Andhra Pradesh, India; [2]Department of Biotechnology, Koneru Lakshmaiah University (KLEF), Vaddeswaram Campus, Green Fields, Guntur 522 302, Andhra Pradesh, India; [3]Department of Academics, Sumandeep Vidyapeeth, Deemed to be University, Vadodara-391760, Gujarat, India; [4]Department of Biotechnology, Graphic Era (Deemed to be University), Dehradun, Uttarakhand-248002, India; [5]Department of Botany, Sanjivani Arts, Commerce and Science College, Kopargoan, Ahmednagar, Maharashtra, India

6.1 Introduction

Synthetic dyes, particularly azo dyes, are major contributors to environmental pollution (Shen *et al.*, 2020). These dyes are widely used as industrial colorants but are notorious for their toxicity and persistence in the environment. Most synthetic dyes are nondegradable, posing serious ecological threats. Their production and usage significantly contribute to water pollution due to the presence of toxic chemicals, which contaminate drinking water sources and aquatic ecosystems, leading to severe damage to aquatic life (Khandel *et al.*, 2019). Furthermore, synthetic dyes can cause adverse health effects, including skin irritation, allergic drug reactions, and in some cases carcinogenicity (Sharma *et al.*, 2020).

Given the growing concerns about environmental sustainability and human health, there is increasing demand for safer alternatives, especially from natural sources. Regulatory guidelines from governments have also prompted industries to seek sustainable options, particularly in the food and beverage sectors. According to Grand View Research (2021), the global market for natural food colorants is expected to reach USD 5.8 billion by 2027. In contrast to synthetic dyes, biopigments are generally recognized as safe (GRAS) and are more appealing from both consumer and regulatory perspectives. Biopigments are nontoxic, biodegradable, and derived from renewable resources. Additionally, they offer health benefits such as antiinflammatory, antioxidant, and antimicrobial properties (Cavalcante *et al.*, 2020). With increasing awareness of environmental pollution and growing consumer preference for eco-friendly products, biopigments present a viable and attractive solution for achieving sustainability (Cai *et al.*, 2021).

*Corresponding author: paramageetham@yahoo.com

 Biopigments of Microbial Origin: Innovations and Applications Across Industries (eds Madan *et al.*)
DOI: 10.1079/ 9781836990505.0006

6.2 Biopigments From Microorganisms

Biopigments are natural pigments synthesized by microorganisms, which have gained attention due to their diverse applications in pharmaceuticals, food, cosmetics, and textiles. Unlike synthetic pigments, these are eco-friendly and biodegradable, and often exhibit bioactive properties such as antioxidant, antimicrobial, and anticancer activities (Venil *et al.*, 2013).

Microbial pigments can be classified based on various criteria, including their chemical structure, solubility, biological function, and color (Fig. 6.1). Based on their chemical structure, biopigments include carotenoids (e.g. β-carotene, astaxanthin, and lycopene), phenazine compounds (e.g. pyocyanin and phenazine-1-carboxylic acid), quinones (e.g. anthraquinones and naphthoquinones), polyketides (e.g. actinorhodin and undecylprodigiosin), indole compounds (e.g. violacein), and tetrapyrroles (e.g. phycocyanin and phycoerythrin) (Hassan *et al.*, 2016). They can also be classified based on solubility into water-soluble pigments, such as phycocyanin and pyoverdine, and lipid-soluble pigments like carotenoids and quinones. Regarding biological function, pigments are categorized as primary pigments, essential for microbial metabolism (e.g. chlorophyll), or secondary pigments, which are nonessential but aid in stress adaptation (e.g. violacein and prodigiosin). Additionally, they can be grouped by color into red pigments (e.g. prodigiosin and lycopene), blue-green pigments (e.g. phycocyanin and pyocyanin), yellow pigments (e.g. flavins and riboflavin), violet pigments (e.g. violacein), and orange pigments (e.g. β-carotene) (Venil *et al.*, 2013; Singh *et al.*, 2020).

Microorganisms that produce pigments include bacteria, fungi, actinomycetes, algae, and yeasts. Specific examples include *Chromobacterium violaceum* (violacein), *Serratia marcescens* (prodigiosin), and *Pseudomonas aeruginosa* (pyocyanin and pyoverdine) among bacteria. Fungi like *Monascus* species produce monascin and ankaflavin, while *Penicillium* species are known for yellow pigments. Actinomycetes such as *Streptomyces* species produce actinorhodin and undecylprodigiosin. Algae, including *Dunaliella salina* and *Spirulina* species, synthesize β-carotene and phycocyanin,

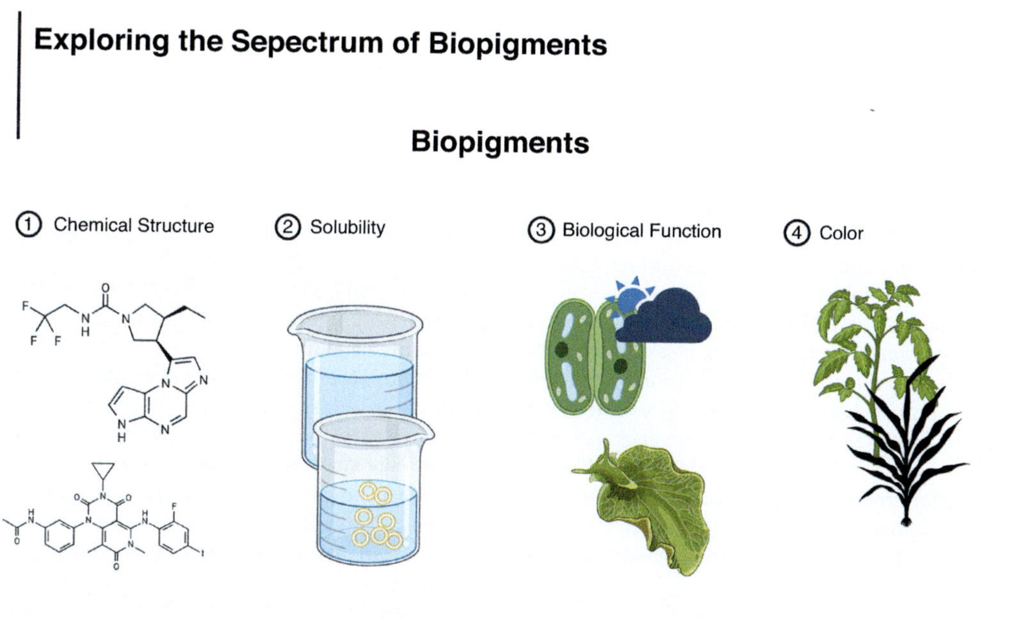

Fig. 6.1. Classification of biopigments.

respectively. Yeasts such as *Rhodotorula* species produce torulene and torularhodin, while *Phaffia rhodozyma* is a source of astaxanthin (Dufossé *et al.*, 2014; Manivasagan *et al.*, 2021).The sustainable production of microbial pigments offers significant potential for industrial applications, although challenges such as scalability, production costs, and regulatory approvals remain to be addressed.

6.2.1 Biopigments from bacteria

Bacteria produce diverse biopigments characterized by their vibrant colors and bioactive properties, making them valuable for sustainable and eco-friendly applications in pharmaceuticals, food, cosmetics, textiles, and bioremediation (Venil *et al.*, 2020). Carotenoids, a group of yellow, orange, and red pigments, are prominent examples, produced by bacteria like *Flavobacterium* spp. (β-carotene), *Dietzia natronolimnaea* and *Brevibacterium* spp. (canthaxanthin), and *Pantoea ananatis* (zeaxanthin). These pigments serve as antioxidants and provide UV protection, with applications as food colorants, cosmetics, and nutraceuticals (Singh *et al.*, 2018).

Prodigiosin, a red pigment synthesized by *Serratia marcescens*, exhibits antimicrobial, anticancer, and immunosuppressive properties, making it useful in pharmaceuticals and biomedicine (Joshi *et al.*, 2021). Violacein, a violet pigment derived from *Chromobacterium violaceum*, demonstrates antimicrobial, anticancer, and antioxidant activities, with applications in pharmaceutical formulations and as research reagents (Kumar *et al.*, 2020).

Pyocyanin, a blue-green pigment produced by *Pseudomonas aeruginosa*, possesses antibiotic properties and plays a role in quorum sensing regulation, while pyoverdine, a yellow-green fluorescent pigment from *Pseudomonas* spp., acts as an iron-chelating siderophore, facilitating bacterial growth under iron-limited conditions. Both pigments have potential applications in antimicrobial research and biosensor development (Choi *et al.*, 2021).

Actinorhodin, a blue pigment with antibiotic properties, is synthesized by *Streptomyces coelicolor* and has pharmaceutical applications (Venil *et al.*, 2020). Phycocyanin and allophycocyanin, blue pigments from cyanobacteria such as *Spirulina* spp. and *Anabaena* spp., are used as light-harvesting pigments and antioxidants in food colorants, nutraceuticals, and cosmetics (Kumar *et al.*, 2020).

Rhodopsins, purple pigments produced by halophilic archaea and *Rhodobacter* spp., function in light absorption for energy production and are applied in optogenetics and bioelectronics (Singh *et al.*, 2018). Flavins, yellow pigments synthesized by *Lactobacillus fermentum* and *Bacillus* spp., are known for their antioxidant and redox-active properties and are used in pharmaceuticals and food fortification (Choi *et al.*, 2021).

Melanins, black or brown pigments produced by *Pseudomonas putida* and *Shewanella colwelliana*, provide UV protection, free radical scavenging, and heavy metal sequestration, with applications in cosmetics, bioremediation, and bioelectronics (Joshi *et al.*, 2021). Bacterial biopigments are highly versatile, serving as antimicrobial and anticancer agents in pharmaceuticals, natural colorants in the food and textile industries, and protective and antioxidant compounds in cosmetics. They also play a crucial role in environmental applications such as heavy metal sequestration and pollutant degradation (Venil *et al.*, 2020).

6.2.1.1 Industrial applications

Bacterial biopigments have gained significant attention due to their versatile applications across various industries. In the pharmaceutical sector, bacterial biopigments such as prodigiosin, violacein, and actinorhodin exhibit antimicrobial, anticancer, and immunosuppressive properties, making them valuable in developing therapeutic agents (Chakraborty *et al.*, 2020; Silva *et al.*, 2021). These pigments have shown promise in addressing issues like multidrug resistance and are being explored for their potential as anticancer drugs due to their cytotoxic effects on cancer cells. In the food industry, bacterial pigments such as carotenoids and phycocyanin are increasingly used as natural colorants, providing a safer alternative to synthetic dyes (Liu *et al.*, 2017). These pigments not only enhance the visual appeal of food products but also offer antioxidant properties, benefiting consumer health.

In the cosmetics industry, bacterial biopigments like melanins and carotenoids are utilized for their UV-protective and antioxidant capabilities, helping to protect skin from oxidative stress and harmful UV radiation (Zhao *et al.*, 2020a). These properties make them valuable ingredients in skincare products. In the textile industry, biopigments like prodigiosin and violacein offer sustainable and natural dyeing alternatives, reducing the environmental impact of synthetic dyes (Hassan *et al.*, 2019). These pigments are also being explored in bioremediation for their ability to sequestrate heavy metals and degrade pollutants, contributing to environmental clean-up efforts (Chakraborty *et al.*, 2020).

The continued exploration of bacterial biopigments will likely lead to new, eco-friendly solutions across these diverse sectors (Table 6.1).

6.2.2 Biopigments from fungi

Fungi are renowned producers of biopigments, which are secondary metabolites exhibiting vivid colors and diverse biological activities. These pigments are highly valued for their applications in the food, pharmaceutical, textile, and cosmetic industries, owing to their natural origin and functional benefits (Venil *et al.*, 2020).

Carotenoids, including β-carotene from *Blakeslea trispora* and *Neurospora crassa*, lycopene from *Blakeslea trispora*, and torulene from *Fusarium* spp. are yellow, orange, or red pigments with antioxidant and UV-protective properties. These are widely used in nutraceuticals, food colorants, and cosmetics (Kumar *et al.*, 2020). Polyketides, such as monascins, ankaflavins, rubropunctatin, and monascorubrin produced by *Monascus* spp., are notable for their antioxidant, antimicrobial, and cholesterol-lowering properties. These pigments find applications in food coloring, especially in fermented rice or beverages, and pharmaceuticals (Singh *et al.*, 2018).

Melanins, produced by fungi such as *Aspergillus niger* and *Alternaria alternata*, are black or brown pigments that provide UV protection, free radical scavenging, and resistance to environmental stress. These are utilized in bioremediation, cosmetics, and UV-protective formulations (Joshi *et al.*, 2021). Quinones, including chrysophanol from *Penicillium* spp. and austinoquinone from *Aspergillus ustus*, are associated with antimicrobial, anticancer, and antiinflammatory activities, making them valuable in pharmaceuticals and natural dye production (Choi *et al.*, 2021).

Azaphilones, characterized by their red, yellow, or orange hues, include pigments such as monascus pigments from *Monascus purpureus* and chaetoviridin from *Chaetomium globosum*. These pigments have antioxidant and antimicrobial properties, with applications in the food and textile industries and as biocontrol agents (Venil *et al.*, 2020). Flavins, produced by fungi like *Aspergillus flavus* and *Trichoderma reesei*, are yellow pigments with redox-active and antioxidant properties, utilized in biocatalysis and as food supplements (Kumar *et al.*, 2020).

Anthraquinones, such as emodin and averufin from *Penicillium* and *Aspergillus* spp., are yellow, red, or orange pigments with antimicrobial and anticancer activities, commonly used in pharmaceuticals and as natural dyes (Joshi *et al.*, 2021). Laccase-derived pigments,

Table 6.1. Notable bacteria and their pigments.

Bacterium	Pigment	Color	Applications
Serratia marcescens	Prodigiosin	Red	Antimicrobial, anticancer
Chromobacterium violaceum	Violacein	Violet	Antioxidant, anticancer
Pseudomonas aeruginosa	Pyocyanin, pyoverdine	Blue-green	Antibiotics, biosensors
Streptomyces coelicolor	Actinorhodin	Blue	Antibiotics
Spirulina platensis	Phycocyanin	Blue	Nutraceuticals, cosmetics
Flavobacterium sp.	β-Carotene	Orange	Food and cosmetics
Dietzia natronolimnaea	Canthaxanthin	Orange-red	Antioxidant, aquaculture feed

ranging in color from brown to dark blue, are enzymatically produced by fungi like *Pleurotus ostreatus* and *Trametes versicolor*, playing roles in enzymatic browning, bioremediation of dyes, and bio-based coloring (Choi *et al.*, 2021).

Fungal biopigments serve as natural colorants in the food industry (e.g. *Monascus* pigments and β-carotene) and functional food ingredients due to their antioxidant properties. In pharmaceuticals, they act as antimicrobial, anticancer, and cholesterol-lowering agents (e.g. anthraquinones and azaphilones). The cosmetic industry benefits from antiaging and UV-protective products containing carotenoids and melanins. Additionally, fungal biopigments are used as sustainable and natural dyes in the textile industry and in environmental applications like dye and pollutant bioremediation (Venil *et al.*, 2020).

6.2.2.1 *Industrial applications*

Fungal biopigments are gaining widespread use across various industries due to their diverse applications and bioactive properties. In the food industry, fungal pigments such as those derived from *Monascus* species, and β-carotene are used as natural colorants and functional food ingredients with antioxidant properties (Zhao *et al.*, 2020b). These pigments not only enhance the visual appeal of food but also offer potential health benefits, promoting consumer wellness. In the pharmaceuticals sector, pigments like anthraquinones and azaphilones have demonstrated significant antimicrobial and anticancer activities, making them valuable candidates for drug development (Chakraborty *et al.*, 2021). Additionally, *Monascus*-derived pigments are known for their cholesterol-lowering effects, providing a potential functional ingredient for heart health products (Liu *et al.*, 2019a). In cosmetics, fungal carotenoids and melanins are utilized for their antiaging and UV-protective properties, playing a critical role in skincare formulations (Zhao *et al.*, 2020c). These pigments help protect the skin from harmful UV radiation and oxidative stress, contributing to healthier, more youthful skin. The textile industry benefits from fungal biopigments as well, with azaphilones and anthraquinones being used as natural and sustainable dyes, offering an eco-friendly alternative to synthetic dyes (Li *et al.*, 2021).

Lastly, fungal pigments also have important environmental applications, such as in bioremediation. Pigments like laccase-derived compounds are capable of degrading dyes and pollutants, contributing to environmental clean-up efforts (Gupta *et al.*, 2020a). The multifunctionality of fungal biopigments across these sectors highlights their growing potential in both commercial and environmental applications (Table 6.2).

6.2.3 Biopigments from actinomycetes

Actinomycetes, particularly from the genus *Streptomyces*, are recognized for their ability to produce biopigments with diverse colors and bioactivities. These secondary metabolites exhibit antimicrobial, antioxidant, and anticancer

Table 6.2. Notable fungi and their pigments.

Fungus	Pigment	Color	Applications
Monascus purpureus	Monascins, rubropunctatin	Yellow, red	Food colorants, nutraceuticals
Blakeslea trispora	β-Carotene, lycopene	Yellow, range	Food and cosmetic industries
Aspergillus niger	Melanin	Black	Bioremediation, UV protection
Penicillium purpurogenum	Chrysophanol, emodin	Yellow, orange	Pharmaceuticals
Fusarium sp.	Torulene	Orange	Nutraceuticals and cosmetics
Trametes versicolor	Laccase-derived pigments	Brown, blue	Bioremediation, textile industries

properties, making them highly valuable in pharmaceuticals, food, cosmetics, and textile industries (Bérdy, 2012; Singh *et al.*, 2019).

One prominent pigment, actinorhodin, is a blue antibiotic with antibacterial properties produced by *Streptomyces coelicolor*. It is primarily utilized in the pharmaceutical sector and as a research tool (Newman and Cragg, 2020). Similarly, prodiginines, such as prodigiosin from *Streptomyces coelicolor* and *S. lividans*, exhibit potent anticancer, antimicrobial, and immunosuppressive properties, making them valuable in pharmaceuticals and natural dye production (Bhute and Bhonde, 2017).

Melanins, black or brown pigments produced by *Streptomyces* and *Nocardia* species, play a crucial role in UV protection, free radical scavenging, and heavy metal sequestration. These pigments find applications in cosmetics, bioremediation, and biomedical devices (Singh *et al.*, 2019). Carotenoids, including β-carotene and lycopene from *Streptomyces griseus* and *Nocardia* spp., are yellow, orange, or red pigments known for their antioxidant and health-promoting properties. These are widely used in the food industry, cosmetics, and nutraceuticals (Bérdy, 2012).

Another important group is anthraquinones, which include pigments such as emodin and alizarin derivatives produced by *Streptomyces* spp. These pigments display antimicrobial, anticancer, and antiinflammatory properties and are employed in pharmaceuticals and natural dyes (Newman and Cragg, 2020). Similarly, phenazines, yellow-to-red pigments such as pyocyanin-like compounds from *Streptomyces*, are notable for their antibiotic and antioxidant activities, with applications in pharmaceuticals and biotechnological innovations (Bhute and Bhonde, 2017).

Finally, lutein-like pigments, yellow pigments produced by *Streptomyces globisporus*, possess antioxidant properties, making them suitable as food colorants and nutraceutical additives (Singh *et al.*, 2019).

6.2.3.1 Industrial applications

Actinomycetes-derived biopigments have a wide range of applications across various industries due to their bioactive properties and natural origins. In the pharmaceutical industry, biopigments such as actinorhodin and prodiginines are highly valued for their antibiotic and anticancer activities, making them potent candidates for drug development (Santos *et al.*, 2021). Additionally, actinomycetes-derived pigments have demonstrated antioxidant and immunosuppressive effects, contributing to the development of therapeutic agents (Thakur *et al.*, 2020). In the **food industry**, pigments like carotenoids and anthraquinones are used as natural colorants, providing vibrant hues to food products, while also offering functional health benefits due to their antioxidant properties (Liu *et al.*, 2019a). These pigments enhance the nutritional value of food, making them desirable for health-conscious consumers.

In the cosmetics industry, pigments such as melanins and carotenoids are incorporated for their UV-protective and antioxidant properties, which help protect the skin from harmful UV radiation and oxidative stress (Tee *et al.*, 2021). These properties make them essential in antiaging and sun-protective cosmetic formulations. The textile industry benefits from actinomycetes-derived biopigments, with natural dyes providing an eco-friendly alternative to synthetic pigments (Gautam *et al.*, 2021).

Lastly, actinomycetes-derived melanins have applications in environmental bioremediation, where they assist in the sequestration of heavy metals, aiding in pollutant degradation and environmental cleanup (Radhakrishnan *et al.*, 2020). These diverse applications demonstrate the versatility and growing importance of actinomycetes-derived biopigments in industrial and environmental sectors (Table 6.3).

6.2.4 Biopigments from algae

Algae are a rich source of diverse biopigments, which play crucial roles in photosynthesis and protection against environmental stress. These pigments are widely utilized in industries such as food, cosmetics, pharmaceuticals, and nutraceuticals due to their functional and bioactive properties (Khan *et al.*, 2018; Ravi *et al.*, 2021).

Chlorophylls, the primary photosynthetic pigments, are responsible for the green coloration in algae. Chlorophyll *a* is universally present in all algae, while chlorophyll *b* is found in green algae (*Chlorophyta*), and chlorophyll *c* occurs in

Table 6.3. Notable actinomycetes and their pigments.

Actinomycete	Pigment	Color	Applications
Streptomyces coelicolor	Actinorhodin	Blue	Pharmaceuticals (antibiotics)
Streptomyces griseus	β-Carotene	Yellow-orange	Food and cosmetic industries
Streptomyces lividans	Prodigiosin	Red	Anticancer and antimicrobial agents
Nocardia sp.	Melanin	Black	Bioremediation, UV protection
Streptomyces sp.	Anthraquinones	Yellow, red	Pharmaceuticals, natural dyes
Streptomyces globisporus	Lutein-like pigment	Yellow	Food and nutraceuticals

brown algae (*Phaeophyta*) and diatoms. Common sources include *Chlorella vulgaris*, *Spirulina platensis*, and *Ulva* species, making these pigments essential for photosynthesis.

Carotenoids, including β-carotene, fucoxanthin, and lutein, are yellow, orange, or red pigments that function as accessory pigments for light harvesting, photoprotection, and antioxidants. Notable sources include *Dunaliella salina* for β-carotene, brown algae such as *Fucus* and *Sargassum* for fucoxanthin, and green algae like *Chlorella* and *Scenedesmus* for lutein (Ravi *et al.*, 2021).

Phycobiliproteins, such as phycocyanin, phycoerythrin, and allophycocyanin, are water-soluble pigments found in cyanobacteria and red algae. They serve as light-harvesting molecules for photosynthesis and are used as natural colorants in food and cosmetics, as well as for their antioxidant properties. For example, phycocyanin is derived from *Spirulina*, while phycoerythrin is sourced from red algae like *Porphyridium* sp. (Khan *et al.*, 2018).

Xanthophylls, including zeaxanthin and violaxanthin, are yellow pigments found in green algae such as *Chlorella* and *Dunaliella*. These pigments protect against photooxidative damage, ensuring algal survival under stress conditions.

Astaxanthin, a red pigment derived from *Haematococcus pluvialis* (a microalga), is a potent antioxidant with antiinflammatory and UV-protective properties. This pigment has widespread applications in nutraceuticals, cosmetics, and aquaculture feed, particularly for the pigmentation of salmon and shrimp (Ravi *et al.*, 2021). Additional pigments like echinenone and canthaxanthin, found in cyanobacteria and red algae respectively, contribute to photoprotection and are used as natural dyes (Khan *et al.*, 2018).

6.2.4.1 *Industrial applications*

Algal biopigments have gained significant attention for their diverse applications in various industries. In the food industry, pigments such as β-carotene and phycocyanin are widely used as natural colorants, providing an eco-friendly alternative to synthetic dyes. Moreover, functional food ingredients like astaxanthin, known for its potent antioxidant properties, are incorporated to enhance the nutritional value of food products (Zhang *et al.*, 2018). In the pharmaceuticals and nutraceuticals sectors, algal biopigments, including astaxanthin and fucoxanthin, are valued for their strong antioxidant activities, which help in mitigating oxidative stress, while phycobiliproteins possess antiinflammatory effects, showing potential for therapeutic applications (Jin *et al.*, 2020).

In cosmetics, carotenoids and astaxanthin are utilized for their UV-protective and antiaging benefits, offering protective effects against environmental damage and contributing to skin health (Neto *et al.*, 2019). The aquaculture industry benefits from algal pigments such as astaxanthin and canthaxanthin, which are used to enhance the pigmentation of fish and crustaceans, improving the aesthetic value and market appeal of aquaculture products (Huang *et al.*, 2019). Furthermore, in biofuels and bioremediation, algal pigments play a role in biofuel production by supporting optimal growth conditions for microalgae, and serve as biomarkers for environmental monitoring, contributing to the sustainability of ecological systems (Zhang *et al.*, 2020).

These applications highlight the wide-ranging potential of algal biopigments in both industrial and environmental fields (Table 6.4).

6.2.5 Biopigments from yeast

Yeasts, unicellular fungi, produce a variety of biopigments with significant industrial and biotechnological applications. These pigments, primarily secondary metabolites, exhibit bioactive properties such as antioxidant, antimicrobial, and anticancer activities. As a result, they are extensively utilized in food, pharmaceutical, and cosmetic industries (Buzzini *et al.*, 2017; Gupta *et al.*, 2020a).

Carotenoids, including β-carotene, torulene, and torularhodin, are major pigments produced by yeasts such as *Rhodotorula glutinis*, *Rhodotorula mucilaginosa*, and *Sporobolomyces* species. These pigments, which range in color from yellow to red, serve as antioxidants and offer photoprotective and health-promoting properties. Their applications span across food supplements, cosmetics, and pharmaceuticals (Buzzini *et al.*, 2017).

Astaxanthin, a red-orange carotenoid, is primarily produced by *Phaffia rhodozyma* (also known as *Xanthophyllomyces dendrorhous*). It is a potent antioxidant and UV protectant, widely used in aquaculture for salmon and shrimp pigmentation, as well as in nutraceuticals (Zhao *et al.*, 2021).

Melanins, black or brown pigments, are synthesized by yeasts such as *Cryptococcus neoformans* and *Saccharomyces cerevisiae* under specific conditions. These pigments provide protection against UV radiation, oxidative stress, and environmental toxins. Melanins find applications in bioremediation, cosmetics, and emerging bioelectronics (Gupta *et al.*, 2020b).

Flavins, yellow pigments produced by *Candida* species and *Saccharomyces cerevisiae*, act as cofactors in enzymatic reactions and antioxidants. They are used in food fortification and pharmaceutical formulations (Zhao *et al.*, 2021).

Pulcherrimin, a reddish-brown pigment produced by *Saccharomyces cerevisiae* and *Metschnikowia pulcherrima*, exhibits iron-chelating properties and antimicrobial activity. This pigment is employed as a biocontrol agent in agriculture and as a natural antimicrobial agent (Buzzini *et al.*, 2017).

Additionally, monascins and ankaflavins, yellow to orange pigments produced by yeast-like *Monascus* species, have antioxidant and cholesterol-lowering properties, making them suitable as food colorants and nutraceuticals (Gupta *et al.*, 2020b).

6.2.5.1 Industrial applications

Yeast-derived biopigments have found diverse applications across several industries due to their bioactive properties and sustainability. In the food industry, pigments such as β-carotene and astaxanthin are utilized as natural colorants, providing an eco-friendly alternative to synthetic dyes, while also serving as functional food ingredients and dietary supplements with antioxidant properties (Yin *et al.*, 2021). In pharmaceuticals, yeast-derived biopigments, including astaxanthin and melanin, offer significant therapeutic potential due to their antioxidant and bioactive effects, contributing to the development of novel drug formulations (Liu *et al.*, 2019b). In cosmetics, carotenoids and melanin are incorporated into anti aging and UV-protective products, leveraging their skin-protective capabilities to address environmental and oxidative stress (Zhao *et al.*, 2020b). In agriculture, pulcherrimin-producing yeasts have gained attention for their biocontrol properties, acting as natural antimicrobial agents to protect crops (Fan *et al.*, 2021).

Table 6.4. Notable alga and their pigments.

Algae	Pigment	Applications
Dunaliella salina	β-Carotene	Food colorant, antioxidant
Haematococcus pluvialis	Astaxanthin	Nutraceuticals, aquaculture feed
Spirulina platensis	Phycocyanin, chlorophyll	Food supplements, cosmetics
Porphyridium sp.	Phycoerythrin	Natural dye, bioactive compounds
Sargassum sp.	Fucoxanthin	Antiobesity and antiinflammatory

Additionally, in biotechnology, melanins are explored for their applications in bioelectronics and environmental monitoring due to their conductive and pollutant-binding properties, highlighting the multifunctional potential of these pigments (Mohammad *et al.*, 2018).

These applications underscore the growing importance of yeast-derived biopigments in various industries, particularly for their eco-friendly and bioactive characteristics (Table 6.5).

6.3 Extremophiles in Pigment Production

Extremophiles, microorganisms that thrive in extreme environmental conditions such as high salinity, temperature, acidity, or alkalinity, have garnered significant attention for their ability to produce bioactive pigments that are both stable and functional. These pigments, adapted to extreme environments, exhibit remarkable stability, making them ideal for various industrial applications, including food, cosmetics, textiles, and bioremediation (Madigan *et al.*, 2015; Fendrihan *et al.*, 2021a).

Halophiles, which thrive in high-salt environments, produce carotenoids like β-carotene, astaxanthin, and canthaxanthin. *Halobacterium salinarum* synthesizes bacterioruberin, a red carotenoid known for its potent antioxidant properties, while *Haloferax alexandrinus* produces zeaxanthin, used in nutraceuticals and cosmetics (Fendrihan *et al.*, 2021b). These pigments are particularly stable in saline conditions, making them ideal for applications in brine environments and high-salt food products.

Thermophiles, microorganisms adapted to high-temperature environments, produce pigments that maintain their functionality under heat. For example, *Thermus thermophilus* produces thermozeaxanthin, a carotenoid utilized in high-temperature industrial processes. Similarly, *Geobacillus* species produce yellow carotenoids, which have potential applications in food and pharmaceuticals (Madigan *et al.*, 2015). These thermostable pigments are valuable in industries that require high-temperature operations, such as food processing and textile dyeing.

Acidophiles and alkaliphiles, which thrive in highly acidic or alkaline environments, produce pigments that can withstand extreme pH levels. For example, *Acidithiobacillus ferrooxidans*, an acidophile, produces melanins with antioxidant properties, while alkaliphilic cyanobacteria such as *Spirulina platensis* produce phycocyanin, a blue pigment widely used in food and cosmetics (Fendrihan *et al.*, 2021b). These pigments are especially useful in formulations that require pH stability, such as those in cosmetics and cleaning agents.

Psychrophiles, adapted to cold environments, produce pigments with antifreeze and photoprotective properties. *Colwellia psychrerythraea* produces carotenoids that protect against oxidative stress at low temperatures, while *Rhodococcus erythropolis* synthesizes pigments for biotechnological applications in cold regions. These pigments are valuable in cryoprotectants and cold-resistant coatings (Madigan *et al.*, 2015).

Table 6.5. Notable yeasts and their pigments.

Yeast	Pigment	Color	Applications
Rhodotorula glutinis	β-Carotene	Yellow- orange	Food, cosmetics, nutraceuticals
Phaffia rhodozyma	Astaxanthin	Red-orange	Aquaculture, pharmaceuticals
Cryptococcus neoformans	Melanin	Black	Bioremediation, UV protection
Saccharomyces cerevisiae	Pulcherrimin, melanin	Reddish-brown	Antimicrobials, cosmetics
Metschnikowia pulcherrima	Pulcherrimin	Reddish-brown	Biocontrol in agriculture

6.3.1 Industrial applications

Pigments derived from extremophiles offer promising applications across several industrial and environmental sectors. In the food industry, these pigments provide stable, natural colorants for processed foods, particularly those requiring stability under extreme pH or temperature conditions, ensuring the preservation of color quality. In cosmetics, extremophile pigments, such as those with UV-protective and antioxidant properties, are incorporated into skincare products, enhancing their effectiveness in protecting against oxidative damage and harmful UV radiation (Madigan *et al.*, 2015). In the textile industry, natural dyes from extremophiles offer enhanced stability under industrial conditions, providing an eco-friendly alternative to synthetic dyes. Additionally, in bioremediation, pigments such as melanins from extremophiles play a significant role in heavy metal sequestration and pollutant degradation, contributing to environmental clean-up efforts (Madigan *et al.*, 2015).

The use of extremophile pigments offers several advantages, including their stability in extreme conditions such as high heat, salinity, and pH, their eco-friendliness due to sustainable and biodegradable production methods, and their bioactivity, which includes enhanced antioxidant, antimicrobial, and UV-protective properties (Madigan *et al.*, 2015) (Table 6.6).

6.4 Advantages of Microbial Pigments Over Plant and Animal Pigments

Microbial pigments offer several advantages over plant and animal pigments, making them a preferred choice in various industries. One of the primary benefits is their high production efficiency. Microorganisms such as bacteria, fungi, and algae can produce pigments in controlled environments, often at a faster rate and with higher yield than plants or animals. This allows for more sustainable and scalable production, particularly in biotechnological applications (Liu *et al.*, 2017). Additionally, microbial pigments are not subject to seasonal or geographic limitations, unlike plant pigments which depend on specific growing conditions (Yuan *et al.*, 2015). Microorganisms can also be cultured in bioreactors, enabling continuous pigment production, which is more consistent and reliable than extraction from plants or animals.

Another significant advantage of microbial pigments is their versatility and diversity. Microbial sources offer a broader range of colors, structures, and bioactivities compared to those derived from plants or animals. These pigments often exhibit antimicrobial, antioxidant, antiinflammatory, and UV-protective properties, making them highly valuable in pharmaceutical, cosmetic, and food industries. Furthermore, microbial pigments are often more stable under extreme environmental conditions such as high temperatures,

Table 6.6. Extremophiles and pigments.

Extremophile	Pigment	Condition	Application
Halobacterium salinarum	Bacterioruberin	High salinity	Antioxidants, cosmetics
Thermus thermophilus	Thermozeaxanthin	High temperature	Food processing, industrial applications
Acidithiobacillus ferrooxidans	Melanin	Acidic pH	Bioremediation, UV protection
Spirulina platensis	Phycocyanin	Alkaline pH	Nutraceuticals, food coloring
Colwellia psychrerythraea	Carotenoids	Low temperature	Cryoprotectants, cold-resistant coatings

pH fluctuations, and salinity, which make them ideal for industrial applications where stability is critical.

Moreover, microbial pigments are more eco-friendly as they can be produced through sustainable and biodegradable processes, reducing the environmental impact compared to plant-based pigment extraction, which can be resource intensive and result in deforestation or overharvesting of natural resources (Becerra *et al.*, 2017). With growing concerns about sustainability and environmental impact, microbial pigments are seen as a more viable alternative to traditional plant and animal-derived pigments.

6.5 Future Scope of Microbial Pigments

The future of microbial pigments holds promising prospects across various sectors, driven by advances in biotechnology, sustainability, and the increasing demand for natural alternatives in industries such as food, cosmetics, pharmaceuticals, and textiles. Several key areas are expected to witness significant growth and innovation.

6.5.1 Sustainable production

With growing environmental concerns, microbial pigments offer an eco-friendly alternative to synthetic dyes and those derived from plants and animals. Future advancements in fermentation technology and metabolic engineering will allow for more efficient and scalable production of microbial pigments. By optimizing microbial strains, improving fermentation conditions, and enhancing pigment yields, industries will benefit from a more sustainable and cost-effective production process (Yuan *et al.*, 2015).

6.5.2 Biotechnological applications

The versatility of microbial pigments in biotechnological applications will continue to expand. In particular, pigments with antioxidant, antimicrobial, antiinflammatory, and UV-protective properties are expected to find increasing use in pharmaceuticals and nutraceuticals. The development of novel microbial strains with enhanced bioactivities will further contribute to their application in therapeutic treatments, cosmetics, and functional foods.

6.5.3 Pigment modification and genetic engineering

Advances in genetic engineering and synthetic biology are likely to enable the modification of microbial pigments for customized properties. For instance, enhancing the stability of pigments under extreme conditions or developing new colorants with unique characteristics could open new avenues in food colorant and textile applications. Additionally, biotransformation techniques could be employed to create pigments with superior bioactive properties for therapeutic applications (Liu *et al.*, 2017).

6.5.4 Natural colorants in the food industry

As consumers increasingly demand natural and clean-label ingredients, microbial pigments will play a vital role in replacing synthetic colorants in food products. The potential to replace artificial food colors with safe, natural pigments from microorganisms offers both health benefits and marketing advantages for food manufacturers. Further research into the safety, stability, and regulatory aspects of microbial pigments in food will enhance their adoption (Becerra *et al.*, 2017).

6.5.5 Bioremediation and environmental applications

Microbial pigments derived from extremophiles, such as halophiles, thermophiles, and psychrophiles, offer unique properties that make them valuable in environmental applications like bioremediation. These pigments can be employed in heavy metal sequestration, pollutant degradation, and environmental monitoring. The future application of microbial pigments in environmental clean-up and sustainable development strategies holds immense potential (Fendrihan *et al.*, 2021a).

6.5.6 Pigment-based bioelectronics and smart materials

The use of microbial pigments in bioelectronics and smart materials is an emerging field. Due to their biocompatibility and functional properties, microbial pigments could be integrated into biosensors, bioelectronics, and wearable devices. Their application in these areas could revolutionize industries like healthcare and electronics, creating more sustainable and bio-friendly technologies (Zhao *et al.*, 2020c).

6.6 Conclusion

In summary, the future scope of microbial pigments is vast, with opportunities for innovation in sustainable production methods, advanced biotechnological applications, and environmental sustainability. The growing interest in natural products and eco-friendly alternatives will drive further research into microbial pigments, leading to novel applications that benefit various industries.

References

Becerra, M., Moya, P. and Medina, M. (2017) Sustainable production of microbial pigments: Biotechnological advances and environmental benefits. *Bioresource Technology* 234, 160–169.

Bérdy, J. (2012) Thoughts and facts about antibiotics: Where we are now and where we are heading. *Journal of Antibiotics* 65(8), 385–395. DOI: 10.1038/ja.2012.27.

Bhute, A.S. and Bhonde, R.R. (2017) Microbial melanins: Applications in bioremediation and medicine. *Applied Microbiology and Biotechnology* 101(9), 3535–3543.

Buzzini, P., Martini, A., Vaughan-Martini, A. and Moraes, P. (2017) Yeasts as sources of biopigments. *Biotechnology Advances* 35(7), 725–742.

Cai, Y., Jiang, J., Zhao, Y. and Li, Z. (2021) Biopigments as sustainable alternatives: Opportunities and challenges. *Sustainable Chemistry* 12(3), 45–60.

Cavalcante, G.R., Santos, D.S. and Oliveira, T.M. (2020) Health benefits of biopigments: Antioxidant, antimicrobial, and anti-inflammatory properties. *Journal of Natural Products* 83(9), 2345–2356.

Chakraborty, S., Sinha, R. and Ghosh, S. (2020) Biopigments from bacteria: Antimicrobial and anticancer potentials. *Frontiers in Microbiology* 11, 567.

Chakraborty, S., Sinha, R. and Ghosh, S. (2021) Fungal biopigments: Applications in pharmaceuticals and food industries. *Applied Microbiology and Biotechnology* 105(3), 1009–1021.

Choi, S., Kim, J.H. and Lee, J. (2021) Advances in microbial pigment production for industrial applications. *Biotechnology Advances* 45, 1–13.

Dufossé, L., Fouillaud, M., Caro, Y., Mapari, S.A.S. and Sutthiwong, N. (2014) Filamentous fungi are large-scale producers of pigments and colorants for the food industry. *Current Opinion in Biotechnology* 26, 56–61. DOI: 10.1016/j.copbio.2013.09.007.

Fan, L., Zhang, Z. and Wang, L. (2021) Pulcherrimin-producing yeasts: A promising tool for biocontrol in agriculture. *Journal of Agricultural and Food Chemistry* 69(4), 1228–1237.

Fendrihan, S., McIntyre, R. and D'Arcy, T. (2021a) Pigments from extremophiles: From production to applications in the food, cosmetic, and pharmaceutical industries. *Applied Microbiology and Biotechnology* 105(4), 1491–1505.

Fendrihan, S., Tămaş, M. and Miclaus, L. (2021b) Pigments from extremophiles: New frontiers in biotechnology and industrial applications. *Journal of Industrial Microbiology & Biotechnology* 48(8), 877–894.

Gautam, N., Yadav, P. and Kumar, A. (2021) Actinomycetes-derived biopigments as natural dyes in the textile industry: Sustainable alternatives. *Industrial Biotechnology* 17(2), 43–50.

Grand View Research (2021) Natural food colorants market size to reach USD 5.8 billion by 2027. Available at: www.grandviewresearch.com (accessed 22 January 2025).

Gupta, A., Singhal, R. and Meena, R. (2020a) Fungal pigments and their bioremediation potential in environmental applications. *Journal of Environmental Management* 258, 110087.

Gupta, N., Prakash, R. and Pathak, N. (2020b) Microbial biopigments: Production, application, and future prospects. *Journal of Applied Microbiology* 129(2), 287–303.

Gupta, S., Abu-Ghannam, N. and Rajauria, G. (2013) Bioactive potential and possible health effects of edible brown seaweeds. *Trends in Food Science & Technology* 22(6), 315–326. DOI: 10.1016/j.tifs.2011.03.011.

Hassan, M., Essam, T. and Yassin, A. (2016) Colouring matters: Production and applications of microbial pigments in food industry. *Journal of Applied Microbiology* 121(5), 1289–1301.

Hassan, M.A., Saeed, M. and Khan, Z. (2019) Sustainable dyeing using bacterial pigments: A natural alternative for the textile industry. *Environmental Sustainability* 2(3), 213–222.

Huang, J., Sun, Y. and Ma, X. (2019) Algal pigments in aquaculture: Enhancing pigmentation of fish and crustaceans. *Aquaculture Nutrition* 25(4), 964–974.

Jin, H., Kim, J. and Lee, S. (2020) Functional properties of algal pigments: Antioxidants and anti-inflammatory agents for pharmaceutical applications. *Phytochemistry Reviews* 19(1), 145–158.

Joshi, V.K., Attri, D., Bala, A. and Bhushan, S. (2021) Microbial pigments. *Indian Journal of Biotechnology* 20(1), 1–15.

Khandel, P., Joshi, B. and Rathi, S. (2019) Environmental and health impacts of synthetic dyes. *Environmental Chemistry Letters* 17(2), 321–335.

Khan, M.I., Shin, J.H. and Kim, J.D. (2018) The promising future of microalgae: Current status, challenges, and optimization of a sustainable and renewable industry for biofuels, feed, and other products. *Microbial Cell Factories* 17(1), 36. DOI: 10.1186/s12934-018-0879-x.

Kumar, S., Singh, D. and Mathur, A. (2020) Recent trends in microbial pigment production and their applications in various industries. *Journal of Applied Microbiology* 129(3), 753–767.

Li, X., Zhang, L. and Liu, Y. (2021) Natural dyes from fungi: Applications in the textile industry. *Textile Research Journal* 91(5), 551–560.

Liu, Q., Wang, X. and Chen, J. (2019a) Antioxidant properties of yeast-derived biopigments and their potential applications in pharmaceuticals. *Antioxidants* 8(5), 132–142.

Liu, Z., Xu, Z. and Cheng, J. (2017) Microbial pigments: sources, production, and applications in food and pharmaceuticals. *Biotechnology Advances* 35(3), 334–343.

Liu, Z., Xu, Z. and Cheng, J. (2019b) Monascus pigments: Production, bioactivity, and applications in food and pharmaceuticals. *Critical Reviews in Food Science and Nutrition* 59(14), 2232–2243.

Liu, Z., Zhang, M. and Wang, S. (2019c) Natural colorants from actinomycetes: Applications in food and health. *Food Research International* 116, 902–912.

Madigan, M.T., Martinko, J.M. and Bender, K.S. (2015) *Brock Biology of Microorganisms*, 14th edn. Pearson, London, UK.

Manivasagan, P., Venkatesan, J. and Kim, S.K. (2021) Biosynthesis, biological, and industrial applications of microbial pigments. *Critical Reviews in Biotechnology* 41(2), 1–23.

Mohammad, N., Ali, M. and Khan, R. (2018) Melanins from yeasts: Applications in bioelectronics and environmental monitoring. *Environmental Science & Technology* 52(3), 1050–1058.

Neto, J., Silva, R. and Rodrigues, A. (2019) Algal carotenoids in cosmetics: Antiaging and UV protection benefits. *Cosmetic Science and Technology* 17(2), 89–101.

Newman, D.J. and Cragg, G.M. (2020) Natural products as sources of new drugs over the nearly four decades from 1981 to 2019. *Journal of Natural Products* 83(3), 770–803. DOI: 10.1021/acs.jnatprod.9b01285.

Radhakrishnan, R., Laiz, L. and Bensalah, M. (2020) Bioremediation potential of melanins from actinomycetes in heavy metal sequestration. *Environmental Science and Pollution Research* 27(15), 18875–18886.

Ravi, M., Kumar, M.N. and Muvva, V. (2021) Algal biopigments: A treasure trove of bioactive compounds. *Journal of Applied Phycology* 33(1), 761–776.

Santos, L., Soares, P. and Castro, F. (2021) Actinomycetes-derived biopigments: Pharmaceuticals, biotechnological, and industrial applications. *Applied Microbiology and Biotechnology* 105(4), 1365–1377.

Sharma, A., Gupta, D. and Mehta, N. (2020) Toxicological impacts of synthetic dyes on human health. *International Journal of Environmental Health Research* 15(5), 123–135.

Shen, W., Li, X. and Zhang, H. (2020) Azo dyes and their role in industrial pollution: Current status and remediation strategies. *Journal of Industrial Pollution Research* 12(4), 455–462.
Silva, M.F., Costa, D.S. and Souza, A.D. (2021) Exploring bacterial pigments for biopharmaceutical applications: From antimicrobial to anticancer potential. *Biotechnology Reports* 28, e00565. DOI: 10.1016/j.btre.2021.e00565.
Singh, P., Patel, A. and Singh, A. (2018) Exploration of microbial pigments: A sustainable and versatile approach. *Frontiers in Microbiology* 9, 1–9. DOI: 10.3389/fmicb.2018.02963.
Singh, P., Patel, A. and Singh, A. (2020) Microbial pigments: An insight into current production trends and future prospects. *Frontiers in Microbiology* 11, 1–15. DOI: 10.3389/fmicb.2020.00763.
Singh, S., Das, S. and Singh, S.K. (2019) Bioprospecting of actinomycetes for secondary metabolites. *Applied Microbiology and Biotechnology* 103(9), 3727–3742. DOI: 10.1007/s00253-019-09883-1.
Tee, Y., Sivaraman, S. and Lee, K. (2021) Actinomycetes pigments for cosmetics and skin care products: Natural UV protection and antioxidant benefits. *Cosmetics* 8(3), 56.
Thakur, A., Yadav, P. and Garg, S. (2020) Immunosuppressive and antioxidant properties of actinomycetes-derived biopigments. *Journal of Biotechnology and Biochemistry* 58(8), 1175–1186.
Venil, C.K., Zakaria, Z.A. and Ahmad, W.A. (2013) Bacterial pigments and their applications. *Process Biochemistry* 48(7), 1065–1079. DOI: 10.1016/j.procbio.2013.06.006.
Venil, C.K., Zakaria, Z.A. and Ahmad, W.A. (2020) Biopigments derived from microbial sources and their industrial applications. *Process Biochemistry* 95, 1–15.
Yin, S., Zhang, F. and Luo, Z. (2021) Applications of yeast-derived biopigments in food and dietary supplements. *Food Research International* 140, 109859.
Yuan, T., Zhang, Y. and Chen, W. (2015) Advances in microbial pigment production and their potential applications. *Journal of Applied Microbiology* 118(3), 587–602.
Zhang, Y., Xu, L. and Wang, X. (2020) Role of algae-derived pigments in biofuels and bioremediation applications. *Environmental Technology and Innovation* 18, 100735.
Zhang, Z., Yang, X. and Zhou, Q. (2018) Algal biopigments as natural colorants in the food industry. *Food Chemistry* 244, 247–254. DOI: 10.1016/j.foodchem.2017.09.123.
Zhao, X., Zhang, L. and Li, Y. (2020a) Fungal biopigments in the cosmetic and food industries: Opportunities and challenges. *Biotechnology Advances* 38, 107355.
Zhao, X., Zhang, L. and Li, Y. (2020b) Microbial pigments in the pharmaceutical and food industries: Current applications and future prospects. *Biotechnology Reports* 28, e00481.
Zhao, Y., Zhang, L. and Zhang, Z. (2020c) Yeast-derived carotenoids and melanins in cosmetic formulations: Antiaging and UV protection. *Journal of Cosmetic Dermatology* 19(9), 2359–2366. DOI: 10.1111/jocd.13279.
Zhao, Y., Wang, L. and Tan, T. (2021) Advances in yeast-based production of carotenoids and astaxanthin. *Critical Reviews in Biotechnology* 41(1), 67–81.

7 Biotechnological Production of Microbial Pigments

Rahul Kumar[1], Sourav Chattaraj[2], Rachan Karmakar[3], Chaithra M.[4], Tanja Vasić[5], Beatriz Elena Guerra Sierra[6] and Debasis Mitra[1*]

[1]Department of Microbiology, Graphic Era (Deemed to be University), Dehradun, Uttarakhand, India; [2]Centre for Industrial Biotechnology Research, School of Pharmaceutical Science, Siksha 'O' Anusandhan University, Kalinga Nagar, Bhubaneswar, Odisha, India; [3]Department of Environmental Science, Graphic Era (Deemed to be University), Dehradun, Uttarakhand, India; [4]ICAR-Central Plantation Crop Research Institute, Research Center, Kahikuchi, Guwahati, Assam, India; [5]Faculty of Agriculture, University of Niš, Niš, Serbia; [6]Facultad de Ciencias Naturales y Agropecuarias, Grupo de Investigación Biotecnología Agroambiente y Salud -Microbiota-Universidad de Santander, Bucaramanga, Colombia

7.1 Introduction

Human psychology, business, and culture are all significantly influenced by color, which influences everything from shopping behavior to food preferences (Casas and Chinoperekweyi, 2019). Due to their immense manufacturing ability, consistent quality, and affordability, artificial dyes are used extensively in industry. Approximately 800,000 metric tons of artificial dyes are manufactured every year, and more than 10,000 different types are being used globally (Albahnasawi *et al.*, 2025). However, due to the toxicity, persistence in the environment, and potential harm to human health of artificial colorants, natural alternatives have become more in demand, particularly in food, fashion, and personal care products. The dye and pigment industry has been reported by the World Health Organization (WHO) to contribute approximately 20% to industrial water pollution worldwide, with most of its chemicals associated with mutagenicity, carcinogenicity, and endocrine disruption (Islam *et al.*, 2023). Natural earth-friendly pigments that are nontoxic to human utilization and environmentally sustainable are increasingly used due to these concerns.

Microbial pigments are increasingly being seen as a helpful alternative. Microorganisms such as microalgae, bacteria, yeast, and fungi can synthesize various pigments of numerous colors and functional properties. They are non-toxic, biodegradable, and often have anticancer, antibacterial, and antioxidant activities. The global market for native pigments was estimated to be $3.6 billion in 2023 and is projected to grow at a compound annual growth rate (CAGR) of 5.1% to reach $5.4 billion by 2027 (Mandree *et al.*, 2023). The increased demand

*Corresponding author: debasismitra3@gmail.com

DOI: 10.1079/ 9781836990505.0007

in the food and beverage sector where natural colorants are used in place of synthetic chemicals is the main driver of this growth. Based on research, 64% of global customers are willing to pay for ingredients produced using natural constituents, pointing to a shift in market trends and consumer behaviors.

Biotechnological innovations have significantly improved the productivity of microbial pigment synthesis. By manipulating factors including nutrient composition, pH, aeration, and temperature, conventional fermentation methods like submerged and solid-state fermentation have been enhanced. Metabolic engineering and synthetic biology tools have been employed to boost pigment biosynthesis (Saini *et al.*, 2020). For instance, CRISPR-Cas9 technology has assisted editing of *Monascus* biosynthetic gene clusters which lowers citrinin production while enhancing yield and purity of the pigment. Technologies including genome editing, metabolomics, transcriptomics, and omics are also being used to unravel and alter pigment biosynthesis processes.

Microbial pigments have broad industrial applications with a growing variety of uses. They provide natural colorants and preservatives for the food industry; some pigments are even nutritionally active. Due to their bioactivities, microbial pigments have found application in pharmaceuticals as lead molecules to develop medicines. Melanin and carotenoids, two pigments that make up approximately 18% of the cosmetics market, are highly valued due to their antioxidant and UV-protective properties (Salian *et al.*, 2021). In addition, microbial pigments have been explored as environmentally friendly textile dyes, especially in view of the harm that synthetic azo dyes cause to the environment.

7.2 Natural Pigments: Plant, Animal, and Microbial Sources

Natural pigments are chemicals that impart color and are derived from biological organisms such as microbes, plants, and animals. Apart from possessing biofunctional properties such as antibacterial, anticancer, and antioxidant activity, they are also aesthetically pleasing. Because of the demand from consumers for clean-label and eco-friendly products, the global natural pigment market was valued at \$3.6 billion in 2023 and is anticipated to grow to \$5.4 billion by 2027 (Budzianowska *et al.*, 2025).

Plant-based pigments such as flavonoids, carotenoids, chlorophylls, and betalains account for 60–65% of the natural pigment market because they are so prevalent and well recognized. However, their usefulness in industry is restricted by their seasonal production, variable yield, and susceptibility to degradation under changes in pH, light, and heat. Pigments derived from animals, e.g. heme, carmine, and melanin, are less commonly used but have potential for drug delivery and photoprotection. Due to their fermentable scale-up, functional diversity, and rapid development, microbial pigments are proving to be excellent alternatives (Usmani *et al.*, 2020). Orange carotenoids from *Rhodotorula glutinis*, blue phycocyanin from *Spirulina platensis*, yellow monascin from *Monascus purpureus*, and red prodigiosin from *Serratia marcescens* are just a few examples. These pigments have shown remarkable bioactivities, including antibacterial activities with minimum inhibiory concentrations (MICs) of 25–100 μg/ml and antioxidant activity with IC_{50} values < 50 μg/ml.

Due to advancements in fermentation technology, metabolic engineering, and synthetic biology, microbial pigments now capture 15–25% of the natural pigment market, and this share is likely to grow (Table 7.1).

7.3 Advantages of Microbial Pigments Over Synthetic Counterparts

Microbial pigments are gaining increasing popularity as superior substitutes to synthetic dyes because of increasing demand for safe, green, and sustainable substitutes (Aman Mohammadi *et al.*, 2022). Due to their consistency, wide range of colors, and affordable cost of production, synthetic dyes have traditionally dominated the market, but environmental and health issues are slowly overtaking them. Among the several advantages of microbial pigments are that they are nontoxic, inexpensive, environmentally sustainable, and scalable (Samal *et al.*, 2025). Compared to petroleum-derived synthetic dyes that contain toxic aromatic amines or heavy

Table 7.1. Natural pigments and their sources.

Source	Examples	Applications	Advantages	Limitations	Market share %	References
Microbial derived	Carotenoids (*Rhodotorula*), prodigiosin (*Serratia*), monascin (*Monascus*), phycocyanin (*Spirulina*)	Textiles, food, pharmaceuticals, cosmetics	Scalable fermentation, nonseasonal, functional biodiversity example antioxidant, antimicrobial	Some yield limitations, regulatory hurdles, downstream processing costs	15–20%	(Rao *et al.*, 2021)
Plant derived	Anthocyanins (*berries*), betalains (beetroot), chlorophyll (spinach), lycopene (tomato), β-carotene (carrot)	Cosmetics, food, beverages, nutraceuticals	Antioxidant properties, widely available, consumer accepted	Variable yields, stability issues (pH, light, heat), seasonal variation	60–65%	(Krishnakripa and Thoppil, 2025)
Animal derived	Heme (hemoglobin), carmine (cochineal insect), melanin (squid ink)	Biomedical use, food coloring, cosmetics	Vibrant colors, high stability	Limited scalability, ethical concerns, allergenicity	<10%	(Scanes, 2018)

metals, microbial pigments are chemically synthesized under natural conditions and are considered safe for use by general recognition of safety (GRAS) guidelines (Intasian *et al.*, 2021). They possess antiinflammatory, anticancer, antibacterial, and antioxidant properties, among other useful health benefits. In addition, microbial culture is more climate change and agricultural constraint friendly because it requires less water and land compared to plant-based options and does not take away from food crops and emerging plant hormones (Jain *et al.*, 2025). Microbial pigments are also argued to be cost-effective and scalable. They can be produced throughout the year under controlled conditions with efficient fermentation processes, converting waste into value-added products. Microbial strains can be genetically engineered to enhance pigment production, reduce byproducts, and tailor specific colors due to advances in metabolic engineering and synthetic biology (Saini *et al.*, 2020).

Microbial pigments also possess significant benefits in consumer acceptability and regulation. Companies are under pressure to eliminate artificial ingredients because of increased customer awareness and the need for clean-label products. Based on surveys, over 60% of consumers globally opt for natural ingredients, and some are even willing to pay extra for naturally colored meals and personal care products.

7.4 Classification of Microbial Pigments

Microorganisms such as fungi, yeasts, bacteria, and microalgae synthesize a variety of biocompounds called microbial pigments (Hamidi *et al.*, 2019; Kamboj *et al.*, 2025). Color, structure, biosynthetic origin, and their biological role are all different from one another. Chemical structure, microbial origin, color, and functional properties are just a few parameters by which the microbial pigments can be grouped (Paillié-Jiménez *et al.*, 2020). Their probable application in food, pharmaceuticals, cosmetics, and textiles industry necessitates recognition of these groups (Table 7.2).

Table 7.2. Classification of microbial pigments.

Classification criteria	Pigment group	Examples	Microbial source	Color	Bioactivity / functional properties	Applications	References
Chemical structure	Carotenoids	β-Carotene, torulene, zeaxanthin, astaxanthin	*Rhodotorula glutinis*, *Blakeslea trispora*, *Haematococcus pluvialis*	Yellow, orange, red	Antioxidant, antiinflammatory	Food colorants, nutraceuticals, cosmetics	(Olaniran *et al*., 2023)
	Polyketides	Monascin, rubropunctatin, prodigiosin	*Monascus purpureus*, *Serratia marcescens*	Red, orange	Antimicrobial, anticancer	Natural food dyes, Pharmaceuticals	(Rao *et al*., 2021)
	Phenazines	Pyocyanin	*Pseudomonas aeruginosa*	Blue-green	Antimicrobial, antiinflammatory	Biomedical, agricultural biocontrol	(Sahoo *et al*., 2023)
	Quinones	Violacein, actinorhodin	*Chromobacterium violaceum*, *Streptomyces coelicolor*	Violet, blue	Antimicrobial, antiparasitic, anticancer	Pharmaceuticals, research	(Kamble *et al*., 2022)
	Melanins	Eumelanin, pyomelanin	*Cryptococcus neoformans*, *Aspergillus niger*, *Streptomyces*	Black, brown	UV-protective, antioxidant, radio-protective	Sunscreens, bio-coatings, Medicine	(Peyrat *et al*., 2019)
	Phycobiliproteins	Phycocyanin, phycoerythrin	*Spirulina platensis*, *Porphyridium cruentum*	Blue, red	Antioxidant, immunomodulatory	Nutraceuticals, food, diagnostics	(Minić *et al*., 2024)
Microbial source	Bacteria	Prodigiosin, violacein, pyocyanin, zeaxanthin	*Serratia marcescens*, *Pseudomonas*, *Micrococcus*, *Chromobacterium*	Red, blue, yellow	Antibacterial, anticancer, antioxidant	Food, textiles, pharmaceuticals	(Devi *et al*., 2024)

Continued

Table 7.2. Continued

Classification criteria	Pigment group	Examples	Microbial source	Color	Bioactivity / functional properties	Applications	References
	Fungi	Monascin, rubropunctatin, citrinin	*Monascus*, *Penicillium*, *Talaromyces*	Red, yellow	Natural dyes, but some mycotoxins present	Food colorants (toxicity monitoring needed)	(Poorniammal *et al.*, 2021)
	Yeasts	β-Carotene, torulene, torularhodin	*Rhodotorula*, *Phaffia rhodozyma*	Red, orange, yellow	Antioxidant, colorant	Food, cosmetics, feed	(Dufossé, 2018)
	Microalgae	Phycocyanin, chlorophyll, astaxanthin	*Spirulina*, *Chlorella*, *Haematococcus*	Blue, green, red	Nutritional, antioxidant	Health drinks, Dietary supplements	(Olaniran *et al.*, 2023)
Colour	Red	Prodigiosin, monascorubrin, astaxanthin	*Serratia*, *Monascus*, *Haematococcus*	Red	Antioxidant, anticancer	Food, pharma, textile	(Rao *et al.*, 2021)
	Yellow	Zeaxanthin, monascin, lutein	*Rhodotorula*, *Monascus*, *Micrococcus*	Yellow	Eye health, antioxidant	Food supplements, colorants	(Rather *et al.*, 2025)
	Orange	β-Carotene, canthaxanthin	*Blakeslea trispora*, *Rhodotorula*, *Phaffia*	Orange	Antioxidant, provitamin A	Nutraceuticals, cosmetics	(Gupta *et al.*, 2022)
	Blue-green	Phycocyanin, pyocyanin, violacein	*Spirulina*, *Pseudomonas*, *Chromobacterium*	Blue, blue-green	Immunomodulatory, antimicrobial	Beverages, skincare, biomedical dyes	(Chwil *et al.*, 2024)
	Black, brown	Melanin	*Cryptococcus*, *Aspergillus*, *Streptomyces*	Brown, black	UV protection, free radical scavenger	Sunscreens, biomedical coatings	(Michael *et al.*, 2023)
Bioactivity	Antioxidant	β-Carotene, astaxanthin, phycocyanin	*Spirulina*, *Rhodotorula*, *Haematococcus*	Varies	Neutralize oxidative stress	Functional foods, antiaging formulations	(Budzianowska *et al.*, 2025)

Continued

Table 7.2. Continued

Classification criteria	Pigment group	Examples	Microbial source	Color	Bioactivity / functional properties	Applications	References
	Antimicrobial	Prodigiosin, violacein, pyocyanin	*Serratia*, *Chromobacterium*, *Pseudomonas*	Red, violet, blue	Kills/inhibits pathogenic microbes	Pharma, food preservation	(Al-Sayaghi *et al*., 2022)
	Anticancer	Prodigiosin, violacein, astaxanthin	*Serratia*, *Chromobacterium*, *Haematococcus*	Red, violet	Inhibits cancer cell growth	Therapeutics, drug development	(Agarwal *et al*., 2023)
	Immunomodulatory	Phycocyanin, astaxanthin	*Spirulina*, *Haematococcus*	Blue, red	Enhances immune system	Health supplements, antiinflammatory drugs	(Omar *et al*., 2022)

7.5 Microorganisms as Pigment Producers

A rich source of natural pigments, microorganisms offer a renewable and eco-friendly alternative to synthetic dyes across a range of industries (Rajendran *et al.*, 2023). Due to their variety in color, structure, bioactive compound, and bioactivity, these pigments are ideal for industrial-scale production (Mahakur *et al.*, 2025).

The principal source of pigments is bacteria, which also produce phenazines, anthraquinones, and carotenoids. *Pseudomonas aeruginosa* produces pyocyanin, a blue-green pigment with antibacterial activity, while *Serratia marcescens* produces prodigiosin, a red pigment (Devi *et al.*, 2024). The purple pigment violacein, which is antibacterial and anticancer in nature, is produced by *Chromobacterium violaceum*. Red pigments named rubropunctatin and monascin, which are commonly utilized for food, cosmetics, and medicine, are yielded by fungi, particularly species of the genus *Monascus*. Typically generated within monitored fermentation systems, fungal pigments offer scalability for commercial applications. Nonetheless, as some fungi produce mycotoxins, careful strain selection is needed to confirm safety (Kagot *et al.*, 2019).

Another important group of microbial pigment manufacturers is yeasts which produce carotenoids such as torulene and β-carotene which are often employed as food color and nutritional supplements. The red-colored pigment astaxanthin with potent antiinflammatory and antioxidant properties can be produced by specific yeasts, e.g. *Phaffia rhodozyma*. Chlorophyll and astaxanthin, which are created by microalgae such as *Spirulina platensis*, *Chlorella*, and *Haematococcus pluvialis*, are extremely valuable in the food industry as natural pigments and in the nutraceutical market due to their health benefits (Sun *et al.*, 2023).

The commercial possibility for producing these pigments on a large scale will increase as methods for microbial pigment production evolve. Ensuring the safety of certain fungal pigments remains experimental, however. Despite these challenges, microbial pigments have been projected to gain increasing prominence in replacing artificial colors and meeting the needs of a greener future.

7.6 Biosynthetic Pathways and Genetic Regulation

The regulation of microbial pigment biosynthesis involves multiple enzyme reactions, gene clusters, and network interactions that produce diverse colors through complex biochemical pathways (Tang *et al.*, 2025). Both internal genetic processes within the microbes and external factors such as pH, temperature, light, and nutrition availability tightly regulate these pathways. While the primary microbial pigments such as carotenoids, phenazines, and melanins are synthesized via distinct biosynthetic pathways that often employ common intermediates or enzymes, each type of pigment is controlled by distinct genes that coordinate the production process (Madan *et al.*, 2025).

The isoprenoid pathway is utilized to synthesize carotenoids, a group of pigments that impart red, yellow, and orange colors to foods (Riaz *et al.*, 2021). Phytoene synthase and phytoene desaturase, carotenoid-specific enzymes, catalyze a series of reactions that yield carotenoids such as β-carotene from GGPP in microbes such as *Rhodotorula* species. In some microorganisms UV light prompts the production of carotenoid pigments (Massoud and Khosravi-Darani, 2017). In addition, certain transcription aspects like CrtE and CrtB control the genetic control of carotenoid biosynthesis in yeast, bacteria, and fungi by promoting or inhibiting genes that synthesize carotenoid constructed on environmental signals. The shikimate pathway primary system for aromatic amino acid biosynthesis is employed to produce phenazine pigments, which provide some bacteria like *Pseudomonas aeruginosa* with their blue-green color. Chorismic acid is the central intermediate here and this is converted through several enzyme-catalyzed conversions into various phenazine products (Huang *et al.*, 2024).

Phenazines are classically regulated by an arrangement of global regulatory organizations such as quorum sensing and two-component signal transduction systems that instruct the bacteria to adjust pigment mixture based on environmental cues or extent. The phz cluster of *Pseudomonas* contains genes for the enzymes contained in phenazine production, and both internal signals (e.g. levels of cyclic AMP) and environmental signals (e.g. availability of oxygen) regulate its appearance (Mavrodi *et al.*, 2013). Another long-standing process is melanin biosynthesis in microorganisms that yields dark pigments like eumelanin. Tyrosinase enzymes catalyze a series of oxidation consequences to convert L-tyrosine to melanin in the fungus *Aspergillus niger*. As an intermediate in the process, dihydroxyindole is formed and then goes through additional polymerization to form melanins (Suthar *et al.*, 2023). The assembly of melanin biosynthesis genes (e.g. the pks genes in *Aspergillus*) affects the genetic regulation of melanin synthesis. These genes are induced by some stimulus, e.g. exposure to UV radiation or nutritional deficiency.

Transcription factors and gene clusters which act as alterations to initiate or suppress the appearance of biosynthetic genes are the primary genetic regulators of color production in microorganisms. Sigma factors and two-component systems in bacteria, for instance, regulate the expression of genes that give color based on stimuli from the environment while regulatory proteins such as bZIP and Zn2Cys6 transcription factors regulate pigment production in fungi and yeasts (Tianqiao *et al.*, 2021). In addition, quorum sensing plays a critical role in regulating pigment production in most bacteria as the accumulation of signaling molecules introduces gene expression that leads to pigment formation at high cell intensity. Pigment production by microorganisms is a composite process regulated by difficult genetic and metabolic networks. Microbial mechanisms for the synthesis of melanin, phenazine, and carotenoid involve complex enzymic processes that are regulated by a variety of genetic factors in response to extracellular cues (Pavan *et al.*, 2020).

In order to maximize microbial pigment production for industrial applications, such as natural pigments, drugs, and other biotechnology, the biosynthetic pathways and their genetic regulation need to be appreciated.

7.7 Biosynthesis of Pigment Involving Genes and Enzymes

A variety of enzymes and genes help to form various pigment mixtures during microbial pigment biosynthesis. Genes for these enzymes are regulated to safeguard effective pigment construction, often in response to clues from the environment, and the enzymes catalyze events in a number of biosynthetic pathways (Saini *et al.*, 2020).

The biosynthesis of carotenoids begins with phytoene synthase enzyme (CrtB), which changes GGPP to phytoene. Phytoene desaturase (CrtI) follows, which converts phytoene to lycopene, a valuable intermediate (Kato *et al.*, 2016). Finally, lycopene is cyclized by lycopene cyclase (CrtY) to yield end carotenoid colors like β-carotene. Genes like crtE and crtI regulate the biosynthesis process in yeasts such as *Rhodotorula*, while the transcription feature CrtR supports the regulation of how genes are extracted in response to environmental cues such as light. Phenazine is often produced by *Pseudomonas aeruginosa* and other bacteria, a process that is mainly shikimate pathway dependent. Pioneers are converted to phenazine-1-carboxylic acid (PCA) by phenazine-1-carboxylic acid synthase (PhzA) during this process. PCA is subsequently further modified to generate colors like pyocyanin (Vilaplana and Marco, 2020). This route is regulated by the phz gene cluster, in which proteins such as PhzR regulate statement of genes due to quorum sensing signals controlling population intensity. *Aspergillus niger* and other fungi employ the enzyme tyrosinase to catalyze conversion of L-tyrosine to dopaquinone, a precursor for melanin.

Tyrosinase is encoded by the *melA* gene while melanin polymerization is encoded by the pks gene. The final steps in melanin progress are facilitated by other enzymes including laccases and peroxidases (Qin and Xia, 2024). Environmental signals such as readiness of nutrients or stressors such as UV radiation often affect the genetic control of melanin development. Expression of pigment biosynthesis genes is monitored by regulatory genes like PhzR in phenazine biosynthesis and CrtR in carotenoid biosynthesis; likewise the enzymes themselves that are directly accountable for pigment synthesis. Such regulatory activities that prompt microbes to modify pigment synthesis involve quorum sensing, sigma factors, and two-component signal transduction systems (McLean *et al.*, 2019). Due to this exceptionally tuned management, pigment biosynthesis occurs under optimal conditions, allowing microorganisms to respond to their environment and generate pigments as necessary.

7.8 Pigment Production and Genetic Engineering Approaches

To improve production of microbial pigments and achieve maximum quantity and stability diversity of the pigments in profitable applications, genetic engineering methods have gained ever-increasing acceptance (Anshi *et al.*, 2024). Such techniques involve modifying microorganisms' gene composition and specific genes controlling pigment production. Microbial strains may be scientifically engineered to construct larger numbers of useful pigments more effectively, often with enhanced activity or novel colors that do not occur naturally. Gene overexpression, one of the leading techniques in genetic modification for pigment improvement, involves the amplification of biosynthetic genes in order to enhance pigment formation (Lagarde *et al.*, 2000). For example, overexpression of the gene *crtE*, which encodes phytoene synthase, could lead to increased phytoene carotenoid precursor synthesis. The production of carotenoids by yeast and bacteria such as *Escherichia coli* has been maximized through coexpression of numerous biosynthetic genes via synthetic pathways.

An alternative widely employed genetic engineering method is heterologous expression of color manufacture pathways in nonnative hosts (Nawab *et al.*, 2020). Scientists can employ more industrially appropriate microorganisms for mass production of pigments by introducing genes that produce pigments into bacteria that do not produce them naturally. Unlike traditional extraction from crustaceans or algae, the highly useful carotenoid antioxidant astaxanthin can now be generated more economically and efficiently with the advent of transgenes from *Xanthophyllomyces dendrorhous* into *Saccharomyces cerevisiae* (Naz *et al.*, 2023). Through diversion of the stream of precursor metabolites, metabolic engineering can be utilized to optimize cell processes favorable for pigment synthesis. With the accurate genetic changes enabled by CRISPR-Cas9 technology, genes that can divert precursor metabolites from pigment production can be knocked out or reduced, leading to improved production efficiency. To design entirely new biosynthetic routes for dyes that are not naturally occurring in the microorganism, synthetic biology strategies have been employed.

7.9 Use of Omics Tools in Pigment Pathway Elucidation

An inclusive understanding of the genetic protein and metabolic pathways in pigment production is assisted by this sophisticated methodology (Lyu *et al.*, 2022). Scientists can regulate relevant proteins, genes, and metabolites that are responsible for the biosynthesis of various microbial pigments melanin, carotenoids, and phenazines. They can study the impact of genetic engineering or environmental influences on pigment production using omics tools, which offer high-quality data and thorough molecular detail. With next-generation sequencing (NGS) technologies exposing gene clusters responsible for pigment production, genomics extracting is pivotal in determining the genes responsible for pigment biosynthesis (Kumar *et al.*, 2025b). For example, genomic studies have identified the genes crtE, crtI, and crtY within carotenoid biosynthesis as responsible for encoding enzymes which catalyze the conversion of precursors into carotenoids (Liang *et al.*, 2018). The synthesis of phenazine pigments is performed by the phz gene cluster that genomics has found in phenazine-producing bacteria *Pseudomonas aeruginosa*. Novel biosynthetic pathways may be identified using qualified genomics to identify new gene clusters in microorganisms that produce characteristic colors.

Illuminating the proteins that take part in pigment production, proteomics extends genomics. Proteins can be detected with techniques including liquid chromatography-tandem mass spectrometry (LC-MS/MS), mass spectrometry (MS), and two-dimensional gel electrophoresis (2D-GE) (Gómez *et al.*, 2021). Metabolomics, or the investigation of small molecule metabolites, allows insights into microbial metabolic actions during pigment production. Scientists can track the flow of metabolites along the biosynthetic pathway and identify metabolic bottlenecks by investigating the levels of intermediates and endproducts (Daly *et al.*, 2022).

7.10 Biotechnological Strategies for Enhanced Production

Strategies employed incorporate the expansion of new bioreactor systems, metabolic engineering, genetic manipulation, and development optimization (Ko *et al.*, 2020). With these approaches, the limitations of natural pigment production can be extended, making it more feasible for mass applications in the textile, food, cosmetics, and pharmaceutical industries. Overexpression of color production-related genes, for example crtE, crtI, and crtY, in microorganisms such as *Saccharomyces cerevisiae* or *Escherichia coli* is referred to as genetic engineering (Choi *et al.*, 2016). One may also synthesize novel pigments such as astaxanthin by expressing heterologous biosynthetic routes in pigment-forming or nonpigment-forming microorganisms. Through the remodelling of metabolite pathways furnishing the precursor needed to produce a pigment, metabolic engineering increases the rate of movement of precursor metabolites toward synthesis of the pigment. By facilitating the formation of entirely new biosynthetic routes for special colors, synthetic biology allows for microorganisms to produce an incredible number of colors (Orlandi *et al.*, 2022).

Increased pigment production can be achieved by optimizing fermentation conditions, such as growing medium, temperature, pH, oxygen levels, and light exposure. To sustain optimal growth conditions during fermentation and ensure a constant nutrient supply, batch or fed-batch fermentation methods are often used to maximize pigment production.

Light plays a fundamental role in improving biosynthetic activities in photobioreactors to produce carotenoid pigments, particularly β-carotene and astaxanthin. Coculturing or microbial consortia is an alternative technique gaining popularity for boosting pigment production (Bhatia *et al.*, 2018). It is possible to create a more beneficial environment for producing pigments by cultivating various microorganisms together. A more effective production process is executed by this method which also takes advantage of synergistic effects of multiple species waste degradation and nutrient utilization.

Encouraging successful translation of laboratory-scale value to industrial-scale pigment production are bioreactor model and scale-up procedures (Abdi *et al.*, 2024). Sophisticated bioreactor systems such as airlift and stirred tank reactors allow regulation of factors such as oxygen transfer, agitation, and nutrition distribution. The repeatability and efficiency of pigment production are also improved through the incorporation of process control and digital sensors.

The increasing demand for natural pigments in different industries can be met using the potential of such biotechnological tools, proposing an ecologically responsive alternative to synthetic pigments (Venil *et al.*, 2020).

7.11 Downstream Processing and Purification

After fermentation or microbial growth, the pigment is removed from the culture broth and purified (Ramesh *et al.*, 2022). The pigment is retained in its pure state with minimal contamination due to competent downstream processing. It can be developed in a range of applications such as fabrics, pharmaceuticals, food, and cosmetics. Cell separation is normally employed as the first step in downstream processing to retrieve biomass from the fermentation broth (Sadare *et al.*, 2021). This can be achieved through centrifugation, flocculation, and filtration depending on the type of broth as well as the type of microbe used. Filtration or microfiltration can also be used for separation of fine pigments. Depending on the solubility and chemical nature of the pigment, elimination from the culture broth is promoted by the second step (Nemer *et al.*, 2021).

To make the pigment acceptable for use in end applications, pollutants, residual solvents, by-harvests, and other contaminants must be removed by purification. Chromatographic techniques such as high-performance liquid chromatography (HPLC) are commonly used for pigment purification (Degano, 2019). Such techniques can further separate pigments based on their chemical properties, including polarity, size, and charge. Thin-layer chromatography (TLC) can also be used to achieve smaller-scale pigment purification and preliminary viewing for large-scale purification procedures like ion exchange chromatography or adsorption chromatography could be commissioned. Formulation and packing are the final steps in downstream processing. To extend shelf-life and prevent degradation during storage, pure pigment is often mixed with stabilizers or carriers.

7.12 Industrial Applications of Microbial Pigments

Due to their natural origins, large range of colors, and environmentally friendly properties, microbial pigments are progressively used as replacements for artificial dyes. Microbial pigments are helpful in both consumer products and industrial processes because they are applied across many industries, including food, cosmetics, textiles, medicines, and agriculture (Venil *et al.*, 2020).

Microbial pigments are used in the food industry as natural colorants instead of artificial dyes, which are associated with health risks such as allergy and hyperactivity in children. Carotenoids, for example astaxanthin and β-carotene, are valued for their antioxidant properties and coloring attributes. Microbial pigments provide an environment-friendly and safer alternative to artificial dyes which have been tainted by health and ecological issues (Dave *et al.*, 2022). Microbial pigments are critical to the formulation of skincare products, hair dyes, and make-up elements in the cosmetics industry. Pigments such as melanin, anthocyanin, and carotenoids are used for their color and potential health benefits. Since it can effectively filter out UV radiation, microbial melanin is used as a natural ingredient in sunscreens, offering a greener means of defending the skin from UV damage (Solano, 2020).

Microbial dyes, such as those made using bacteria such as *Chromobacterium violaceum*, yield rich colors that are ecological and nontoxic (Rather *et al.*, 2025). They assist in establishing a more eco-friendly and sustainable textile industry by substituting for destructive artificial dyes. Due to the increase in demand for sustainable fashion, the application of microbial pigments in textiles will significantly reduce environmental pollution by removing harmful chemicals from traditional dyeing processes.

Microbial pigments are also applied in the pharmaceutical industry as they possess antioxidant, antibacterial, and antiinflammatory properties (Agarwal *et al.*, 2023). Drug discovery is aided by the therapeutic value of pigments like pyocyanin from *Pseudomonas aeruginosa* and anthraquinones from fungi. Quinones and indigo are being explored as drugs for cancer therapy and have been promising.

Microbial pigments are used as natural biocontrol agents in agricultural settings (Elnahal *et al.*, 2022). Certain pigments, including phenazines produced by microbes such as *Pseudomonas*, possess antibacterial properties that protect crops against disease-causing microbes and provide a greener alternative to traditional pesticides (Table 7.3).

7.13 Safety, Toxicity, and Regulatory Considerations

Food, cosmetics, textiles, and pharmaceuticals are only a few of the industries that rely significantly on microbial pigments. Their application demands thorough assessment of their toxicity, safety, and regulatory status. Certain pigments, like pyocyanin, are toxic to human health as well as the environment, while others, like melanin, carotenoids, and anthocyanins, are generally considered harmless (Gonçalves and Vasconcelos, 2021).

To determine the safety of any pigment, one must know its toxicity, especially in the case of cosmetics or edible product formulations. Since contaminations like endotoxins or heavy metals can cause severe harm to health, purity is also important. To ensure that microbial pigments are not contaminated, purification processes must be carefully structured. Another important aspect of microbial pigment application is protection in the environment. In contrast to synthetic dyes, they can be biodegradable and less harmful to the environment (Tkaczyk *et al.*, 2020). Reducing the environmental footprint of microbial pigment production is realized through sustainable synthesis processes with appropriate waste management and water recycling (Gupta *et al.*, 2025).

Prior to their approval for application in food, cosmetics, and pharmaceuticals, monitoring bodies like the FDA and EFSA evaluate microbial pigments against welfare data and toxicity standards (Poorniammal *et al.*, 2021). Prior to authorization as food flavors, pigments are required to undergo demanding testing within the food industry incorporating toxicological and allergenicity aspects (Mitra *et al.*, 2024).

Table 7.3. Industrial applications of microbial pigments.

Industry	Pigments	Applications	References
Food industry	Carotenoids (e.g. β-carotene, astaxanthin)	Natural colorants in juices, candies, dairy products; antioxidant supplements; food coloring alternatives to synthetic dyes	(Ghosh *et al.*, 2022)
Cosmetics industry	Melanin, carotenoids, anthocyanins	Pigments in make-up (lipsticks, foundations, eye shadows); sunscreen formulations; skincare products for pigmentation and UV protection	(Li *et al.*, 2023)
Textile industry	Violacein, carotenoids, anthocyanins	Natural dyes for fabric coloring, providing eco-friendly alternatives to synthetic dyes that pollute water bodies; sustainable and biodegradable textile production	(Renita *et al.*, 2023)
Pharmaceuticals	Pyocyanin, anthraquinones, carotenoids	Pigments with antimicrobial, antiinflammatory, antioxidant, and anticancer properties for drug development and therapeutic applications	(Agarwal *et al.*, 2023)
Agriculture	Phenazines, carotenoids	Biocontrol agents for protecting crops from pathogens; natural pesticide alternatives in integrated pest management systems	(Shahid *et al.*, 2017)

7.14 Safety and Toxological Studies Assessment

In deciding if microbial pigments can be used in a wide range of industrial methods, toxicological studies and safety estimations must be performed. Through these examinations, microbial pigments are investigated for environment and human health by testing any potential risks correlated with experience (Nigam and Luke, 2016). The acute toxicity test is the first phase in toxicological testing. A model organism grouping, such as rats or water species, is given a one-time dose of the pigment and the effects are followed over a precise amount of time. The results of the test indicate the likelihood of acute injury from effects including organ poisoning, respiratory distress or skin irritation (Ramadori, 2023). Genotoxicity testing discovers whether a substance can modify DNA integrity, induce chromosome damage, or create genetic mutations (Mohamed *et al.*, 2017). Examinations of ecological safety are required to recognize the impact of microbial pigments on the environment (Chatragadda and Dufossé, 2021). For industrial use, specifically in the textile industry where wastewater discharge can have a substantial environmental impact, biodegradable and nonhazardous pigments are favored (Kumar *et al.*, 2025a).

Microbial pigments used in products must be authorized and sold in line with established guidelines like those of the FAO/WHO Expert Committee on Food Additives (JECFA) (World Health Organization, 2019).

7.15 Guidelines in Different Countries

Safety guidelines, testing rules, and approval protocols for the relevance of microbial pigments in different industries are expressed in regulatory guidelines that differ across countries. To ensure the safety and value of microbial pigments in their various jurisdictions, crucial monitoring bodies like the Food Safety and Standards Authority of India (FSSAI), European Food Safety Authority (EFSA), and US Food and Drug Administration (FDA) are vital (Dhara *et al.*, 2021) (Fig. 7.1). These bodies offer precise standards and guidelines for assessing microbial pigments before they are commercialized for human use. Global regulations are necessary to ensure the safety of microbiological pigments applied in food, cosmetics, and pharmaceuticals (Dave *et al.*, 2022).

7.16 Challenges and Future Prospects

Large-scale production is one of the greatest challenges due to specific cultivation restrictions and the massive cost of fermentation developments (da Silva and Reis, 2015). One substantial hindrance is posed by genetic diversity which causes concerns regarding biosafety, ethics, and regulatory sanction (van Belkum *et al.*, 2019). Regulatory bodies might limit the use of evident microbial strains or require general testing before granting approval. Since microbial pigments are generated as complex fermentation broths containing other metabolic products that must be removed to obtain pure pigment, purification and downstream processing are also challenging. They often involve solvents

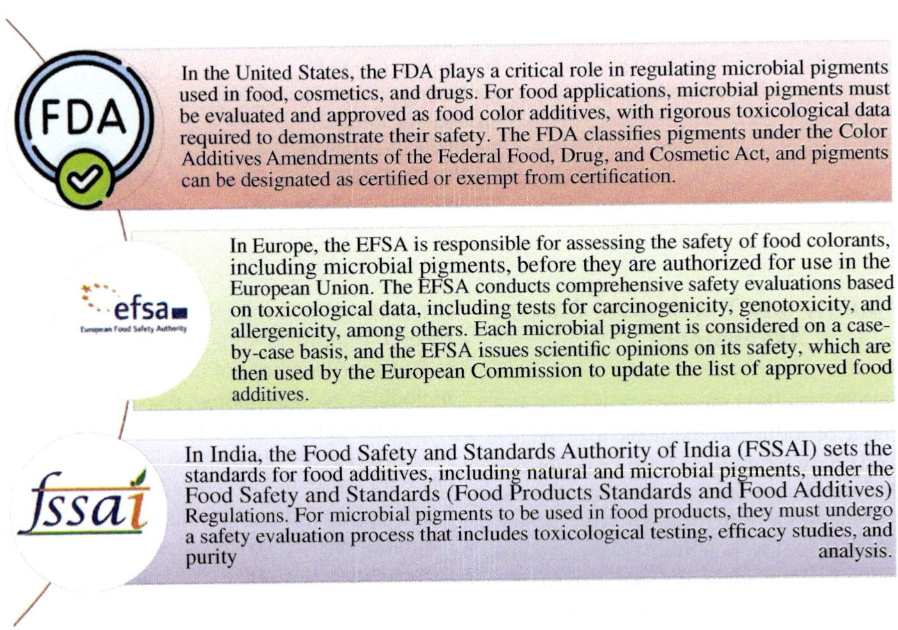

Fig. 7.1. Regulatory guideline of different countries.

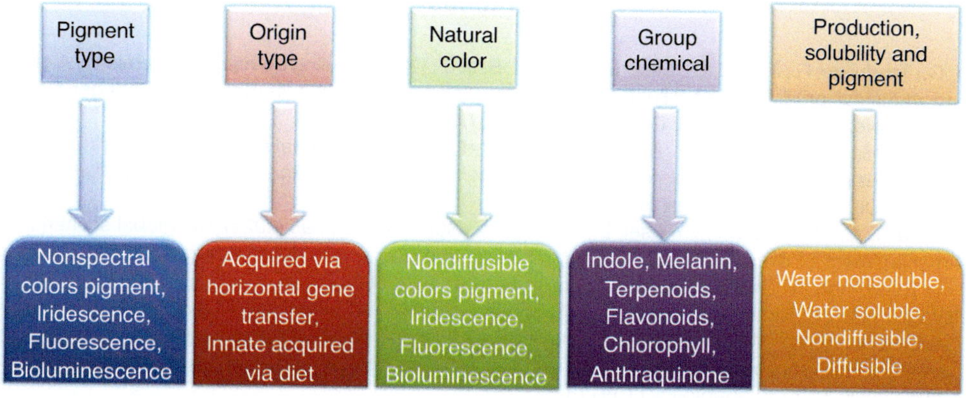

Fig. 7.2. Future challenges.

or chromatographic procedures that may not be cost-effective or sustainable for large-scale applications (Heidari *et al.*, 2021).

As the demand for natural colorants in food, cosmetics, and fabrics continues to grow, microbial pigments have a promising future despite these challenges. The yield and properties of microbial pigments could be substantially improved by advances in metabolic engineering and synthetic biology. To make beneficial use of microbial pigments in several industrial applications, innovative research and development will be required (Fig. 7.2).

7.17 Conclusion

Microbial pigments hold great promise as alternatives to synthetic products for various industries including textiles, food, cosmetics, and pharmaceuticals. Compared to synthetic dyes, they offer sustainability, biodegradability, and lower environmental impact. Microbial pigments have gained prominence with the increased requirement for natural and environmentally friendly products, making them a sustainable and safe option for different applications.

However, there are challenges to large-scale production and commercialization of microbial pigments, such as downstream processing, purification, genetic engineering, and process optimization. It is difficult to efficiently detect and isolate pigments because the fermentation broths are complex. To minimize cost and reduce environmental impact, advances in green chemistry and sustainable purification methods, including solvent-free extraction and membrane filtration, are critical. By integrating bioprocess intensification with optimization, the overall production pipeline can become more scalable, efficient, and environmentally benign. Existing limitations in microbial pigment production could be addressed through application of biotechnological approaches like metabolic engineering, synthetic biology, and omics technologies. To optimize the production of complex pigments, enhance pigment titers, and engineer microbial strains for the production of specific colorants, scientists may modify microbial processes. These advances can also facilitate the production of new pigments possessing novel characteristics, such as greater stability, more vibrant color, or resistance to degradation in different environmental conditions.

The regulatory environment for microbial pigments is evolving as countries including the US, EU, and India establish protocols for the safe use of such colorants in medicines, food, and cosmetics. Global regulatory harmonization remains a problem, however, and routine safety assessments must be conducted to ensure microbial pigments meet the standards for environmental sustainability and human health.

References

Abdi, G., Tendulkar, R., Thatte, C., Mishra, S., Desai, B. *et al.* (2024) Scaling up nature's chemistry: A guide to industrial production of valuable metabolites. In: *Advances in Metabolomics*. Springer Nature Singapore, Singapore, pp. 331–375. DOI: 10.1007/978-981-97-7459-3_15.

Agarwal, H., Bajpai, S., Mishra, A., Kohli, I., Varma, A. *et al.* (2023) Bacterial pigments and their multifaceted roles in contemporary biotechnology and pharmacological applications. *Microorganisms* 11(3), 614. DOI: 10.3390/microorganisms11030614.

Albahnasawi, A., Alazaiza, M.Y., Yüksel, E., Tekbaş, M., Gürbulak, E. *et al.* (2025) The global scenario for management of dyes: Recent advances in natural and man-made dyes. *Advances in Natural Dyes for Environmental Protection* 279–305. DOI: 10.1201/9781003473466.

Al-Sayaghi, A.M., Al-Kabsi, A.M., Abduh, M.S., Saghir, S.A.M. and Alshawsh, M.A. (2022) Antibacterial mechanism of action of two types of honey against *Escherichia coli* through interfering with bacterial membrane permeability, inhibiting proteins, and inducing bacterial DNA damage. *Antibiotics* 11(9), 1182. DOI: 10.3390/antibiotics11091182.

Aman Mohammadi, M., Ahangari, H., Mousazadeh, S., Hosseini, S.M. and Dufossé, L. (2022) Microbial pigments as an alternative to synthetic dyes and food additives: A brief review of recent studies. *Bioprocess and Biosystems Engineering* 45(1), 1–12. DOI: 10.1007/s00449-021-02621-8.

Anshi, G., Kapil, S., Goswami, L. and Sharma, V. (2024) Unveiling the intricacies of microbial pigments as sustainable alternatives to synthetic colorants: Recent trends and advancements. *Micro* 4(4), 621–640. DOI: 10.3390/micro4040038.

Bhatia, S.K., Bhatia, R.K., Choi, Y.K., Kan, E., Kim, Y.G. *et al.* (2018) Biotechnological potential of microbial consortia and future perspectives. *Critical Reviews in Biotechnology* 38(8), 1209–1229. DOI: 10.1080/07388551.2018.1471445.

Budzianowska, A., Banaś, K., Budzianowski, J. and Kikowska, M. (2025) Antioxidants to defend healthy and youthful skin—current trends and future directions in cosmetology. *Applied Sciences* 15(5), 2571. DOI: 10.3390/app15052571.

Casas, M.C. and Chinoperekweyi, J. (2019) Color psychology and its influence on consumer buying behavior: A case of apparel products. *Saudi Journal of Business and Management Studies* 4(5), 441–456. DOI: 10.21276/sjbms.2019.4.5.8.

Chatragadda, R. and Dufossé, L. (2021) Ecological and biotechnological aspects of pigmented microbes: A way forward in development of food and pharmaceutical grade pigments. *Microorganisms* 9(3), 637. DOI: 10.3390/microorganisms9030637.

Choi, K.R., Shin, J.H., Cho, J.S., Yang, D. and Lee, S.Y. (2016) Systems metabolic engineering of escherichia coli. *EcoSal Plus* 7(1). DOI: 10.1128/ecosalplus.ESP-0010-2015.

Chwil, M., Mihelič, R., Matraszek-Gawron, R., Terlecka, P., Skoczylas, M.M. *et al.* (2024) Comprehensive review of the latest investigations of the health-enhancing effects of selected properties of *Arthrospira* and *Spirulina microalgae* on skin. *Pharmaceuticals* 17(10), 1321. DOI: 10.3390/ph17101321.

Daly, G., Ghini, V., Adessi, A., Fondi, M., Buchan, A. *et al.* (2022) Towards a mechanistic understanding of microalgae-bacteria interactions: Integration of metabolomic analysis and computational models. *FEMS Microbiology Reviews* 46(5), fuac020. DOI: 10.1093/femsre/fuac020.

da Silva, T.L. and Reis, A. (2015) Scale-up problems for the large scale production of algae. In: *Algal Biorefinery: An Integrated Approach*. Springer, Cham, Switzerland, pp. 125–149. DOI: 10.1007/978-3-319-22813-6_6.

Dave, S., Das, J., Varshney, B. and Sharma, V.P. (2022) Dyes and pigments: Interventions and how safe and sustainable are colors of life. In: *Trends and Contemporary Technologies for Photocatalytic Degradation of Dye*. Springer, Cham, Switzerland, pp. 1–20. DOI: 10.1007/978-3-031-08991-6_1.

Degano, I. (2019) Liquid chromatography: Current applications in heritage science and recent developments. *Physical Sciences Reviews* 4(5), 20180009. DOI: 10.1515/psr-2018-0009.

Devi, M., Ramakrishnan, E., Deka, S. and Parasar, D.P. (2024) Bacteria as a source of biopigments and their potential applications. *Journal of Microbiological Methods* 219, 106907. DOI: 10.1016/j.mimet.2024.106907.

Dhara, D., Biswas, S., Das, S.K. and Biswas, O. (2021) Status of food safety and food security in india in the perspective of FSSAI. *Indian Journal of Animal Health* 60(2), 167–173. DOI: 10.36062/ijah.2021.spl.01821.

Dufossé, L. (2018) Microbial pigments from bacteria, yeasts, fungi, and microalgae for the food and feed industries. In: *Natural and Artificial Flavoring Agents and Food Dyes*. Academic Press, New York, pp. 113–132. DOI: 10.1016/B978-0-12-811518-3.00004-1.

Elnahal, A.S.M., El-Saadony, M.T., Saad, A.M., Desoky, E.-S.M., El-Tahan, A.M. *et al.* (2022) The use of microbial inoculants for biological control, plant growth promotion, and sustainable agriculture: A review. *European Journal of Plant Pathology* 162(4), 759–792. DOI: 10.1007/s10658-021-02393-7.

Ghosh, S., Sarkar, T., Das, A. and Chakraborty, R. (2022) Natural colorants from plant pigments and their encapsulation: An emerging window for the food industry. *LWT* 153, 112527. DOI: 10.1016/j.lwt.2021.112527.

Gómez, B.B., López-Cortés, R., Casas-Nebra, F.J., Vázquez-Estévez, S., Pérez-Fentes, D. *et al.* (2021) Detection of circulating serum protein biomarkers of non-muscle invasive bladder cancer after protein corona-silver nanoparticles analysis by SWATH-MS. *Nanomaterials* 11(9), 2384. DOI: 10.3390/nano11092384.

Gonçalves, T. and Vasconcelos, U. (2021) Colour me blue: The history and the biotechnological potential of pyocyanin. *Molecules* 26(4), 927. DOI: 10.3390/molecules26040927.

Gupta, I., Adin, S.N., Panda, B.P. and Mujeeb, M. (2022) β-Carotene-production methods, biosynthesis from Phaffia rhodozyma, factors affecting its production during fermentation, pharmacological properties: A review. *Biotechnology and Applied Biochemistry* 69(6), 2517–2529. DOI: 10.1002/bab.2301.

Gupta, N., Wijenayake, W.P.T., Roy, D., Kumar, R., Rangot, M. *et al.* (2025) Smart recycling and sustainable lignocellulosic waste management. In: *Value Addition and Utilization of Lignocellulosic Biomass: Through Novel Technological Interventions*. Springer Nature Singapore, Singapore, pp. 221–250. DOI: 10.1007/978-981-96-2786-8_9.

Hamidi, M., Kozani, P.S., Kozani, P.S., Pierre, G., Michaud, P. *et al.* (2019) Marine bacteria versus microalgae: Who is the best for biotechnological production of bioactive compounds with antioxidant properties and other biological applications? *Marine Drugs* 18(1), 28. DOI: 10.3390/md18010028.

Heidari, S., Sadatshojaei, E. and Wood, D.A. (2021) Large-scale molecular solvents for environmentally sustainable applications. In: *Green Sustainable Process for Chemical and Environmental Engineering and Science*. Elsevier, Amsterdam, The Netherlands, pp. 267–282. DOI: 10.1016/B978-0-12-821884-6.00001-2.

Huang, W., Wan, Y., Su, H., Zhang, Z., Liu, Y. *et al.* (2024) Recent advances in phenazine natural products: Biosynthesis and metabolic engineering. *Journal of Agricultural and Food Chemistry* 72(39), 21364–21379. DOI: 10.1021/acs.jafc.4c05294.

Intasian, P., Prakinee, K., Phintha, A., Trisrivirat, D., Weeranoppanant, N. *et al.* (2021) Enzymes, *in vivo* biocatalysis, and metabolic engineering for enabling a circular economy and sustainability. *Chemical Reviews* 121(17), 10367–10451. DOI: 10.1021/acs.chemrev.1c00121.

Islam, T., Repon, M.R., Islam, T., Sarwar, Z. and Rahman, M.M. (2023) Impact of textile dyes on health and ecosystem: A review of structure, causes, and potential solutions. *Environmental Science and Pollution Research International* 30(4), 9207–9242. DOI: 10.1007/s11356-022-24398-3.

Jain, S., Kumar, R., Gunsola, D., Živković, S., Vasić, T. *et al.* (2025) Strigolactones and control of parasitic weeds. *Strigolactones: Emerging Plant Hormones* 261–279. DOI: 10.1002/9781394302826.ch12.

Kagot, V., Okoth, S., De Boevre, M. and De Saeger, S. (2019) Biocontrol of *Aspergillus* and *Fusarium* mycotoxins in Africa: Benefits and limitations. *Toxins* 11(2), 109. DOI: 10.3390/toxins11020109.

Kamble, G.R., Hiremath, G.B., Hiremath, S.V. and Hiremath, M.B. (2022) Bacterial pigments: An untapped colorful microbial world. In: *Enzymes for Pollutant Degradation*. Springer Nature Singapore, Singapore, pp. 285–307. DOI: 10.1007/978-981-16-4574-7_15.

Kamboj, N., Chugh, P., Wijenayake, W.P.T., Mitra, D., Panneerselvam, P. *et al.* (2025) Multifunctionality and diversity of antagonistic potential fungi as biocontrol agent. In: Mitra, D., de los Santos Villalobos, S., Rani, A., Elena Guerra Sierra, B. and Andjelković, S. (eds) *Bio-Control Agents for Sustainable Agriculture*. Springer, Singapore. DOI: 10.1007/978-981-96-3232-9_9.

Kato, S., Takaichi, S., Ishikawa, T., Asahina, M., Takahashi, S. *et al.* (2016) Identification and functional analysis of the geranylgeranyl pyrophosphate synthase gene (crtE) and phytoene synthase gene (crtB) for carotenoid biosynthesis in Euglena gracilis. *BMC Plant Biology* 16, 4. DOI: 10.1186/s12870-015-0698-8.

Ko, Y.S., Kim, J.W., Lee, J.A., Han, T., Kim, G.B. *et al.* (2020) Tools and strategies of systems metabolic engineering for the development of microbial cell factories for chemical production. *Chemical Society Reviews* 49(14), 4615–4636. DOI: 10.1039/d0cs00155d.

Krishnakripa, P.K. and Thoppil, J.E. (2025) Beyond a pigment – a review on the beneficial aspects of anthocyanins, chlorophylls, carotenoids, and betalains. *Vegetos* 1–11. DOI: 10.1007/s42535-025-01166-5.

Kumar, R., Gupta, G., Hussain, A., Rani, A., Thapliyal, A. *et al.* (2025a) Pioneering zero-waste technologies utilization and its framework on sustainable management: International, national and state level. *Discover Applied Sciences* 7(3), 224. DOI: 10.1007/s42452-025-06693-z.

Kumar, R., Lakhera, A., Gulati, M., Rani, A., Thapliyal, A. *et al.* (2025b) Genome mining of biocontrol agents for bioprospecting action modes. In: Mitra, D., de los Santos Villalobos, S., Rani, A., Elena Guerra Sierra, B. and Andjelković, S. (eds) *Bio-Control Agents for Sustainable Agriculture*. Springer, Singapore. DOI: 10.1007/978-981-96-3232-9_3.

Lagarde, D., Beuf, L. and Vermaas, W. (2000) Increased production of zeaxanthin and other pigments by application of genetic engineering techniques to Synechocystis sp. strain PCC 6803. *Applied and Environmental Microbiology* 66(1), 64–72. DOI: 10.1128/AEM.66.1.64-72.2000.

Liang, M.H., Zhu, J. and Jiang, J.G. (2018) Carotenoids biosynthesis and cleavage related genes from bacteria to plants. *Critical Reviews in Food Science and Nutrition* 58(14), 2314–2333. DOI: 10.1080/10408398.2017.1322552.

Li, X., Liu, C., Li, Y., Yuan, K., Zhang, W. *et al.* (2023) Bioactivity and application of anthocyanins in skin protection and cosmetics: An extension as a functional pigment. *Phytochemistry Reviews* 22(6), 1441–1467. DOI: 10.1007/s11101-023-09890-z.

Lyu, X., Lyu, Y., Yu, H., Chen, W., Ye, L. *et al.* (2022) Biotechnological advances for improving natural pigment production: A state-of-the-art review. *Bioresources and Bioprocessing* 9(1), 8. DOI: 10.1186/s40643-022-00497-4.

Madan, A., Kumar, R., Garg, R., Chugh, P., Chattaraj, S. *et al.* (2025) Revolutionizing insights from genes fundamental role of data science in bioinformatics and healthcare. *Indian Journal of Microbiology* 12(1), 21–33. DOI: 10.18231/j.ijmr.2025.004.

Mahakur, B., Bedaprabha, A., Barik, D.P., Gupta, S., Kumar, R. *et al.* (2025) Hairy root culture: An effective method for production of plant bioactive compounds. In: *Genetic Improvement and Conservation Practices of Medicinal Plants*. CRC Press, Boca Raton, FL, p. 9.

Mandree, P., Thopil, G.A. and Ramchuran, S. (2023) Potential opportunities to convert waste to bio-based chemicals at an industrial scale in South Africa. *Fermentation* 9(10), 908. DOI: 10.3390/fermentation9100908.

Massoud, R. and Khosravi-Darani, K. (2017) A review on the impacts of process variables on microbial production of carotenoid pigments. *Food Biosynthesis* 183–211. DOI: 10.1016/B978-0-12-811372-1.00006-3.

Mavrodi, D.V., Parejko, J.A., Mavrodi, O.V., Kwak, Y.S., Weller, D.M. *et al.* (2013) Recent insights into the diversity, frequency and ecological roles of phenazines in fluorescent *Pseudomonas* spp. *Environmental Microbiology* 15(3), 675–686. DOI: 10.1111/j.1462-2920.2012.02846.x.

McLean, T.C., Lo, R., Tschowri, N., Hoskisson, P.A., Al Bassam, M.M. *et al.* (2019) Sensing and responding to diverse extracellular signals: An updated analysis of the sensor kinases and response regulators of streptomyces species. *Microbiology* 165(9), 929–952. DOI: 10.1099/mic.0.000817.

Michael, H.S.R., Subiramanian, S.R., Thyagarajan, D., Mohammed, N.B., Saravanakumar, V.K. *et al.* (2023) Melanin biopolymers from microbial world with future perspectives–a review. *Archives of Microbiology* 205(9), 306. DOI: 10.1007/s00203-023-03642-5.

Minić, S., Gligorijević, N., Veličković, L. and Nikolić, M. (2024) Narrative review of the current and future perspectives of Phycobiliproteins' applications in the food industry: From natural colors to alternative proteins. *International Journal of Molecular Sciences* 25(13), 7187. DOI: 10.3390/ijms25137187.

Mitra, D., Kumar, R. and Kamboj, N. (2024) Microbial innovations in agriculture: Interdisciplinary approaches to leveraging microbes for food sustainability and security. *Indian Journal of Microbiology Research* 11(3), 129–139. DOI: 10.18231/j.ijmr.2024.026.

Mohamed, S.A.K.S., Sabita, U., Rajendra, S. and Raman, D. (2017) Genotoxicity: Mechanisms, testing guidelines and methods. *Global Journal of Pharmacy & Pharmaceutical Sciences* 1(5), 133–138. DOI: 10.19080/GJPPS.2017.02.555575.

Nawab, S., Wang, N., Ma, X. and Huo, Y.X. (2020) Genetic engineering of non-native hosts for 1-butanol production and its challenges: A review. *Microbial Cell Factories* 19(1), 79. DOI: 10.1186/s12934-020-01337-w.

Naz, T., Ullah, S., Nazir, Y., Li, S., Iqbal, B. *et al.* (2023) Industrially important fungal carotenoids: Advancements in biotechnological production and extraction. *Journal of Fungi* 9(5), 578. DOI: 10.3390/jof9050578.

Nemer, G., Louka, N., Vorobiev, E., Salameh, D., Nicaud, J.M. *et al.* (2021) Mechanical cell disruption technologies for the extraction of dyes and pigments from microorganisms: A review. *Fermentation* 7(1), 36. DOI: 10.3390/fermentation7010036.

Nigam, P.S. and Luke, J.S. (2016) Food additives: Production of microbial pigments and their antioxidant properties. *Current Opinion in Food Science* 7, 93–100. DOI: 10.1016/j.cofs.2016.02.004.

Olaniran, A.F., Folorunsho, J.O., Akinsanola, B.A., Taiwo, A.E., Iranloye, Y.M. *et al.* (2023) Application of astaxanthin and carotenoids derived from algae for the production of nutraceuticals, pharmaceuticals, additives, food supplement and feed. *Next-Generation Algae* 2, 95–124. DOI: 10.1002/9781394194704.

Omar, A.E., Al-Khalaifah, H.S., Osman, A., Gouda, A., Shalaby, S.I. *et al.* (2022) Modulating the growth, antioxidant activity, and immunoexpression of proinflammatory cytokines and apoptotic proteins in broiler chickens by adding dietary *spirulina platensis* phycocyanin. *Antioxidants* 11(5), 991. DOI: 10.3390/antiox11050991.

Orlandi, V.T., Martegani, E., Giaroni, C., Baj, A. and Bolognese, F. (2022) Bacterial pigments: A colorful palette reservoir for biotechnological applications. *Biotechnology and Applied Biochemistry* 69(3), 981–1001. DOI: 10.1002/bab.2170.

Paillié-Jiménez, M.E., Stincone, P. and Brandelli, A. (2020) Natural pigments of microbial origin. *Frontiers in Sustainable Food Systems* 4, 590439. DOI: 10.3389/fsufs.2020.590439.

Pavan, M.E., López, N.I. and Pettinari, M.J. (2020) Melanin biosynthesis in bacteria, regulation and production perspectives. *Applied Microbiology and Biotechnology* 104(4), 1357–1370. DOI: 10.1007/s00253-019-10245-y.

Peyrat, L.A., Tsafantakis, N., Georgousaki, K., Ouazzani, J., Genilloud, O. *et al.* (2019) Terrestrial microorganisms: Cell factories of bioactive molecules with skin protecting applications. *Molecules* 24(9), 1836. DOI: 10.3390/molecules24091836.

Poorniammal, R., Prabhu, S., Dufossé, L. and Kannan, J. (2021) Safety evaluation of fungal pigments for food applications. *Journal of Fungi* 7(9), 692. DOI: 10.3390/jof7090692.

Qin, Y. and Xia, Y. (2024) Melanin in fungi: Advances in structure, biosynthesis, regulation, and metabolic engineering. *Microbial Cell Factories* 23(1), 334. DOI: 10.1186/s12934-024-02614-8.

Rajendran, P., Somasundaram, P. and Dufossé, L. (2023) Microbial pigments: Eco-friendly extraction techniques and some industrial applications. *Journal of Molecular Structure* 1290, 135958. DOI: 10.1016/j.molstruc.2023.135958.

Ramadori, G.P. (2023) Organophosphorus poisoning: Acute Respiratory Distress Syndrome (ARDS) and cardiac failure as cause of death in hospitalized patients. *International Journal of Molecular Sciences* 24(7), 6658. DOI: 10.3390/ijms24076658.

Ramesh, C., Prasastha, V.R., Venkatachalam, M. and Dufossé, L. (2022) Natural substrates and culture conditions to produce pigments from potential microbes in submerged fermentation. *Fermentation* 8(9), 460. DOI: 10.3390/fermentation8090460.

Rao, A.S., Deka, S.P., More, S.S., Nair, A., More, V.S. *et al.* (2021) A comprehensive review on different microbial-derived pigments and their multipurpose activities. *Microbial Polymers: Applications and Ecological Perspectives* 479–519. DOI: 10.1007/978-981-16-0045-6_20.

Rather, L.J., Mir, S.S. and Islam, Z. (2025) Introduction to microbial colorants. *Microbial Colorants: Chemistry, Biosynthesis and Applications* 1–32. DOI: 10.1002/9781394287888.

Renita, A.A., Gajaria, T.K., Sathish, S., Kumar, J.A., Lakshmi, D.S. *et al.* (2023) Progress and prospective of the industrial development and applications of ECO-friendly colorants: An insight into environmental impact and sustainability issues. *Foods* 12(7), 1521. DOI: 10.3390/foods12071521.

Riaz, M., Zia-Ul-Haq, M. and Dou, D. (2021) Chemistry of carotenoids. In: *Carotenoids: Structure and Function in the Human Body*. Springer, Cham, Switzerland, pp. 43–76. DOI: 10.1007/978-3-030-46459-2_2.

Sadare, O.O., Ejekwu, O., Moshokoa, M.F., Jimoh, M.O. and Daramola, M.O. (2021) Membrane purification techniques for recovery of succinic acid obtained from fermentation broth during bioconversion of lignocellulosic biomass: Current advances and future perspectives. *Sustainability* 13(12), 6794. DOI: 10.3390/su13126794.

Sahoo, S.R., Pradhan, A.K., Das, R.P., Panigrahi, L.L. and Arakha, M. (2023) Pyocyanin is the microbial blue-green pigment: A review on its history, virulence, and therapeutic use. *Current Bioactive Compounds* 19(6), 62–76. DOI: 10.2174/1573407219666221209092023.

Saini, D.K., Chakdar, H., Pabbi, S. and Shukla, P. (2020) Enhancing production of microalgal biopigments through metabolic and genetic engineering. *Critical Reviews in Food Science and Nutrition* 60(3), 391–405. DOI: 10.1080/10408398.2018.1533518.

Salian, A., Dutta, S. and Mandal, S. (2021) A roadmap to UV-protective natural resources: Classification, characteristics, and applications. *Materials Chemistry Frontiers* 5(21), 7696–7723. DOI: 10.1039/D1QM00741F.

Samal, S., Kumar, R., Thapliyal, M., Gunsola, D., Apaarna, K. *et al.* (2025) Insight into an effective development of biocontrol agent formulation for commercial production. In: Mitra, D., de los Santos Villalobos, S., Rani, A., Elena Guerra Sierra, B. and Andjelković, S. (eds) *Bio-Control Agents for Sustainable Agriculture*. Springer, Singapore, pp. 385–407. DOI: 10.1007/978-981-96-3232-9_16.

Scanes, C.G. (2018) Invertebrates and their use by humans. In: *Animals and Human Society*. Academic Press, New York. DOI: 10.1016/B978-0-12-805247-1.00012-5.

Shahid, M., Zaidi, A., Khan, M.S., Rizvi, A., Saif, S. *et al.* (2017) Recent advances in management strategies of vegetable diseases. *Microbial Strategies for Vegetable Production* 197–226. DOI: 10.1007/978-3-319-54401-4_9.

Solano, F. (2020) Photoprotection and skin pigmentation: Melanin-related molecules and some other new agents obtained from natural sources. *Molecules* 25(7), 1537. DOI: 10.3390/molecules25071537.

Sun, H., Wang, Y., He, Y., Liu, B., Mou, H. *et al.* (2023) Microalgae-derived pigments for the food industry. *Marine Drugs* 21(2), 82. DOI: 10.3390/md21020082.

Suthar, M., Dufossé, L. and Singh, S.K. (2023) The enigmatic world of fungal melanin: A comprehensive review. *Journal of Fungi* 9(9), 891. DOI: 10.3390/jof9090891.

Tang, Q., Li, Z., Chen, N., Luo, X. and Zhao, Q. (2025) Natural pigments derived from plants and microorganisms: Classification, biosynthesis, and applications. *Plant Biotechnology Journal* 23(2), 592–614. DOI: 10.1111/pbi.14522.

Tianqiao, S., Xiong, Z., You, Z., Dong, L., Jiaoling, Y. *et al.* (2021) Genome-wide identification of Zn2Cys6 class fungal-specific transcription factors (ZnFTFs) and functional analysis of UvZnFTF1 in ustilaginoidea virens. *Rice Science* 28(6), 567–578. DOI: 10.1016/j.rsci.2021.03.001.

Tkaczyk, A., Mitrowska, K. and Posyniak, A. (2020) Synthetic organic dyes as contaminants of the aquatic environment and their implications for ecosystems: A review. *Science of the Total Environment* 717, 137222. DOI: 10.1016/j.scitotenv.2020.137222.

Usmani, Z., Sharma, M., Sudheer, S., Gupta, V.K. and Bhat, R. (2020) Engineered microbes for pigment production using waste biomass. *Current Genomics* 21(2), 80–95. DOI: 10.2174/1389202921999200330152007.

van Belkum, A., Bachmann, T.T., Lüdke, G., Lisby, J.G., Kahlmeter, G. *et al.* (2019) Developmental roadmap for antimicrobial susceptibility testing systems. *Nature Reviews Microbiology* 17(1), 51–62. DOI: 10.1038/s41579-018-0098-9.

Venil, C.K., Dufossé, L. and Renuka Devi, P. (2020) Bacterial pigments: Sustainable compounds with market potential for pharma and food industry. *Frontiers in Sustainable Food Systems* 4, 100. DOI: 10.3389/fsufs.2020.00100.

Vilaplana, L. and Marco, M.P. (2020) Phenazines as potential biomarkers of pseudomonas aeruginosa infections: Synthesis regulation, pathogenesis and analytical methods for their detection. *Analytical and Bioanalytical Chemistry* 412(24), 5897–5912. DOI: 10.1007/s00216-020-02696-4.

World Health Organization (2019) *Evaluation of Certain Food Additives: Eighty-sixth Report of the Joint FAO/WHO Expert Committee on Food Additives*, Vol. 965. World Health Organization, Geneva, Switzerland.

8 Isolation and Purification Techniques for Microbial Pigments

Mary Malilah, Uma Thakur, Mohammedfaizan Shaikh and Krishan Kumar*
Department of Life Sciences, Parul Institute of Applied Sciences, Parul University, Vadodara, Gujarat, India

8.1 Introduction

The term "microbial pigments" refers to bioactive, natural pigments produced by microorganisms, such as bacteria, fungi, and algae, which have functional characteristics and economic value. Microbial pigments are green, nontoxic, and biodegradable substitutes for synthetic pigments that are harmful to human health and the environment (Venil *et al.*, 2020).

Microbial pigments can be categorized according to their biological sources and chemical compositions. Along with their colorant properties, these pigments are also known for their antioxidant and antimicrobial activities (Venil *et al.*, 2020). Fungal pigments, such as melanins, flavins, and azaphilones, are found in common organisms such as *Monascus* and *Aspergillus* and are beneficial in food items and drugs (Lagashetti *et al.*, 2019). Algal pigments, such as chlorophylls, carotenoids, and phycobiliproteins, are primarily synthesized by microalgae, such as *Spirulina* and *Dunaliella*, and are commonly used in the food, textile, cosmetic, and pharmaceutical industries (Patel *et al.*, 2022). Microbial pigments have a variety of biological activities, such as antibacterial, antioxidant, and anticancer activities, and their application in various industries such as food, healthcare, textiles, and cosmetics is increasing at a very rapid rate (Soliev *et al.*, 2011).

The growing demand for natural pigments has led to radical advancements in the methods applied for their extraction and purification to improve their purity, stability, and application in various industries. The variable chemical composition of the pigments influences their solubility, stability, and functional properties, and hence extraction and purification methods are required using appropriate extraction and purification methods appropriate to the specific microbial origin. Raw pigment extracts contain impurities such as cellular detritus, proteins, and secondary metabolites, which may affect the functionality, stability, and safety of products (Venil *et al.*, 2020). The food and pharmaceutical industries require high purity levels of pigments according to regulatory criteria, whereas textile industries can sustain lower purity pigments (Tiwari, 2022). Additionally, the purified pigments were found to be least affected by external variables (light, oxygen, and temperature) along with extended shelf-life and bioactivity of products (Dufossé and Tiwari, 2024).

Various methods for the extraction and purification of microbial pigments have

*Corresponding author: krishan.kumar20140@paruluniversity.ac.in

DOI: 10.1079/ 9781836990505.0008

been reported. Solvent extraction, ultrasonic extraction, microwave-assisted extraction, supercritical fluid extraction, enzymatic-assisted extraction, ionic liquid-assisted extraction, and pulsed electric field-assisted extraction are commonly used extraction techniques for pigments.

Extraction techniques depend on the source and type of pigment used to maximize the yield and quality of the pigments. For instance, intracellular bacterial pigments require cell disruption methods, such as enzymatic lysis or sonication, for efficient recovery (Nemer *et al.*, 2021). Pigments from algae, such as carotenoids and phycobilins, can be extracted using organic solvents or pressurized liquid extraction (Pagels *et al.*, 2021). Fungal pigments are commonly extracellular in nature so they can be extracted through simple filtration and solvent partitioning (Sharma *et al.*, 2012; Wang *et al.*, 2018).

Along with the extraction techniques, this chapter also discusses the various chromatographic techniques used in the purification of pigments. Preparative thin-layer chromatography (PTLC), high-performance liquid chromatography (HPLC), ion exchange chromatography, and size exclusion chromatography are used to purify the pigments. In large-scale industrial applications, nonchromatographic purification techniques are preferred because of their simplicity, economy, and scalability. Methods such as crystallization, precipitation, and liquid-liquid partitioning are elaborated upon, with emphasis placed on their concepts and practical applications. For instance, crystallization has been highlighted for its ability to purify pigments, such as violacein and astaxanthin, by exploiting differences in solubility (Rettori and Durán, 1998; Sun *et al.*, 2014). The ammonium sulfate precipitation method is explored for use in pigment purification and concentration, e.g. for phycobiliproteins (Sonani *et al.*, 2014). Liquid-liquid partitioning, as a solvent extraction procedure, is a flexible technique for pigment separation based on differences in solubility.

By combining the development of these extraction and purification methods, this chapter provides an exhaustive manual for researchers, students, and industries. It is not only a discussion of the theoretical principle of each technique but also demonstrates their practical application through examples. Whether the aim is carotenoid extraction from algal material, violacein purification from bacterial culture, or astaxanthin refining for pharmaceutical applications, this chapter provides readers with the knowledge to make an informed decision and to optimize the most appropriate techniques for their specific projects.

Finally, this chapter acts as a bridge connecting traditional practices with the latest breakthroughs in microbial pigment processing. By describing emerging technologies, it puts into context the advancements that make the future of natural pigment manufacturing feasible. The general overview described here will be a guide for individuals in this rapidly changing field of activity, calling for further research and innovation of eco-friendly pigment substitutes.

8.2 Pigment Extraction Techniques

Microbial pigment recovery is an important process for obtaining bio-based colorants for a number of applications. Microbial cell disruption releases pigments from microbial cells to maintain their stability and functional characteristics. The recovery process is influenced by parameters including solubility, polarity, and tolerance to environmental factors, such as temperature and pH. The different pigment extraction techniques with their optimized conditions and yields from various microbes are listed in Table 8.1.

8.2.1 Solvent liquid extraction

Solvent extraction is one of the most studied techniques for the extraction of microbial pigments. It works on the principle of the differential solubility of pigments between two immiscible phases: the aqueous and organic phases (Hessel *et al.*, 2022). Microbial pigments are usually lipophilic in nature so they prefer an organic solvent over the aqueous phase that contains the microbial biomass (Fig. 8.1A). The partitioning of pigments between two solvents depends on the Nernst Distribution Law, which states that in equilibrium, the ratio of the concentration of a solute in two immiscible solvents

Table 8.1. Ccomparative analysis of various pigments obtained using different extraction techniques from various microbes.

Pigment	Microorganism	Name	Extraction technique	Condition	Yield	Reference
Carotenoids	Bacteria	*Arthrospira platensis*	Microwave-assisted extraction	P: 400 W; Pr: 1 bar; S1: ammonia acetate, ethanol; S2: limonene, ethyl acetate	123.6	(Esquivel-Hernández *et al*., 2017)
		Synechococcus spp.	Supercritical fluid extraction	S: SC-CO_2; Pr: 300 bar; T: 500C; t: 3 hour	1.511	(Macías- Sánchez *et al*., 2010)
		Gordonia alkanivorans	Ionic liquid extraction	S: tributyl phosphonium diethyl phosphate, ethyl acetate; T: 220C; t: overnight	171	(Silva *et al*., 2024)
		Streptomyces hyderabadensis	Ionic liquid extraction	S: TBAOH; T: room temp; t: 1 hour	5548	(Ghadge *et al*., 2022)
	Fungi	*Rhodotorula glutinis*	Ultrasound-assisted extraction	S: water; t: 180 min; T: 300C; frequency: 20 kHz		(Martínez *et al*., 2020a)
		Rhodotorula glutinis CCT- 2186	Ionic liquid extraction	S: ammonium-based PIL; T: 250–650C; t: 1 hour	28.97	(Mussagy *et al*., 2019)
	Algae	*Spirulina platensis*	Ultrasound-assisted extraction	S: methanol; t: 8 min; T: 300C; I: 167 W/cm^2	47.10	(Dey and Rathod, 2013)
		Chlorella vulgaris	Microwave-assisted extraction	P: 300 W; t: 14 min; T: 600C; S: ethanol 63.36	24.88	(Georgiopoulou *et al*., 2023)
		Dunaliella salina	Supercritical fluid extraction	S: SC-CO_2; P: 443 bar; T: 27.50C; t: 20 min	0.429	(Jaime *et al*., 2007)
		Nannochloropsis oculata	Supercritical fluid extraction	S: SC-CO_2, ethanol; P: 350 bar; T: 323K	7.61	(Liau *et al*., 2010)
		Scenedesmus almeriensis	Supercritical fluid extraction	S: SC-CO_2; P: 400 bar; T: 600C	1.5	(Macías- Sánchez *et al*., 2010)
		Porphyridium cruentum	Pulsed electric field extraction	e: 1 Kv/cm, 80 pulses, 400 kJ/kg	7.69	(Martínez *et al*., 2019a)

Continued

Table 8.1. Continued

Pigment	Microorganism	Name	Extraction technique	Condition	Yield	Reference
Astaxanthin	Fungi	*Xanthophyllomyces dendrorhous*	Ultrasound-assisted extraction	S: DMSO; T: 300C; t: 56 min	82	(Unagul *et al.*, 2005)
		P. rhodozyma	Ionic liquid extraction	S: cholinium- based ionic liquids; t: 1 hour; T: 650C	47.9	(Mussagy *et al.*, 2021)
		Xanthophyllomyces dendrorhou s	Pulsed electric field extraction	e: 20 25 kv/cm, 135 μs; T: 250C; S: ethanol	2.4	Aguilar-Machado *et al.*, 2020)
	Algae	*H. pluvialis*	Ultrasound-assisted extraction	S: ethanol/ethyl acetate (1/1, v/v); t: 16 min; T: 41.10C; P: 200 W	27.58	(Zou *et al.*, 2013)
		H. pluvialis	Supercritical fluid extraction	S: SC-CO_2; P: 550bar; T: 500C; t: 100 min	237.4	(Di Sanzo *et al.*, 2018)
		H. pluvialis	Pulsed electric field extraction	ten pulses of 5μs; S: ethanol; T: 200C	18.3	(Di Sanzo *et al.*, 2018)
Chlorophyll	Bacteria	*Phormidium autumnale*	Ultrasound-assisted extraction	S: ethanol; t: 24.5 min; T: 44.40C	46.3	(Zhu *et al.*, 2017)
		Spirulina platensis	Microwave-assisted extraction	P: 560 W; t: 52.8 min; T: 55.10C; S: ethanol	1.27	(Tong *et al.*, 2012)
	Algae	*Chlorella vulgaris*	Microwave-assisted extraction	P: 300 W; t: 14 min; T: 600C; S: ethanol 63.36	24:88	(Georgiopoulou *et al.*, 2023)
Phycobiliproteins	Bacteria	*Arthrospira platensis*	Microwave-assisted extraction	P:400 W; Pr: 1 bar; S1: ammonia acetate, ethanol; S2: limonene, ethyl acetate	2.28	(Esquivel-Hernández *et al.*, 2017)
		A. platensis	Enzyme-assisted extraction	S: lysozyme, surfactants; T: 370C; t: 20 h	32.27	(Tavanandi *et al.*, 2019)
		Nostoc commune	Pulsed electric	F: 2 Hz; T: 400C, 15000	29.66	(Chittapun *et al.*, 2020)

Continued

Table 8.1. Continued

Pigment	Microorganism	Name	Extraction technique	Condition	Yield	Reference
			field extraction	Pulses of 1 μs; e: 5kv; S: water		
	Algae	*Porphyridium purpureum*	Microwave-assisted extraction	P: 300 W; t: 10 sec; T: 400C	73.7	(Juin *et al*., 2015)
		Porphyra	Enzyme-assisted extraction	S: distilled water; T: 260C; t: 72 h; E: MA103, MAEF108	33.94	(Huang *et al*., 2021)
Torularhodin	Fungi	*Rhodotoru a* spp.	Supercritical fluid extraction	S: SC-CO_2, ethanol; P: 400 bar; T: 600C; t: 90 min	65.8	(Larocca *et al*., 2023)
		Rhodotorula glutinis	Pulsed electric field extraction	e: 15 kV/cm, 84 kJ/kg; S: ethanol		(Martínez *et al*., 2019b)
α-Tocopherol	Bacteria	*Arthrospira platensis*	Microwave-assisted extraction	P: 400 W; Pr: 1 bar; S1: ammonia acetate, ethanol; S2: limonene, ethyl acetate	37.87	(Esquivel-Hernández *et al*., 2017)
Thiamine	Bacteria	*Arthrospira platensis*	Microwave-assisted extraction	P: 400 W; Pr: 1 bar; S1: ammonia acetate, ethanol S2: limonene, ethyl acetate	845.57	(Esquivel-Hernández *et al*., 2017)

Continued

Table 8.1. Continued

Pigment	Microorganism	Name	Extraction technique	Condition	Yield	Reference
Riboflavin	Bacteria	*Arthrospira platensis*	Microwave-assisted extraction	P: 400 W; Pr: 1 bar; S1: ammonia acetate, ethanol S2: limonene, ethyl acetate	101.09	(Esquivel-Hernández *et al.*, 2017)
Melanin	Fungi	*Cryptococcocus neoformans*	Enzyme-assisted extraction	S: 1.0M sodium citrate, 0.1M sorbit ol; T: 300C; t: 12 h E: Trichoderma harzianum enzymes	Not specified	(Dadachova *et al.*, 2007)
Natural red	Fungi	*Penicillium purpurogenum*	Ionic liquid extraction	S: potassium citrate buffer; T: 250C; t: 12 h	4.64	(Ventura *et al.*, 2013)
Green pigment	Algae	*Spirulina platensis*	Ultrasound-assisted extraction	S: glycerol/water; t: 20 min; T: 600C; P: 240 W	165.19	(Martins *et al.*, 2023)
Fucoxanthin	Algae	*Undaria pinnatifida*	Enzyme-assisted extraction	S: dimethyl ether, ethanol; T: 370C; t: 19.5 min	0.7	(Billakanti *et al.*, 2013)
Lutein	Algae	*Scenedesmus almeriensis*	Supercritical fluid extraction	S: SC-CO2; P: 400 bar; T: 600C	1.5 0.0466	(Macías-Sánchez *et al.*, 2010)

S, solvent;T, temperature;t, time;P, power;F, frequency;DW, dry weight;S1, first solvent;S2, second solvent;Pr, pressure;E, enzymes;e, energy;

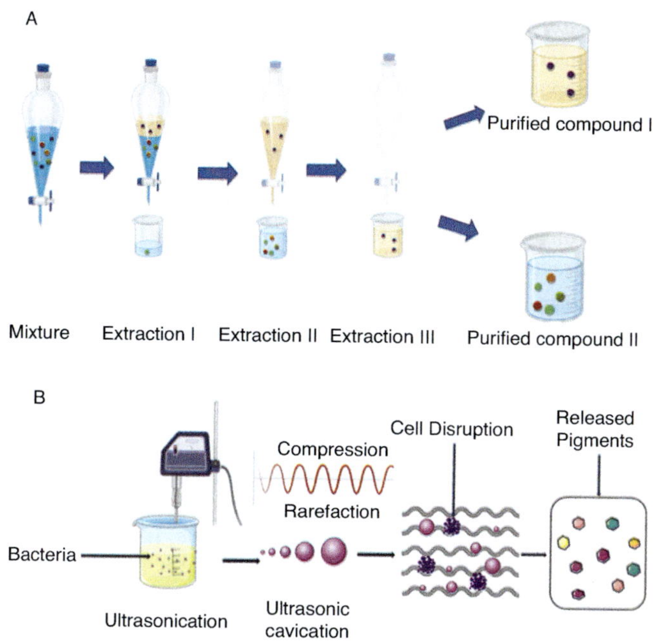

Fig. 8.1. (A) Solvent extraction: sequential partitioning of microbial pigments into an organic solvent using a separatory funnel. (B) Ultrasonic-assisted extraction: ultrasonic cavitation disrupts microbial cells, releasing intracellular pigments.

is constant (Arce *et al.*, 2022). Thus, organic solvents are chosen based on the higher Kd for the target pigment to maximize extraction.

Solvent liquid-assisted extraction (SLE) has been used for pigment extraction from bacteria, fungi, and microalgae. Phycobiliproteins were isolated from *Arthrospira platensis* using phosphate buffer at 25°C for 4 hour. Carotenoids were recovered from *Rhodotorula glutinis* and *Chlorella vulgaris* by using 1-butyl-3-methylimidazolium chloride as solvent at 53°C for 57 min and produced a yield of 1.311 mg/gDW (Zhiping *et al.*, 2014). Carotenoids from *Puccinia distincta* were extracted using ethyl acetate and cyclohexane as solvents, with a yield of 28.97 μg/gDW of β-carotene (Davoli, 2002). Astaxanthin was extracted from *P. rhodozyma* with cholinium-based ionic liquids solvent at 65°C for 1 hour and yielded 47.9μg/gDW (Mussagy *et al.*, 2021). The fermented broth of *Penicillium purpurogenum* was treated with potassium citrate ionic liquids at 25°C for 12 hour and yielded 4.64 μg/gDW for red colorants (Ventura *et al.*, 2013). The algal pigment fucoxanthin was extracted from *Isochrysis galbana* microalgae at room temperature using ethanol as a solvent, with a yield of 18.3 mg/gDW of fucoxanthin (Kim *et al.*, 2012). The astaxanthin pigment of *Haematococcus pluvialis* was extracted with ethanol:ethyl acetate (1:1, v/v) as the solvent at room temperature and incubated for 90 min to give a yield of 16.48 mg/gDW (Zou *et al.*, 2013).

8.2.2 Ultrasonic extraction

Ultrasonic extraction, also known as ultrasonication or sonication, is a modern, nonthermal technique that employs high-frequency sound waves to enhance the release of bioactive compounds from plant, animal, or microbial sources. The procedure generates cavitation and small bubbles in a liquid that rupture abruptly, releasing intense energy via ultrasonic sound waves (usually between 20 and 100 kHz) (Fig. 8.1B). This energy upsets cell structures and releases intracellular chemicals into the solvent for collection (Zhao *et al.*, 2019; Shen *et al.*, 2023).

Pigment extraction using ultrasound-assisted extraction (UAE) is affected by multiple factors such as ultrasound frequency, extraction time, solvent choice, temperature, and pH. Ultrasound frequency (typically 20–40 kHz) impacts cavitation intensity; extraction time affects the yield, where longer periods enhance yields but may lead to compound degradation; and solvent choice and solvent-to-sample ratio affect the solubility and effectiveness of the extracted pigment Kalra *et al.* (2020). Temperature regulation is essential for optimizing yield and stability, as ultrasound generates heat throughout the extraction process. pH adjustment is necessary for stabilizing target microbial pigment compounds, and a sonication mode (continuous or pulsed) is essential for managing the temperature. Controlling these factors can boost outputs and reduce processing duration (Palma *et al.*, 2021; Martins *et al.*, 2023).

Ultrasound-assisted extraction has already been used to extract various intracellular pigments from different microbes, such as bacteria, microalgae, and fungi. An intracellular pink pigment from a novel psychrotolerant *Paenibacillus* sp. BPW19 was extracted using 100% methanol (v/v) as the solvent at 25°C with a sonication time of 30 min. This resulted in the extraction of 5.41% crude pink pigments (Padhan *et al.*, 2021). A green pigment from *Spirulina platensis* (Martins *et al.*, 2023) was extracted using UAE at 60°C for 20 min with glycerol and water as the solvents at 240 watts, resulting in the extraction of 165.19 mg/g pigments compounds. Astaxanthin pigments of 27.58 mg/g yield were extracted from *H. pluvialis* at 41.1°C for 16 min with 200 watts using ethanol and ethyl acetate as solvents (Zou *et al.*, 2013) and in another study, UAE was used to extract β-carotenes from spirulina at 30°C for 8 min with acoustic intensity of 167 W/cm^2 using methanol as solvent, which resulted in a 47.10 mg/g yield (Dey and Rathod, 2013). Another study showed that 159.3 mg/g of chlorophyll pigment was extracted from *Phormidium autumnale* at 44.4°C for 24.5 min using 46.3% ethanol as solvent (Zhu *et al.*, 2017). Eighty-two percent of carotenoid content was extracted from the yeast *Rhodotorula glutinis* at 30°C with a frequency of 20 kHz (Aguilar-Machado *et al.*, 2020). Astaxanthin pigments from *Xanthophyllomyces dendrorhous* were extracted by UAE at temperature 30°C for 56 min using DMSO as the solvent (Unagul *et al.*, 2005). Torularhodin pigments from *Rhodotorula glutinis* were extracted using UAE at 200 kPa hydrostatic pressure and 120 μm amplitude for 180 min (Aguilar-Machado *et al.*, 2020).

8.2.3 Microwave-assisted extraction

Microwave-assisted extraction (MAE) is a method of extracting bioactive compounds, lipids, and pigments from microalgae, bacteria, and fungi using microwave energy (Jhanji *et al.*, 2024). The MAE method uses microwave energy to generate heat in a sample and solvent to increase the extraction of target compounds (Fig. 8.2A). Microwaves induce the rotation of the dipole in polar molecules and ionic motion in charged particles, generating fast and specific heating. Internal heating disturbs the sample matrix, weakens the cell walls, and induces the extrusion of intracellular contents into the solvent (Pagels *et al.*, 2021).

Microwave-assisted extraction is influenced by numerous factors such as temperature, extraction time, microwave power, nature of the solvent, and volume (Fomo *et al.*, 2020). Temperature plays a major role in MAE by increasing solvent penetration and the solubility of target compounds, and breaking down cell walls, which in turn increases extraction efficiency and yield (Jhanji *et al.*, 2024). The extraction time affects the efficacy, yield, and quality of the extracted compound. Prolonged extraction time increases the contact between the sample and solvent, with a greater release of targeted compounds.

Microwave-assisted extraction is a low-cost, environment-friendly, and energy-efficient approach for obtaining bioactive compounds from pharmaceuticals, food, and environmental applications. Increased power levels generate greater heat and localized pressure in the sample matrix, enhancing cell wall disruption, solvent infiltration, and mass transfer of desired chemicals. However, excessive power may lead to overheating, thermal damage to heat-sensitive materials, and a reduction in volatile substances. Inadequate power can result in incomplete extraction and lower yield. Hence, the power should be chosen correctly to avoid overheating, which may lead to degradation of the pigment compounds (Jhanji *et al.*, 2024), and solvent nature and volume have a substantial impact on MAE efficiency, selectivity, and yield, with the optimal solvent volume minimizing overheating and energy consumption (Sparr Eskilsson and Björklund, 2000).

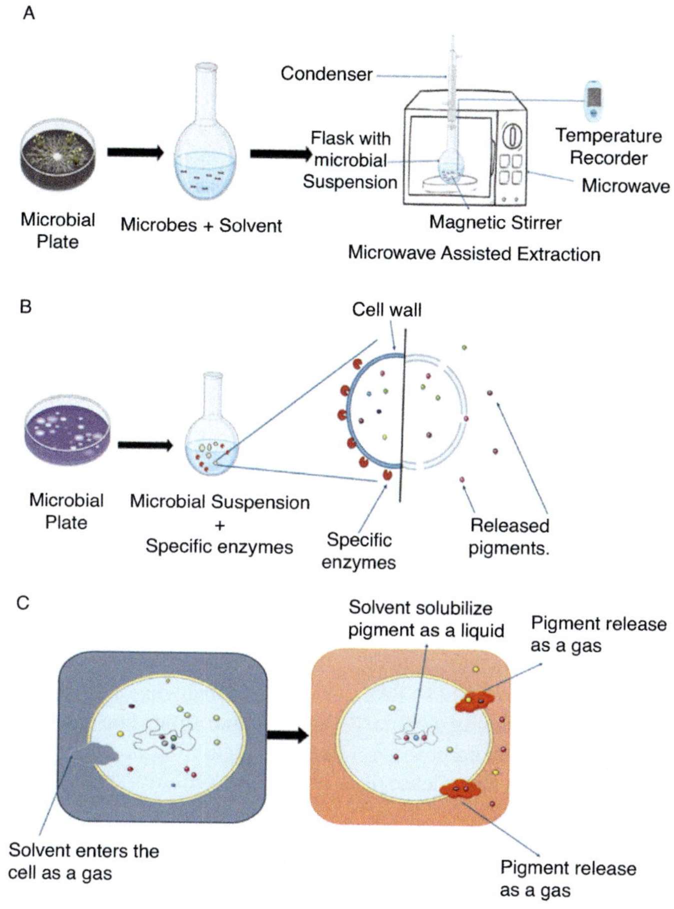

Fig. 8.2. (A) Microwave-assisted extraction: quick pigment release by microwave heating to disrupt microbial cells in a solvent. (B) Enzyme-assisted extraction: degradation of the microbial cell wall by specific enzymes to allow pigment release. (C) Supercritical fluid extraction: supercritical solvent penetrates microbial cells, dissolves pigments, and releases as a gas upon decompression.

Microwave-assisted extraction minimizes extraction times, enhances efficiency, and enables selective extraction with increased yields (Jhanji *et al.*, 2024) but it has numerous flaws that must be accounted for. Subsequent separation procedures such as centrifugation or filtration may be required after extraction (Mandal *et al.*, 2007). High temperatures can destroy thermolabile compounds and alter their chemical structure, potentially reducing their production. MAE is less efficient for nonpolar chemicals or solvents and can extract untargeted molecules, requiring subsequent purification. Optimization of parameters such as temperature, time, and microwave power is essential, and these may vary depending on the sample (Sparr Eskilsson and Björklund, 2000; Jhanji *et al.*, 2024).

Microwave-assisted extraction has been used in the extraction of various pigment compounds from different microbes such as bacteria, fungi, and microalgae. Bacterial pigments from *Arthrospira platensis* were extracted at 400 W power and 1 bar, using two types of solvents: polar solvents including ammonium acetate 10 mM and ethanol to extract thiamine, riboflavin, C-phycocyanin, and A-phycocyanin; and nonpolar solvents to extract β-carotene and α-tocopherol (Esquivel-Hernández *et al.*, 2017). Phycoerythin (PE) from the microalga *Porphyridium purpureum* was successfully extracted using this technique at 400°C for 10 s (flash irradiation) and resulted in a 73 μg/mg yield (Juin *et al.*, 2015). Chlorophyll was extracted from *Spirulina platensis* using MAE under 60 s of radiation, 560 W power, 52.8 min of extraction time, and a temperature of 55.1°C, with ethanol as the solvent, yielding 1.27% (Tong *et al.*, 2012). Chlorophylls and carotenoids were extracted from *Chlorella vulgaris* using this technique at 60°C for 14 min, with 300 W microwave power and ethanol as the solvent, resulting in a total chlorophyll yield of 63.36 mg/gDW and a carotenoid yield of 24.88 mg/gDW (Georgiopoulou *et al.*, 2023).

8.2.4 Enzyme-assisted extraction

Enzyme-assisted extraction (EAE) utilizes specific hydrolytic enzymes to break down cell walls, membranes, or complex molecules in raw materials to enhance extraction and recovery of the desired component (Pagels *et al.*, 2021). EAE functions by enzymatically degrading cellular structures, such as cell walls and membranes, as shown in Fig. 8.2B, facilitating the release of intracellular substances (Nadar *et al.*, 2018).

The efficiency of EAE of microbial pigments depends on various factors such as enzyme type and concentration, temperature, incubation time, substrate composition, and enzyme-to-substrate ratio (Picot-Allain *et al.*, 2021). The enzymes used in this process are cellulases, proteases, and pectinases, which break cell walls and release pigments. The operation of these factors and enzymes improves not only the efficiency of extraction but also the integrity and bioactivity of the pigment (Nadar *et al.*, 2018; Picot-Allain *et al.*, 2021).

Enzyme-assisted extraction has been used for the extraction of microbial pigments from various sources, such as bacteria, fungi, and microalgae. Melanin pigment was extracted from *Cryptocococus neoformans* by suspending the cells in 1.0 M sorbital and 0.1 M sodium citrate at pH 5.5 along with lyzing enzymes from *Trichoderma harzarium* (Dadachova *et al.*, 2007). Phycobiliproteins from *Porphyra* were extracted at 26°C for 72 hour using the crude enzymes prepared from *Pseudomonas vesicularis* and *Aeromonas salmonicida*, which yielded 33.94 mg/gDW (Huang *et al.*, 2021). Fucoxanthin was isolated from *Undaria pinnatifida* with dimethyl ether and ethanol as cosolvents at 37°C for 19.5 min incubation which returned 0.7 mg/g wet weight of the fucoxanthin (Billakanti *et al.*, 2013). EAE has been utilized to obtain allophycocyanin from *A. platensis* with the help of lysozyme and surfactants at 37°C for 20 hour, resulting in 32.27 mg/gDW allophycocyanin yield (Tavanandi *et al.*, 2019).

8.2.5 Supercritical fluid extraction

Supercritical fluid extraction (SFE) is a method for isolating and extracting a compound from a material using a solvent in the supercritical state. In such a condition, the fluid functions above its critical point in temperature and pressure, combining the properties of gases and liquids (Pagels *et al.*, 2021). This dual nature allows the fluid to effectively penetrate the matter while simultaneously dissolving the desired chemicals at the same time. The most commonly applied supercritical fluid is carbon dioxide (CO_2) due to its safety, nonflammability, and eco-friendliness (Guedes *et al.*, 2011).

Supercritical fluid extraction utilizes a fluid in its supercritical state with liquid-like density and gas-like viscosity to penetrate materials successfully and dissolve particular chemicals or pigment compounds, as shown in Fig. 8.2C. This technique allows for precise control over solubility through the adjustment of temperature and pressure. The dissolved compounds are separated by lowering the pressure of the solution, allowing them to precipitate as the fluid is either recycled or discharged (Ahmad *et al.*, 2019). SFE is determined by parameters such as the temperature, pressure, and nature of the supercritical fluid used (Sharif *et al.*, 2014). Higher temperatures increase diffusivity but lower fluid density, whereas higher pressures increase solvent efficiency by increasing fluid density. The choice of supercritical fluid is most commonly carbon dioxide, but cosolvents, such as ethanol, can be added to improve the solubility of polar molecules. Ethanol was used as the cosolvent in this study. Furthermore, the rate of flow for the supercritical fluid and the time of extraction, as well as the separator parameters for compound precipitation and recycling of the fluid, influence the operation. Maximizing these parameters ensures that the extractions are efficient, selective, and provide a high output (Abou Elmaaty *et al.*, 2022).

Supercritical fluid extraction has been applied to the extraction of pigment compounds from various microbes, such as bacteria, fungi, and microalgae. Carotenoids and chlorophyll *a* of the cyanobacterium *Synechococcus* spp. were extracted using SFE, where maximum carotenoid yield was obtained at 300 bar and 50°C. Interestingly, the temperature that maximized chlorophyll a production also coincided with a substantial torularhodin yield of 65.8 μg/g DW, suggesting a potential overlap in the optimal extraction conditions for these metabolites(Larocca *et al.*, 2023). SC-CO_2 was effectively employed to obtain β-carotene from red yeast cells at 300 bar pressure, 40°C, and CO_2 as the solvent for 90 min of extraction time which yielded 94 μg/gDW of β-carotene (Wang *et al.*, 2012). The algal pigment astaxanthin was extracted using SC-CO_2 from *H. pluvialis* at 550 bar pressure, 50°C temperature for 100 min, resulting in 237.4 mg/gDW (Di Sanzo *et al.*, 2018). A study found that a pressure of 443 bar and temperature of 27.5°C are optimum for the extraction of β-carotene from *Dunaliella salina* (Jaime *et al.*, 2007). Another study found that the use of ethanol as a cosolvent raised carotenoid yields of *Nannochloropsis oculata* at 350 bar pressure, 323 K temperature, and 16.7% ethanol, resulting in 7.61 mg/gDW carotenoid recovery (Liau *et al.*, 2010). Lutein and β-carotene from *Scenedesmus almeriensis* were effectively recovered by this method at 400 bar and 60°C, yielding 1.50 mg/gDW of β-carotene and 0.0466 mg/gDw of lutein (Macías-Sánchez *et al.*, 2010).

8.2.6 Ionic liquid extraction

Ionic liquid-assisted extraction (ILAE) uses ionic liquids (ILs) as cosolvents or solvents to extract pigments from various microbes such as fungi, bacteria, and microalgae (Fig. 8.3A) (Ventura *et al.*, 2017). Ionic liquids consist of organic cations and organic/inorganic anions that have low volatility, heat stability, and controllable solubility. ILAE can also involve microwave or ultrasonic extraction to facilitate the extraction of pigments from microbial sources (Rajendran *et al.*, 2023). Ionizable ligands also aid in extraction by protonation or deprotonation. These processes utilize the solubility, ion exchange ability, and controllable interactions of ILs for efficient and selective extraction processes (Janssen *et al.*, 2015).

Carotenoids from *Gordonia alkanivorans* bacterial masses were effectively extracted at 22°C with tributyl phosphonium diethyl phosphate and ethyl acetate as cosolvents during overnight incubation, yielding 171 μg/gDW of carotenoids (Silva *et al.*, 2024). Another study explored the extraction of carotenoid pigments from the endophytic *Streptomyces hyderabadensis* 7VPT5-5R by employing tetrabutylammonium hydroxide (40% w/w TBAOH in water) as the solvent at room temperature with an incubation time of 196 hour, yielding 5548 μg/gDW (Ghadge *et al.*, 2022). Intracellular carotenoids were isolated from the yeast *Rhodotorula glutinis* CCT-2186 between 25°C and 65°C using ammonium-derived protic ionic liquids, with a yield of 28.97 μg/gDW of β-carotene (Mussagy *et al.*, 2019). Another study utilized cholinium ionic liquids as solvents in the extraction of astaxanthin from *P. rhodozyma* with the conditions of 65°C for 1 hour with a yield of 47.9 μg/gDW (Mussagy *et al.*, 2021). Red natural pigment of *Penicillium purpurogenum* fermented broth was prepared using potassium citrate ionic liquids at 25°C for 12 hour and gave 4.64 μg/gDW (Ventura *et al.*, 2013). *Chlorella sorokiniana* carotenoids were obtained using an ionic liquid and ultrasound apparatus. With a 1:10 solid-to-liquid ratio, an IL-cosolvent ratio of 1:4 and doubling up on 7.5 min of extraction, the process yielded 1.29 mg/gDW of carotenoids (Bianchini *et al.*, 2024). Similarly, optimized extraction of *Rhodotorula glutinis* sludge and *Chlorella vulgaris* sludge was subjected to treatment with 1-butyl-3-methylimidazolium chloride as the IL at 53°C for 57 min at an IL concentration of 1 mol/L, yielding carotenoids at 1.311 mg/gDW (Zhiping *et al.*, 2014).

The above research confirms the effectiveness and versatility of ILAE for the extraction of microbial pigments from a vast array of sources.

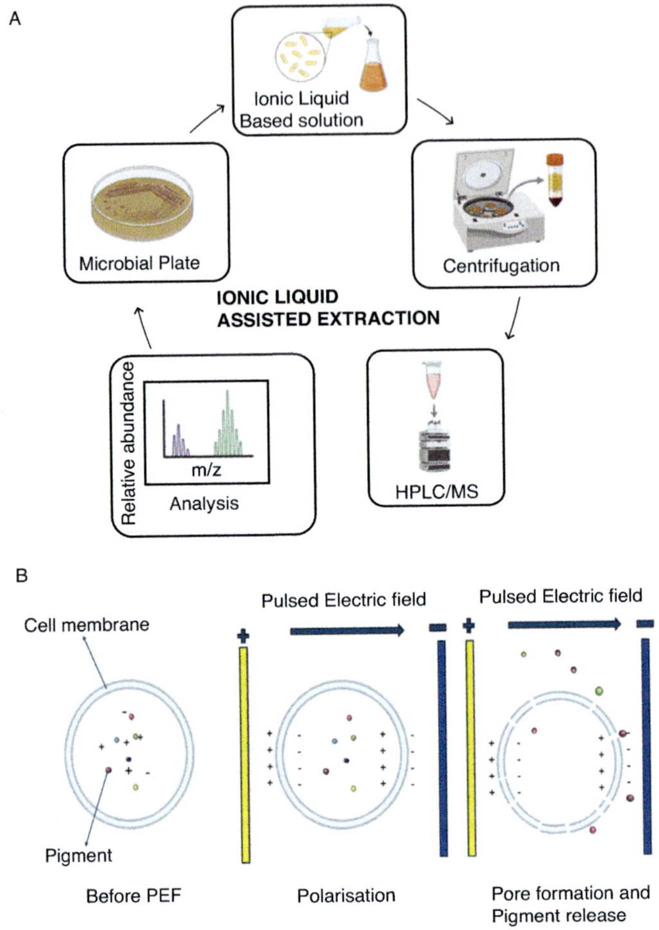

Fig. 8.3. Extraction methods. (A) Ionic liquid-assisted extraction: microbial pigments are extracted with the help of ionic liquids as green solvents and then centrifuged followed by analysis with HPLC/MS. (B) Pulsed electric field-assisted extraction: cell membrane polarization is caused by high-voltage electric pulses, which results in the creation of pores and the release of pigments.

8.2.7 Pulsed electric field-assisted extraction

Pulsed electric field (PEF) is a nonthermal process that targets the cytoplasmic membrane of microbes, such as bacteria, fungi, and microalgae, to increase the selective discharge of intracellular compounds (e.g. pigment compounds) while reducing the negative effects of heat (Martínez *et al.*, 2020b). Pulsed electric fields utilize repeated high-intensity pulses, typically in the range of a few nanoseconds to a few milliseconds, which raises cytoplasmic membrane permeability; this is also referred to as electroporation, as indicated by Fig. 8.3B. PEF pretreatment is low in energy, has high extraction yields, and maintains the properties and purity of the extracts (Kalra *et al.*, 2020; Rajendran *et al.*, 2023).

Pulsed electric field extraction increases yield with a decrease in extraction time by disrupting membrane structures to allow mass transfer. PEF treatment efficacy is influenced by a number of factors such as field strength, pulse frequency, energy input, and treatment temperature. PEF extraction is a nonthermal process so it is used to preserve thermally sensitive compounds (Zhang *et al.*, 2018). C-phycocyanin pigments from *Nostoc commune* were extracted using PEF at 40°C, 2 Hz, 1500 pulses of 1 μs, water solvent, and 5 kV, with a yield of 29.66 mg/gDW (Chittapun *et al.*, 2020). Torularhodin, a red pigment, was extracted from *Rhodotorula glutinis* using PEF at 15 kV/cm and ethanol as the solvent (Martínez *et al.*, 2020a). This technique was also used for the extraction of pigment from microalgae; carotenoids were extracted from *Porphyridium mcruentum* using PEF under field strength of 1 kV/cm, 80 pulses, 400 kJ/kg, and pulse lengths of 150 μs in the presence of ethanol, which yielded 7.69 mg/gDW (Martínez *et al.*, 2019a). Carotenoids and chlorophylls have been extracted from *Chlorella vulgaris* using PEF, which improved intracellular pigment extraction (Luengo *et al.*, 2014). Another study employed PEF permeabilization for the extraction of astaxanthin from *Haematococcus pluvialis* at 20°C under 10 pulses of 5 μs, using ethanol as a solvent and yielding 18.3 mg/gDW (Martínez *et al.*, 2019b).

8.3 Chromatographic Techniques for Purification

Chromatography is a method applied to separate components within a mixture based on the variation in distribution between a mobile phase and a stationary phase. The method takes advantage of the physical or chemical characteristics of components, such as polarity, size, or charge, to carry out separation effectively (Wilson, 2010).

8.3.1 Preparative thin-layer chromatography

Preparative thin-layer chromatography (PTLC) is a rapid and universal chromatography technique employed for the separation, purification, and isolation of compounds from mixtures. It is a high-performance version of standard analytical thin-layer chromatography (TLC), which is adapted to handle large quantities of samples for preparative purposes. PTLC is widely used in many fields, such as natural product chemistry, pharmaceutical development, and microbiology, primarily for the purification of bioactive metabolites, pigments, and other small molecules (Kumar and Lenhoff, 2020).

The method distinguishes compounds based on their differential interactions with stationary and mobile phases. The stationary phase is silica gel coated onto a glass or plastic plate. The mobile phase (solvent system) travels through the stationary phase via capillary action. With increasing solvent ascent, the components of the mixture interact with the two phases in a different way. Pigments with a higher affinity for the stationary phase move less rapidly, whereas those with a preferential interaction with the mobile phase migrate more quickly (Fig. 8.4A). For instance, carotenoids that are hydrophobic in nature require non-polar solvents for efficient separation (Saini and Keum, 2018). Separation leads to the isolation of pigments into individual bands, and hence easy purification. Pigments derived from bacteria, fungi, and microalgae have different chemical structures. This method is very flexible in choosing the solvent, as illustrated in Table 8.2, and is also very adaptable to different classes of compounds; thus, it is a useful tool for the purification of microbial pigments.

Pigment samples are prepared by solvent extraction using ethanol, methanol, or acetone based on the solubility of the pigments. For example, *Monascus purpureus* pigments were extracted with ethanol before separation on PTLC plates (Mukherjee and Singh, 2011). The stationary phase material utilized for PTLC, such as silica gel, alumina, or other materials, is generally mixed with water or an appropriate binding agent in the desired proportions to prepare a homogeneous slurry. This slurry is evenly applied to clean glass plates to obtain a uniform layer (0.5–2 mm). Then, the coated plates are air dried, followed by activation at 100–120°C for 30 min to evaporate water. The activated plates are kept in a desiccator until use to avoid rehydration (Sherma and Fried, 2003). The pigments are spotted on PTLC plates and developed with a specific mobile phase. The migration of pigments on PTLC depends on their affinity for the stationary phase. The separated pigments can be seen using UV light or by identifying their natural coloration; for example, the prodigiosin pigment from *Serratia marcescens* is red (Shaikh, 2016). After separation, the bands of pigment are gently

Table 8.2. Chromatographic purification of various microbial pigments from different microbes utilizing PTLC.

Pigments	Adsorbent	Solvent system	Microorganism	Reference
Astaxanthin	Silica gel	Acetone: petroleumether: chloroform (2:2:1 v/v)	*Haematococcus pluvialis*	(Elumalai et al., 2014)
Prodigiosin	Silica gel	Methanol: ethyl acetate: chloroform (6:3:1 v/v)	*Serratia marcescens*	(Pore *et al.*, 2016)
Pyocyanin	Silica gel/ alumina	Chloroform: methanol(9:1 v/v)	*Pseudomonas aeruginosa*	(Abdul-Hussein and Atia, 2016)
Violacein	Silica gel	Benzene: acetone(2:1 v/v)	*Chromobacteriumviolaceum*	(Durán *et al.*, 1994)
Melanin	Silica gel	Petroleum ether: ethyl acetate: 95 ethanol: conc. ammonia (4:4:6:1 v/v)	*Streptomyces djakartensis NSS3*	(El-Zawawy *et al.*, 2024)
Phycocyanin	Silica gel	Petroleum ether: acetone (7:3 v/v)	*Synechococcus* sp.	(Elumalai et al., 2014)

scraped off the silica gel layer and eluted in suitable solvents such as acetone or methanol. For example, monascin from *Monascus purpureus* was washed out using acetone after PTLC separation (Lin *et al.*, 2008). The refined pigments are concentrated by removing residual solvents, yielding high-purity compounds for subsequent use.

Preparative thin-layer chromatography is widely used to determine the identification and qualitative purity of microbial pigment extracts and fractions, assisting in the optimization of solvent systems for efficient separation (Rabel and Sharma, 2017).

8.3.2 High-performance liquid chromatography

High-performance liquid chromatography (HPLC) is a robust analytical method used for the separation, identification, and quantitative determination of individual components of a complex mixture. High pressure is used to force the solution through a packed column containing stationary phase materials (Fig. 8.4B). HPLC is efficient for the separation of complex mixtures because it yields high sensitivity and resolution.

High-performance liquid chromatography is generally supplemented by more recently developed detection techniques such as fluorescence detection (FLD), diode array detection (DAD), and mass spectrometry (MS) for the purification of microbial pigments. For instance, HPLC-FLD has been used to examine pigments such as monascin, rubropunctatin, ankaflavin, astaxanthin, and monascorubin (Elumalai *et al.*, 2014; Shi *et al.*, 2017), where aqueous solutions of phosphoric acid and acetonitrile were employed as the mobile phase, flow rate of 1 ml/min, and fluorescence parameters (excitation at 360 nm and emission at 565 nm) enhanced sensitivity and specificity for low concentrations of pigments. HPLC-DAD has been used to determine and quantify the content of fucoxanthin, β-carotene, and chlorophyll *a* in brown algal extracts from *Laminaria saccharina* and *Undaria pinnatifada*, employing ammonium acetate in water, ammonium acetate in methanol, and ethyl acetate in the optimized ratio as the mobile phase (Lourenço-Lopes *et al.*, 2022) and C18 silica columns as the stationary phase (Žuvela *et al.*, 2019). Fucoxanthin exhibits its characteristic broad absorbance band at 453 nm (Erdoğan *et al.*, 2022), chlorophyll *a* exhibits its expected

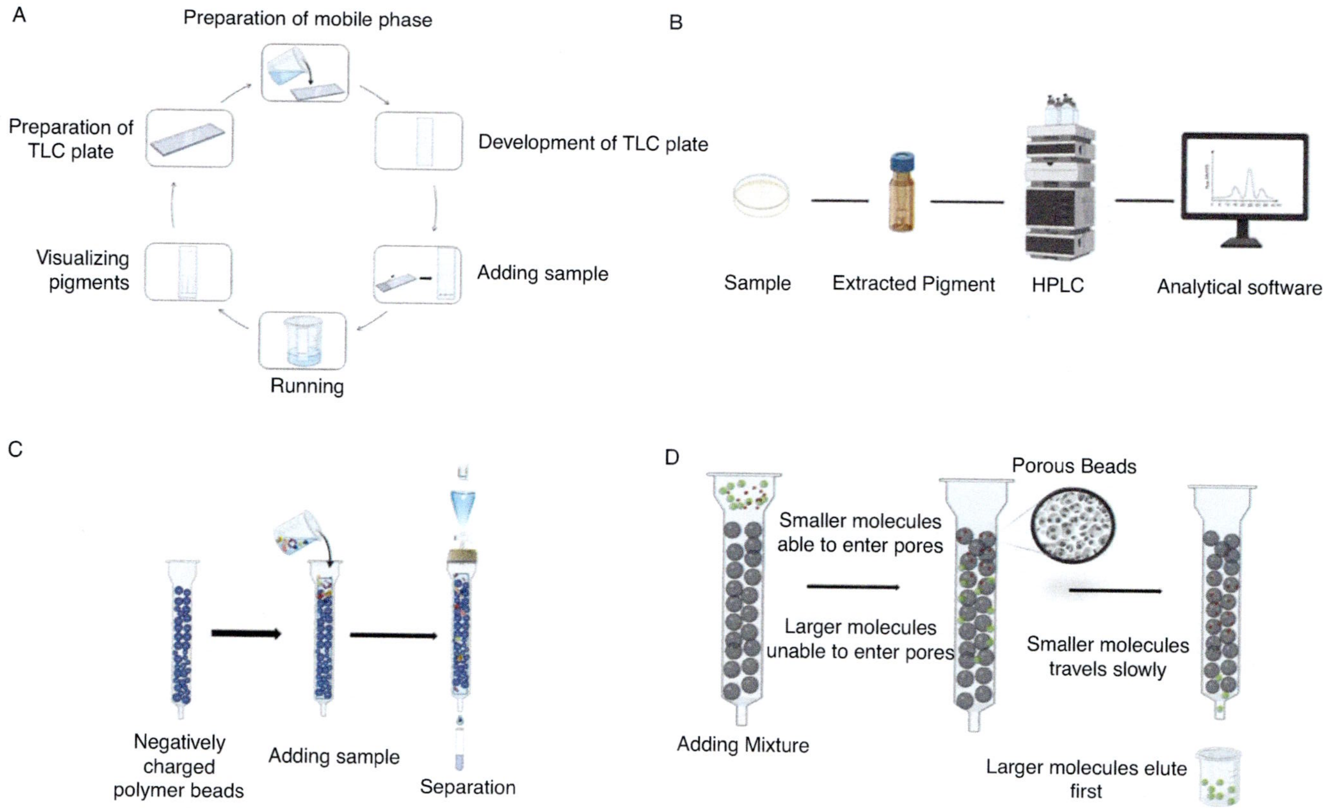

Fig. 8.4. Methods of purification of microbial pigments. (A) Preparative thin-layer chromatography (PTLC): pigment separation on the basis of differential migration over a stationary phase. (B) High-performance liquid chromatography (HPLC): delicate pigment separation and identification by high-pressure liquid flow through a chromatographic column. (C) Ion exchange chromatography: separation on the basis of charge interaction between the pigments and negatively charged polymer beads. (D) Size exclusion chromatography: separation of molecules occurs based on their size. Large molecules pass through the column rapidly compared to small molecules.

peak band at 662 nm (Singh *et al.*, 2020) while β-carotene exhibits two-peak spectra with peak wavelengths of 454 nm and 480 nm (Rodríguez-Rodríguez *et al.*, 2020). Because of pigment spectrophotometric properties, diode array detectors (DAD) are commonly used for analytical detection, offering high resolution, low noise, reliable identification and quantification, reproducibility, and ease of use (Swartz, 2010).

Moreover, HPLC-MS/MS enables precise quantification and structural analysis of pigments, as demonstrated in the study of metaprodigiosin from *Streptomyces spectabilis*, offering structural insights through mass fragmentation, making it ideal for complex pigment mixtures (Meng-xi *et al.*, 2021). Some of the optimum conditions employed for the purification of microbial pigments using HPLC are listed in Table 8.3. The use of HPLC-DAD and FLD is even more effective for detection because fluorescent and nonfluorescent pigments can now be detected simultaneously. All of these advances enhance pigment purification by maximizing resolution, sensitivity, and identification.

8.3.3 Ion exchange chromatography

Ion exchange chromatography (IEC) is a technique through which a mixture of comparably charged ions can be separated using an ion exchange resin that exchanges ions based on their relative affinities, as depicted in Fig. 8.4C. The technique is based on the variation in the net charge of compounds under certain pH values and is used in the isolation of ionic pigments from intricate biological mixtures. In the recent past, IEC has become an essential method in pharmaceutical, biotechnology, food and beverage, environmental, chemical, agriculture, cosmetics, and research fields for the separation of charged compounds such as biomolecules, pigments, and chemicals (Wallace and Rochfort, 2023).

The process of separation in IEC is based on charge-based interactions between the pigments and the stationary phase.IEC distinguishes compounds based on their ionic characteristics using resins functionalized with charged groups. For example, carboxymethyl groups are used for cation exchange and quaternary ammonium groups for anion exchange (Aasim *et al.*, 2022). Interactions between pigments and the stationary phase are facilitated by the mobile phase, which is typically an aqueous buffer. Depending on the charge, pigments attach to the resin and are selectively eluted by progressively altering the pH or ionic strength of the buffer. The negatively charged C-phycocyanin (C-PC) pigments were purified from *Limnothrix* sp. NS01 extract using Q-Sepharose resin with elution at 80mM NaCl in 10mM sodium phosphate buffer (pH 7.5), which resulted in a yield of 85.31% (Safaei *et al.*, 2019). Similarly, Gorgich *et al.* (2020) used DEAE-Sepharose resin to purify phycocyanin. C-phycoerythrin (C-PE) was purified from *Pseudanabaena* sp. using DEAE-cellulose with a sodium acetate buffer (pH 5.5) and a NaCl gradient (0–0.5 M), which yielded 47% (Mishra *et al.*, 2011).

Table 8.4 illustrates different optimum conditions for the purification of various microbial pigments using the IEC technique.

8.3.4 Size exclusion chromatography

Size exclusion chromatography (SEC) or gel filtration chromatography is used to separate molecules according to their size (Barth *et al.*, 1996). Porous beads form a stationary phase in which smaller molecules penetrate the pores and are retained for a longer time, and larger molecules elute sooner, as illustrated in Fig. 8.4D. Separation occurs without chemical interactions, which makes it rely solely on the physical differences between the molecules in the mixture. SEC runs under mild, nondenaturing conditions so it is best suited for the purification of delicate molecules, such as proteins, polysaccharides, and pigments (Hatti-Kaul and Mattiasson, 2003). It has broad applications in biochemistry and biotechnology, where highly pure and intact molecules are required.

Size exclusion chromatography is utilized for the purification of microbial pigments according to their molecular weight, providing gentle, nondenaturing conditions that are suitable for fragile pigments. It has been used successfully to purify many pigments, with modifications adapted to increase efficiency (Table 8.5). Phycobiliproteins were purified in three successive purification steps: ammonium sulfate precipitation, anion exchange chromatography using DEAE cellulose columns, and size exclusion chromatography using Sephadex G100, which provided effective separation from proteins (Kannaujiya *et al.*, 2017).

Table 8.3. Summary of different HPLC parameters used for the separation and detection of various microbial pigments.

Pigment	Microbial source	Stationary phase	Mobile phase	Flow rate	Detection wavelength (nm)	Retention time (min)	References
Monashin	*Monascs purpureus*	C18column	Acetonitrile-water (0.1% TFA)	1.0ml/min	464	16.37	(Mondal *et al*., 2019)
Actinorhodin	*Streptomyces coelicolor*	C18column	Acetonitrile-water (0.1%formicacid)	1.0ml/min	520	14.02	(Taguchi *et al*., 2000)
Phycocyanin	*Arthrospira platensis*	Zorbax Eclipse C18column	Acetonitrile-water (0.1% TFA)	0.8ml/min	620	9.98	(Kissoudi *et al*., 2017)
Prodigiosin	*Serratia marcescens*	C18column	Methanol-water (7:3 v/v)	20 ml/min	535	10.1	(Song *et al*., 2006)
Carotenoids	*Rhodotorula glutinis*	C18reverse-phase column	Methanol: acetonitrile: dichloromethane (60:10:30v/v/v)	1.0ml/min	450	5.5	(Mussagy *et al*., 2021)
Chlorophyll	*Chlorella vulgaris*	C18reverse-phase column	Methanol- tert-butyl methyl ether (80:20)	1.2ml/min	660	6.8	(Cha *et al*.,2010)
Violacein	*Chromobacterium violaceum*	C18column	Methanol- water (75:25)	7.0ml/min	230	9.2	(Rettori and Durán, 1998)

Table 8.4. Summary of microbial pigment purification utilizing IEC from various microbial sources. Techniques include both anion and cation exchange chromatography optimized for specific pigments to achieve high purity and recovery, suitable for industrial and research applications.

Pigment name	Microbial source	Type of IEC	Optimized conditions	Purity/yield	Reference
Astaxanthin	*Haematococcus pluvialis*	Anion exchange	pH 7.0, salt gradient elution	High purity, ~90% recovery	(Dong *et al.*, 2014)
Carotenoids	*Monascus purpureus*	Cation exchange	pH 6.5, stepwise increase in ionic strength	High purity, industrial scale	(Lakshman and Rao, 1999)
Prodigiosin	*Serratia marcescens*	Cation exchange	pH 7.2, phosphate buffer, gradient NaCl elution	High purity	(Juang and Yeh, 2014)
Anthocyanins	*Monarda fistulosa*	Anion exchange	pH 5.5, multicolumn system, gradual ionic strength increase	Industrial-scale purity	(Shanaida *et al.*, 2021)
Chlorophyll	*Chlorella vulgaris*	Anion exchange	pH 8.0, buffer system for selective separation	High purity	(Hynstova *et al.*, 2018)

Several studies have successfully used SEC to purify microbial pigments. For instance, researchers have purified phycoerythrin from *Pseudanabaena* sp. using a Sephadex G-150 column with 50 mM phosphate buffer (pH 7.2) (Mishra *et al.*, 2011). Other studies have included the separation of anthocyanins from *Monascus purpureus* on a Bio-Gel P-2 column using water as the mobile phase (Luzardo-Ocampo *et al.*, 2021), separation of phycocyanin from *Spirulina platensis* using a Sephadex G-25 column with phosphate buffer (pH 6.8) (Bhaskar *et al.*, 2005), and successful separation of prodigiosin from *Serratia marcescens* with a Sephadex LH-20 column and methanol-water eluent (Sherma, 2017).

The above examples demonstrate the adaptability of SEC for various pigments by selecting appropriate resins and mobile phases. The versatility of SEC makes it valuable for pigment purification in the food, pharmaceutical, and cosmetic industries.

8.4 Nonchromatographic Techniques for Purification

Nonchromatographic purification processes for microbial pigments rely on basic physicochemical features such as solubility, polarity, and phase separation. Nonchromatographic approaches are usually preferred because of their simplicity, ease of scaling up for industrial production, and cost-effectiveness. By handling important parameters such as pH, temperature, ionic strength, and solvent composition, pigment contamination can be minimized by the unique properties of the pigment, such as its solubility in various solvents and durability under various conditions, which determine the purification procedure to use. These strategies provide effective substitutes for chromatographic techniques.

8.4.1 Ammonium sulfate precipitation

Ammonium sulfate precipitation is commonly used for protein and microbial pigment purification because of its solubility in the presence of high levels of ammonium sulfate. It is a type of salting out, and the solubility of biomolecules decreases as salt concentration increases, resulting in their precipitation from the solution (Green and Hughes, 1955). Ammonium sulfate is the salt of choice because of its high solubility, ability to stabilize proteins during processing, and compatibility with biological molecules (Goldring, 2019). Ammonium sulfate precipitation works

Table 8.5. Comparison of various features of different chromatographic techniques for the purification of microbial pigments.

Feature	Size exclusion chromatograph y (SEC)	Ion exchange chromatograph y (IEC)	Thin-layer chromatograph y (TLC)	High- performance liquid chromatography (HPLC)
Separation basis	Molecular size	Charge differences	Polarity or solubility differences	Polarity, size, or specificinteractions
Stationary phase	Porous beads (e.g. Sephadex, Bio-Gel)	Charged resins (cationic or anionic)	Adsorbent laye (silica gel r or alumina)	Packed columns with materials like C18, silica, orpolymeric resins
Mobile phase	Neutral or buffer solutions	Buffers with controlled pH andionic strength	Solvent mixtures	Liquid solvent or buffer systemsunder pressure
Conditions	Nondenaturing, mild (e.g. neutral pH, ambienttemp)	Requires controlled pH and ionic strength	Operates at ambient conditions	Operates under high pressure and controlledconditions
Applications in pigments	Purifies pigments like carotenoids, anthocyanins, and phycobiliproteins by separating them from macromolecules (e.g. proteins, polysaccharides)	Separates charged pigments like prodigiosin, astaxanthin, and phycocyanin by exploiting their ionic properties	Preliminary pigment profiling and identification. Used to separate mixtures into individual components	High-purity separation of complex pigment mixtures such as carotenoids, astaxanthin, and chlorophyll
Advantages	Nondenaturing, ideal for sensitive pigment molecules	High resolution for charged pigments	Simple, cost effective, and quick	High resolution, speed, and reproducibility for pigments
Limitations	Low resolution for similar-sized pigment molecules	Requires precise buffer and pH optimization	Limited resolution for complex mixtures	Expensive set-up and maintenance
Pigment examples	Carotenoids from Rhodotorula glutinis, phycobiliproteins from Spirulina platensis	Prodigiosin from Serratia marcescens, anthocyanins from Monascus purpureus	Separation of carotenoid mixture or simple pigment profiling	High-resolution purification of astaxanthin, chlorophyll, and carotenoids
Separation	Easily scalable for large-scale pigment production	Moderately scalable for industrial use	Poor scalability	Highly scalable and compatible with automated process
Detection	Requires external detectors (e.g. UV/Vis)	Requires external detectors (e.g. UV/Vis)	Visual (UV light, stains)	Integrated detectors (UV/Vis, fluorescence, etc.)
Cost	Moderate	Moderate to high	Low	High

on the simple principle of decreasing the water solubility of protein pigments by increasing the ionic strength of the solution. Salting out is a simple method, in which salt is added to a protein solution to reduce the hydration layer, which ultimately leads to precipitation (Rupley *et al.*, 1983). The quantity of ammonium sulfate required for precipitation depends on the characteristics of the target compound (e.g. size, charge, and hydrophobicity) (Timasheff and Arakawa, 1997).

Phycocyanin, phycoerythrin, and allophycocyanin were purified from *Lyngbya* sp. A09DM, using ammonium sulfate precipitation, was amended with Triton X-100 (Sonani *et al.*, 2014). C-phycoerythrin was isolated from marine cyanobacterium *Pseudanabaena* sp. using 60% ammonium sulfate saturation at 4°C (Mishra *et al.*, 2011).

8.4.2 Crystallization

Crystallization is defined as a process of transition solidification, whereby molecules are arranged into a highly ordered form referred to as a crystalline structure (Armitstead *et al.*, 1992; Zhang *et al.*, 2016). The principle of crystallization during purification depends on the variation in the solubility of a compound under varying temperatures or solvent conditions. When the solution is cooled or the solvent is evaporated, the solute loses its solubility and starts precipitating in the form of solid crystals. The more purified compound is precipitated from the solution first, with impurities remaining in the leftover solution. By regulating the temperature, solvent, and other parameters very carefully, the crystalline form of the pure compound is separated and can be harvested, as shown in Fig. 8.5A. It utilizes the properties of most compounds, including microbial pigments, to obtain differentiated solubility characteristics that can be utilized for selective separation. It is also used in chemical and petroleum industries to refine compounds.

Violacein isolated from *Chromobacterium violaceum* was purified by the initial evaporation of ethanol from the semi-purified extract under reduced pressure, yielding approximately 40 mg of crude pigment. The pigment was further purified by crystallization using an ethanol/water solvent pair. Harvested crystals were obtained by centrifugation and dried at 100°C for 24 hour to obtain purified pigments for subsequent application (Rettori and Durán, 1998). Crystallization was employed to purify astaxanthin from a sample of *Haematococcus pluvialis* by dissolving 5g of all-trans-astaxanthin in 100 mt of acetone at room temperature. The mixture was incubated for 72 hour at 4°C for the crystallization of astaxanthin. The purity of final crystalline powder was verified by HPLC.

8.4.3 Liquid-liquid partitioning

Liquid-liquid partitioning (LLP) is a separation technique based on the partitioning of a solute between two immiscible liquids (Fig. 8.5B). These liquids are typically organic and aqueous solutions. As the solute is more soluble in one liquid, various chemicals, including microbial pigments, can be separated from one another based on their different affinities for different solvent phases. The partition coefficient, which determines the solubility of pigments across the two phases, is essential for the liquid-liquid partitioning mechanism (Berthod and Carda-Broch, 2004). Some pigments remain in the aqueous phase, but more soluble pigments enter the organic phase. The pigment is concentrated and undesirable contaminants are removed using this selective separation process. Using LLP, all-trans-astaxanthin was isolated from *Haematococcus pluvialis*. The pigment extracts were mixed with an equal amount of n-hexane and distilled water and centrifuged. The n-hexane phase was rinsed in water until neutral pH was achieved, and the procedure was repeated to remove contaminants (Sun *et al.*, 2014).

8.5 Conclusion

Microbial pigments have been proven to be lucrative, green and sustainable alternatives to synthetic dyes, with enormous potential in various industries such as food, pharmaceuticals, cosmetics, and textiles. Their widespread application depends on efficient extraction and purification methods that result in high purity,

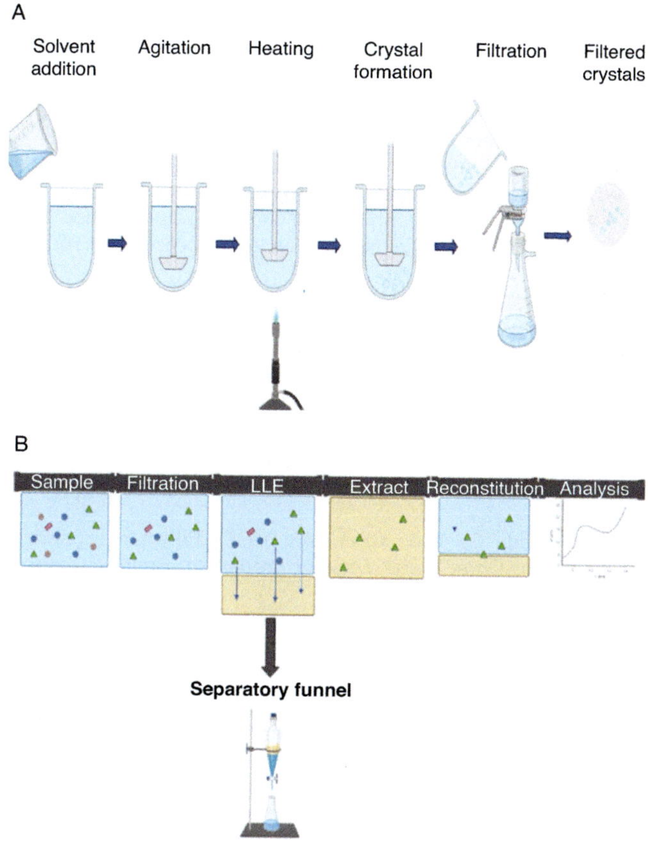

Fig. 8.5. Separation technique.

stability, and yield. This chapter has surveyed a wide range of pigment extraction techniques, equally emphasizing traditional and modern techniques. After extraction, purification plays a key role in pigment cleaning, which makes it suitable for research and industrial applications. Chromatographic methods achieve high-resolution separations that are ideal for the analytical isolation of pigments. In contrast, nonchromatographic methods are cost-effective and scalable, and hence an effective approach for bulk use. Improvements in extraction and purification technologies not only make microbial pigment recovery more efficient but also open up new avenues for commercial exploitation. Through continuous research, the implementation of such methods into green technologies and bioprocess optimization will continue to promote productivity and sustainability.

The future of microbial pigments lies in optimizing such processes to the industrial level without a reduction in economic and environmental efficiency. Through continuous research, microbial pigments have the potential to revolutionize industries by providing natural, nontoxic, high-quality colorants for several applications. The development of extraction and purification technologies not only enhances the recovery yield of microbial pigments, but also creates new areas of industrial applications. With progressive research, the integration of these techniques with future green technology and bioprocess optimizations will further improve sustainability and yield. The future of microbial pigments lies in enhancing these processes to meet industrial requirements while not harming the environment and economy.

References

Aasim, M., Khan, M.H., Bibi, N.S. and Fernandez-Lahore, M. (2022) Understanding the interaction of proteins to ion exchange chromatographic supports: A surface energetics approach. *Biotechnology Progress* 38(2). DOI: 10.1002/btpr.3232.

Abdul-Hussein, Z.R. and Atia, S.S. (2016) Antimicrobial effect of pyocyanin extracted from *Pseudomonas aeruginosa*. *European Journal of Experimental Biology* 6(6).

Abou Elmaaty, T., Sayed-Ahmed, K., Elsisi, H. and Magdi, M. (2022) Optimization of extraction of natural antimicrobial pigments using supercritical fluids: A review. *Processes* 10(10), 2111. DOI: 10.3390/pr10102111.

Aguilar-Machado, D., Delso, C., Martinez, J.M., Morales-Oyervides, L., Montañez, J. *et al.* (2020) Enzymatic processes triggered by PEF for astaxanthin extraction from *Xanthophyllomyces dendrorhous*. *Frontiers in Bioengineering and Biotechnology* 8. DOI: 10.3389/fbioe.2020.00857.

Ahmad, T., Masoodi, F.A., Rather, S., Wani, S.M. and Gull, A. (2019) Supercritical fluid extraction: A review. *Journal of Biological and Chemical Chronicles* 5(1), 114–122. DOI: 10.33980/jbcc.2019.v05i01.019.

Arce, A., Arce, A. and Rodriguez, O. (2022) Revising concepts on liquid–liquid extraction: data treatment and data reliability. *Journal of Chemical & Engineering Data* 67(1), 286–296. DOI: 10.1021/acs.jced.1c00778.

Armitstead, K., Goldbeck-Wood, G. and Keller, A. (1992) Polymer crystallization theories. In: *Macromolecules: Synthesis, Order and Advanced Properties*. Springer-Verlag, Berlin/Heidelberg, pp. 219–312.

Barth, H.G., Boyes, B.E. and Jackson, C. (eds) (1996) Size exclusion chromatography. *Analytical Chemistry* 68(12), 445–466. DOI: 10.1021/a19600193.

Berthod, A. and Carda-Broch, S. (2004) Determination of liquid-liquid partition coefficients by separation methods. *Journal of Chromatography. A* 1037(1–2), 3–14. DOI: 10.1016/j.chroma.2004.01.001.

Bhaskar, S.U., Gopalaswamy, G. and Raghu, R. (2005) A simple method for efficient extraction and purification of C-phycocyanin from *Spirulina platensis* Geitler. *Indian Journal of Experimental Biology* 43(3), 277–279.

Bianchini, C.B., Ramos-Souza, C., Schappo, F.B., Farina, M., De Rosso, V.V. *et al.* (2024) Ionic liquid and ultrasound as a fast and innovative combination for improved extraction of chlorella sorokiniana-derived carotenoids. *Algal Research* 82, 103650. DOI: 10.1016/j.algal.2024.103650.

Billakanti, J.M., Catchpole, O.J., Fenton, T.A., Mitchell, K.A. and MacKenzie, A.D. (2013) Enzyme-assisted extraction of fucoxanthin and lipids containing polyunsaturated fatty acids from *Undaria pinnatifida* using dimethyl ether and ethanol. *Process Biochemistry* 48(12), 1999–2008. DOI: 10.1016/j.procbio.2013.09.015.

Cha, K.H., Kang, S.W., Kim, C.Y., Um, B.H., Na, Y.R. and Pan, C.H. (2010) Effect of pressurized liquids on extraction of antioxidants from *Chlorella vulgaris*. *Journal of Agricutlure and Food Chemistry* 58(8):4756–61. DOI: 10.1021/jf100062m.

Chittapun, S., Jonjaroen, V., Khumrangsee, K. and Charoenrat, T. (2020) C-phycocyanin extraction from two freshwater cyanobacteria by freeze thaw and pulsed electric field techniques to improve extraction efficiency and purity. *Algal Research* 46, 101789. DOI: 10.1016/j.algal.2020.101789.

Dadachova, E., Bryan, R.A., Huang, X., Moadel, T., Schweitzer, A.D. *et al.* (2007) Ionizing radiation changes the electronic properties of melanin and enhances the growth of melanized fungi. *PLoS ONE* 2(5), e457. DOI: 10.1371/journal.pone.0000457.

Davoli, P. (2002) Identification and quantification of carotenoid pigments in aeciospores of the daisy rust fungus, *Puccinia distincta*. *Phytochemistry* 60(3), 309–313. DOI: 10.1016/s0031-9422(02)00120-6.

Dey, S. and Rathod, V.K. (2013) Ultrasound assisted extraction of β-carotene from *Spirulina platensis*. *Ultrasonics Sonochemistry* 20(1), 271–276. DOI: 10.1016/j.ultsonch.2012.05.010.

Di Sanzo, G.D., Mehariya, S., Martino, M., Larocca, V., Casella, P. *et al.* (2018) Supercritical carbon dioxide extraction of astaxanthin, lutein, and fatty acids from *Haematococcus pluvialis* microalgae. *Marine Drugs* 16(9), 334. DOI: 10.3390/md16090334.

Dong, S., Huang, Y., Zhang, R., Wang, S. and Liu, Y. (2014) Four different methods comparison for extraction of astaxanthin from green alga *Haematococcus pluvialis*. *Scientific World Journal* 1–7. DOI: 10.1155/2014/694305.

Dufossé, L. and Tiwari, P. (2024) Editorial: Emerging trends and advances in the socioeconomic applications of beneficial microbes. *Frontiers in Microbiology* 15. DOI: 10.3389/fmicb.2024.1523569.

Durán, N., Antonio, R.V., Haun, M. and Pilli, R.A. (1994) Biosynthesis of a trypanocide by *Chromobacterium violaceum*. *World Journal of Microbiology & Biotechnology* 10(6), 686–690. DOI: 10.1007/BF00327960.

Elumalai S., Infant B.S., Rajesh K.G. (2014) Extraction of carotenoid and thin layer chromatography (TLC), GC- MS, FT- IR and HPLC analysis of pharmaceutically important pigment astaxanthin from a new strain of *Haematococcus pluvialis*. *Weekly Science Research Journal* 2(8), 2321–7871. DOI: 10.9780/2321-7871/1202013/53.

El-Zawawy, N.A., Kenawy, E.-R., Ahmed, S. and El-Sapagh, S. (2024) Bioproduction and optimization of newly characterized melanin pigment from *Streptomyces djakartensis* NSS-3 with its anticancer, antimicrobial, and radioprotective properties. *Microbial Cell Factories* 23(1). DOI: 10.1186/s12934-023-02276-y.

Erdoğan, A., Karataş, A.B., Demirel, Z. and Dalay, M.C. (2022) Purification of fucoxanthin from the diatom *Amphora capitellata* by preparative chromatography after its enhanced productivity via oxidative stress. *Journal of Applied Phycology* 34(1), 301–309. DOI: 10.1007/s10811-021-02625-7.

Esquivel-Hernández, D.A., Rodríguez-Rodríguez, J., Rostro-Alanis, M., Cuéllar-Bermúdez, S.P., Mancera-Andrade, E.I. *et al.* (2017) Advancement of green process through microwave-assisted extraction of bioactive metabolites from *Arthrospira platensis* and bioactivity evaluation. *Bioresource Technology* 224, 618–629. DOI: 10.1016/j.biortech.2016.10.061.

Fomo, G., Madzimbamuto, T.N. and Ojumu, T.V. (2020) Applications of nonconventional green extraction technologies in process industries: Challenges, limitations and perspectives. *Sustainability* 12(13), 5244. DOI: 10.3390/su12135244.

Georgiopoulou, I., Tzima, S., Louli, V. and Magoulas, K. (2023) Process optimization of microwave-assisted extraction of chlorophyll, carotenoid and phenolic compounds from chlorella vulgaris and comparison with conventional and supercritical fluid extraction. *Applied Sciences* 13, 2740. DOI: 10.3390/app13042740.

Ghadge, V., Kumar, P., Maity, T.K., Prasad, K. and Shinde, P.B. (2022) A facile alternative sustainable process for the extraction and purification of microbial melanin. *ACS Sustainable Chemistry and Engineering* 10(8). DOI: 10.1101/2021.05.10.443376.

Goldring, J.D. (2019) Concentrating proteins by salt, polyethylene glycol, solvent, SDS precipitation, three-phase partitioning, dialysis, centrifugation, ultrafiltration, lyophilization, affinity chromatography, immunoprecipitation or increased temperature for protein isolation, drug interaction, and proteomic and peptidome evaluation. In: *Electrophoretic Separation of Proteins: Methods and Protocols*. Humana Press, New York, pp. 41–59. DOI: 10.1007/978-1-4939-8793-1_4.

Gorgich, M., Passos, M.L., Mata, T.M., Martins, A.A., Saraiva, M.L.M. *et al.* (2020) Enhancing extraction and purification of phycocyanin from *Arthrospira* sp. with lower energy consumption. *Energy Reports* 6, 312–318. DOI: 10.1016/j.egyr.2020.11.151.

Green, A.A. and Hughes, W.L. (1955) Protein fractionation on the basis of solubility in aqueous solutions of salts and organic solvents. *Methods in Enzymology* 1, 67–90.

Guedes, A.C., Amaro, H.M. and Malcata, F.X. (2011) Microalgae as sources of carotenoids. *Marine Drugs* 9(4), 625–644. DOI: 10.3390/md9040625.

Hatti-Kaul, R. and Mattiasson, B. (2003) Isolation and purification of proteins. In: Ratledge, C. and Kristiansen, B. (eds) *Basic Biotechnology*. Cambridge University Press, Cambridge, MA. DOI: 10.1201/9780203911686.

Hessel, V., Tran, N.N., Asrami, M.R., Tran, Q.D., Van Duc Long, N. *et al.* (2022) Sustainability of green solvents – review and perspective. *Green Chemistry* 24(2), 410–437. DOI: 10.1039/D1GC03662A.

Huang, C.-H., Chen, W.-C., Gao, Y.-H., Chen, G.-W., Lin, H.-T.V. *et al.* (2021) Enzyme-assisted method for phycobiliproteins extraction from *Porphyra* and evaluation of their bioactivity. *Processes* 9(3), 560. DOI: 10.3390/pr9030560.

Hynstova, V., Sterbova, D., Klejdus, B., Hedbavny, J., Huska, D. *et al.* (2018) Separation, identification and quantification of carotenoids and chlorophylls in dietary supplements containing *Chlorella vulgaris* and *Spirulina platensis* using High Performance Thin Layer Chromatography. *Journal of Pharmaceutical and Biomedical Analysis* 148, 108–118. DOI: 10.1016/j.jpba.2017.09.018.

Jaime, L., Mendiola, J.A., Ibáñez, E., Martin-Álvarez, P.J., Cifuentes, A. *et al.* (2007) β-carotene isomer composition of sub- and supercritical carbon dioxide extracts Antioxidant Activity Measurement. *Journal of Agricultural and Food Chemistry* 55(26), 10585–10590. DOI: 10.1021/jf0711789.

Janssen, C.H.C., Macías-Ruvalcaba, N.A., Aguilar-Martínez, M. and Kobrak, M.N. (2015) Metal extraction to ionic liquids: The relationship between structure, mechanism and application. *International Reviews in Physical Chemistry* 34(4), 591–622. DOI: 10.1080/0144235X.2015.1088217.

Jhanji, P., Jhanji, P. and Sharma, Dr.N. (2024) Microwave assisted extraction of food colorants: Principle, mechanism, extraction technique and applications. *International Journal of Advanced Biochemistry Research* 8(8), 1202–1213. DOI: 10.33545/26174693.2024.v8.i8o.1950.

Juang, R.-S. and Yeh, C.-L. (2014) Adsorptive recovery and purification of prodigiosin from methanol/water solutions of *Serratia marcescens* fermentation broth. *Biotechnology and Bioprocess Engineering* 19(1), 159–168. DOI: 10.1007/s12257-013-0547-2.

Juin, C., Chérouvrier, J.-R., Thiéry, V., Gagez, A.-L., Bérard, J.-B. *et al.* (2015) Microwave-assisted extraction of phycobiliproteins from *Porphyridium purpureum*. *Applied Biochemistry and Biotechnology* 175(1), 1–15. DOI: 10.1007/s12010-014-1250-2.

Kalra, R., Conlan, X.A. and Goel, M. (2020) Fungi as a potential source of pigments: Harnessing filamentous fungi. *Frontiers in Chemistry* 8, 369. DOI: 10.3389/fchem.2020.00369.

Kannaujiya, V.K., Sundaram, S. and Sinha, R.P. (2017) Advances and strategies of purification technology. In: *Phycobiliproteins: Recent Developments and Future Applications*. Springer Singapore, Singapore, pp. 99–120. DOI: 10.1007/978-981-10-6460-9_7.

Kim, S.M., Kang, S., Kwon, O., Chung, D. and Pan, C. (2012) Fucoxanthin as a major carotenoid in *Isochrysis* aff. Galbana: Characterization of extraction for commercial application. *Journal of the Korean Society for Applied Biological Chemistry* 55(4), 477–483. DOI: 10.1007/s13765-012-2108-3.

Kissoudi, M., Sarakatsianos, I. and Samanidou, V. (2017) Isolation and purification of food-grade C-phycocyanin from *Arthrospira platensis* and its determination in confectionery by HPLC with diode array detection. *Journal of Separation Science* 41(4), 975–981. DOI: 10.1002/jssc.201701151.

Kumar, V. and Lenhoff, A.M. (2020) Mechanistic modeling of preparative column chromatography for biotherapeutics. *Annual Review of Chemical and Biomolecular Engineering* 11(1), 235–255. DOI: 10.1146/annurev-chembioeng-102419-125430.

Lagashetti, A.C., Dufossé, L., Singh, S.K. and Singh, P.N. (2019) Fungal pigments and their prospects in different industries. *Microorganisms* 7(12), 604. DOI: 10.3390/microorganisms7120604.

Lakshman, M.R. and Rao, M.N. (1999) Purification and characterization of cellular carotenoid-binding protein from mammalian liver. In: *Methods in Enzymology*, Vol. 299. Academic Press, San Diego, CA, pp. 441–456.

Larocca, V., Martino, M., Trupo, M., Magarelli, R.A., Spagnoletta, A. *et al.* (2023) Evaluation of carbon dioxide supercritical fluid extraction (CO2-SFE) on carotenoids recovery from red yeast cells. *Biomass Conversion and Biorefinery*. DOI: 10.1007/s13399-023-04434-z.

Liau, B.-C., Shen, C.-T., Liang, F.-P., Hong, S.-E., Hsu, S.-L. *et al.* (2010) Supercritical fluids extraction and anti-solvent purification of carotenoids from microalgae and associated bioactivity. *Journal of Supercritical Fluids* 55(1), 169–175. DOI: 10.1016/j.supflu.2010.07.002.

Lin, Y., Wang, T., Lee, M. and Su, N. (2008) Biologically active components and nutraceuticals in the monascus-fermented rice: A review. *Applied Microbiology and Biotechnology* 77(5), 965–973. DOI: 10.1007/s00253-007-1256-6.

Lourenço-Lopes, C., Fraga-Corral, M., Garcia-Perez, P., Carreira-Casais, A., Silva, A. *et al.* (2022) A HPLC-DAD method for identifying and estimating the content of fucoxanthin, β-carotene and chlorophyll a in brown algal extracts. *Food Chemistry Advances* 1, 100095. DOI: 10.1016/j.focha.2022.100095.

Luengo, E., Condón-Abanto, S., Álvarez, I. and Raso, J. (2014) Effect of pulsed electric field treatments on permeabilization and extraction of pigments from *Chlorella vulgaris*. *Journal of Membrane Biology* 247(12), 1269–1277. DOI: 10.1007/s00232-014-9688-2.

Luzardo-Ocampo, I., Ramírez-Jiménez, A.K., Yañez, J., Mojica, L. and Luna-Vital, D.A. (2021) Technological applications of natural colorants in food systems: A review. *Foods* 10, 634. DOI: 10.3390/foods10030634.

Macías-Sánchez, M.D., Fernandez-Sevilla, J.M., Fernández, F.G.A., García, M.C.C. and Grima, E.M. (2010) Supercritical fluid extraction of carotenoids from *Scenedesmus almeriensis*. *Food Chemistry* 123(3), 928–935. DOI: 10.1016/j.foodchem.2010.04.076.

Mandal, V., Mohan, Y. and Hemalatha, S. (2007). Microwave assisted extraction—An innovative and promising extraction tool for medicinal plant research. *Pharmacognosy Reviews*, 1(1), 7–18.

Martins, R., Mouro, C., Pontes, R., Nunes, J. and Gouveia, I. (2023) Ultrasound-assisted extraction of bioactive pigments from *Spirulina platensis* in natural deep eutectic solvents. *Bioresources and Bioprocessing* 10(1), 88. DOI: 10.1186/s40643-023-00692-x.

Martínez, J.M., Delso, C., Álvarez, I. and Raso, J. (2019a) Pulsed electric field permeabilization and extraction of phycoerythrin from *Porphyridium cruentum*. *Algal Research* 37, 51–56. DOI: 10.1016/j.algal.2018.11.005.

Martínez, J.M., Delso, C., Aguilar, D.E., Gómez, P.F. and López, M.T. (2020a) Organic-solvent-free extraction of carotenoids from yeast *Rhodotorulaglutinis* by application of ultrasound under pressure. *Ultrasonics Sonochemistry* 61, 104833. DOI: 10.1016/j.ultsonch.2019.104833.

Martínez, J.M., Delso, C., Álvarez, I. and Raso, J. (2020b) Pulsed electric field-assisted extraction of valuable compounds from microorganisms. *Comprehensive Reviews in Food Science and Food Safety* 19(2), 530–552. DOI: 10.1111/1541-4337.12512.

Martínez, J.M., Gojkovic, Z., Ferro, L., Maza, M., Álvarez, I. *et al.* (2019b) Use of pulsed electric field permeabilization to extract astaxanthin from the nordic microalga *Haematococcus pluvialis*. *Bioresource Technology* 289, 121694. DOI: 10.1016/j.biortech.2019.121694.

Meng-xi, L.I., Hui-bin, H., Jie-yun, L., Jing-xiao, C.A.O. and Zhen-wang, Z. (2021) Antibacterial performance of a *Streptomyces spectabilis* strain producing metacycloprodigiosin. *Current Microbiology* 78(7), 2569–2576. DOI: 10.1007/s00284-021-02513-w.

Mishra, S.K., Shrivastav, A. and Mishra, S. (2011) Preparation of highly purified C-phycoerythrin from marine cyanobacterium *Pseudanabaena* sp. *Protein Expression and Purification* 80(2), 234–238. DOI: 10.1016/j.pep.2011.06.016.

Mondal, S., Pandit, S.G., Puttananjaiah, M.H., Harohally, N.V. and Dhale, M.A. (2019) Structural and functional characterization of new pigment molecule *Monashin* from *Monascus purpureus* CFR410-11. *Process Biochemistry* 82, 173–178. DOI: 10.1016/j.procbio.2019.04.004.

Mukherjee, G. and Singh, S.K. (2011) Purification and characterization of a new red pigment from *Monascus purpureus* in submerged fermentation. *Process Biochemistry* 46(1), 188–192. DOI: 10.1016/j.procbio.2010.08.006.

Mussagy, C.U., Santos-Ebinuma, V.C., Gonzalez-Miquel, M., Coutinho, J.A.P. and Pereira, J.F.B. (2019) Protic ionic liquids as cell-disrupting agents for the recovery of intracellular carotenoids from yeast *Rhodotorula glutinis* CCT-2186. *ACS Sustainable Chemistry & Engineering* 7(19), 16765–16776. DOI: 10.1021/acssuschemeng.9b04247.

Mussagy, C.U., Guimarães, A.A.C., Rocha, L.V.F., Winterburn, J., Santos-Ebinuma, V. de C. *et al.* (2021) Improvement of carotenoids production from *Rhodotorula glutinis* CCT-2186. *Biochemical Engineering Journal* 165, 107827. DOI: 10.1016/j.bej.2020.107827.

Nadar, S.S., Rao, P. and Rathod, V.K. (2018) Enzyme assisted extraction of biomolecules as an approach to novel extraction technology: A review. *Food Research International* 108, 309–330. DOI: 10.1016/j.foodres.2018.03.006.

Nemer, G., Louka, N., Vorobiev, E., Salameh, D., Nicaud, J.-M. *et al.* (2021) Mechanical cell disruption technologies for the extraction of dyes and pigments from microorganisms: A review. *Fermentation* 7(1), 36. DOI: 10.3390/fermentation7010036.

Padhan, B., Poddar, K., Sarkar, D. and Sarkar, A. (2021) Production, purification, and process optimization of intracellular pigment from novel *Psychrotolerantpaenibacillus* sp BPW19. *Biotechnology Reports* 29, e00592. DOI: 10.1016/j.btre.2021.e00592.

Pagels, F., Pereira, R.N., Vicente, A.A. and Guedes, A.C. (2021) Extraction of pigments from microalgae and cyanobacteria—A review on current methodologies. *Applied Sciences* 11(11), 5187. DOI: 10.3390/app11115187.

Palma, A., Díaz, M.J., Ruiz-Montoya, M., Morales, E. and Giráldez, I. (2021) Ultrasound extraction optimization for bioactive molecules from *Eucalyptus globulus* leaves through antioxidant activity. *Ultrasonics Sonochemistry* 76, 105654. DOI: 10.1016/j.ultsonch.2021.105654.

Patel, A.K., Albarico, F.P.J.B., Perumal, P.K., Vadrale, A.P., Nian, C.T. *et al.* (2022) Algae as an emerging source of bioactive pigments. *Bioresource Technology* 351, 126910. DOI: 10.1016/j.biortech.2022.126910.

Picot-Allain, C., Mahomoodally, M.F., Ak, G. and Zengin, G. (2021) Conventional versus green extraction techniques — A comparative perspective. *Current Opinion in Food Science* 40, 144–156. DOI: 10.1016/j.cofs.2021.02.009.

Pore, T.S., Khanolkar, A.B. and Nadaf, N.H. (2016) Production, purification, identification of prodigiosin from *Serratia* sp. and its antimicrobial activity. Available at: www.rjlbpcs.com/article-pdf-downloads/2016/6/37.pdf (accessed 12 August 2025).

Rabel, F. and Sharma, J. (2017) Review of the state of the art of preparative thin-layer chromatography. *Journal of Liquid Chromatography & Amp Related Technologies* 40(4), 165–176. DOI: 10.1080/10826076.2017.1294081.

Rajendran, P., Somasundaram, P. and Dufossé, L. (2023) Microbial pigments: Eco-friendly extraction techniques and some industrial applications. *Journal of Molecular Structure* 1290, 135958. DOI: 10.1016/j.molstruc.2023.135958.

Rettori, D. and Durán, N. (1998) Production, extraction and purification of violacein: An antibiotic pigment produced by *Chromobacterium violaceum*. *World Journal of Microbiology and Biotechnology* 14(5), 685–688. DOI: 10.1023/A:1008809504504.

Rodríguez-Rodríguez, E., Sánchez-Prieto, M. and Olmedilla-Alonso, B. (2020) Assessment of carotenoid concentrations in red peppers (*Capsicum annuum*) under domestic refrigeration for three weeks as determined by HPLC-DAD. *Food Chemistry* X 6, 100092. DOI: 10.1016/j.fochx.2020.100092.

Rupley, J.A., Gratton, E. and Careri, G. (1983) Water and globular proteins. *Trends in Biochemical Sciences* 8(1), 18–22. DOI: 10.1016/0968-0004(83)90063-4.

Safaei, M., Maleki, H., Soleimanpour, H., Norouzy, A., Zahiri, H.S *et al.* (2019) Development of a novel method for the purification of C-phycocyanin pigment from a local cyanobacterial strain *Limnothrix* sp. NS01 and evaluation of its anticancer properties. *Scientific Reports* 9(1). DOI: 10.1038/s41598-019-45905-6.

Saini, R.K. and Keum, Y.S. (2018) Carotenoid extraction methods: A review of recent developments. *Food Chemistry* 240, 90–103. DOI: 10.1016/j.foodchem.2017.07.099.

Shaikh, Z. (2016) Biosynthesis of prodigiosin and its applications. *Journal of Pharmacy and Biological Sciences* 11(6), 01–28. DOI: 10.9790/3008-1106050128.

Shanaida, M., Klochkov, V. and Kovalev, G. (2021) Polyphenols and pharmacological screening of a *Monarda fistulosa* L. dry extract based on a hydrodistilled residue by-product. *Frontiers in Pharmacology* 12. DOI: 10.3389/fphar.2021.563436.

Sharif, K.M., Rahman, M.M., Azmir, J., Mohamed, A., Jahurul, M.H.A *et al.* (2014) Experimental design of supercritical fluid extraction – A review. *Journal of Food Engineering* 124, 105–116. DOI: 10.1016/j.jfoodeng.2013.10.003.

Sharma, D., Singh, S. and Gupta, R. (2012) Pigment extraction from fungus for textile dyeing. *Indian Journal of Fibre & Textile Research* 37, 68–73.

Sherma, J. (2017) Review of thin-layer chromatography in pesticide analysis: 2014–2016. *Journal of Liquid Chromatography & Related Technologies* 40(5–6), 226–238. DOI: 10.1080/10826076.2017.1298024.

Shen, L., Pang, S., Zhong, M., Sun, Y., Qayum, A. *et al.* (2023) A comprehensive review of ultrasonic assisted extraction (UAE) for bioactive components: Principles, advantages, equipment, and combined technologies. *Ultrasonics Sonochemistry* 101, 106646. DOI: 10.1016/j.ultsonch.2023.106646.

Sherma, J. and Fried, B. (eds) (2003) *Handbook of Thin-Layer Chromatography*. CRC Press, Boca Raton, FL. DOI: 10.1201/9780203912430.

Shi, K., Tang, R., Huang, T., Wang, L. and Wu, Z. (2017) Pigment fingerprint profile during extractive fermentation with *Monascus anka* GIM 3.592. *BMC Biotechnology* 17. DOI: 10.1186/s12896-017-0366-1.

Silva, T.P., Alves, L., Salgado, F., Roseiro, J.C., Łukasik, R.M. *et al.* (2024) Ionic liquids toward enhanced carotenoid extraction from bacterial biomass. *Molecules* 29(17), 4132. DOI: 10.3390/molecules29174132.

Singh, A.K., Rana, H.K. and Pandey, A.K. (2020) Analysis of chlorophylls. In: *Recent Advances in Natural Products Analysis*. pp. 635–650.

Soliev, A.B., Hosokawa, K. and Enomoto, K. (2011) Bioactive pigments from marine bacteria: Applications and physiological roles. *Evidence-Based Complementary and Alternative Medicine* 670349. DOI: 10.1155/2011/670349.

Sonani, R.R., Singh, N.K., Kumar, J., Thakar, D. and Madamwar, D. (2014) Concurrent purification and antioxidant activity of phycobiliproteins from *Lyngbya* sp. A09DM: An antioxidant and anti-aging potential of phycoerythrin in *Caenorhabditis elegans*. *Process Biochemistry* 49(10), 1757–1766. DOI: 10.1016/j.procbio.2014.06.022.

Song, M.-J., Bae, J., Lee, D.-S., Kim, C.-H., Kim, J.-S. *et al.* (2006) Purification and characterization of prodigiosin produced by integrated bioreactor from *Serratia* sp KH-95. *Journal of Bioscience and Bioengineering* 101(2), 157–161. DOI: 10.1263/jbb.101.157.

Sparr Eskilsson, C. and Björklund, E. (2000) Analytical-scale microwave-assisted extraction. *Journal of Chromatography A* 902(1), 227–250. DOI: 10.1016/S0021-9673(00)00921-3.

Sun, W., Lin, H., Zhai, Y., Cao, L., Leng, K. *et al.* (2014) Separation, purification, and identification of (3S,3′s)-trans-astaxanthin from *Haematococcus pluvialis*. *Separation Science and Technology* 50, 1377–1383. DOI: 10.1080/01496395.2014.976873.

Swartz, M. (2010) HPLC detectors: A brief review. *Journal of Liquid Chromatography &Amp Related Technologies* 33(9–12), 1130–1150. DOI: 10.1080/10826076.2010.484356.

Taguchi, T., Itou, K., Ebizuka, Y., Malpartida, F., Hopwood, D.A. *et al.* (2000) Chemical characterisation of disruptants of the *Streptomyces coelicolor* A3(2). Actvi genes involved in actinorhodin biosynthesis. *Journal of Antibiotics* 53, 144–152. DOI: 10.7164/antibiotics.53.144.

Tavanandi, H.A., Vanjari, P. and Raghavarao, K.S.M.S. (2019) Synergistic method for extraction of high purity Allophycocyanin from dry biomass of *Arthrospira platensis* and utilization of spent biomass for recovery of carotenoids. *Separation and Purification Technology* 225, 97–111. DOI: 10.1016/j.seppur.2019.05.064.

Timasheff, S.N. and Arakawa, T. (1997) Stabilization of protein structure by solvents. In: Creighton, T.E. (ed.) *Protein Structure: A Practical Approach*, 2nd edn. Oxford University Press, Oxford, UK, pp. 349–364. DOI: 10.1093/oso/9780199636198.003.0014.

Tiwari, K. (2022) A potential application of microbial pigment: An alternative to synthetic dye & colourants. *International Journal of Current Science Research and Review* 05. DOI: 10.47191/ijcsrr/V5-i9-07.

Tong, Y., Gao, L., Xiao, G. and Pan, X. (2012) Microwave pretreatment-assisted ethanol extraction of chlorophylls from *Spirulina platensis*. *Journal of Food Process Engineering* 35(5), 792–799. DOI: 10.1111/j.1745-4530.2010.00629.x.

Unagul, P., Wongsa, P., Kittakoop, P., Intamas, S., Srikitikulchai, P. *et al.* (2005) Production of red pigments by the insect pathogenic fungus *Cordyceps unilateralis* BCC 1869. *Journal of Industrial Microbiology & Biotechnology* 32(4), 135–140. DOI: 10.1007/s10295-005-0213-6.

Venil, C.K., Velmurugan, P., Dufossé, L., Devi, P.R. and Ravi, A.V. (2020) Fungal pigments: Potential coloring compounds for wide ranging applications in textile dyeing. *Journal of Fungi* 6(2), 68. DOI: 10.3390/jof6020068.

Ventura, S.P.M., Santos-Ebinuma, V.C., Pereira, J.F.B., Teixeira, M.F.S., Pessoa, A. *et al.* (2013) Isolation of natural red colorants from fermented broth using ionic liquid-based aqueous two-phase systems. *Journal of Industrial Microbiology and Biotechnology* 40(5), 507–516. DOI: 10.1007/s10295-013-1237-y.

Ventura, S.P.M., Silva, F.A., Quental, M.V., Mondal, D., Freire, M.G. *et al.* (2017) Ionic-liquid-mediated extraction and separation processes for bioactive compounds: Past, present, and future trends. *Chemical Reviews* 117(10), 6984–7052. DOI: 10.1021/acs.chemrev.6b00550.

Wallace, R.G. and Rochfort, K.D. (2023) Ion-exchange chromatography: Basic principles and application. In: *Protein Chromatography*, Vol. 2699. Springer Nature, Dordrecht, The Netherlands, pp. 161–177.

Wang, E., Dong, C., Park, R.F. and Roberts, T.H. (2018) Carotenoid pigments in rust fungi: Extraction, separation, quantification and characterisation. *Fungal Biology Reviews* 32(3), 166–180. DOI: 10.1016/j.fbr.2018.02.002.

Wang, S.L., Liu, W., Wang, H.X. and Lv, C.H. (2012) The extraction of β-carotene from red yeast cells by supercritical carbon dioxide technique. *Advanced Materials Research* 554–556, 949–952. DOI: 10.4028/www.scientific.net/AMR.554-556.949.

Wilson, K. (2010) Chromatographic techniques. In: Wilson, K. and Walker, J. (eds) *Principles and Techniques of Biochemistry and Molecular Biology*. Cambridge University Press, Cambridge, UK, pp. 433–476. DOI: 10.1017/cbo9780511841477.012.

Zhang, M., Guo, B.-H. and Xu, J. (2016) A review on polymer crystallization theories. *Crystals* 7(1), 4. DOI: 10.3390/cryst7010004.

Zhang, Q.W., Lin, L.G. and Ye, W.C. (2018) Techniques for extraction and isolation of natural products: A comprehensive review. *Chinese Medicine* 13(1), 20. DOI: 10.1186/s13020-018-0177-x.

Zhao, Y., Guo, L., Xia, Y., Zhuang, X. and Chu, W. (2019) Isolation, identification of carotenoid-producing *Rhodotorula* sp. from marine environment and optimization for carotenoid production. *Marine Drugs* 17. DOI: 10.3390/md17030161.

Zhiping, Z., Ji, H., Gong, G., Zhang, X. and Tan, T. (2014) Synergistic effects of oleaginous yeast *Rhodotorula glutinis* and microalga *Chlorella vulgaris* for enhancement of biomass and lipid yields. *Bioresource Technology* 164, 93–99. DOI: 10.1016/j.biortech.2014.04.039.

Zhu, Z., Wu, Q., Di, X., Li, S., Barba, F.J. *et al.* (2017) Multistage recovery process of seaweed pigments: Investigation of ultrasound assisted extraction and ultra-filtration performances. *Food and Bioproducts Processing* 104, 40–47. DOI: 10.1016/j.fbp.2017.04.008.

Zou, T.-B., Jia, Q., Li, H.-W., Wang, C.-X. and Wu, H.-F. (2013) Response surface methodology for ultrasound-assisted extraction of astaxanthin from *Haematococcus pluvialis*. *Marine Drugs* 11(5), 1644–1655. DOI: 10.3390/md11051644.

Žuvela, P., Skoczylas, M., Jay Liu, J., Ba Czek, T., Kaliszan, R. *et al.* (2019) Column characterization and selection systems in reversed-phase high-performance liquid chromatography. *Chemical Reviews* 119(6), 3674–3729. DOI: 10.1021/acs.chemrev.8b00246.

9 Structural Characterization and Analysis of Microbial Pigments

Rashid Abbas Khan[1], Tauqeer Ahmed Qadri[2], Amjid Khan[3] and Muhammad Anas[4*]

[1]Department of Biology, Faculty of Science, Prince of Songkhla University, Hat Yai, Songkhla, Thailand; [2]Department of Biosciences, COMSATS University Islamabad, Islamabad, Pakistan; [3]UNESCO UNISA Africa Chair in Nanoscience and Nanotechnology, College of Graduate Studies, University of South Africa, Pretoria, South Africa; [4]Department of Soil Science & Plant Nutrition, Selçuk University, Konya, Türkiye

9.1 Introduction

Microbial pigments have attracted increasing research interest because of their structural diversity, ecological value, and broad industrial applications. Derived from bacteria, fungi, and algae, these natural pigments offer key advantages over their synthetic counterparts: they are biodegradable, environmentally safe, and often exhibit bioactive properties, such as antimicrobial effects and potential anticancer activity (Jorge and Pereira, 2025). The growing demand for sustainable, eco-friendly materials driven by environmental concerns and stricter regulations on synthetic pigments has further fueled interest in their production (Grewal *et al.*, 2022).

Chemically, bacterial pigments display a remarkable variety of properties, including aromatic chromophores, complex polyketide frameworks, and nitrogen-containing heterocycles. Each microbial species synthesizes pigments through unique metabolic pathways that are influenced by nutrient availability and environmental conditions. Because pigment performance is closely tied to structural features, detailed structural characterization is essential to optimize stability, bioactivity, and industrial usability (Ramesh *et al.*, 2019).

Advances in analytical science in recent years have contributed significantly to the expansion of the repertoire of microbial pigment characterization. Conventional spectroscopic and chromatographic techniques, such as UV-Vis spectroscopy, FTIR, Raman spectroscopy, HPLC, NMR, and mass spectroscopy-based methods, provide a detailed understanding of the functional groups, molecular make-up, and design of chromophores. New technologies such as X-ray diffraction, scanning and transmission electron microscopy (SEM/TEM), and many hyphenated chromatographic-spectroscopic platforms provide higher resolution and more advanced analytical characterization. Simultaneously, *in silico* methods have become

*Corresponding author: muhammed.anas@selcuk.edu.tr

DOI: 10.1079/ 9781836990505.0009

available for predicting pigment structures, performing *in silico* simulations of pigment bioactivity, and designing pigment biosynthesis (Figon and Casas, 2021; Mummaleti *et al.*, 2024).

SEM/TEM integrates a wide range of analytical techniques, including both established methods (UV-vis, FTIR, Raman, HPLC, MS, and NMR) and recent advancements (X-ray diffraction, SEM/TEM, hyphenated chromatographic-spectroscopic), along with computational tools to predict structures, evaluate bioactivity, and enhance biosynthesis (Valenzuela-Gloria *et al.*, 2021). Practical examples illustrate that enhanced structural elucidation affects pigment stability, biological activity, and industrial applications. This comprehensive approach, merging chemical and methodological insights, serves as a powerful research tool for scientists and industry professionals, thereby advancing the global understanding of microbial pigment structures and their functional potential (Singhal *et al.*, 2024).

9.2 Source Microorganisms and Pigment Classes

Microbial pigments represent a diverse array of secondary metabolic products with varied biosynthetic origins and are present in bacteria, fungi, algae, and archaea. These pigments display structural motifs, ranging from conjugated polyenes to complex heteroaromatic polymers, shaped by the organism's metabolic capabilities and adaptations for survival in different environments. Understanding the origins and classification of these pigments is essential for their structural characterization, as the chemical source and biosynthetic context often determine their molecular structure, chromophore nature, and functional stability.

9.2.1 Pigment-producing microorganisms: bacteria, fungi, algae, and archaea

Bacteria form a class of pigment-rich factors over a wide taxonomic range. Examples of pigment-producing organisms include representatives of the high-latitude genera *Serratia*, *Chromobacterium*, *Pseudomonas*, *Streptomyces*, and the extremophilic *Halobacterium*, which form pigments, such as prodigiosin, violacein, pyocyanin, carotenoids, and bacterioruberin, respectively. Many of these pigments act as antioxidants, photoprotectants, virulence factors, and electron carriers. A good example is *Pseudomonas aeruginosa*, which produces a blue-green pigment called pyocyanin phenazine that is essential for redox cycling and other antimicrobial effects (Nawaz *et al.*, 2020; Himani Agarwal *et al.*, 2023).

Fungal pigments also produce pigments such as melanin, azaphilones, anthraquinones, and polyketide derivatives. These pigments have ecological roles such as UV protection, spore pigmentation, and grazing or microbial competitor defense. The red pigments from *Monascus* species via the production of red polyketides, such as monascorubrin, have long been used in traditional food. *Aspergillus* and *Alternaria* species produce complex melanin (Lin and Xu, 2022; Afroz Toma *et al.*, 2023).

Algae, especially microalgae (*Dunaliella salina*, *Chlorella vulgaris*) and cyanobacteria (*Spirulina platensis*), provide rich sources of carotenoids (β-carotene, lutein, and astaxanthin), chlorophylls, and phycobiliproteins (phycocyanin and phycoerythrin). Photosynthetic light harvesting and photoprotection strongly depend on these pigments, which have highly conjugated chromophores tuned to the light wavelength of interest (Chen *et al.*, 2023; Fatima *et al.*, 2024).

Archaea, which are less well understood, have their own special pigments that are suited to extreme conditions. Halophilic strains produce bacterioruberin and other C50 carotenoids, which are long conjugated systems that provide color and membrane stability in elevated high-salt, high-radiation environments (Grivard *et al.*, 2022; Martínez-Espinosa, 2024).

Nearly every type of microbe creates biosynthesized colors through biological processes, with each microbial group producing special color types. Under natural stress, microorganisms create specific pigment molecules as secondary products. Pigment-producing microbes create aromatic molecules all the way up to sophisticated heterocyclic and polyketide arrangements that display many biological functions and produce different colors.

9.2.2 Bacteria-derived pigments

Several bacterial genera produce biopigments. The red tripyrrole pigment prodigiosin, synthesized by *Serratia marcescens*, is known to have anticancer and antibacterial properties. *Streptomyces* species are known producers of melanin, actinorhodin (blue), and other polyketide pigments, which are often associated with soil ecology and UV protection. *Chromobacterium violaceum* produces the purple pigment violacein, which has antioxidant, antiprotozoal, cytotoxic, and other activities.

The chemical architecture of each pigment is defined based on its distinct nonribosomal peptide synthesis or polyketide biosynthetic origin (Mantri *et al.*, 2022; Huang *et al.*, 2024).

9.2.3 Fungal pigments

Fungi produce stable, high-yield pigments that are used in the food, cosmetics, and textiles industries. Among the known polyketide-based azaphilones, the pigments used in food industries in Asia are monascorubrin and ankaflavin produced by *Monascus purpureus*. *Penicillium* and *Aspergillus* spp. synthesize many anthraquinones and naphthoquinones with antioxidant and antimicrobial properties. *Alternaria* and *Cladosporium* produce melanin for stress protection and metal tolerance, apart from pathogenicity factors, acetyl-CoA precursors and multiple enzymatic oxidations (Christiansen *et al.*, 2021; Lin and Xu, 2022).

9.2.4 Algal pigments

Cyanobacteria and chlorophytes are especially useful microalgae as they produce pigments such as phycocyanin, phycoerythrin, chlorophylls, and carotenoids (such as astaxanthin in *Haematococcus pluvialis*). Phycobiliproteins from cyanobacteria can serve as fluorescent markers and natural food colorants. These water-soluble pigments are derived from proteinaceous substances, exhibit good bioavailability, and are easily extracted under mild conditions. Structural features include linear tetrapyrrole chromophores covalently attached to apoproteins; thus, they differ from purely organic microbial pigments (Chini Zittelli *et al.*, 2023; Fatima *et al.*, 2024).

9.3 Classification Based on Chemical Structure

Microbial pigments can be grouped based on their core chemical skeleton. Historically, major classes have included pyrroles (e.g. prodigiosin), quinones (e.g. actinorhodin), phenazines (e.g. pyocyanin), carotenoids (e.g. β-carotene), and azaphilones. This helps in predicting the physicochemical behavior, solubility, stability, and interaction potential of pigments. The criteria for classification also determine the steps in structural elucidation that should be taken. Conjugated double-bond systems of carotenoids are ideal for UV-Vis spectral analysis, while quinones are ideal because of the presence of a large variety of functional groups in the FT-IR and MS techniques. The origin and chemical nature of microbial pigments are used as a basis for their structural characterization. The final product has complexity, yield, and application spectrum determined by the biosynthetic capabilities of each pigment-producing organism (Valenzuela-Gloria *et al.*, 2021; Barreto *et al.*, 2023).

9.4 Extraction, Purification, and Sample Preparation

Purification and subsequent extraction are essential steps that must be performed prior to structural analysis of pigments in microbes. The choice of an appropriate extraction procedure depends mainly on the chemical properties of the pigment, such as polarity, solubility, chromophore type, and stability (Table 9.1). To perform comprehensive research applications, these procedures should be effective in removing contaminants without damaging the structure so that it is easy to obtain precise characterization (Cuellar-Bermudez *et al.*, 2015).

9.4.1 Solvent-based extraction methods

The main method of microbial pigment isolation from biomass is solvent extraction, and

Table 9.1. Summary of common extraction and purification techniques used for microbial pigments, their target pigment classes, advantages, and supporting literature. The table highlights both conventional solvent-based methods and emerging green technologies relevant for high-purity pigment isolation.

Technique	Description	Suitable pigments	Advantages	References
Solvent extraction (methanol)	Biomass treated with methanol to dissolve intracellular pigments	Prodigiosin, violacein, melanin	High extraction efficiency, simple protocol	(Sarmiento-Tovar *et al.*, 2022)
Solvent extraction (acetone)	Disruption of cells followed by pigment solubilization in acetone	Carotenoids, xanthophylls	Suitable for lipophilic pigments, fast evaporation	(Butnariu, 2016)
Methanol:chloroform mixture	Biphasic solvent system used for enhanced extraction of nonpolar pigments	Prodigiosin, actinorhodin	Increases solubility and selectivity	(Fatima *et al.*, 2024)
Ethanol extraction	Food-grade solvent used under agitation or sonication	Azaphilones, violacein	GRAS status, applicable for food and pharma industries	(Aman Mohammadi *et al.*, 2022)
Supercritical CO_2 extraction	Nontoxic, tunable CO_2 used under pressure to extract pigments	Carotenoids, astaxanthin	Green method, no solvent residue, high purity	(Yan *et al.*, 2024)
Microwave-assisted extraction (MAE)	Rapid heating breaks cells, releasing pigments	Melanin, azaphilones	Time-saving, energy-efficient	(Rajendran *et al.*, 2023)
Ultrasound-assisted extraction (UAE)	Acoustic cavitation enhances solvent penetration	Anthraquinones, phycocyanin	Increases yield, reduces solvent and energy consumption	(Bhosle *et al.*, 2015)
Ionic liquid-based extraction	Novel solvents used to dissolve complex pigments	Melanin, anthocyanins	Recyclable, low vapor pressure, customizable solvents	(Saha *et al.*, 2022)
Deep eutectic solvents (DES)	Biodegradable solvent systems for sustainable extraction	Anthraquinones, flavins	High stability, low cost, eco-friendly	(Chan, 2020)
Thin layer chromatography (TLC)	Analytical technique to monitor pigment purity	Prodigiosin, violacein	Quick screening of compound mixtures	(Ramesh *et al.*, 2020)

Continued

Table 9.1. Continued

Technique	Description	Suitable pigments	Advantages	References
Column chromatography	Separates based on polarity using silica/C18 columns	Xanthones, quinones	Simple lab-scale separation	(El-Gazzar *et al.*, 2025)
High-performance liquid chromatography (HPLC)	High-resolution purification using pressure-driven columns	Violacein, deoxyviolacein	High accuracy, simultaneous quantification and analysis	(Boysen and Hearn, 2011)
Size exclusion chromatography (SEC)	Separation based on molecular size	Phycocyanin, phycobiliproteins	Useful for protein-bound pigments	(Park *et al.*, 2021)
Flash chromatography	Pressurized column chromatography with rapid elution	Carotenoids, flavins	Faster and more scalable than gravity-fed methods	(Bhusal *et al.*, 2017)
Preparative TLC	Large-scale version of TLC for semi-purification	Violacein, prodigiosin	Inexpensive and efficient for small samples	(Derguine-Mecheri *et al.*, 2021)
Solid phase extraction (SPE)	Concentration and purification on cartridge-based stationary phases	Xanthophylls, chlorophylls	Efficient clean-up prior to HPLC	(Dugheri *et al.*, 2021)
Liquid-liquid extraction (LLE)	Two immiscible solvents used to separate pigment from aqueous matrix	Water-insoluble pigments	Selective phase partitioning	(Martins *et al.*, 2021)
Crystallization	Precipitation from solvents based on solubility differences	Prodigiosin	High-purity product, no need for expensive equipment	(Castro *et al.*, 1959)

the choice of a solvent class is based on the polarity and stability of the target compound. Carotenoids and other lipophilic pigments are always extracted with nonpolar solvents (e.g. hexane and petroleum ether), whereas water-soluble flavins and phycobiliproteins are traditionally extracted using polar solvents (methanol, ethanol, and acetone) (Fatima and Anuradha, 2022). Sequential solvent extraction has the potential to dramatically increase the recovery of several classes of pigments from a sample.

The efficiency of the extraction process and the stability of chromophores can be affected by factors such as the solvent-to-biomass ratio, the method of agitation (including ultrasonication and homogenization), and temperature, which must be performed in a manner that minimizes thermal degradation or photodegradation. For structural analysis, it is important to minimize solvent impurities and coadsorbed metabolites as they may interfere with spectral readings (Pagels *et al.*, 2021).

9.4.2 Chromatographic purification techniques

One method for refining postisolation pigments from crude extracts involves purifying the extract to remove lipids, proteins, polysaccharides, and secondary metabolites, which may complicate subsequent analysis. High-performance liquid chromatography (HPLC) is widely recognized as the gold standard for pigment separation, as it effectively differentiates pigments based on their polarity and chromophore absorbance properties, particularly when combined with diode array detection. Carotenoids and phenazines are typically separated using reversed-phase C18 columns, whereas charged pigments, such as phycobiliproteins, are more appropriately separated by ion exchange chromatography (Kiani *et al.*, 2022; Fatima *et al.*, 2024). Thin-layer chromatography (TLC) is a rapid and cost-effective method for the initial separation and preliminary purity assessment. For thermo-labile pigments, oxidative degradation can be significantly minimized using flash chromatography under an inert atmosphere. Overall, the purification strategy should yield spectroscopically pure pigments suitable for further spectroscopic, mass spectrometric, and crystallographic analyses (Colyer *et al.*, 2005).

9.4.3 Green and sustainable extraction technologies

Green bioprocessing has advanced significantly, resulting in increased interest in the adoption of environmentally sustainable technologies for pigment extraction. There is particular emphasis on economic scalability and applications subject to stringent regulations. Supercritical fluid extraction (SFE) utilizing carbon dioxide is effective in recovering nonpolar pigments, offering solvent-free purification with adjustable selectivity by manipulating the pressure and cosolvent composition (Abou Elmaaty *et al.*, 2022; Tzima *et al.*, 2023). An additional increase in the yield and a significant decrease in the solvent volume were obtained using pressurized liquid extraction and ultrasound-assisted extraction. Enzyme-assisted extraction also offers selective breakdown of microbial cellular walls under relatively physiologically benign conditions, and hence retains the labile pigment structure. From a structural characterization perspective, these methods reduce the chemical alteration and degradation of the pigment, and such analytical data reflect the natural pigment structure (Fatima *et al.*, 2024; Morón-Ortiz *et al.*, 2024).

9.4.4 Prevention of degradation during handling and storage

Pigments may also degrade owing to light, oxygen, temperature, and pH changes. It is necessary to ensure structural integrity by implementing strategies to prevent such changes. To minimize damage to light-sensitive pigments, work should be done in light-reduced circumstances or under the protection of amber glasses (Zhao *et al.*, 2021). Samples that contain oxygen-sensitive extracts should undergo processing under inert gas atmospheres, and thermo-labile molecules should be refrigerated or stored at low temperatures (Shkolnikov Lozober *et al.*, 2023). The addition of antioxidants, such as butylated hydroxytoluene (BHT), is optional. In instances when pigments are extremely susceptible to oxidation, as in the case of carotenoids, this can be done during the extraction process. Adjusting the pH can mitigate hydrolytic breakdown, particularly for esterified or glycosylated pigments. Adhering to strict sample handling protocols is essential to ensure that the structural results obtained through spectroscopy or crystallography accurately reflect the original unaltered pigment (Zhao *et al.*, 2021).

9.4.5 Sample preparation for spectroscopy, chromatography, and crystallography

The preparation protocols vary according to the analytical method employed; however, they universally aim to yield pigment samples that are clean, concentrated, and stable. For spectroscopic analyses, such as UV-Vis, FTIR, and Raman spectroscopy, it is imperative to dissolve the pigments in a solvent that does not exhibit absorption or emission bands overlapping with those of the analyte. In the context of mass spectrometry and NMR experiments, it is essential to remove residual salts and solvents that may impede ionization or obscure spectral lines. Typical NMR preparation involves placing a lyophilized material in a deuterated solvent to achieve a solubilized sample (Saif *et al.*, 2021; Malissa *et al.*, 2023). In X-ray crystallography, methods such as slow solvent evaporation, vapor diffusion, and cocrystallization with inert carriers are utilized to produce diffraction-quality crystals, which can be particularly challenging with flexible or amorphous pigments. The thoroughness of preparation dictates the answers provided by an examination of the shape of the resolution, accuracy, and interpretability of the obtained structural data (Li *et al.*, 2025).

9.5 Analytical Techniques for Structural Characterization (Fig. 9.1)

9.5.1 Spectroscopic approaches

Spectroscopy techniques offer a source of information regarding pigment chromophores, functional groups, and the entire design of molecular organic molecules without being destructive in nature. Identification of the type of chromophore and length of conjugation is regularly carried out by UV-Vis absorption spectroscopy by correlating the characteristic lambda$_{max}$ values of all corresponding chromophore types; UV-Vis spectral markers may also be used to denote structural modification or degradation reactions (Mahabaduge, 2025). Electronic transitions are linked with each other, provided by fluorescence spectroscopy, which allows study of the intrinsic emissive properties, as well as conformational assessments (Tamburini *et al.*, 2024). Fourier transform infrared spectroscopy (FTIR) offers the possibility of identifying functional groups such as hydroxyl, carbonyl, or aromatic groups and offers vibrational fingerprints that complement other structural information. Raman spectroscopy is also highly suitable for the characterization of carotenoids and pigments with very large conjugated systems owing to its increased sensitivity to vibrations of C = C bonds, and spectroscopies can be easily correlated with molecular physics (Longoni and Bruni, 2021).

9.5.2 Chromatography

Chromatographic separation is a preproduct of the structural characterization of pigments identified with a complex microbial origin. HPLC, particularly when used in conjunction with diode-array detection, offers the potential to separate the isomers of pigments and derivatives and record the ultraviolet–visible absorption spectra of each fraction as it elutes. Conversely, there is a quick and economic way to verify the purity of the pigments and the category of the pigments on the retention factor and the sequential variation of colors accompanied by the predicted standard materials that can be administered by TLC. Less commonly used in the case of nondegraded pigments, gas chromatography (GC) may analyze the volatile-derivatized pigments in this respect, supporting the evidence of structures (Venil *et al.*, 2021; Aman Mohammadi *et al.*, 2022).

9.5.3 Mass spectrometry

Mass spectrometry is a key component of pigment structure studies because it can provide acceptable molecular mass information and elemental compositions (Sousa, 2022). Electrospray ionization (ESI)-MS can provide molecular ions of unstable pigments, even when they are predominantly characterized by fragmentation. Very high molecular weight pigments and pigment protein complexes can be examined without excessive fragmentation through matrix-assisted laser desorption/ionization time-of-flight (MALDI-TOF). In an instrument featuring high-resolution mass spectrometry (HRMS), molecular formula determination is further developed, where differences between isomers can be distinguished, and the nature of the substituents and how they are chain-linked can be resolved in more detail through rigorous fragmentation studies (Garcia-Lopez *et al.*, 2021).

9.5.4 NMR spectroscopy

Characterization of the connectivity and stereochemistry of pigment molecules is provided by NMR spectroscopy. One-dimensional data (^1 hour and ^13C) present a chemical environment of protons and carbons, whereas two-dimensional experiments (COSY, HSQC,

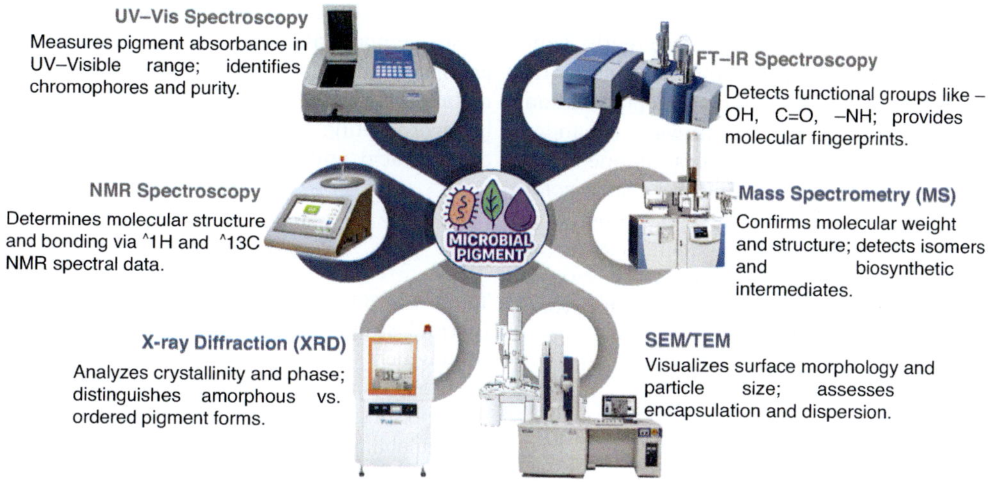

Fig. 9.1. Key analytical techniques used for structural characterization of microbial pigments, highlighting their primary function and relevance to pigment identification, purity assessment, and formulation suitability.

and HMBC) present information on spin–spin coupling and heteronuclear correlation and are indispensable for deriving the entire molecular structure (Khan *et al.*, 2021). Combining the new cryoprobe technology and refinement of solvents will enable greatly enhanced spectral resolution of even unrefined pigments, and solid-state NMR-based methods can be applied to investigate insoluble pigments (e.g. melanin).

Thus, NMR spectroscopy has been used as an essential tool for carrying out investigations into the structural complexity of pigment molecules (Valenzuela-Gloria *et al.*, 2021).

9.5.5 Scanning and transmission electron microscopy (SEM/TEM)

It is possible to image pigment particles to nanometer dimensions using SEM and TEM and to determine their morphology, crystal habit, and aggregate shapes. These characteristics can influence pigment photostability, light interactions, and practical use, thereby offering a valuable context for spectroscopic and crystallographic findings (Falsafi *et al.*, 2020).

9.6 Stability and Functional Group Analysis

9.6.1 Thermal and pH stability linked to structural features (Table 9.2)

9.6.1.1 Thermal stability

The thermal resilience of microbial pigments is determined by molecular rigidity, conjugation, and aromaticity. Carotenoids, a class of polyene pigments, are particularly prone to thermal isomerization and oxidative cleavage of carbon–carbon double bonds (Abdollahi *et al.*, 2021). Conversely, aromatic heterocyclic pigments, including melanin and phenazines, demonstrate considerable thermal stability owing to their extensive two-dimensional and three-dimensional conjugation and intermolecular polymeric cross-linking. Furthermore, stabilizing intramolecular hydrogen bonds enhances the heat resistance of various flavin and anthraquinone derivatives (Venil *et al.*, 2013).

9.6.1.2 pH stability

Ionizable functional groups play a crucial role in influencing structural alterations in response to pH fluctuations. Pigments containing phenolic, carboxylic, and amino groups frequently demonstrate pH-sensitive color changes because the balance between protonation and deprotonation modifies the conjugation pathways (Chandra *et al.*, 2021). For example, anthraquinone-based fungal pigments can undergo reversible color changes under acidic or alkaline conditions, whereas protein-bound pigments, such as phycobiliproteins, may experience irreversible denaturation when exposed to pH levels beyond their optimal range (Kumar *et al.*, 2024).

9.6.2 Reactive functional groups and chemical modifiability

9.6.2.1 Functional group profiling

Defining the location and availability of functional groups such as hydroxyl, carbonyl, amine, and halogen substituents determines the reactivity and stability of pigments. These moieties determine the degradation routes. As such, glycosylation is possible due to the presence of hydroxyl groups in flavonoids that increase water solubility, and conjugated aldehydes and ketones in carotenoids can be considered electrophilic sites that are susceptible to nucleophilic attack (Zhang *et al.*, 2023).

9.6.2.2 Chemical modifications for stability enhancement

Pigment stability may be enhanced by a selective derivatization process that does not affect color properties. Photostability and resistance to oxidative degradation have been achieved through esterification, etherification, and metal complexation. Melanin analogues can achieve structural rigidity and protection against photobleaching by being oxidatively polymerized or cross-linked with biopolymers (Wang *et al.*, 2024).

9.6.3 Stability in formulations and environmental conditions

9.6.3.1 Formulation stability

In industrial products, microbial pigments are usually confronted by the formulation of ingredients (surfactants, emulsifiers, and pH adjusters). The combination of pigments and these components may cause aggregation, precipitation, or fading of color. Oxidative and metal-catalyzed degradation can be alleviated by incorporating stabilizers such as tocopherols, ascorbic acid, and EDTA (Jurić *et al.*, 2022).

9.6.3.2 Encapsulation and protective delivery systems

These pigments, as well as light, oxygen, and thermal stress, are shielded by encapsulating the pigments in lipid vesicles, polymeric nanoparticles, or protein matrices. This method not only prolongs shelf-life but may also increase bioavailability for nutraceutical and biomedical applications (Jurić *et al.*, 2022; Singh *et al.*, 2023).

9.6.3.3 Environmental stress resistance

Exposure to UV light, in conjunction with variations in humidity and temperature, significantly affects pigment performance. Enhancing the aromatic characteristics and increasing the number of metal coordination sites appear to enhance the photostability of pigments. Conversely, pigments characterized by extended polyene chains require more robust stabilization strategies (Randhawa, 2024).

9.7 Computational and *In Silico* Approaches

The financial implications of computational tools are evaluated to understand the synthesis of microbial pigments and their interactions with biological systems (Fig. 9.2). This evaluation also considers potential modifications to enhance pigment properties. These methods are particularly advantageous when pigment quantities are limited or when new structural variants are developed to improve performance (Venil *et al.*, 2021).

9.7.1 Molecular modeling and docking

Molecular modeling uses the structures of pigments (3D) that display the conformational aspects and active sites of the pigment.

Table 9.2. Stability profiles and functional group composition of selected microbial pigments, along with their application potential and literature references. The table outlines thermal and pH stability, key chemical functionalities, and relevance to food, pharmaceutical, cosmetic, and environmental industries.

Pigment	Stability characteristic	Functional groups present	Application implications	References
Prodigiosin	Unstable at high pH; degrades above 60°C	Pyrrole rings, methoxy, imine	Suitable for slightly acidic formulations; used in anticancer and antimicrobial products	(Rapson, 2004)
Violacein	Stable across pH 4–10; tolerates moderate heat	Indole, hydroxyl, ketone	Applicable in diverse formulations including pharmaceuticals and biosensors	(Del Hoyo, 2007)
Melanin	Extremely heat stable; UV resistant	Phenolic, amine, quinone	Acts as a natural UV shield; used in sunscreens and environmental coatings	(Salian *et al.*, 2021)
Carotenoids	Heat sensitive; degraded by UV and oxygen	Polyene chain, methyl, epoxide	Encapsulation required for food and cosmetic stability	(Aslam *et al.*, 2021)
Phycocyanin	pH sensitive below pH 4; thermo-labile above 50°C	Tetrapyrrole chromophore, carboxyl, amine	Requires cold storage and stabilizers for use in food and diagnostics	(Fabre *et al.*, 2022)
Astaxanthin	Oxidatively unstable; sensitive to light and heat	Keto, hydroxyl	Encapsulation in liposomes enhances use in supplements and feed	(Martínez-Álvarez *et al.*, 2020)
Monascorubrin	Stable in pH 3–7; loses activity above 70°C	Ketone, ester	Widely used in fermented food industries	(Wang and Lin, 2007)
Actinorhodin	Stable under mild acidic conditions; degrades at high temp	Anthraquinone, hydroxyl	Applied in antibacterial coatings; limited by thermal instability	(Abdel-Rahiem, 2022)
Anthraquinones	Stable at pH 4–9; moderate heat resistance	Carbonyl, aromatic ring	Textile dyes and antioxidant additives	(Marzec, 2014)
Xanthophylls	Easily degraded in light and heat	Hydroxyl, epoxy, ester	Often coformulated with antioxidants in cosmetics	(Bunea *et al.*, 2014)
Azaphilones	Sensitive to UV and alkaline pH	Lactone, pyran ring, ketone	Require pH buffers and UV protectants in formulations	(Zabala *et al.*, 2012)

Continued

Table 9.2. Continued

Pigment	Stability characteristic	Functional groups present	Application implications	References
Pyocyanin	Moderately heat stable; pH sensitive	Phenazine, hydroxyl	Used in biosensors and redox applications	(Keerthana *et al.*, 2025)
Flavins	Light sensitive; stable in neutral pH	Isoalloxazine ring, ketone, nitrogen	Requires UV shielding in formulations for pharmaceutical use	(Beztsinna *et al.*, 2016)
Deoxyviolacein	Similar to violacein but slightly more heat stable	Indole, ketone	Extended use in pigment-based drug development	(Yusof, 2010)
Indirubin	Light sensitive; moderate thermal stability	Indole, ketone	Applied in anticancer and antiinflammatory drug design	(Serafini *et al.*, 2024)
Chlorophylls	Unstable in acidic and light conditions	Porphyrin ring, magnesium ion	Requires encapsulation for cosmetics and nutraceuticals	(Borah and Bhuyan, 2017)

Programs such as Chem3D, Gaussian, and Avogadro can reduce the molecular energies to determine the shape that is energetically stable (Prasetiya *et al.*, 2023). Computational docking is a technique used to predict the affinity of pigments for enzymes, DNA, and membrane receptors to identify their binding sites and possible bioactivities. Docking studies of prodigiosin with apoptotic proteins revealed important interhydrophobic hydrogen bonds associated with anticancer activity, whereas docking of violacein has proven to have antibacterial and antiprotozoal activity (Maranata *et al.*, 2024).

9.7.2 Structure–activity relationship studies

Structure–activity relationship (SAR) research has established connections between the molecular characteristics of pigments and their biological and physical properties. By employing computational tools, such as QSARINS, PASS, and MOE, researchers can identify pharmacophores and evaluate the effects of chemical modifications, including hydroxyl, methoxy, and halogen substitutions, on activity. These models facilitate the development of safer, more effective pigments for applications in pharmaceuticals, cosmetics, and food products (Masi *et al.*, 2018; Ishfaq *et al.*, 2025).

9.7.3 Predictive toxicology and ADMET profiling

Toxtree, ProTox-II, and SwissADME are computational tools designed to predict pigment toxicity and evaluate ADMET (Absorption, Distribution, Metabolism, Excretion, Toxicity) properties. These predictive capabilities facilitate the early elimination of unsuitable candidates and inform the development of safe and bioavailable substances. For example, melanin analogues assessed using ADMET demonstrated low toxicity and effectively provided protection against skin toxicity (Rizkuloh *et al.*, 2021; Khan *et al.*, 2024).

9.7.4 Integration with omics and network analysis

The integration of computational algorithms with omics technologies, including genomics, transcriptomics, and metabolomics, facilitates the mapping of pigment biosynthetic

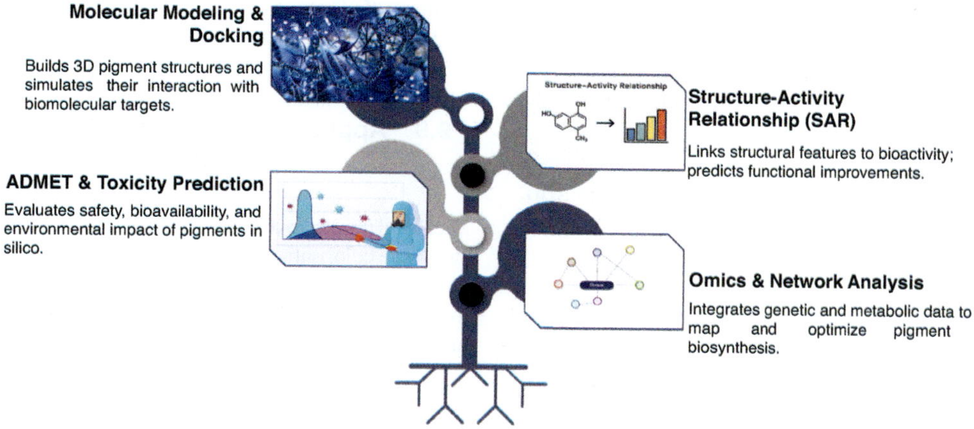

Fig. 9.2. Computational and *in silico* approaches for microbial pigment analysis, highlighting their role in structural prediction, bioactivity screening, toxicological assessment, and biosynthetic optimization.

pathways and identification of regulatory nodes. Metabolic engineering techniques, such as the enhancement of violacein production in *Chromobacterium violaceum* through selectively optimized pathways, are supported by software tools, such as Cytoscape and KEGG Mapper (Chen *et al.*, 2022; Wan *et al.*, 2023).

9.8 Case Studies of Structurally Characterized Pigments

Well-characterized microbial pigments are discussed in detail with regard to their biosynthesis, structure, functional properties, and industrial relevance. The process of isolation for the application of these representative examples, however, also provides models for classes of pigments broader than these. We focused on three well-studied microbial pigments, prodigiosin, violacein, and melanin, for which significant analytical, biochemical, and computational studies have been conducted.

9.8.1 Prodigiosin from *Serratia marcescens*: origin, structure, functional properties, stability, applications

Prodigiosin is a red tripyrrole pigment found in *Serratia marcescens* and some species of *Streptomyces*. Structurally, it is made up of two pyrrole rings with a middle methoxy-substituted pyrrole that makes up its conjugated system, which gives it intense color and bioactive properties. UV rhodic, NMR, and HR/MS, together with X-ray crystallography analysis, have provided an analytical explanation of a flat conformation that enables planar stacking interactions (Paul *et al.*, 2022). Prodigiosin has antibacterial, anticancer, and immunosuppressive effects, and has been suggested to act through DNA intercalation and membrane proton gradients. Prodigiosin is easily affected in alkaline media and when exposed to light so stabilization is needed in pharmaceutical and manufacturing processes (Araújo *et al.*, 2022).

9.8.2 Violacein from *Chromobacterium violaceum*: structural features and bioactivity

Violacein is a purple bisindole obtained through the oxidative condensation of two L-tryptophan molecules catalyzed by the Vio ABCDE gene cluster. It is a structurally simple, centrally oxo-indole-core-based hydroxy indole-stabilized hydrogen-bonded, conjugated molecule. NMR, MS, and crystallographic analyses have played a significant role in the structural characterization of this compound, showing that certain functional groups are vital to its wide-spectrum antimicrobial, antiparasitic, and anticancer activities. The hydrophobic nature of the pigment provides stability in lipid-based formulations but also affects water solubility, a feature that determines the delivery and bioavailability of therapeutic products (Durán *et al.*, 2012, 2021).

9.8.3 Melanin from *Streptomyces* spp. and fungi: structural complexity and functional roles

Melanin is a heterogeneous polymer with a high molecular weight that is prepared by oxidative polymerization of phenolic or indolic precursors (e.g. L-DOPA or tyrosine). Because of its structure, structural characterization is based on indirect methods such as FTIR spectroscopy, solid-state NMR, electron paramagnetic resonance (EPR), and X-ray photoelectron spectroscopy. The resultant data suggested that the network could have a highly cross-linked aromatic system with nitrogen and oxygen functional groups. The architecture has unprecedented photostability, thermal stability, and radical scavenging potential, which have made melanin of interest for UV protective coatings, cosmetics, and biomedical implants (Tran-Ly *et al.*, 2020; Muñoz-Torres *et al.*, 2024).

9.8.4 Carotenoids from extremophiles: structural adaptation to harsh environments

Extremophilic microorganisms (*Halobacterium salinarum* and thermophilic cyanobacteria) produce carotenoids with unusual structural adaptations that increase stability in harsh environments, such as extremely high temperatures, salinity levels, and radiation. Some variants, such as bacterioruberin, bear longer conjugation forms with an extra number of hydroxyl groups that enhance photostability and antioxidant capacity. These modifications have been shown to increase the rigidity of the pigment and decrease its conformational flexibility, based on structural elucidation using HPLC-MS, FTIR, and resonance Raman spectroscopy to reduce degradation under extreme conditions. Such modifications have led to the creation of synthetic analogues that can be used in the development of photoprotective agents and antioxidants in food, cosmetics, and aerospace products (Lafarga *et al.*, 2021; Bouhamed *et al.*, 2024).

9.9 Applications of Characterized Pigments

Microbial pigments are safe alternatives to harmful synthetic colorants owing to their stability, bioactivity, and compatibility. They are used in many industrial applications owing to their unique attributes (Aman Mohammadi *et al.*, 2022).

9.9.1 Pharmaceutical applications

Microbial pigments are attractive because they exhibit high bioactivity; hence, they are of interest in the pharmaceutical industry. Prodigiosin has strong cytotoxic activity against tumor cells that disrupts the integrity of mitochondria and triggers apoptosis. Violacein has antibacterial and antiparasitic activities, particularly against *Staphylococcus aureus* and *Leishmania* species, and acts on redox balance and DNA synthesis. Melanin can be applied as a radioprotectant and antioxidant, as a drug delivery vehicle, and in photothermal therapy (Manivasagan *et al.*, 2018). Through their structures that contain hydroxyls, carbonyls, and nitrogen-based heterocycles, these pigments allow interactions with biomolecular targets in the form of hydrogen bonding and elliptic covering. Computational docking, ADMET modeling, encapsulation, and surface modification are used to increase bioavailability and therapeutic impact. This increases biocompatibility and limits the long-term toxicity risk (Celedón and Díaz, 2021).

9.9.2 Food and cosmetics industry

Recently, the market for natural colorants has experienced rapid growth owing to the combination of consumer demand and stringent restrictions on synthetic dyes. Monascorubrin, carotenoids, and phycocyanin are produced by microbes, giving rise to natural food colorants with antioxidant properties, which increase nutritional value. Studies on the pH stability of microbial pigments and thermal reactivity contribute to adjusting pigment processing stability (Venil *et al.*, 2021). Violacein in cosmetics works as an antimicrobial and protects melanin against UV light. UV radiation and free radicals are inhibited by products that contain melanin as a sunscreen and antiaging factor. It has consistent aromatic stability, making it effective under harsh conditions to extend its application to skincare. The carotenoid astaxanthins are also commercially added to antiwrinkle and skin brightening products to protect the fragile polyene backbone (Salvo *et al.*, 2023).

9.9.3 Environmental and textile use

Microbial pigments are renewable greener successors to nonrenewable natural pigments, and are useful in environmental and industrial applications. They have metal-chelating and antioxidant effects and are thus suitable for wastewater treatment, biosensing, and dyeing of textiles because of their photostability (Agarwal *et al.*, 2023). For example, melanin and violacein have the capacity to bind heavy metals, such as lead and cadmium, in contaminated habitats. Pigments such as violacein, prodigiosin, and monascus are not only aesthetically pleasing in fabrics but also possess antimicrobial effects, which makes them applicable in medical and active wear. Their resistance to washing and UV, and colorfastness have been confirmed through structural analyses. They can be used in the dyeing process with natural mordants or green solvents and therefore help ensure sustainable dyeing (Toki *et al.*, 2025).

Microbial pigments are versatile in industry and are characterized by chemical stability, biological activity, and ecological safety. Their use will be extended through further development of fermentation, metabolic engineering, and nanoformulation (Barreto *et al.*, 2023).

9.10 Future Perspectives and Research Gaps

Standardization of the structural characterization protocol is essential for advancing microbial pigment research and addressing current inconsistencies in solvent systems, chromatographic techniques, and spectral calibration, which hinder cross-study comparisons. To enhance data reliability, the establishment of internationally recognized reference standards and interlaboratory proficiency tests is necessary for developing pigment-specific spectral libraries and structural databases. Integrating multi-omics approaches with advanced analytical tools such as cryo-electron microscopy, ultra-high-resolution mass spectrometry, and solid-state NMR enables comprehensive insights into pigment biosynthesis and structure–function relationships, accelerating the identification of novel pigments with optimized properties. High-throughput and automated pigment analysis utilizing microfluidic separation, machine learning for spectral analysis, and automated sampling processing offer significant potential for efficient pigment discovery by enabling rapid screening of microbial libraries for desirable traits.

Transitioning microbial pigment production on an industrial scale requires stringent structural quality control. Employing real-time, nondestructive spectroscopic monitoring during fermentation, extraction, and formulation and delivery systems, such as encapsulation in biodegradable polymers or lipid carriers, is key to improving pigment stability, solubility, and targeted delivery while minimizing environmental impact. Comprehensive structural data will be vital for regulatory approval in food, cosmetics, and pharmaceutical applications, focusing on purity, stability, toxicological risks, and microbial pigment characterization. *In silico* toxicology aids in the identification of structural motifs associated with adverse effects, thus ensuring safety and expediting regulatory reviews.

Despite considerable progress, numerous critical research gaps remain. The lack of standardized analytical procedures undermines reproducibility, and the instability of the pigments used for extraction and characterization impedes effective structural analysis. Furthermore, computational modeling, high-throughput screening, and integration of multi-omics approaches are underutilized in predicting and optimizing pigment properties.

Emerging opportunities lie in the integration of experimental and computational methods, adoption of sustainable extraction and formulation techniques, and implementation of structural quality control in industrial processes.

Addressing these gaps will not only enhance fundamental knowledge but also facilitate the development of microbial pigments into reliable, safe, and sustainable solutions for global use.

9.11 Conclusion

Investigating the structural characteristics of microbial pigments represents an interdisciplinary pursuit that integrates analytical chemistry, microbiology, and material science. The central focus of this research is to obtain structural insights that elucidate the molecular underpinnings of the stability, bioactivity, and functional versatility of these pigments. Techniques such as high-resolution spectroscopy, various forms of chromatography (including gas, liquid, supercritical fluid, and capillary), mass spectrometry (encompassing powder, gas, ion trap, and analytical methods), and crystallography collectively facilitate the precise determination of pigment structures, ranging from simple aromatic compounds to highly conjugated polyenes and complex polymers.

The correlation between pigment structure and function is evident in their diverse applications: the chemistry of conjugated systems influences optical absorption and redox potential; the chemistry of chemical functional groups affects solubility, stability, and reactivity; and the supramolecular organization dictates interactions with biological membranes, light-harvesting complexes, and active formulation matrices. Understanding these structure–function relationships is essential for designing and engineering pigments with specific structures and functions tailored to meet industrial, biomedical, and environmental demands.

References

Abdel-Rahiem, M. (2022) Microbial pigments: Sources and applications in marine environments. *Egyptian Journal of Aquatic Biology and Fisheries* 26(1), 99–124. DOI: 10.21608/ejabf.2022.215163.

Abdollahi, F., Jahadi, M. and Ghavami, M. (2021) Thermal stability of natural pigments produced by *Monascus purpureus* in submerged fermentation. *Food Science & Nutrition* 9(9), 4855–4862. DOI: 10.1002/fsn3.2425.

Abou Elmaaty, T., Sayed-Ahmed, K., Elsisi, H. and Magdi, M. (2022) Optimization of extraction of natural antimicrobial pigments using supercritical fluids: A review. *Processes* 10(10), 2111. DOI: 10.3390/pr10102111.

Afroz Toma, M., Rahman, M.H., Rahman, M.S., Arif, M., Nazir, K.N.H. *et al.* (2023) Fungal pigments: Carotenoids, riboflavin, and polyketides with diverse applications. *Journal of Fungi* 9(4), 454. DOI: 10.3390/jof9040454.

Agarwal, H., Bajpai, S., Mishra, A., Kohli, I., Varma, A. *et al.* (2023) Bacterial pigments and their multifaceted roles in contemporary biotechnology and pharmacological applications. *Microorganisms* 11(3), 614. DOI: 10.3390/microorganisms11030614.

Aman Mohammadi, M., Ahangari, H., Mousazadeh, S., Hosseini, S.M. and Dufossé, L. (2022) Microbial pigments as an alternative to synthetic dyes and food additives: A brief review of recent studies. *Bioprocess and Biosystems Engineering* 45(1), 1–12. DOI: 10.1007/s00449-021-02621-8.

Araújo, R.G., Zavala, N.R., Castillo-Zacarías, C., Barocio, M.E., Hidalgo-Vázquez, E. *et al.* (2022) Recent advances in prodigiosin as a bioactive compound in nanocomposite applications. *Molecules* 27, 15. DOI: 10.3390/molecules27154982.

Aslam, S., Ahmad, M. and Riaz, M. (2021) Stability of carotenoids. In: *Carotenoids: Structure and Function in the Human Body*. Springer, Dordrecht, The Netherlands, pp. 251–315. DOI: 10.1007/978-3-030-46459-2_8.

Barreto, J.V. d. O., Casanova, L.M., Junior, A.N., Reis-Mansur, M.C.P.P. and Vermelho, A.B. (2023) Microbial pigments: Major groups and industrial applications. *Microorganisms* 11(12), 2920. DOI: 10.3390/microorganisms11122920.

Beztsinna, N., Solé, M., Taib, N. and Bestel, I. (2016) Bioengineered riboflavin in nanotechnology. *Biomaterials* 80, 121–133. DOI: 10.1016/j.biomaterials.2015.11.030.

Bhosle, D., Janghel, A., Deo, S., Raut, P., Verma, C. *et al.* (2015) Emerging Ultrasound Assisted Extraction (UAE) techniques as innovative green technologies for the effective extraction of the active phytopharmaceuticals. *Research Journal of Pharmacy and Technology* 8(7), 963–970. DOI: 10.5958/0974-360X.2015.00161.4.

Bhusal, R.D., Nahar, D.M. and Dalvi, P.B. (2017) Review on: Flash column chromatography. *Indo American Journal of Pharmaceutical Research* 7, 1. DOI: 10.5281/zenodo.1006789.

Borah, K.D. and Bhuyan, J. (2017) Magnesium porphyrins with relevance to chlorophylls. *Dalton Transactions* 46(20), 6497–6509. DOI: 10.1039/C7DT00236A.

Bouhamed, S.B.H., Chaari, M., Baati, H., Zouari, S. and Ammar, E. (2024) Extreme halophilic *Archaea*: *Halobacterium salinarum* carotenoids characterization and antioxidant properties. *Heliyon* 10, 17. DOI: 10.1016/j.heliyon.2024.e30556.

Boysen, R.I. and Hearn, M.T. (2011) High-performance liquid chromatography of peptides and proteins. In: Hughes, A. (ed.) *Analysis and Function of Amino Acids and Peptides*, Vol. 5. John Wiley & Sons, Chichester, UK, pp. 167–210. DOI: 10.1002/9783527631803.ch5.

Bunea, A., Socaciu, C. and Pintea, A. (2014) Xanthophyll esters in fruits and vegetables. *Notulae Botanicae Horti Agrobotanici Cluj-Napoca* 42(2), 310–324. DOI: 10.15835/nbha4229631.

Butnariu, M. (2016) Methods of analysis (extraction, separation, identification and quantification) of carotenoids from natural products. *Journal of Ecosystem & Ecography* 6(2), 1–19. DOI: 10.4172/2157-7625.1000193.

Castro, A., Corwin, A., Waxham, F. and Beilby, A. (1959) Products from *Serratia marcescens*. *Journal of Organic Chemistry* 24(4), 455–459. DOI: 10.1021/jo01086a004.

Celedón, R.S. and Díaz, L.B. (2021) Natural pigments of bacterial origin and their possible biomedical applications. *Microorganisms* 9(4). DOI: 10.3390/microorganisms9040739.

Chan, J.C. (2020) *Interactions between deep eutectic solvent and laccase*. ProQuest Dissertations & Theses, Washington State University.

Chandra, R.D., Prihastyanti, M.N.U. and Lukitasari, D.M. (2021) Effects of pH, high pressure processing, and ultraviolet light on carotenoids, chlorophylls, and anthocyanins of fresh fruit and vegetable juices. *EFood* 2(3), 113–124. DOI: 10.2991/efood.k.210630.001.

Chen, D., Yuan, X., Zheng, X., Fang, J., Lin, G. *et al.* (2022) Multi-omics analyses provide insight into the biosynthesis pathways of fucoxanthin in *Isochrysis galbana*. *Genomics, Proteomics & Bioinformatics* 20(6), 1138–1153. DOI: 10.1016/j.gpb.2022.05.010.

Chen, Z., Wu, W., Wen, Y., Zhang, L., Wu, Y. *et al.* (2023) Recent advances of natural pigments from algae. *Food Production, Processing and Nutrition* 5(1), 39. DOI: 10.1186/s43014-023-00155-y.

Chini Zittelli, G., Lauceri, R., Faraloni, C., Silva Benavides, A.M. and Torzillo, G. (2023) Valuable pigments from microalgae: Phycobiliproteins, primary carotenoids, and fucoxanthin. *Photochemical & Photobiological Sciences* 22(8), 1733–1789. DOI: 10.007/s43630-023-00407-3.

Christiansen, J.V., Isbrandt, T., Petersen, C., Sondergaard, T.E., Nielsen, M.R. *et al.* (2021) Fungal quinones: Diversity, producers, and applications of quinones from *Aspergillus, Penicillium, Talaromyces, Fusarium, and Arthrinium*. *Applied Microbiology and Biotechnology* 105(21), 8157–8193. DOI: 10.1007/s00253-021-11597-0.

Colyer, C.L., Kinkade, C.S., Viskari, P.J. and Landers, J.P. (2005) Analysis of cyanobacterial pigments and proteins by electrophoretic and chromatographic methods. *Analytical and Bioanalytical Chemistry* 382(3), 559–569. DOI: 10.1007/s00216-004-3020-4.

Cuellar-Bermudez, S.P., Aguilar-Hernandez, I., Cardenas-Chavez, D.L., Ornelas-Soto, N., Romero-Ogawa, M.A. *et al.* (2015) Extraction and purification of high-value metabolites from microalgae: Essential lipids, astaxanthin and phycobiliproteins. *Microbial Biotechnology* 8(2), 190–209. DOI: 10.1111/1751-7915.12167.

Del Hoyo, C. (2007) Layered double hydroxides and human health: An overview. *Applied Clay Science* 36(1–3), 103–121. DOI: 10.1016/j.clay.2006.06.010.

Derguine-Mecheri, L., Kebbouche-Gana, S. and Djenane, D. (2021) Biosurfactant production from newly isolated *Rhodotorula* sp. YBR and its great potential in enhanced removal of hydrocarbons from contaminated soils. *World Journal of Microbiology and Biotechnology* 37(1), 18. DOI: 10.1007/s11274-020-02983-3.

Dugheri, S., Marrubini, G., Mucci, N., Cappelli, G., Bonari, A. *et al.* (2021) A review of micro-solid-phase extraction techniques and devices applied in sample pretreatment coupled with chromatographic analysis. *Acta Chromatographica* 33(2), 99–111. DOI: 10.1556/1326.2020.00790.

Durán, M., Ponezi, A.N., Faljoni-Alario, A., Teixeira, M.F.S., Justo, G.Z. *et al.* (2012) Potential applications of violacein: A microbial pigment. *Medicinal Chemistry Research* 21(7), 1524–1532. DOI: 10.1007/s00044-011-9654-9.

Durán, N., Nakazato, G., Durán, M., Berti, I.R., Castro, G.R. *et al.* (2021) Multi-target drug with potential applications: Violacein in the spotlight. *World Journal of Microbiology and Biotechnology* 37(9), 151. DOI: 10.1007/s11274-021-03120-4.

El-Gazzar, N., Said, L., Al-Otibi, F.O., AbdelGawwad, M.R. and Rabie, G. (2025) Antimicrobial and cytotoxic activities of natural (z)-13-docosenamide derived from *Penicillium chrysogenum*. *Frontiers in Cellular and Infection Microbiology* 15, 1529104. DOI: 10.3389/fcimb.2025.1529104.

Fabre, J.-F., Niangoran, N. g. U.F., Gaignard, C., Buso, D., Mouloungui, Z. *et al.* (2022) Extraction, purification and stability of c-phycocyanin from *Arthrospira platensis*. *European Food Research and Technology* 248(6), 1583–1599. DOI: 10.1007/s00217-022-03987-z.

Falsafi, S.R., Rostamabadi, H., Assadpour, E. and Jafari, S.M. (2020) Morphology and microstructural analysis of bioactive-loaded micro/nanocarriers via microscopy techniques; CLSM/SEM/TEM/AFM. *Advances in Colloid and Interface Science* 280, 102166. DOI: 10.1016/j.cis.2020.102166.

Fatima, M. and Anuradha, K. (2022) Isolation, characterization, and optimization studies of bacterial pigments. *Journal of Pure & Applied Microbiology* 16(2). DOI: 10.22207/JPAM.16.2.28.

Fatima, I., Munir, M., Qureshi, R., Hanif, U., Gulzar, N. *et al.* (2024) Advanced methods of algal pigments extraction: A review. *Critical Reviews in Food Science and Nutrition* 64(27), 9771–9788. DOI: 10.1080/10408398.2023.2216782.

Figon, F. and Casas, J. (2021) The integrative biology of pigment organelles, a quantum chemical approach. *Integrative and Comparative Biology* 61(4), 1490–1501. DOI: 10.1093/icb/icab045.

Garcia-Lopez, E., Alcazar, P. and Cid, C. (2021) Identification of biomolecules involved in the adaptation to the environment of cold-loving microorganisms and metabolic pathways for their production. *Biomolecules* 11(8), 1155. DOI: 10.3390/biom11081155.

Grewal, J., Wołacewicz, M., Pyter, W., Joshi, N., Drewniak, L. *et al.* (2022) Colorful treasure from agro-industrial wastes: A sustainable chassis for microbial pigment production. *Frontiers in Microbiology* 13, 832918. DOI: 10.3389/fmicb.2022.832918.

Grivard, A., Goubet, I., Duarte Filho, L.M. d. S., Thiéry, V., Chevalier, S. *et al.* (2022) Archaea carotenoids: Natural pigments with unexplored innovative potential. *Marine Drugs* 20(8), 524. DOI: 10.3390/md20080524.

Huang, X., Gan, L., He, Z., Jiang, G. and He, T. (2024) Bacterial pigments are promising alternatives to synthetic colorants, from fundamentals to applications. *Journal of Microbiology and Biotechnology* 34(11), 2153. DOI: 10.4014/jmb.2404.04018.

Ishfaq, M., Shah, S.W.A., Li, S., Bilawal, A., Shah, Z. *et al.* (2025) QSAR and pharmacophore modeling in computational drug design. In: *Computational Methods in Medicinal Chemistry, Pharmacology, and Toxicology*. Elsevier, St Louis, MO, pp. 99–118. DOI: 10.1016/B978-0-443-33024-7.00011-4.

Jorge, A.M. and Pereira, J.F. (2025) Microbial colorants for ecofriendly textile colorations. In: *Microbial Colorants: Chemistry, Biosynthesis and Applications*. Scrivener, Beverly, MA, pp. 397–415. DOI: 10.1002/9781394287888.ch18.

Jurić, S., Jurić, M., Król-Kilińska, V., Vlahoviček-Kahlina, K., Vinceković, M. *et al.* (2022) Sources, stability, encapsulation, and applications of natural pigments in foods. *Food Reviews International* 38(8), 1735–1790. DOI: 10.1080/87559129.2020.1837862.

Keerthana, S., Divya, K., Rajapriya, A., Viswanathan, C. and Ponpandian, N. (2025) Nanodiamond-embedded tungsten disulfide nanosheets for electrochemical detection of the bacterial metabolite pyocyanin. *Mikrochimica Acta* 192, 621. DOI: 10.21203/rs.3.rs-5406178/v1.

Khan, A.A., Alshabi, A.M., Alqahtani, Y.S., Alqahtani, R., Bennur, I.A. *et al.* (2021) Extraction and identification of fungal pigments from *Penicillium europium* using different spectral studies. *Journal of King Saud University-Science Science* 33(4), 101437. DOI: 10.1016/j.jksus.2021.101437.

Khan, S., Ul Haq, I., Ali, I., Rehman, A., Almehmadi, M. *et al.* (2024) Antibacterial and antibiofilm potential of chlorophyllin against *Streptococcus* mutans *in vitro* and *in silico*. *Antibiotics* 13(9), 899. DOI: 10.3390/antibiotics13090899.

Kiani, H., Aznar, R., Poojary, M.M., Tiwari, B.K. and Halim, R. (2022) Chromatographic techniques to separate and identify bioactive compounds in microalgae. *Frontiers in Energy Research* 10, 904014. DOI: 10.3389/fenrg.2022.904014.

Kumar, Y., Bist, Y., Thakur, D., Nagar, M. and Saxena, D.C. (2024) Review of the role of pH-sensitive natural pigments in biopolymer-based intelligent food packaging films. *International Journal of Biological Macromolecules* 276, 133869. DOI: 10.1016/j.ijbiomac.2024.133869.

Lafarga, T., Sanchez-Zurano, A., Morillas-España, A. and Acien-Fernandez, F.G. (2021) Extremophilic microalgae as feedstock for high-value carotenoids: A review. *International Journal of Food Science and Technology* 56(10), 4934–4941. DOI: 10.1111/ijfs.15069.

Li, Q., Ju, X., Hu, X., Ma, P., McClements, D.J. *et al.* (2025) Improvement of the blue color and structural stability of phycocyanin by forming soluble complexes with algae oligosaccharides. *Food Hydrocolloids* 167, 111452. DOI: 10.1016/j.foodhyd.2025.111452.

Lin, L. and Xu, J. (2022) Production of fungal pigments: Molecular processes and their applications. *Journal of Fungi* 9(1), 44. DOI: 10.3390/jof9010044.

Longoni, M. and Bruni, S. (2021) Identification of synthetic organic pigments in contemporary artists' paints using FT-IR and FT-Raman: An advanced analytical experiment. *Journal of Chemical Education* 98(3), 966–972. DOI: 10.1021/acs.jchemed.0c00875.

Mahabaduge, S.P. (2025) Harnessing light via sea-derived pigments: A solvent-based study of marine macroalgae towards bio-inspired photovoltaics. PhD thesis, Mechanical and Sustainability Engineering. Available at: www.theseus.fi/handle/10024/892214

Malissa, A., Cappa, F., Schreiner, M. and Marchetti-Deschmann, M. (2023) Spectral features differentiate aging-induced changes in parchment–a combined approach of UV/VIS, μ-ATR/FTIR, and μ-Raman spectroscopy with multivariate data analysis. *Molecules* 28(12), 4584. DOI: 10.3390/molecules28124584.

Manivasagan, P., Bharathiraja, S., Santha Moorthy, M., Mondal, S., Seo, H. *et al.* (2018) Natural marine pigments are potential sources of therapeutic applications. *Critical Review in Biotechnology* 38(5), 745–761. DOI: 10.1080/07388551.2017.1398713.

Mantri, S., Dondapati, M., Ramakrishna, K., Audipudi, A.V. and Srinath, B.S. (2022) Production, characterization, and applications of bacterial pigments: A decade of review. *Biomedicine* 42(3), 434–440. DOI: 10.51248/.v42i3.1449.

Maranata, G.J., Megantara, S. and Hasanah, A.N. (2024) An update in computational methods for environmental monitoring: Theoretical evaluation of the molecular and electronic structures of natural pigment-metal complexes. *Molecules* 29(7), 1680. DOI: 10.3390/molecules29071680.

Martins, M., Mesquita, L.M. d., Vaz, B.M., Dias, A.C., Torres-Acosta, M.A. *et al.* (2021) Extraction and fractionation of pigments from *Saccharina latissima* (Linnaeus, 2006) using an ionic liquid+ oil+ water system. *ACS Sustainable Chemistry & Engineering* 9(19), 6599–6612. DOI: 10.1021/acssuschemeng.0c09110.

Martínez-Álvarez, Ó., Calvo, M.M. and Gómez-Estaca, J. (2020) Recent advances in the micro/nanoencapsulation of astaxanthin have improved its stability and functionality as a food ingredient. *Marine Drugs* 18(8), 406. DOI: 10.3390/md18080406.

Martínez-Espinosa, R.M. (2024) Halocins and C50 carotenoids from haloarchaea: Potential natural tools against cancer. *Marine Drugs* 22(10), 448. DOI: 10.3390/md22100448.

Marzec, A. (2014) The effect of dyes, pigments and ionic liquids on the properties of elastomer composites. Doctoral dissertation, Université Claude Bernard-Lyon I; Uniwersytet łódzki. Available at: https://theses.hal.science/tel-01166530/file/TH2014MarzecAnna.pdf

Masi, M., Nocera, P., Reveglia, P., Cimmino, A. and Evidente, A. (2018) Fungal metabolite antagonists for plant pests and human pathogens: Structure–activity relationship studies. *Molecules* 23(4), 834. DOI: 10.3390/molecules23040834.

Morón-Ortiz, Á., Mapelli-Brahm, P. and Meléndez-Martínez, A.J. (2024) Sustainable green extraction of carotenoid pigments: Innovative technologies and bio-based solvents. *Antioxidants* 13(2), 239. DOI: 10.3390/antiox13020239.

Mummaleti, G., Udo, T., Mohan, A. and Kong, F. (2024) Synthesis, characterization, and application of microbial pigments in foods as natural colors. *Critical Reviews in Food Science and Nutrition* 1–30. DOI: 10.1080/10408398.2024.2417802.

Muñoz-Torres, P., Cárdenas-Ninasivincha, S. and Aguilar, Y. (2024) Exploring agricultural applications of microbial melanin. *Microorganisms* 12(7), 1352. DOI: 10.3390/microorganisms12071352.

Nawaz, A., Chaudhary, R., Shah, Z., Dufossé, L., Fouillaud, M. *et al.* (2020) Overview on industrial and medical applications of bio-pigments synthesized by marine bacteria. *Microorganisms* 9(1). DOI: 10.3390/microorganisms9010011.

Pagels, F., Pereira, R.N., Vicente, A.A. and Guedes, A.C. (2021) Extraction of pigments from microalgae and cyanobacteria: A review of current methodologies. *Applied Sciences* 11(11), 5187. DOI: 10.3390/app11115187.

Park, J.H., Song, H.G., Kim, Y.K. and Shin, H.S. (2021) Differential spectroscopic analysis of size-dependent phycobilisomes from *Spirulina maxima*. *Biotechnology and Applied Biochemistry* 68(3), 669–675. DOI: 10.1002/bab.1978.

Paul, T., Bandyopadhyay, T.K., Mondal, A., Tiwari, O.N., Muthuraj, M. *et al.* (2022) A comprehensive review of recent trends in the production, purification, and application of prodigiosin. *Biomass Conversion and Biorefinery* 12(4), 1409–1431. DOI: 10.1007/s13399-020-00928-2.

Prasetiya, F.S., Destiarani, W., Prihastaningtyas, I.R.C., Agung, M.U.K. and Yusuf, M. (2023) Computational simulations of microalgae-derived bioactive compounds as novel inhibitors of B-Raf V600E driven melanoma. *Journal of Applied Pharmaceutical Science* 13(6), 068–086. DOI: 10.7324/JAPS.2023.10012.

Rajendran, P., Somasundaram, P. and Dufossé, L. (2023) Microbial pigments: Eco-friendly extraction techniques and industrial applications. *Journal of Molecular Structure* 1290, 135958. DOI: 10.1016/j.molstruc.2023.135958.

Ramesh, C., Vinithkumar, N.V., Kirubagaran, R., Venil, C.K. and Dufossé, L. (2019) Multifaceted applications of microbial pigments: Current knowledge, challenges, and future directions for public health implications. *Microorganisms* 7(7), 186. DOI: 10.3390/microorganisms7070186.

Ramesh, C., Vinithkumar, N.V., Kirubagaran, R., Venil, C.K. and Dufossé, L. (2020) Applications of prodigiosin extracted from the marine red-pigmented bacteria *Zooshikella* sp. and actinomycete *Streptomyces* sp. *Microorganisms* 8(4), 556. DOI: 10.3390/microorganisms8040556.

Randhawa, K.S. (2024) Synthesis, properties, and environmental impact of hybrid pigments. *Scientific World Journal* 2024(1), 2773950. DOI: 10.1155/tswj/2773950.

Rapson, T.D. (2004) Bioactive 4-methoxypyrrolic natural products from two South African marine invertebrates. MSc thesis, Rhodes University, South Africa.

Rizkuloh, L.R., Pratita, A.T.K., Mediana, M. and Yuliana, A. (2021) In silico study of toxicity parameters of pigment-derived compounds of *Monascus* sp. mold as cervical anti-cancer drug candidates. *Jurnal Teknologi Laboratorium* 10(2), 93–100. DOI: 10.29238/teknolabjournal.v10i1.238.

Saha, N., Sarkar, B. and Sen, K. (2022) Aqueous biphasic systems: A robust platform for the green extraction of biomolecules. *Journal of Molecular Liquids* 363, 119882. DOI: 10.1016/j.molliq.2022.119882.

Saif, F., Yaseen, S., Alameen, A., Mane, S. and Undre, P. (2021) Identification and characterization of *Aspergillus* species of fruit rot fungi using microscopy, FT-IR, Raman spectroscopy, and UV–Vis spectroscopy. *Spectrochimica Acta Part A: Molecular and Biomolecular Spectroscopy* 246, 119010. DOI: 10.1016/j.saa.2020.119010.

Salian, A., Dutta, S. and Mandal, S. (2021) A roadmap for UV-protective natural resources: Classification, characteristics, and applications. *Materials Chemistry Frontiers* 5(21), 7696–7723. DOI: 10.1039/D1QM00741F.

Salvo, E., Lo Vecchio, G., Pasquale, R., Maria, L., Tardugno, R. *et al.* (2023) Natural pigments production and their application in food, health and other industries. *Nutrients* 15(8). DOI: 10.3390/nu15081923.

Sarmiento-Tovar, A.A., Silva, L., Sanchez-Suarez, J. and Diaz, L. (2022) Streptomyces-derived bioactive pigments: Ecofriendly sources of bioactive compounds. *Coatings* 12(12), 1858. DOI: 10.3390/coatings12121858.

Serafini, I., Ciccola, A., Curini, R., Favero, G., Kavich, G.M. *et al.* (2024) Revealing the unknown: How can a multi-technical approach be crucial in the identification of dyes and proteins in archeological remains? In: *Multidisciplinary Approaches for the Investigation of Textiles and Fibres in the Archaeological Field*. Springer, Berlin, Germany, pp. 71–122. DOI: 10.1007/978-3-031-73812-8_6.

Shkolnikov Lozober, H., Okun, Z., Parvari, G. and Shpigelman, A. (2023) Effects of storage and pasteurization (thermal and high-pressure) conditions on the stability of phycocyanobilin and phycobiliproteins. *Antioxidants* 12(3), 568. DOI: 10.3390/antiox12030568.

Singh, S., Aeri, V. and Sharma, V. (2023) Encapsulated natural pigments: Techniques and applications. *Journal of Food Process Engineering* 46(12), e14311. DOI: 10.1111/jfpe.14311.

Singhal, A., Saini, U., Chopra, B., Dhingra, A.K., Jain, A. *et al.* (2024) UV-visible spectroscopy: A review on its pharmaceutical and bio-allied science applications. *Current Pharmaceutical Analysis* 20(3), 161–177. DOI: 10.2174/0115734129300562240408042614.

Sousa, C. (2022) Anthocyanins, carotenoids, and chlorophylls in edible plant leaves were identified using tandem mass spectrometry. *Foods* 11, 13. DOI: 10.3390/foods11131924.

Tamburini, D., Sabatini, F., Berbers, S., van Bommel, M.R. and Degano, I. (2024) Introduction and recent advances in the analytical study of early synthetic dyes and organic pigments in cultural heritage sites. *Heritage* 7(4), 1969–2010. DOI: 10.3390/heritage7040094.

Toki, G.F.I., Mia, R. and Alave, R.K.K. (2025) Extraction and purification of microbial pigments: Eco-friendly techniques and applications. In: *Microbial Colorants: Chemistry, Biosynthesis and Applications*. John Wiley & Sons, Oxford, UK, pp. 193–221. DOI: 10.1002/9781394287888.ch8.

Tran-Ly, A.N., Reyes, C., Schwarze, F.W.M.R. and Ribera, J. (2020) Microbial production of melanin and its various applications. *World Journal of Microbiology and Biotechnology* 36(11), 170. DOI: 10.1007/s11274-020-02941-z.

Tzima, S., Georgiopoulou, I., Louli, V. and Magoulas, K. (2023) Recent advances in supercritical CO2 extraction of pigments, lipids, and bioactive compounds from microalgae. *Molecules* 28(3), 1410. DOI: 10.3390/molecules28031410.

Valenzuela-Gloria, M.S., Balagurusamy, N., Chávez-González, M.L., Aguilar, O., Hernández-Almanza, A. *et al.* (2021) Molecular characterization of fungal pigments. *Journal of Fungi* 7(5), 326. DOI: 10.3390/jof7050326.

Venil, C.K., Zakaria, Z.A. and Ahmad, W.A. (2013) Bacterial pigments and their applications. *Process Biochemistry* 48(7), 1065–1079. DOI: 10.1016/j.procbio.2013.06.006.

Venil, C.K., Dufossé, L., Velmurugan, P., Malathi, M. and Lakshmanaperumalsamy, P. (2021) Extraction and application of pigment from *Serratia marcescens* SB08, an insect enteric gut bacterium, for textile dyeing. *Textiles* 1(1), 21–36. DOI: 10.3390/textiles1010003.

Wan, S., Liu, X., Sun, W., Lv, B. and Li, C. (2023) Current advances in omics-guided process optimization of microbial manufacturing. *Bioresour Bioprocess* 10(1), 30. DOI: 10.1186/s40643-023-00647-2.

Wang, T.H. and Lin, T.F. (2007) *Monascus* rice products. *Advances in Food and Nutrition Research* 53, 123–159. DOI: 10.1016/S1043-4526(07)53004-4.

Wang, Y., Zhong, J., McClements, D.J., Zhang, Z., Zhang, R. *et al.* (2024) Stability enhancement methods for natural pigments in intelligent packaging: A review. *Critical Reviews in Food Science and Nutrition* 1–16. DOI: 10.1080/10408398.2024.2437570.

Yan, B., Hu, Y., Wang, J., Tao, J., Xia, S. *et al.* (2024) State-of-the-art conceptual design of supercritical carbon dioxide as a green technology involved in bioresource conversion processes. *Chemical Engineering Journal* 486, 150166. DOI: 10.1016/j.cej.2024.150166.

Yusof, N.Z. (2010) Isolation and application of violet pigment extracted from *Chromobacterium violaceum*. Doctoral dissertation, Universiti Teknologi Malaysia.

Zabala, A.O., Xu, W., Chooi, Y.-H. and Tang, Y. (2012) Characterization of a silent azaphilone gene cluster from *Aspergillus niger* ATCC 1015 revealed hydroxylation-mediated pyran-ring formation. *Chemistry & Biology* 19(8), 1049–1059. DOI: 10.1016/j.chembiol.2012.07.004.

Zhang, Y., Zhang, L., Hu, J., Wang, Z., Meng, D. *et al.* (2023) Structural characterization and color stabilization of pigment protein-phycoerythrin glycosylated with oligochitosan. *Food Hydrocolloids* 136, 108241. DOI: 10.1016/j.foodhyd.2022.108241.

Zhao, R., Zhang, B., Sun, J., Zheng, Z. and Qiao, X. (2021) Evaluation of degradation of pigments formed during garlic discoloration at different pH. *Food Research International* 140, 109957. DOI: 10.1016/j.foodres.2020.109957.

10 Biopigments as Natural Colorants in the Food Industry

Abdullah Ahmed Butt*
Department of Food Science and Technology, Government College University, Faisalabad, Pakistan

10.1 Introduction

Natural colorants are pigments which are bio-derived from plants, microorganisms, and other natural sources that are sustainable health-friendly alternatives to synthetic dye. These pigments are vital for sophisticated clean-label food product development, as they are needed for both functional and nutritional benefits and for taking product visual presentation to the next level. Natural colorants work differently from synthetic versions, as they decompose in the environment and support food safety without any harmful outcomes to the consumer demand for environment-friendly and healthier dietary alternatives.

Significant market expansion has been seen globally for natural pigments due to growing consumer awareness of synthetic colorant risks and the current demand for food made from natural ingredients and minimal processing technologies. The market has a preference for bio-based pigments, particularly carotenoids, anthocyanins, and phycobiliproteins, which manufacturers increasingly use across various sectors like beverages, dairy products, and confectionery (Kuddus *et al.*, 2024). The efficient production and enhancement of natural colorants are possible by accounting for responsible scientific progress in biotechnology as well as improvements in extraction techniques, strengthening their emerging marketplace value.

The trend for natural colorants in the food industry is due to regulatory frameworks and consumer preferences for traceability. Synthetic dyes derived from petroleum sources present allergenic potential and exhibit toxic, possibly carcinogenic effects (Banc *et al.*, 2024). Manufacturers now exchange synthetic additives for naturally derived options to keep pace with escalating food safety requirements while reaching health-conscious consumers in expanding markets. The food industry is undergoing a fundamental transformation where natural colorants lead to sustainability and innovation in future food development (Fig. 10.1).

10.2 Classification of Biopigments

10.2.1 Chlorophylls

Plants, algae, and cyanobacteria use green chlorophylls as critical light-capturing pigments during photosynthesis to convert carbon dioxide and water into organic compounds. Of all the chlorophyll molecules (chlorophyll a, chlorophyll

*Corresponding author: Abdullahahmad100233@gmail.com

 Biopigments of Microbial Origin: Innovations and Applications Across Industries (eds Madan *et al.*)
DOI: 10.1079/ 9781836990505.0010

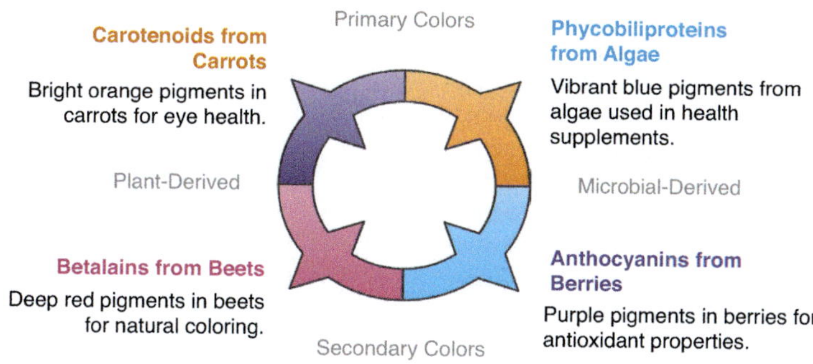

Fig. 10.1. Biopigments commonly used in food the industry.

b, chlorophyll c, and chlorophyll d), chlorophyll a represents the most widespread chemical form. As structural components, chlorophylls use a magnesium-centered porphyrin ring coordinated with a hydrophobic phytol tail to anchor them within biological membranes.

Food manufacturers use chlorophylls to create natural green food coloring that follows food standards, such as E140 for chlorophyll and E141 for chlorophyllin. Food manufacturers incorporate chlorophyll pigments into many products, including beverages, ice creams, confectioneries, and processed snack foods. Due to their better stability and improved water compatibility, chlorophyllin derivatives surpass chlorophylls in aqueous food formulation applications. Alongside their coloring functions, chlorophylls demonstrate multiple bioactive properties, including antioxidants and antiinflammatory elements, while providing detoxification benefits (Martins *et al.*, 2023). Their advantageous characteristics enable their use in functional food manufacturing processes. Under light exposure, unfavorable pH values, and high temperatures, chlorophylls display increased sensitivity, leading to degradation and fading color. Scientists have developed encapsulation methods and stabilizer techniques to enhance chlorophyll stability within food products, therefore maintaining their natural colorant properties.

10.2.2 Carotenoids (e.g., β-carotene, lutein, lycopene)

Carotenoids produce the yellow, orange, and red colors within fruits, vegetables, algae, and microorganisms. The tetraterpenoid composition of 40 carbon backbones resulting in conjugated double bonds provides both vibrant coloration and antioxidant function to carotenoids (Lu *et al.*, 2021). These pigments are classified into two main groups; features of carotenoids include β-carotene as a hydrocarbon and lutein and zeaxanthin as xanthophylls with the additional presence of oxygenated functional groups.

10.2.2.1 β-Carotene

As a significant vitamin A precursor, β-carotene exists abundantly in carrots, sweet potatoes, and certain microalgae (Zia-Ul-Haq, 2021). The natural food industry relies on β-carotene for its orange coloring function. Due to its antioxidant qualities, β-carotene defends cells against oxidative stress while enabling the immune system to function better and protect eyesight (Mitra *et al.*, 2021).

10.2.2.2 Lutein

The antioxidant xanthophyll lutein appears primarily in yellow-colored vegetables, including spinach and kale, and doctors use it in functional food products and dietary supplements. Eye health depends heavily on lutein because it contains cells that collect blue light to prevent macular degeneration from causing vision loss (Arunkumar *et al.*, 2020). Food manufacturers add lutein as a colorant to bakery goods, beverages, and dairy products.

10.2.2.3 Lycopene

The antioxidant effectiveness of lycopene, the edible carotene that colors tomatoes and watermelons, helps protect cardiovascular health and reduce cancer risks. Food manufacturers commonly include this natural component (E160d) in colored beverages, processed meat, and sauce-based foods. The lipophilic nature of carotenoids makes these compounds vulnerable to environmental factors including light, oxygen, and heat, which promote oxidative breakdown that reduces their functionality in food systems (Polyakov *et al.*, 2023). Microencapsulation technology combined with stabilizers helps food systems achieve improved stability for these components.

10.2.3 Anthocyanins

The water-soluble flavonoid pigments called anthocyanins provide red, purple, and blue coloration to fruits, vegetables, and flowers. Anthocyanins represent glycosylated anthocyanidins with an anthocyanidin core bonded to sugar chains that number one or more per molecule. Anthocyanins exist throughout nature and appear naturally in berries, including blueberries and blackberries, red grapes, cabbage, and eggplants (Qi *et al.*, 2023).

10.2.3.1 Chemical structure and variability

Anthocyanins use the C6-C3-C6 backbone structure to derive their compounds from basic anthocyanidin frameworks. Cyanidin stands alongside six additional anthocyanidins known as delphinidin, pelargonidin, peonidin, malvidin, and petunidin. Anthocyanins demonstrate structural variations in two key categories: through sugar add-ons, such as glucose and galactose, and through organic acid attachments, including acetic acid and ferulic acid. These anthocyanins react to change color expression, affecting solubility and stability levels.

10.2.3.2 Color expression and pH sensitivity

Anthocyanins are used as multipurpose ingredients in natural colorants due to the change of color and pH. In acidic conditions, such as a pH of less than 3, the anthocyanins are almost entirely in flavylium cation structures and produce a good red color (Fenger, 2020). Purple or blue color development occurs due to structural modifications from neutral or anionic quinoidal bases at elevated pH values. Due to their pH sensitivity, anthocyanins are helpful in acid foods, but they present problems in systems where constant colors are desired in neutral or alkaline systems.

10.2.3.3 Stability challenges

Anthocyanins demonstrate variable stability patterns based on two main elements: the amount of light and temperature and the presence of oxygen, enzymes, and minerals. When light degrades anthocyanins, there is normally a loss of color and less antioxidant functionality. Anthocyanin protection during processing and storage has been achieved by encapsulation, through copigmentation, or stabilized with ascorbic or citric acid (Jurić *et al.*, 2022).

10.2.3.4 Health benefits

Anthocyanins have powerful antioxidant properties that include canceling reactive free radicals and fighting oxidative stress. They are also antiinflammatory, can protect the heart, and fight cancer cells. According to evidence-based epidemiological research, eating substances rich in anthocyanins lowers the risk of chronic conditions such as type 2 diabetes, hypertension, and neurodegenerative disorders (Gonçalves *et al.*, 2021).

10.2.3.5 Applications and advances in the food industry

Natural anthocyanins are standard in beverages, confectionery, dairy products, and sauces. Food manufacturers use red anthocyanin extracts from red cabbage and purple sweet potato to make confectionery and baked goods (Luzardo-Ocampo *et al.*, 2021). Their ability to serve simultaneously as coloring agents while offering functional properties makes them highly attractive for clean-label and health-focused formulation products. Research into biotechnological anthocyanin production examines scalable fermentation methods that use engineered yeast and bacteria to meet rising demands. The new methods demonstrate their ability to achieve affordable production costs and abundant output, as well as uniform quality results.

10.2.4 Betalains (betacyanins, betaxanthins)

Plants within the Caryophyllales order produce water-soluble nitrogen-containing pigments known as betalains, leading to red, pink, purple, and yellow plant colorations (Roriz, 2021). These pigments are categorized into two main groups: the red to purple betacyanins and the yellow to orange betaxanthins. Betalains demonstrate dual value as vibrant natural pigments for color selection and beneficial biological compounds appropriate for functional food development.

10.2.4.1 Chemical structure and classification

Betalains generate colors ranging from red to violet because of their molecular properties. Betaxanthins result when balsamic acid combines with either amino acids or amines. Vulgaxanthin, miraxanthin, and indicaxanthin represent three betaxanthin compounds that produce yellow to orange colors (Coy-Barrera, 2020). The combination of structural changes and molecule linkages in betalains creates variations that modify their strength of color, chemical stability, and biological effects. *Beta vulgaris* is the principal natural source of betalains because it contains significant levels of betanin. Betalains also occur in prickly pears, amaranth, and Swiss chard. Plants that produce betalains instead of anthocyanins create a distinctive identifying signature among natural pigment-producing species.

Betalains are sensitive to environmental factors such as pH, temperature, light, and oxygen, which can influence their stability and color integrity. Betalain pigments demonstrate maximum stability when pH measures between 4–6 (Calva-Estrada *et al.*, 2022). High levels of acidic or alkaline conditions will trigger pigment destruction and browning discoloration. High temperatures create conditions that damage betalains, thereby decreasing their color intensity. Due to this sensitivity, betalains require processing at low temperatures through encapsulation techniques that have become standard practices within the industry (Zannou *et al.*, 2024). The sensitivity of betalains to light degradation requires protective packaging solutions combined with stabilization through ascorbic acid.

10.2.4.2 Applications and recent advances in the food industry

Food manufacturers take advantage of betalains as natural coloring agents for numerous food applications, including drinks, dairy products, confectionery foods, and baked goods. The colorant properties of betanin red make it an ideal natural substitute for market dyes, including Allura Red and carmine pigments (Renita *et al.*, 2023). Betaxanthins represent emerging opportunities in development of yellow-colored functional foods even though their use remains less prevalent. Betalain synthesis advances are emerging through biotechnological methods that overcome the constraints of natural extraction methods. Microbial fermentation supported by genetically altered yeast and bacterial systems creates sustainable betalain production at scale while maintaining stable quality standards. Betalain yields continue to attract research interest because scientists are utilizing metabolic engineering techniques to boost betalain production in crops.

10.2.5 Phycobiliproteins (phycocyanin, phycoerythrin)

Cyanobacteria, red algae, and some cryptophytes carry water-soluble protein–pigment complexes known as phycobiliproteins, which efficiently transfer harvested light energy to chlorophyll during photosynthesis. The molecules possess stunning colors and robust fluorescence, making them important as natural food additives and biotechnological and biomedical tools. Biotechnologists focus on the phycobiliproteins phycocyanin and phycoerythrin because phycocyanin shows blue coloration, whereas phycoerythrin displays red coloration.

Due to its easy commercial acquisition and GRAS status, phycocyanin from cyanobacteria *Spirulina platensis* is the most studied type of phycobiliprotein. Phycocyanin provides products with a bright blue color used widely in beverages, confectionery, and dairy products, including yogurts and ice creams. Scientists obtain phycoerythrin chiefly from red algae species *Porphyra* and *Gracilaria*. The demand for clean-label, plant-based, and allergen-free products has created high approval levels for both pigments.

A protein structure links phycobiliproteins to linear tetrapyrrole chromophores through covalent bonds, which result in phycobilins. The linear tetrapyrrole chromophores phycocyanobilin, found in phycocyanin, and phycoerythrobilin, within phycoerythrin, produce characteristic colors and fluorescence in these proteins. Under moderate conditions, phycobiliproteins maintain their brightness and structural stability but degrade when exposed to extreme light and pH conditions. Encapsulation methods like spray drying and liposomal encapsulation help improve phycobiliprotein stability in food products (Haghdoost *et al.*, 2023).

Apart from acting as food color components, phycobiliproteins exhibit valuable health properties through their antioxidant, antiinflammatory, and immunomodulatory effects. Phycocyanin effectively scavenges free radicals while lowering oxidative stress to help protect against chronic diseases, including cardiovascular disorders and cancer (Ziyaei *et al.*, 2023; Husain *et al.*, 2024). These pigments demonstrate both bioavailability and nontoxic properties, which qualify them for inclusion in functional food products.

Phycobiliprotein production encounters problems because of its scarcity in nature and vulnerability to environmental conditions. Biotechnology advancements through controlled bioreactor cultivation of cyanobacteria and recombinant production system development now enable much more effective phycobiliprotein extraction with better scalability options. The recent advances maintain a steady quality pigment supply by fulfilling food manufacturers' growing demand for eco-friendly natural alternatives to synthetic colorants (Table 10.1).

10.3 Microbial-Derived Biopigments

10.3.1 Pigments from algae

Food, pharmaceutical, and cosmetic industries show significant interest in microbe-sourced colorants obtained from algae species, particularly *Spirulina platensis* and *Dunaliella salina*. Algae produce multiple pigments alongside phycobiliproteins (like phycocyanin and phycoerythrin), several kinds of carotenoids (for example, β-carotene and astaxanthin), and chlorophylls. Their vibrant coloration and bioactive properties enable these pigments to fulfill the rising consumer demand for functional natural additives.

The aquatic microorganism *S. platensis* is recognized as the principal producer of phycocyanin since this water-soluble blue pigment offers multiple functional applications (Roy *et al.*, 2024). The structure of phycocyanin involves protein building blocks bound through covalent bonds to phycocyanobilins, which function as chromophores to create a distinctive blue color. Food regulatory authorities such as the FDA and EFSA have approved this pigment under its designation E18 as a food additive. Phycocyanin serves as a natural hue source for confectionery products, dairy items, beverages, and dietary supplements. The main reasons for phycocyanin's growing use include its clean-label status along with vegan-friendly creation through sustainable practices.

Algae also produce β-carotene, a visual nutrient for algae such as *D. salina*. In addition to phycobiliproteins, algal carotenoid production can act as a orange to yellow colorant and provitamin for foods and beverages. High algal carotenoid production can be attained via controlled cultivation when the environment controls the hereditary basis for pigmentation synthesis, such as extreme light, nutrient deficiency, or high salinity conditions.

Similarly, microalgae like *Haematococcus pluvialis* produce astaxanthin, a red-orange carotenoid with exceptional antioxidant properties (Patil *et al.*, 2022), which is used in aquaculture to enhance the pigment coloration of salmon and trout. Green food product coloration results from chlorophyll pigments obtained from algae-based producers. Colorants in beverages, sauces, and dietary supplements benefit from pigments obtained from *Chlorella vulgaris* species.

Both chlorophyll and chlorophyllin demonstrate antioxidant and detoxifying actions that supplement their primary function as natural colors. Their method of sustainable fabrication provides significant benefit to algae-based pigments. Microalgae cultivation occurs in several methodologies, including

Table 10.1. Key biopigments in food industry, chemical properties, and applications.

Biopigment	Chemical structure	Stability	Functional properties	Food applications
Carotenoids	Tetraterpene (C_4OH_{56})	Sensitive to light, heat	Antioxidant, provitamin A activity	Beverages, dairy, confectionery
Anthocyanins	Flavonoid Glycoside ($C_{15}H_{11}O_6$)	Sensitive to pH, heat	Antioxidant, antiinflammatory	Juices, sauces, desserts
Betalains	Betacyanins ($C_{15}H_{11}N_2O_6$) Betaxanthins ($C_{15}H_{11}N_3O_5$)	Sensitive to pH, heat	Antioxidant, antiinflammatory	Processed foods, beverages
Phycobiliproteins	Phycocyanin ($C_{34}H_{47}N_3O_{14}$) Phycoerythrin ($C_{34}H_{48}N_3O_{14}$)	Stable in aqueous solutions	Antioxidant, antiinflammatory	Dairy products, beverages
Chlorophylls	Porphyrin Ring with Mg^{+2} at center	Sensitive to light, heat	Antioxidant, detoxification	Dairy products, beverages

open ponds and photobioreactors, which demand considerably fewer land area, water, and energy resources than traditional land plants (Peter *et al.*, 2022). Control of light exposure, nutrient supply, and thermal conditions boosts pigment production by inducing superior outputs of desired compounds. Through genetic engineering and metabolic pathway optimization, scientists have improved microalgae pigment production, which enables tailored pigment profiles to meet precise industrial requirements. Researchers use spray-drying, freeze-drying, and liposomal encapsulation to extend the shelf-life and stability of such products. Scientists create mixtures of pigments obtained from algae together with natural stabilizers to improve pigment performance in various environmental settings.

10.3.2 Fungal pigments

Pigments derived from microorganisms collected from *Monascus* and *Penicillium* species serve important roles across food production, pharmaceutical, and textile industries. Fungi generate these secondary metabolites as pigments that present a bright spectrum of colors, from red to blue, in their natural composition. Fungal pigments have attracted much attention as natural compounds that are easily manufactured and scalable while providing the necessary functional features to replace synthetic dyes.

Fungal pigments from *Monascus* species earn the title of *Monascus* pigments (MPs) through their position as some of the most researched fungal pigment types. Historically connected to *Monascus purpureus*, these natural pigments derive from the filamentous fungus used throughout East Asia to create red yeast rice. *Monascus* pigments include six main compounds: three yellow pigments (monascin, ankaflavin, and monascinol) and three red pigments (rubropunctatin, monascorubrin, and rubropunctamine). Industry professionals value these pigments because they dissolve well in organic solvents and maintain stability throughout various pH levels, which enables their use in food and beverage applications. *Monascus* pigments are used in fermented rice production, sauces, meat products, and beverages to create attractive presentations.

In addition to coloring substances, MPs have many bioactive properties, including antioxidant and antimicrobial effects, and a reduction in cholesterol levels. *Monascus* cultures synthesize lovastatin, an HMG-CoA reductase inhibitor that inhibits hydroxymethylglutaryl coenzyme A reductase (HMG-CoA reductase) (Mohankumari *et al.*, 2021), and thus controls hypercholesterolemia. Monascus pigments possess dual functionality for food coloring and nutraceutical applications. There are many obstacles in the biosynthesis of MPs. *Monascus* species occur in conditions favorable to the production of a harmful mycotoxin, called citrinin, that is harmful to the kidney and liver (Zhang *et al.*, 2021). Food-grade applications pose safety concerns when citrinin exists. Optimized fermentation of *Monascus* strains has been developed to prepare strains that lack citrinin via fungal cultivation and genetic engineering improvement for secure pigment production. Due to systematic regulatory supervision and rigorous safety checks, MP global sales potential is restricted.

Most fungal pigments produced in the genus *Penicillium* lie in the green and blue ranges. The most well known of these pigments is blue. It gives color to blue-veined cheeses such as Roquefort and Gorgonzola due to the presence of green and blue pigments, respectively, produced by fungi *Penicillium roqueforti* and *Penicillium glaucum*. MPs deliver excellent colors, flavors, and textures that make each cheese unique (López-Díaz *et al.*, 2023; Ferroukhi *et al.*, 2024). *Penicillium* species also generate rubratoxins and citreorosein, a promising natural colorant for use in foods and textiles.

Solid-state or submerged fermentation systems enable scalable fungal pigment production. In submerged fermentation, where fungi grow in a liquid medium with controllable pH and temperature, scientists can optimize environmental parameters including temperature and pH, maximizing pigment production. Natural fungal habitat conditions are reproduced in solid-state fermentation processes on solid materials such as wheat bran and rice bran for cost-efficient large-scale production of pigments (pigment specification, solubility properties, and intended application define which fermentation system is used).

Fungal pigments have an adjustable chemical structure that metabolic engineering processes and fermentation optimization can manipulate. The adaptable production system enables scientists to make specially designed pigments with better stability, solubility, and enhanced bioactive properties for specific industrial use. In coculture systems, several fungi or fungi and bacteria combinations are metabolized under different wavelengths to obtain further pigments which are combined after biotechnology research.

Lack of stability and short shelf-life continue to limit fungal pigment applications. The functionality of numerous pigments declines when exposed to light and thermal variations or oxidation because these lead to both physical degradation and diminished color intensity. Encapsulation methods, including spray-drying and freeze-drying combined with nanoencapsulation, help preserve pigments when stored and applied. Applying these methods helps create more stable pigments, while the new formulation makes them applicable across various food products and processing methods.

10.3.3 Bacterial pigments

Bacterial sources of pigments include *Serratia marcescens* and *Streptomyces* species. Bacteria produce these secondary metabolite pigments, which vary in color from red to green, under particular growth conditions. Scientists increasingly value bacterial pigments because they arise naturally and require climate-friendly production processes while serving numerous industries, including food production, pharmaceutical development, and cosmetics and textiles.

Serratia marcescens synthesizes prodigiosin, a distinct crimson red pigment. This tripyrrolic pigment demonstrates significant bioactivities, including antimicrobial properties, anticancer effects, and immunosuppression. Researchers focus on prodigiosin's antimicrobial properties because they show great potential for food conservation and protecting natural products from harmful microbes. Results from preclinical studies indicated that prodigiosin also had anticancer activity by causing cancer cells to undergo apoptosis additional (Shafqat, 2024). Production rates of prodigiosin depend on multiple factors that can be adjusted for optimal yield improvement: nutrient presence, acidity or alkalinity (pH), temperature, and oxygen concentration levels. *S. marcescens* is a good source of this pigment under stress and is a dependable microbial source.

Bioactive secondary metabolites and colorful pigments, such as actinorhodin (blue), undecylprodigiosin (red), and virginiamycin (yellow), are generally produced by the *Streptomyces* species. In most cases, *Streptomyces* strains generate pigments with antibiotic biosynthesis. Actinorhodin is a blue pigment produced in research by *Streptomyces coelicolor*, a model microorganism. Actinorhodin is the polyketide regulator of actinorhodin synthesis in streptomycete species. Researchers have focused on the pigment due to its unique stability, ease of production, and utilization in colorant creation. Undecylprodigiosin is part of the prodigiosin chemical group but demonstrates distinct structural properties despite its bacterial origin. Research has shown that this pigment functions as an antimicrobial against several pathogens and may even have some cytotoxic characteristics, which could provide new sources for pharmacological development and extended therapeutic practice (Cuozzo *et al.*, 2023).

Preservation of bacterial pigment stability is essential for the functional application of bacterial pigments in food products and consumer goods. Degradation may be caused by light, heat, and oxidation conditions for the bacterial pigments prodigiosin, actinorhodin, and others. Encapsulation techniques reduce these significant limitations. Bacterial pigments are preserved from environmental factors by microencapsulation, spray-drying, and extended shelf-life, and they enhance function when used in food ingredient mixes (Singh *et al.*, 2023). With the high production of bacterial pigments through metabolic engineering and synthetic biology, novel methods are being developed to enhance yield and stability. These factors account for genetic manipulation research on *S. marcescens*, *Streptomyces*, and other bacterial strains, such as increased pigment production capacity and more excellent pigment durability with reduced byproduct creation.

10.4 Extraction and Production Techniques

10.4.1 Conventional solvent extraction

Solvent extraction is the leading traditional approach for pigment retrieval from natural materials, including plants and specific microorganisms. This technique employs organic solvents to effectively separate target pigment compounds from natural biological substances. Solvent extraction stands out as a vital pigment separation method because of its straightforward set-up, which enables effective extraction across multiple pigment types. The solvent extraction process faces difficulties because toxic solvents affect health and the environment while still needing refinement to deliver high-yield and purity results.

Conventional solvent extraction begins with biomass material preparation. During plant-based pigment extraction, scientists need to grind or crush plant material to increase surface area, which aids in extracting the pigments. The production of microbial pigments requires growing cultures in environments that promote pigment production before biomass removal at the peak time of pigment deposition. After preparation, the material is mixed with the proper solvent that extracts the pigment. Selecting the appropriate solvent defines the success rate and precision of the pigment extraction methods. Standard techniques for extracting biopigments use ethanol or methanol and acetone as solvents, while ethyl acetate, chloroform, and sometimes water serve similar purposes. Extraction scientists utilize polar solvents such as ethanol and methanol to obtain hydrophilic pigments, including anthocyanins and phycocyanins, because nonpolar solvents, including chloroform and hexane, perform best in extracting lipophilic compounds such as carotenoids and chlorophylls (de Souza Mesquita *et al.*, 2021). When selecting a solvent, developers must analyze its pigment compatibility and evaluate its pigment dissolution power alongside environmental safety standards.

Researchers apply solvent to the biological mass during standard extraction procedures and mix it over a defined timeframe to enable pigment dissolution. Agitation methods for extracting biomass pigments include basic stirring and shaking and mechanical devices such as ball mills and rotary evaporators. Process efficiency depends on extraction duration, solvent-to-biomass ratio, and control. Heat-sensitive pigments remain intact during extraction at room temperature or slightly elevated temperatures, which is the standard method, while specific pigments need higher temperatures to achieve efficient extraction.

After extraction, solvent pigment mixtures must be filtered or centrifuged to remove solid biomass particles. The extraction product, dissolved pigments, is finally extracted and further refined or concentrated. Using rotary evaporators and alternative distillation methods, the solvent extract of the pigment is rendered as a solvent-free extract by evaporation under reduced pressure. The manufacturers dry the leftover pigment matter and make it ready for use in varied applications, where it gets powdered and finished. The conventional solvent extraction approach produces high-yield and high-purity pigments more efficiently than mechanical pressing and enzymatic extraction methods (López-Cruz *et al.*, 2023). However, some disadvantages need consideration in the process. A significant concern of conventional solvent extraction is the danger that toxic solvents pose to environmental ecosystems and human safety (Thakur and Modi, 2024). Flammable and volatile, methanol, acetone, and chloroform can harm health if not properly dealt with. Solvent waste that is not handled through reliable regulatory compliance procedures can result in environmental contamination.

However, traditional solvent extraction methods involve the risk of pigment degradation during the extraction operation. Structurally degrading pigments, which return a different color when exposed to light or oxygen, may degrade under high-temperature conditions or longer extraction times. The most effective levels of pigment extraction are determined by adjusting the extraction variables to balance the yield vs the quality of the extracted pigment. Degradation of pigments in extraction processes may be reduced by fine-tuning solvent concentration, extraction time, temperature parameters, or the addition of antioxidant substances (Miranda *et al.*, 2021; Molina *et al.*, 2023). However, even though such extractions are complex, solvent extraction as a pigment extraction method is standard because it is highly convenient for large-scale demands.

Our growing understanding of the environmental health impact of solvents calls for the development of green extraction approaches that align with environmental health standards. Alternative processes to conventional solvent extraction, including supercritical fluid extraction, ultrasonic-assisted extraction, and enzymatic extraction, are promising, with higher efficiency and better selectivity, and are also environmentally sustainable (Fig. 10.2, Table 10.2).

10.4.2 Enzymatic extraction

The enzymatic approach represents a greener, more directed route than conventional solvent-based pigment extraction methods. Enzymatic action is precise and mild; it dissociates pigment compounds by dissolving biological material cell walls or membranes. Its ability to generate high amounts of pigments with minimum use of solvent is increasingly being observed as a method to decrease its environmental footprint (Morón-Ortiz *et al.*, 2024). Enzymatic extraction provides a superior harvest of plant- or microorganism-based pigments due to the bioactive compound release without degradation.

Enzymatic extraction from plant tissues, algae, fungi, and bacteria includes physical treatment, grinding, crushing, or homogenizing to increase surface area so that the enzymes can reach the target material. After preparation, the enzymes targeting cell walls or membrane elements such as cellulose, pectin, hemicellulose, or chitin should be applied based on the pigment source. Cellulases, pectinases, xylanases, and proteases function as standard enzymes to decompose proteins, lipids, and complex polysaccharides that make up cell walls and membrane structures. The cell walls of plants mostly contain cellulose, fruits contain pectin, and fungi and algae contain chitin. These molecular targets allow for enzymes to break down either cell walls or membranes, which results in the efflux of cellular pigmentation compounds, including anthocyanins, carotenoids, chlorophylls, and additional pigments. Enzymatic pigment release happens under moderate temperature and pH conditions, which protect pigments from degradation in reactive environments.

The enzymatic extraction method combines raw material with enzymes before incubation for several minutes to a few hours based on material characteristics and target pigment. This is because enzyme activity decreases at temperatures above 30–60°C as they lose their structure at that temperature. Scientists optimize pH levels for extraction at the optimal

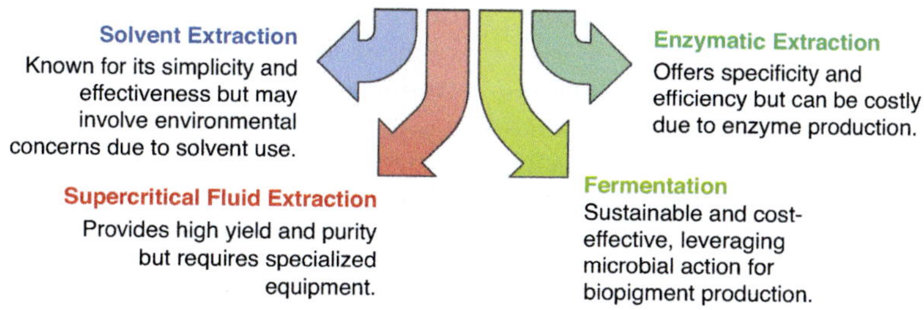

Fig. 10.2. Overview of different extraction techniques.

pH levels for enzyme function. After the enzymatic process, the pigment is separated from the remaining solid biomass through filtration and centrifugation. The benefit of enzymatic extraction is its mild operation, which allows delicate pigments to be processed without degradation (Pagels *et al.*, 2021). Biopigments such as anthocyanins and carotenoids degrade when exposed to high heat or prolonged exposure to strong solvents. Since processing risks are minimal, precious pigments maintain their natural color and biological activity during enzymatic extraction. Specific to the targeted components, enzymes produce target-specific activity and break them down into purer extracts, free from side-by-products.

Table 10.2. Comparison of different extraction techniques with potential advantages and disadvantages.

Extraction methods	Biopigments extracted	Advantages	Disadvantages
Solvent extraction	Carotenoids, anthocyanins, betalains	Simple, cost-effective	Requires organic solvents, potential toxicity
Enzymatic extraction	Carotenoids, phycobiliproteins	Environmentally friendly, specific	Slow process, expensive enzymes
Supercritical fluid extraction	Carotenoids, betalains, phycobiliproteins	High selectivity, no solvents required	Expensive equipment, complex operation
Fermentation (microbial)	Carotenoids, phycobiliproteins	Sustainable, scalable	Requires optimal microbial conditions

Enzymatic extraction minimizes reliance on toxic and volatile organic solvents, which remain standard practices in traditional processes for compound extraction. Producing pigments employing enzymatic extraction yields sustainable and environmentally friendly methods of pigment yield. Increasing use of natural source enzymes from fungi and bacteria for green technologies in food production and pharmaceutical sectors drives demand. With the enzymatic extraction method, producers can further improve their processes via optimization capacity (Fatima *et al.*, 2024). Researchers can design the enzymes to be more efficient in the desired applications. Particle wall components become more vulnerable to enzyme cocktails because these mixtures of various enzymes extend their functional range towards different cell wall components, leading to enhanced pigment extraction. This method is especially advantageous during pigment extraction from complex biological materials such as algae or fungi with their multicomponent resistant cell walls. Careful adjustment of enzyme concentration, extraction time, temperature, and pH produces an optimal extraction process for top-level pigment recovery.

10.4.3 Supercritical fluid extraction

Modern supercritical fluid extraction (SFE) for pigment extraction employs supercritical fluids, wherein carbon dioxide is a preferred extraction solvent (Herzyk *et al.*, 2024). Supercritical fluid is a term that indicates that a substance has surpassed its critical temperature and pressure to produce unique gas and liquid properties. A substance in its supercritical state allows CO_2 to demonstrate high diffusive power and minimize viscosity while maintaining liquid-like density, enabling the efficient separation of numerous compounds, including pigments.

Supercritical fluid extraction delivers better results than solvent and enzymatic extraction because it extracts biopigments selectively and with minimal solvent usage under gentle operating parameters. Through SFE, pigments dissolve from raw biomass into CO_2, and when pressure is decreased, the CO_2 separates from the pigments, creating a nontoxic, environmentally friendly extraction method. Because SFE technology effectively minimizes the chemical burden on industries, it presents a viable environmentally friendly production approach.

Following biomass preparation through drying or grinding for enhanced surface area contact, the process continues with extraction through supercritical CO_2 application. An extraction chamber receives raw materials before supercritical CO_2 applies pressure for extraction. Nitrogenous substances like carotenoids, chlorophylls, and lipophilic pigments, together with nonpolar and medium-polar compounds, dissolve in supercritical CO_2 in a chamber at operational temperatures (30–80°C) under pressure (80–350 bar). The CO_2 absorption capacity can be controlled by adjusting temperature and pressure to extract specific fatty pigments (Cvejić *et al.*, 2022). Through its interaction with biomass, the supercritical CO_2 enters cell structures to break binding forces, resulting in pigment transfer to the fluid within the structure. In a separator vessel, pressure reduction drives CO_2 into its gaseous form while leaving the extracted pigments as a concentrated solution or solid residue. The compressed CO_2 returns to service for new extraction cycles, creating an efficient system with sustainable operation. The separation method excels due to the fast transition of CO_2 from liquid to gas, resulting in clean biopigment retrieval without residual solvent.

Supercritical fluid extraction performance depends on fluid selection, combined with temperature, pressure conditions, and the properties of the biomass material. Supercritical CO_2 is the primary extraction fluid because of its affordability, harmless properties, and simple extraction from the endproduct (Elmaaty and Sorour, 2024). Adding ethanol or water as modifiers in CO_2 enhances polarity, resulting in better solubility of anthocyanins and similar polar pigments. Researchers have successfully applied this CO_2 cosolvent combination to extract more pigments from different natural materials.

Among the advantages of SFE is its outstanding ability to maintain intact the chemical properties of extracted pigments. Low-temperature operations characteristic of SFE cause minimal harm to temperature-labile pigments such as carotenoids and anthocyanins (Goñi *et al.*, 2022). This process generates pigments with enhanced color properties, improved stability, and bioactivity effects. The nontoxic nature of CO_2 allows extracted pigments to be utilized directly in food products, pharmaceuticals, and cosmetics without additional purification or solvent removal procedures.

10.4.4 Fermentation

The extraction and manufacturing of microbial pigments rely on fermentation as a sustainable method gaining increasing application. The system grows microorganisms such as bacteria, fungi, and yeast under specific laboratory constraints to create important pigments. Microbial fermentation demonstrates multiple key benefits compared to traditional extraction methods through its high scalability, low environmental impact, and ability to utilize cheap raw materials as substrates. The fermentation process for microbial pigment creation shows exceptional value in industries requiring organic pigments free of synthetic additives, especially in food production, cosmetics development, and pharmaceutical manufacturing.

Researchers employ microbial fermentation systems to cultivate microorganisms in supplemental media that support peak pigment production levels within favorable cultivation conditions. Engineered gene-modified microbial strains that produce specified pigments join naturally pigment-containing microorganisms as viable fermentation producers. Among microbial pigments, *Monascus* spp. yield monascorubrin red pigments, *Penicillium* spp. manufacture yellow pigments, *Streptomyces* spp. yield both yellow and red pigments, and *Spirulina* spp. generate blue pigments through phycocyanin production (Meruvu and Dos Santos, 2021; Rao *et al.*, 2021). During fermentation, the microorganisms are inoculated into media containing carbon and nitrogen elements with additional growth-promoting minerals and components for pigment synthesis.

The process starts by creating media solutions that meet all nutritional needs of the target microorganism. To support growth, the culture medium requires carbon (glucose, sucrose) and nitrogen (peptones and ammonium salts) sources, vitamins, trace minerals, and pH regulators (Mummaleti *et al.*, 2024). The exact composition of fermentation media plays a key role in determining which pigments will be produced alongside their overall production yields because different microorganisms function through different metabolic pathways that respond to available nutrients. Control of fermentation medium temperature and pH is vital for optimizing pigment production output.

After medium preparation, researchers add microorganisms along with parameter control to promote proper microorganism growth while creating conditions supportive of pigment development. The production method works as batch, fed-batch, or continuous fermentation according to the desired manufacturing scope and microbial culture properties. While operating batch fermentation, microorganisms grow autonomously within a limited tank, which maintains its status as a closed system until the fermentation period ends. Industrial production of microbial pigments mainly employs this basic method, which remains the most common practice.

Fed-batch fermentation enables additional nutrient delivery during operation to support optimal growth while maximizing production time (Villegas-Méndez *et al.*, 2023). Microorganisms receive continuous nutrient flow through continuous fermentation, and culture broth extraction maintains a steady production speed. The microorganisms transform metabolic routes toward pigment synthesis pathways while simultaneously creating pigments that function as their secondary metabolic products during fermentation. *Monascus* spp. generate red monascorubrin pigments by fermenting carbohydrates while blue phycocyanin forms through photosynthetic pigment biosynthetic processes in *Spirulina* spp. (Meruvu and Dos Santos, 2021). Genetic engineering techniques can improve pigment output or create synthetic cellular pigments that do not occur naturally. Microbiological pigments can be genetically engineered through targeted overexpression of biosynthetic genes and by adding foreign genes that control pigment synthesis.

Fermentation duration depends on the microorganism used and the specific concentration of pigment desired. The duration of microorganism growth with pigment release spans from a few days to many weeks until microorganisms achieve their maximum growth and pigment synthesis rate. When fermentation ends, the culture solution holds pigment compounds that remain dissolved yet bound to cellular materials. Industrial applications require the extraction and purification of the pigment through end-of-fermentation processing, including mechanical separation techniques and chemical extraction processes. The biomass retrieval process using centrifugation followed by filtration separates the pigmented material from the fermentation media. The extraction process of pigments from biomass requires solutions such as ethanol, acetone, or methanol for lipid-soluble pigments (including carotenoids and chlorophylls), followed by aqueous solutions for water-soluble pigments like anthocyanins and phycocyanins. Cell disruption technologies (sonication and high-pressure homogenization) open microbial cells, and pigments will be trapped inside free cells and are secured by enzymes.

After pigment extraction, liquid chromatography, adsorption, and precipitation are used to purify the products. Then the pigment is dried commonly using freeze- or spray-drying to produce a powdery product stable for storage and transport procedures. This pigment comes from natural sources and has no synthetic chemicals, so it can be used for several commercial applications.

The kinetic model is most commonly used for analyzing biopigment production during fermentation.

$$\mu = \mu_{max} * \frac{S}{Ks+S}$$

where μ is the growth rate of the microorganism
μ_{max} is the maximum specific growth rate
S is the concentration of limiting substrate, and
K_s is the half-saturation constant.

The production of microbial pigments through fermentation encounters multiple obstacles in the process. Maximizing pigment production during fermentation requires the maintenance of optimal conditions because biopigments respond strongly to factors like pH and oxygen levels, nutrient concentrations, and fermentation temperatures. Large-scale production of fermented products faces challenges when processes are shifted from laboratory to industrial applications because it becomes harder to keep production consistent while maintaining efficiency. The pigment manufacturing process requires downstream activities, including extraction, purification, and drying, which add expense and processing complexity to the overall production. The commercial use of genetically modified organisms (GMOs) for pigment production encounters regulatory barriers that affect their acceptance in some markets worldwide.

10.5 Applications in the Food Industry (Fig. 10.3)

10.5.1 Beverages

The beverage industry relies on pigments as natural solutions for coloring drinks because they enhance visual appeal and deliver functional advantages to products, including juices and alcoholic drinks. The superior quality that consumers seek and sustainable ingredient preferences drive the growing use of natural pigments in various products. Beverages containing pigments gain visual appeal from their natural color and possible health improvements because of the active biocompounds in specific pigment sources. Below are some key applications of pigments in beverages.

10.5.1.1 Juices and smoothies

Biopigments are primarily applied in fruit juices and smoothies because consumers regularly choose these beverages for their wholesome natural ingredients. Carotenoids such as β-carotene and lycopene, anthocyanins, and chlorophylls intensify juice colors through their rich pigment content while delivering appealing hues that match the ingredient's natural profiles (Nabi *et al.*, 2023). Carotenoids in carrots, tomatoes, and orange pulp give juices orange to red shades, whereas anthocyanins from berries like blueberries deliver colors such as red and purple. Live smoothies are the leading option because biopigments make them look better and deliver additional nutritional value to consumers. People like green smoothies because spirulina or chlorella give these options a beautiful green look while at the same time providing antioxidant and detoxing benefits. The natural red pigments extracted from beetroot and red amaranth show a deep red hue in smoothies. This attracts health-conscious consumers who want antioxidant qualities and antiinflammatory power. Several biopigments, such as anthocyanins, are recognized for their antioxidant actions, which protect against oxidative stress and promote general health. Carotenoids help immune system function and eyesight.

10.5.1.2 Alcoholic drinks

The use of pigments in alcoholic drinks helps to give coloration, which heightens market interest in natural and organic drink lines. The popularity of pigments in alcoholic products stems not just from craft beverages but also from their organic status. Anthocyanins from grapes and berries are used by winemakers to naturally color wine purple, and red, and even cocktails and spirits.

Cocktails utilize pigments such as beet-derived betalains along with spirulina-extracted

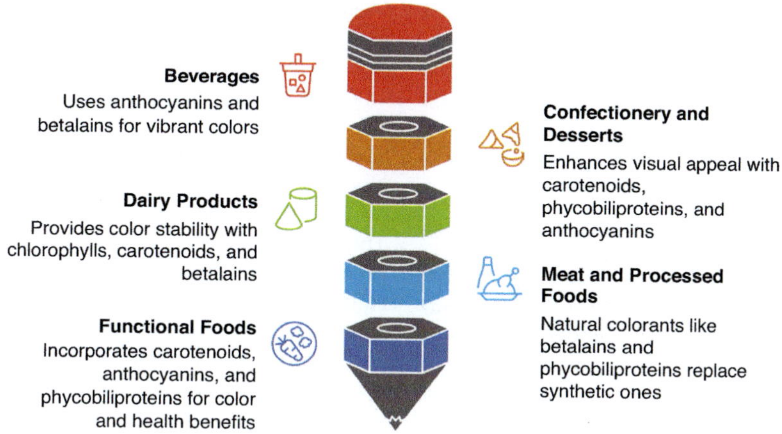

Fig. 10.3. Different applications of biopigments in different industries.

phycocyanin to create striking color options that pierce through marketplace competition. Bar industry experts use phycocyanin to achieve a brilliant blue or green color, essential for crafting modern cocktails such as blue margaritas. Biopigments deliver additional functional benefits when they become part of alcoholic products. Biopigments currently have mainly aesthetic functions in alcoholic beverages, but their bioactive characteristics will attract more research due to rising consumer awareness of drink ingredients.

10.5.1.3 Consumer preferences and market trends

The increasing application of pigments in beverages reflects a market tendency that supports plant-derived natural ingredients. Health concerns about linkages to children's hyperactivity and other issues make consumers avoid synthetic colorants and artificial additives (Miller *et al.*, 2022; de Oliveira *et al.*, 2024). The health and wellness market strongly prefers natural colorants since consumers value transparency and sustainability while maintaining product nutritional integrity. The market transition to biopigments has dual drivers: consumers demand natural products, and strict new regulations by governments and food safety agencies limit artificial food additives. The beverage industry now prefers biopigments to satisfy consumers and adhere to updated regulations.

10.5.1.4 Environmental and economic considerations

The beverage industry benefits environmentally when it adopts pigments for sustainability support. Algae-based, fungi-derived, and bacterial pigments generated through fermentation processes use sustainable resources to achieve minimal environmental impact, unlike synthetic pigments which depend heavily on petrochemical ingredients. Microbial-derived pigments serve as environmentally responsible choices that meet the beverage industry's increased interest in sustainable production standards. Biopigment prices show variability when different sources and extraction methods are used. Biopigments will become more affordable because advanced production methods and scaling up production will reduce manufacturing costs, making them competitive with synthetic colorants. Beverage brands can attract specific segments within busy markets with pigments serving as distinctive attributes, particularly in organic plant-based and health-oriented drink sectors.

10.5.2 Confectionery and desserts

Concerns about clean-label requirements and demands for natural components drive manufacturers to use pigments to improve the aesthetic, taste, and nutritional value of dessert products. Natural biopigments draw the attention of confectionery manufacturers because they create vivid colors without added artificial chemicals. Below are some key applications of biopigments in confectionery and desserts.

10.5.2.1 Candies and sweets

Biopigments are common coloring agents that impart bright, attractive colors into various candy products, including hard candies, gummies, licorice sections, and marshmallows. Natural beetroot betalains supply red color, anthocyanins from berries achieve purple shades, and carotenoids from carrots provide orange and yellow colors. Food manufacturers apply anthocyanin solutions from blackberries, blueberries, and raspberries to add red, purple, and blue colors to fruit-flavored confections. Beetroot betalains give candies intense red and purple colors, substitutes for commercial red dyes such as carmine or Red 40. Hard candies and jellies are colored with octagon orange from carotenoid β-carotene obtained from carrots and pumpkins. These pigments add color and help improve product appearance while possibly contributing vitamin A to support optimal immune system health and antioxidant activity. When food producers choose to use biopigments, they make sweets that are visually appealing, attractive to both adult and child taste, and which also answer consumer demand for healthy products (Algarni, 2020).

10.5.2.2 Chocolates and sugar-coated confectionery

Pigments are showing strong market growth in producing chocolate and sugar-coated candy products. Chlorophyll from spinach or spirulina gives chocolate confections or truffles a natural green coloring, bringing new life to familiar chocolate shades through unique visual appeal (Mouritsen *et al.*, 2024). Anthocyanins in chocolate products produce deep purple and red tones, while carotenoids produce yellow to orange colors. Candy or nut coatings that integrate pigment into coating mixtures render appealing colors. Industrial turmeric has produced the natural yellow base for sugar-coated products, similar to synthetic coloring competitors. Beetroot and hibiscus anthocyanins supply red and purple candy coatings.

10.5.2.3 Ice cream and frozen desserts

The ice cream and frozen dessert sectors utilize biopigments as important visual elements because they naturally color different flavors. The food industry is transforming ice cream products by using biopigments from natural sources instead of artificial colors to meet customer demand for natural components. The carotenoids β-carotene and lutein provide orange and yellow tints to vanilla ice cream and mango- and citrus-flavored sorbets. Bio-extracted anthocyanins from berries and red cabbage allow berry-flavored ice creams to be colored purple and red. Phycocyanin extracted from spirulina gives ice cream products a bright blue color.

Biopigment technology in frozen desserts meets consumer clean-label demands and creates an improved sensory experience for consumers. Product designers use natural purple and red melanins derived from anthocyanin-rich fruits to improve the appearance of fruit-flavored frozen desserts. Carotenoids help create appetizing appearances for tropical-flavored desserts. Multiple biopigments that function as antioxidants make frozen desserts appear more healthful to consumers.

10.5.2.4 Cakes, pastries, and puddings

In cakes, pastries, and puddings, biopigments add color through icings, fillings, and batters. Icing and frosting creations achieve visually striking results through natural food coloring extracted from sources like hibiscus (anthocyanins), turmeric (curcumin), and matcha (chlorophyll), eliminating the need for artificial coloring. Anthocyanins added to cake icings turn purple to pink and chlorophyll used for cupcakes and pastries lends natural green tones. Turmeric gives an exotic, sophisticated yellow color when added to pastry fillings or desserts (Ravindran, 2023).

Biopigments are applied to dairy-based products, such as puddings and mousse desserts, to enhance color-layer design and visual appeal. Lutein and zeaxanthin carotenoids provide rich yellow and orange hues; berry extract-derived anthocyanins can produce striking purple and red pipings in layered puddings (Maoka, 2020). Confectionery pigments are used to impart color while also providing functional health benefits to consumers. Electrochemically generated tris-carotene is a carotenoid variety investigated by researchers because it serves an antioxidant role to safeguard cells against oxidative damage (Wei *et al.*, 2023). Anthocyanins have antioxidant antiinflammatory properties and the ability to oppose cancer growth. Manufacturers are now marketing their confectionery and desserts as dual-purpose products with desirable visual appearance and health claims from adding natural compounds.

10.5.3 Dairy products

Natural pigments can replace synthetic colorants to enhance color perception and the nutritional benefits of color attributes in dairy products such as yogurt, ice cream, and cheese. Vivid colors are brought to products by natural pigments, with antioxidant activity from natural pigments helping to enhance properties and provide nutritional benefits. Below are some key applications of pigments in dairy products.

10.5.3.1 Yogurt

Biopigments allow producers to create an attractive look without employing artificial colors. Yogurt products can obtain rich, vibrant colors reflecting fruit ingredients through natural pigments such as anthocyanins, carotenoids, and chlorophylls. These pigments are visually attractive to consumers, but they also provide antioxidant benefits. Carrots, pumpkins, and other orange fruits and vegetables can be used to make β-carotene products to add an orange or yellow coloring, for example, in tropical or citrus-flavored yogurt products. The green color in yogurt results from chlorophyll when paired with matcha or spirulina, which enhances presentation and additional health benefits as it detoxifies and provides antioxidant support (Ramadhani *et al.*, 2024). Anthocyanins, among many other pigments, improve the health benefits of yogurt with antioxidant and antiinflammatory properties. The carotenoids lutein and zeaxanthin used in yogurt can also deliver functional health advantages, as they may include eye health-promoting agents (Wu *et al.*, 2022).

10.5.3.2 Ice cream

Pigments provide attractive natural colors that perfectly fit the consumer preference for organic ingredients. β-Carotene gives ice cream a bright orange hue, which enriches mango, orange, or pumpkin-flavored creations. Lutein, a yellow pigment found in corn and egg yolks, is a good match for sweet flavors like vanilla, butter pecan, and banana.

Strawberries, raspberries, and cherries contain red-producing anthocyanins, which produce pink and red color in many berry-flavored ice creams. Beetroot and different fruit varieties are used to produce ice cream flavors with vibrant red or purple colors. Grown from spirulina blue-green algae, phycocyanin delivers a blue color and is used in blue raspberry and blueberry ice cream to achieve vivid azure shades (Bause *et al.*, 2024).

Biopigments provide more than visual appeal when added to ice cream. Numerous natural pigments contain antioxidant properties that benefit the product's health profile. Carotenoids eliminate free radicals, while anthocyanins and betalains provide antiinflammatory protection and anticancer benefits. Biopigments enhance products nutritionally and functionally since health-aware individuals demand indulgent items enriched with nutrients.

10.5.3.3 Cheese

Natural cheese has little color until producers apply pigments, transforming cheeses to exhibit vibrant and attractive coloring. Annatto, among carotenoids, is a primary choice because it imparts yellow and orange colors to standard cheese types like cheddar, Colby, and Gouda (Mohd Shukri *et al.*, 2022). Pigments from marigold flowers and paprika that contain capsanthin are also now used by cheese manufacturers as alternatives to traditional sources of pigments, such as annatto (Rodriguez-Amaya *et al.*, 2023). Using chlorophyll from algae and plants makes producing green-colored cheeses easy. Natural coloring ingredients in green pesto cheese include chlorophyll-rich spirulina and chlorella, both of which enhance antioxidant health benefits. Some artisanal and novelty cheeses that have a blue or greenish color effect contain natural blue phycocyanine pigment derived from spirulina.

Biopigments from natural sources are used to increase food sensory quality and develop the color diversity of cheese varieties. Making biopigments a feature of cheese products extends clean-label products for buyers who want to use naturally colored ingredients rather than opting for synthetic additives in dairy products.

10.5.4 Meat and processed foods

Biopigments are starting to replace synthetic colorants in processed food products, especially meat applications. Natural pigments increase meat's appeal while delivering health benefits, which matches the consumer trend toward more sustainable food components. Below are some key applications of biopigments in meat and processed foods.

Biopigments are frequently used in meat processing as color enhancers in products like sausages, ham, and plant-based meat substitutes. The appearance of meat color affects consumer judgment because consumers link color directly to product freshness, quality standards, and flavor expectations. Traditional artificial color agents, nitrate, and nitrites for curing meats provide pink and red hues, while biopigments naturally accomplish similar color enhancement without chemical additives. β-Carotene, among other carotenoids, offers a natural way to improve red and orange coloring of processed meat products while eliminating artificial food dyes (Farkas *et al.*, 2020). Orange-pigmented plants like carrots and paprika provide β-carotene. Lutein creates a yellowish color when added to processed meats, which helps maintain a golden appearance in smoked meats and prepared meal products (Zia-Ul-Haq, 2021).

Food manufacturers apply anthocyanins derived from fruits like berries and red cabbage to give processed meats a natural red or purple shade. Producers can use red cabbage anthocyanins to achieve a richly pigmented coating on sausages, pâtés, and various deli meats. Betalains add visually striking red or purple colors to meat products while supplying antioxidant benefits. The expanding market for plant-based meat alternatives depends heavily on pigments for better visual quality because plant-based meats lack natural animal meat colors. Vegan burgers and sausages achieve a realistic meat appearance with natural pigments from beetroot-derived betalains and red cabbage anthocyanins. These natural pigments improve product appearance, which helps plant-based meat alternatives meats compete with their traditional counterparts.

Ready-to-eat meals and snacks use biopigments to maintain attractive appearance standards and enhance visual quality. Pigments like carotenoids from paprika and curcumin from turmeric give sauces, gravies, and dips their natural orange and yellow tones. These pigments make food products attractive while providing antioxidant and antiinflammatory health benefits. Curcumin is an excellent source of antiinflammatory nutrients (Sharifi-Rad *et al.*, 2020).

Processed food producers use anthocyanins to add red, purple, and blue coloration to products such as jams, jellies, and sauces. Biopigments have been used to enhance the visual appeal of chip and cracker colors. People who prefer natural food often see colorful snacks made from beetroot betalains and spinach chlorophyll as healthier than classic choices. Beetroot extract is used for pink/red hues in restaurant potato chip processing, and spinach chlorophyll is used for green hues in snacks that meet plant-based and transparent labeling requirements.

10.5.5 Functional foods

The natural origins and antioxidant capabilities of biopigments give them distinct health advantages within functional foods. These pigments make functional foods more attractive and provide nutritional benefits and health-enhancing properties. Current consumer demands for functional foods go beyond basic nutritional offerings. Hence, the functional food market relies heavily on pigments. Below are some key applications of pigments in functional foods.

10.5.5.1 Antioxidant-rich functional foods

Pigments are the leading natural antioxidants when used in functional food generation. Pigments such as carotenoids, like β-carotene, lutein, zeaxanthin, anthocyanins, and betalains protect the body against oxidative stress. Oxidative stress contributes to the development of chronic diseases such as cardiovascular disease, cancer, and neurodegenerative disease (Teleanu *et al.*, 2022).

β-Carotene provides us with vitamin A. It is essential for shielding cells against oxidative harm and maintaining visual function. Lutein and zeaxanthin, carotenoid nutrients found in leafy greens and corn, protect eye health and reduce age-related macular degeneration risk. Carotenoids are increasingly found in fortified juices and snacks, aiding in eye health maintenance and whole-body wellness.

Water-dispersible natural colors can be found in berries, grapes, and red cabbage anthocyanins with an antioxidant advantage. These pigments have been shown to help with brain function and heart health while lowering the risk of inflammation. Natural anthocyanin supplements are added to yogurt, breakfast cereals, and energy bars to improve color while enhancing antioxidant effects.

10.5.5.2 Immune support and antiinflammatory effects

Functional foods utilize fluorescent plant pigments, which enhance immune system performance while decreasing inflammatory reactions. Natural antiinflammatory pigments can help control the immune system and decrease chronic disease risk to improve general health and wellbeing. Bioactive compounds contained within pigments have been shown to affect inflammatory pathways while minimizing cytokine output and safeguarding immune function against oxidative stress damage.

Functional foods containing carotenoids such as lutein and zeaxanthin enhance the immune system. Carotenoids promote immune response regulation alongside their established roles in fighting oxidative damage. They apper in immune-boosting products from the functional foods industry and dairy varieties. Research shows that anthocyanins extracted from blueberries, elderberries, and dark-colored fruits have antioxidant and antiinflammatory properties and optimize immune system performance. Anthocyanins are key ingredients in functional foods such as drinks and supplements used for immune defense or to prevent or protect against infection and inflammation. Betalains extracted from beets, according to research, aid in maintaining good liver health and reduce inflammatory markers (Bangar *et al.*, 2022). Betalains are the perfect component for functional food products designed to assist in detoxification and liver health and prevent inflammatory disorders. Thus, beetroot powder is the perfect addition to functional smoothies, juices, or supplements because of its detoxification and health benefits.

10.5.5.3 Cognitive health and brain function

Biopigments have important potential for cognitive health benefits and are thus an important functional food component. Anthocyanins and carotenoids with antioxidant function identified by scientists are pigments that strengthen brain health and improve cognitive function. They are also desirable compounds for functional foods developed for the elderly and people who want to improve their mental performance. Research into anthocyanins has shown how they could slow down brain aging in older adults while improving memory function, strategic memory function, and neuroplasticity ability. Functional foods, such as fortified breakfast cereals, energy bars, and anthocyanin-rich smoothies, can be used by people to improve brain health and cognitive performance. Brain function relies on important carotenoids, including lutein and zeaxanthin. New evidence from emerging research shows that carotenoids within the eye's retina support brain health through neural tissue protection against oxidative damage and cognitive function improvement. Food manufacturers are turning to these specific carotenoids as additive elements for brain-enhancing snacks, fortified liquid products, and supplementary nutrition capsules.

10.6 Stability and Health Benefits of Biopigments

10.6.1 Effect of pH, temperature and light exposure

Environmental conditions, such as pH balance, temperature changes, and light exposure, determine pigment stability. External factors affect pigments' functional properties and chemical structures, substantially changing their stability and health-related advantages. Understanding how environmental factors affect pigments used in food applications is critical because proficiency in employing pigments depends on color strength, stability, and health benefit properties (Fig. 10.4, Table 10.3).

10.6.1.1 Effect of pH on biopigment stability and health benefits

Changes in pH heavily influence pigment stability. During pH variations, multiple pigments, including anthocyanins, betalains, and carotenoids, display high sensitivity to changes in color, chemical structure, and bioactivity.

Anthocyanins show high sensitivity to pH levels of stability. Anthocyanins are pale purple to red anthocyanins in acidic environments above pH 3, and red and blue or purple in basic and neutral environments above pH 7. At higher pH values, they can become colorless. This is because when anthocyanins are exposed to high pH environments, they lose much of their stability and, when structure changes, they lose their antioxidant capabilities (Etxabide *et al.*, 2021).

Betalain contains two substances, betacyanins and betaxanthins, which are pH sensitive but more stable than anthocyanins. Betalains have excellent color stability in acid environments and are therefore the pigment choice for acidic drinks and acidic processed food products. Betalains show stability loss under alkaline environments, which results in both faded pigmentation and a decline in antioxidant effectiveness. Betalains' antiinflammatory and antioxidative effects have maximum potency when preserved under acidic pH conditions.

Research shows that pigments such as β-carotene, luteone, and lysine achieve better stability across different pH levels than other pigments, including anthocyanins and betalains. Slightly acidic to neutral pH levels sustain carotenoid bioactivity, but high extremes of acidic or alkaline environments destabilize carotenoids, affecting their possible health benefits.

The model for estimating color stability model is:

$$\text{ColorChange} = (\text{Abs}_{\text{initial}} - \text{Abs}_{\text{Initial}}) * 100$$

where $\text{Abs}_{\text{initial}}$ is absorbance of biopigment before exposure to pH, and
$\text{Abs}_{\text{final}}$ is absorbance after exposure to pH.

10.6.1.2 Effect of temperature on biopigment stability and health benefits

The stability of biopigments depends critically on temperature variations. The thermal sensitivity of pigments causes them to degrade when exposed to high heat conditions through color loss and bioactivity reduction.

Anthocyanins lose stability when subjected to elevated temperatures. Their stability breaks down faster through heat, which stimulates hydrolysis and polymerization reactions. Food product manufacturing, which requires prolonged heating or pasteurization, leads to a breakdown of anthocyanins that eliminates their coloring and antioxidant abilities and thereby reduces their health advantages (Enaru *et al.*, 2021; Oancea, 2021). In juice and jam products, excessive heating leads to anthocyanin loss, which reduces their nutritional functionality.

Heat exposure affects betalains in the same way that it damages anthocyanins. Both betacyanins and betalains denature when exposed to high temperatures because betacyanins demonstrate a greater sensitivity to heat. Betalain stability is affected when temperatures exceed 70°C, resulting in diminished color and antioxidant features (Esquivel, 2024). Bioactivity and color stability remain intact in some vegetables with high betalain content, such as beets, during processing at reduced temperatures. Functional foods maintain the health qualities of betalains

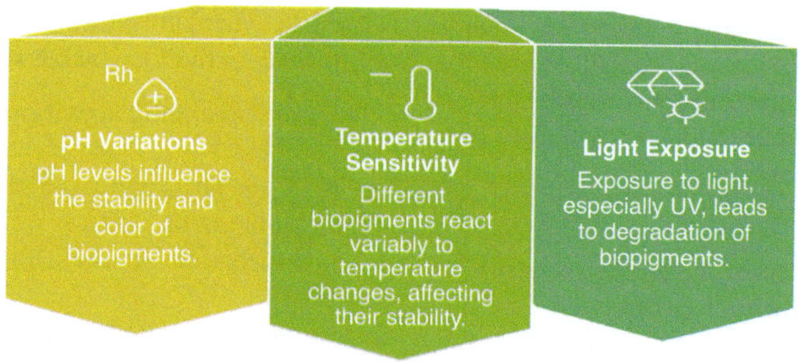

Fig. 10.4. Effect of pH, temperature and light on stability of biopigments.

Table 10.3. Stability of biopigments under different conditions.

Biopigment	pH	Temperature	Light sensitivity	Shelf-life
Carotenoids	4–7	Sensitive to heat (>40°C)	Sensitive to UV light	Six months (under optimal conditions)
Anthocyanins	3–5	Sensitive to high temperatures	Highly sensitive to light	3–4 months
Betalains	4–7	Stable at low temperatures	Sensitive to light	4–5 months
Phycobiliproteins	6–8	Stable at room temperature	Stable in dark storage	6–12 months

better when processing methods limit their heat exposure.

When exposed to high temperatures and food processing conditions, carotenoids remain more stable than anthocyanins and betalains. Bioactive compounds remain stable but endure degradation from prolonged exposure to extreme heat. β-Carotene begins to degrade when temperatures go above 60°C, which results in the loss of both color and its provitamin A activity (Akram *et al.*, 2021). Some food processing techniques that carotenoids undergo, such as blanching and pasteurization, need strict heat management to maintain active biologically functional properties.

10.6.1.3 Effect of light exposure on biopigment stability and health benefits

When anthocyanins are exposed to UV light, they break down faster due to structural damage. The pigments' structural changes lead to diminished color intensity and lowered antioxidant effectiveness. The stability of anthocyanins in food items requires packaging that blocks light using UV-resistant materials or opaque options (Xu *et al.*, 2024). The color quality and health benefits of fruit juices diminish during storage because light exposure induces substantial anthocyanin content loss.

Ultraviolet radiation causes especially harmful effects on betalains. When exposed to light, betalain pigments suffer color loss and antioxidant reduction (Prieto-Santiago *et al.*, 2020). Light protection of betalain-rich products throughout storage and distribution maintains their stability and health advantages, much like anthocyanin protection. Food products like powders and supplements use dark and UV-protective containers to stop light-induced degradation affecting betalains.

Although carotenoids exhibit higher stability than anthocyanins and betalains when exposed to light sources, their long-term exposure to light radiation, especially UV, can result in degradation. Photochemical breakdown removes carotenoid pigmentation and reduces antioxidant function and associated health advantages. Verifying carotenoid resilience requires reduced light exposure while storing nutrients in products such as fortified oils, smoothies, and snacks.

10.6.2 Antioxidant, antiinflammatory, and anticancer properties

Biopigments occur naturally in fungi, bacteria, algae, and plants because of their striking shades and specific functional characteristics. They deliver outstanding bioactive properties that function as antioxidants while providing antiinflammatory and anticancer protection. The health-promoting bioactivities in pigments make them very important for designing functional foods, nutraceutical products, and treatments (Table 10.4).

10.6.2.1 *Antioxidant properties of biopigments*

The antioxidant capability of biopigments stands out as their best-known function since they protect organisms from oxidative stress by destroying free radicals. Multiple chronic diseases, such as cardiovascular conditions, diabetes, and neurodegenerative disorders, develop because of oxidative stress. Biopigments reduce disease risk by scavenging free radicals and protecting lipids, proteins, and DNA from oxidative damage. Researchers recognize biological compounds as antioxidants, including β-carotene, lutein, and lycopene. The carotenoids lutein and zeaxanthin derived from green leafy vegetables defend the retinal structure from oxidative harm and limit age-related macular degeneration (AMD) development. Tomatoes contain lycopene, which successfully acts as an antioxidant that protects cells from oxidative stress and improves their health.

The water-soluble pigments known as anthocyanins, which produce red, purple, and blue colors in fruits and vegetables, demonstrate potent antioxidant capabilities. The pigments show exceptional capabilities in neutralizing reactive oxygen species (ROS) while blocking lipid peroxidation. By shielding cells against oxidative damage, anthocyanins become essential for preventing age-related degenerative diseases, cancer, and cardiovascular diseases. The DPPH assay is used to calculate radical scavenging activity (antioxidant activity), which is represented as:

$$\text{Radical scavenging activity} = \frac{A0-At}{A0} * 100$$

where A_0 is absorbance of the DPPH before adding biopigment, and
A_t is absorbance of the DPPH after adding biopgment.
Certain vegetables and fruits contain betalains along with their subgroups betacyanins and betaxanthins. As potent antioxidants, these pigments shield cells against oxidative damage while working to decrease inflammation. Recent studies reveal that betalain protects cells against damage that causes chronic diseases, including cancer, cardiovascular diseases, and diabetes, while neutralizing reactive oxygen species. The algal-derived pigments phycocyanin and phycoerythrin demonstrate antioxidant activity. Research shows that phycocyanin is a potent free radical scavenger that shields cells from oxidative damage while decreasing the risk of cancer and heart disease (Campos *et al.*, 2024). Functional food products and supplements use these pigments to achieve antioxidant and health improvement benefits.

10.6.2.2 *Antiinflammatory activity of biopigments*

While inflammation functions as a standard defensive reaction of the immune system against injury and infections, natural body processes sometimes produce chronic inflammation, which leads to severe conditions like arthritis, cardiovascular diseases, and cancer. Multiple studies demonstrate that biopigments activate antiinflammatory properties through inflammatory pathway regulation alongside proton antiinflammatory cytokine reduction paired with inflammatory enzyme inhibition. A few carotenoids possess antiinflammatory attributes, among which lutein, zeaxanthin, and β-carotene are important examples. The carotenoids block nuclear factor-XB (NF-XB) activation, which controls proinflammatory cytokine gene expression (Akkara and Sabina, 2020). Carotenoids act against NF-XB to decrease inflammatory molecule production, which helps to reduce inflammation. Research demonstrates how lutein and zeaxanthin lower

Table 10.4. Potential health benefits of different biopigments.

Biopigment	Health benefits	Mechanism of action	Potential applications
Carotenoids	Eye health, cardiovascular health	Antioxidant activity, protect against oxidative stress, reduce LDL	Functional foods, dairy, beverages, supplements
Anthocyanins	Antioxidant, antiinflammatory, eye health	Scavenge free radicals, improve blood circulation, reduce inflammation	Juices, snacks, sauces, desserts
Betalains	Antiinflammatory, antioxidant	Reduce oxidative damage, modulate inflammatory pathways	Processed foods, beverages, supplements
Phycobiliproteins	Antioxidant, antiinflammatory	Scavenge reactive oxygen species, reduce cell damage	Dairy, beverages, functional foods

inflammation levels to combat eye conditions such as AMD, where inflammation is a principal mechanism of disease advancement.

Research indicates that blueberry- and elderberry-derived anthocyanins suppress the production of proinflammatory cytokines like IL-6, IL-1β, and TNF-α. Anthocyanins display antiinflammatory effects to help reduce symptoms in people with arthritis, cardiovascular diseases, and inflammatory bowel ailments. Research confirms that betalains produce notable antiinflammatory benefits by blocking the generation of inflammation markers including TNF-α and IL-6 (Martinez *et al.*, 2024; Nirmal *et al.*, 2024). Ingesting betalains from beets reduces inflammation and improves high blood pressure plus metabolic disorders through interactions with inflammatory pathways. Through their antiinflammatory mechanism, phycobiliproteins such as phycocyanin help lessen the functioning mode of proinflammatory cytokines and enzymes. Phycocyanin prevents nitric oxide (NO) production and adjusts gene expression related to inflammatory processes. Due to the antiinflammatory properties of phycobiliproteins, pharmaceutical treatment can benefit patients with rheumatoid arthritis and inflammatory bowel disease.

10.6.2.3 Anticancer activity of biopigments

The potential anticancer properties of pigments have an expanding focus within scientific research. Through their antioxidant and antiinflammatory properties, biopigments demonstrate the power to restrict cancer cell development while blocking metastatic spread and activating programmed cell death mechanisms. The anticancer potential of β-carotene, lutein, and lycopene has been validated by multiple *in vitro* and *in vivo* studies. Through its role in modulating cancer signaling and promoting cell death, lycopene effectively reduces prostate cancer risk while simultaneously obstructing cancer cell multiplication (Tufail *et al.*, 2024). Dietary intake of lutein and zeaxanthin creates a protective effect against women developing breast and ovarian cancer (Sanlier *et al.*, 2024).

Anticancer potential in anthocyanins arises through the mechanisms of tumor cell growth inhibition and metastasis prevention. These pigments control multiple molecular pathways within cancer cells that control cell division, programmed cell death, and new blood vessel formation. Research shows that the anthocyanins in blackberries, blueberries, and other fruit species restrain cancer cell growth for several primary types of cancer affecting the breast, colon, and lungs. These compounds decrease the spread of tumor cells by inhibiting matrix metalloproteinases (MMPs), which enable cancer cell invasion.

Research reveals that betalains, especially betacyanins, demonstrate anticancer properties through their ability to stop tumor cell growth and trigger cancer cell death by apoptosis. Betalains extracted from beets possess properties that both stop colon cancer cell growth and decrease cancer cell movement and invasion (Fu *et al.*, 2020). Studies demonstrate that betalains prevent cancer progression by stopping new cell blood vessels from forming because this angiogenesis process supports cancer spread.

The algal pigment phycocyanin acts against cancer by blocking cancer cell growth and slowing tumor expansion. Research demonstrates that phycocyanin induces apoptosis in blood, lung, and breast cancer cells while blocking tumor growth in animal studies. Because phycocyanin can control signaling routes that result in cell cycle stasis alongside apoptosis development therapy, research cannot ignore its potential as an anticancer treatment and preventive measure.

10.6.3 Role in eye health and cardiovascular function

Bioactive pigments naturally present within plants and various microorganisms, including algae, and draw widespread attention for their striking colors and functional properties. Biopigments demonstrate multiple health advantages, especially in maintaining eye health and cardiovascular function. The primary health-related effects of pigments result from their combined antioxidant abilities to protect cells from damage and their capacity to reduce inflammation. In-depth research shows that carotenoids, anthocyanins, and betalains deliver essential health benefits because they promote healthy eyesight and strengthen the cardiovascular systems.

Age-related macular degeneration, cataracts, and other vision-related diseases develop substantially because of oxidative stress and inflammation in our eyes. Research shows that carotenoids exhibit excellent natural defenses against eye damage from harmful factors. The retina maintains high levels of lutein and zeaxanthin within the macular area, where these two carotenoids perform protective filtering actions. These pigments capture dangerous blue light before it reaches deep into the retina, where it creates oxidative damage (Kumar *et al.*, 2024). These light-filtering nanopigments work alongside antioxidant effects to shield retinal tissues from oxidative harm while diminishing AMD formation risk. Research shows that lutein and zeaxanthin enhance visual acuity and contrast sensitivity in older adults, and their deficiency during advanced age hastens the development of visual health problems.

The carotenoids protect retinal cells from oxidative damage by neutralizing ROS. The compound β-carotene is a significant carotenoid that converts into vitamin A, necessary to maintain perfect vision. Rhodopsin production in the retina depends on vitamin A because rhodopsin helps humans achieve low-light vision. When vitamin A levels drop too low, people develop night blindness. While β-carotene is less specialized than lutein or zeaxanthin for eye health, its antioxidants preserve retinal cells, which helps prevent degenerative diseases such as cataracts from arising (Rodella *et al.*, 2023).

Red, purple, and blue fruits such as berries and grapes support healthy vision. The color-causing chemicals offer eye protection against oxidative damage induced by UV light, which is responsible for cataract formation. Retinal health is supported by anthocyanins in boosting retinal oxygen supply and increasing blood flow. By comparing the effects of familiar pigments with those known to be toxic to succulent plants, researchers found these pigments have positive qualities like improved night vision and reduced eye strain. Anthocyanins aid diabetic retinopathy, a disease of diabetic retinal blood vessels.

Research demonstrates that biopigments improve heart health while protecting blood vessel networks. Lycopene and β-carotene have received research attention because carotenoids show strong antioxidant abilities along with antiinflammatory activity. Lycopene is a potent antioxidant because red fruits provide protection against LDL cholesterol oxidation. Atherosclerosis development hinges on when LDL cholesterol oxidates because plaque accumulation follows, leading to heart attacks and strokes. Lycopene prevents LDL from oxidizing, lowering atherosclerosis risk and improving cardiovascular health. Research indicates that consuming foods high in lycopene lowers blood pressure, reduces harmful cholesterol levels, and improves artery function leading to improved heart health.

The antioxidant capability of β-carotene lowers oxidative stress, thus helping to stop the development of many cardiovascular diseases. The protective effect of β-carotene against oxidative vessel damage and its antiinflammatory properties preserve optimal endothelial function that protects blood vessel linings and ensures proper circulation throughout the body. The carotenoid group displays improved vascular health outcomes through its ability to boost blood flow and reduce blood clot factors while extending protection against hypertension and cardiovascular events like strokes and heart disease.

10.7 Conclusion

Natural pigments derived from plants and microorganisms offer a dual sustainable benefit in food industries: good coloration and improvement in public health. Research into natural pigments, such as carotenoids, anthocyanins, betalains, and phycobiliproteins, supports their antioxidant properties, such as protection from cancer, cardioprotection, and antiinflammatory advantages. Natural compounds protect the retina and promote visual function improvement. Besides their multiple benefits for vision health, they are functional foods that help fight oxidative stress and improve blood vessel function, leading to a lower risk of heart disease. In the future, pigment extraction will benefit from enzymatic and solvent extraction, supercritical fluid extraction, and fermentation in sustainable production. The rise in demand for natural and health-improving foods means that food manufacturers prefer natural colorants. Consumer concern regarding synthetic colorants is prompting a more significant application of pigments across food sectors, including drinks, dairy products, and specialized functional foods.

References

Akkara, P.J. and Sabina, E.P. (2020) Pre-treatment with beta carotene gives protection against nephrotoxicity induced by bromobenzene via modulation of antioxidant system, pro-inflammatory cytokines and pro-apoptotic factors. *Applied Biochemistry and Biotechnology* 190(2), 616–633. DOI: 10.1007/s12010-019-03111-0.

Akram, S., Mushtaq, M. and Waheed, A. (2021) β-carotene: Beyond provitamin A. In: *A Centum of Valuable Plant Bioactives*. Elsevier, New York, pp. 1–31. DOI: 10.1016/B978-0-12-822923-1.00008-X.

Algarni, E.H.A. (2020) Extraction of natural pigments from food-industrial waste and their use in the manufacture of jelly candy for a child. *World Journal of Environmental Biosciences* 9(4), 33–40. DOI: 10.51847/AWRihX9.

Arunkumar, R., Gorusupudi, A. and Bernstein, P.S. (2020) The macular carotenoids: A biochemical overview. *Biochimica et Biophysica Acta (BBA)-Molecular and Cell Biology of Lipids* 1865(11), 158617. DOI: 10.1016/j.bbalip.2020.158617.

Banc, R., Filip, L., Cozma-petruț, A., Ciobârcă, D. and Miere, D. (2024) Yellow and red synthetic food dyes and potential health hazards: A mini review. *Bulletin of University of Agricultural Sciences and Veterinary Medicine Cluj-Napoca. Food Science and Technology* 81(1), 1–17. DOI: 10.15835/buasvmcn-fst:2024.0005.

Bangar, S.P., Sharma, N., Sanwal, N., Lorenzo, J.M. and Sahu, J.K. (2022) Bioactive potential of beetroot (*Beta vulgaris*). *Food Research International* 158, 111556. DOI: 10.1016/j.foodres.2022.111556.

Bause, K., Bußler, S., de Munnik, M., Reineke, K., Berg-Stolp, F. *et al.* (2024) Challenges and opportunities in the use of natural colors in foods and beverages: An industrial perspective. *Handbook on Natural Pigments in Food and Beverages* 577–604. DOI: 10.1016/B978-0-323-99608-2.00013-6.

Calva-Estrada, S.J., Jiménez-Fernández, M. and Lugo-Cervantes, E. (2022) Betalains and their applications in food: The current state of processing, stability and future opportunities in the industry. *Food Chemistry. Molecular Sciences* 4, 100089. DOI: 10.1016/j.fochms.2022.100089.

Campos, J., Fernandes, R. and Barros, A.N. (2024) Phycocyanin, the microalgae. *Functional Food: Upgrading Natural and Synthetic Sources* 11, 125. DOI: 10.5772/intechopen.115108.

Coy-Barrera, E. (2020) Analysis of betalains (betacyanins and betaxanthins). In: *Recent Advances in Natural Products Analysis*. Elsevier, New York, pp. 593–619. DOI: 10.1016/B978-0-12-816455-6.00017-2.

Cuozzo, S., de LeBlanc, A. d. M., de LeBlanc, J., Hoffmann, N. and Tortella, G.R. (2023) *Streptomyces* genus as a source of probiotics and its potential for its use in health. *Microbiological Research* 266, 127248. DOI: 10.1016/j.micres.2022.127248.

Cvejić, J., Atanacković Krstonošić, M., Mikulić, M. and Miljić, U. (2022) Polyphenols. In: Galanakis, C.M. (ed.) *Nutraceutical and Functional Food Components*, 2nd edn. Academic Press, Washington, DC, pp. 243–312. DOI: 10.1016/B978-0-323-85052-0.00012-X.

de Oliveira, Z.B., Costa, D.V., Silva Dos Santos, A.C., Silva Júnior, A.Q., Lima Silva, A. *et al*. (2024) Synthetic colors in food: A warning for children's health. *International Journal of Environmental Research and Public Health* 21(6), 682. DOI: 10.3390/ijerph21060682.

de Souza Mesquita, L.M., Martins, M., Pisani, L.P., Ventura, S.P.M. and Rosso, V.V. (2021) Insights on the use of alternative solvents and technologies to recover bio-based food pigments. *Comprehensive Reviews in Food Science and Food Safety* 20(1), 787–818. DOI: 10.1111/1541-4337.12685.

Elmaaty, T.A. and Sorour, H. (2024) Extraction of natural pigments using supercritical carbon dioxide and its application on different fabrics. In: *Natural Dyes and Sustainability*. Springer, Dordrecht, The Netherlands, pp. 231–264. DOI: 10.1007/978-3-031-47471-2_11.

Enaru, B., Drețcanu, G., Pop, T.D., Stănilă, A. and Diaconeasa, Z. (2021) Anthocyanins: Factors affecting their stability and degradation. *Antioxidants* 10(12), 12. DOI: 10.3390/antiox10121967.

Esquivel, P. (2024) Betalains. In: *Handbook on Natural Pigments in Food and Beverages*. Elsevier, New York, pp. 147–167. DOI: 10.1016/B978-0-323-99608-2.00002-1.

Etxabide, A., Kilmartin, P.A. and Maté, J.I. (2021) Color stability and pH-indicator ability of curcumin, anthocyanin and betanin containing colorants under different storage conditions for intelligent packaging development. *Food Control* 121, 107645. DOI: 10.1016/j.foodcont.2020.107645.

Farkas, Á., Bencsik, T. and Deli, J. (2020) Carotenoids as food additives. In: *Pigments from Microalgae Handbook*. Springer, Dordrecht, The Netherlands, pp. 421–447. DOI: 10.1007/978-3-030-50971-2_17.

Fatima, I., Munir, M., Qureshi, R., Hanif, U., Gulzar, N. *et al*. (2024) Advanced methods of algal pigments extraction: A review. *Critical Reviews in Food Science and Nutrition* 64(27), 9771–9788. DOI: 10.1080/10408398.2023.2216782.

Fenger, J.-A. (2020) Acylated anthocyanins as natural colors: Reactivity in aqueous solution, metal binding, stabilization for food applications. Université d'Avignon. Available at: https://theses.hal.science/tel-03356341v1/file/PhD_Manuscript_Julie-Anne_Fenger_VF.pdf (accessed 21 August 2025).

Ferroukhi, I., Chassard, C. and Mardon, J. (2024) A comprehensive overview of blue-veined cheeses. *International Dairy Journal* 154, 105926. DOI: 10.1016/j.idairyj.2024.105926.

Fu, Y., Shi, J., Xie, S.-Y., Zhang, T.-Y., Soladoye, O.P. *et al*. (2020) Red beetroot betalains: Perspectives on extraction, processing, and potential health benefits. *Journal of Agricultural and Food Chemistry* 68(42), 11595–11611. DOI: 10.1021/acs.jafc.0c04241.

Gonçalves, A.C., Nunes, A.R., Falcão, A., Alves, G. and Silva, L.R. (2021) Dietary effects of anthocyanins in human health: A comprehensive review. *Pharmaceuticals* 14(7), 690. DOI: 10.3390/ph14070690.

Goñi, M.G., Pellegrini, M.C. and Ponce, A.G. (2022) Effect of thermal treatments on the properties of natural food additives. In: *Natural Additives in Foods*. Springer, Dordrecht, The Netherlands, pp. 251–267. DOI: 10.1007/978-3-031-17346-2_9.

Haghdoost, A., Golestan, L., Hasani, M., Shahidi Noghabi, M. and Shahidi, S. (2023) Encapsulation of the extracted phycobiliprotein from *Gracilaria gracilis* in nanoliposomes: Physicochemical, structural and stability properties. *Iranian Journal of Fisheries Sciences* 22(5), 1018–1038. DOI: 10.22092/ijfs.2023.130549.

Herzyk, F., Piłakowska-Pietras, D. and Korzeniowska, M. (2024) Supercritical extraction techniques for obtaining biologically active substances from a variety of plant byproducts. *Foods* 13(11), 1713. DOI: 10.3390/foods13111713.

Husain, A., Khanam, A., Alouffi, S., Shahab, U., Alharazi, T. *et al*. (2024) C-phycocyanin from cyanobacteria: A therapeutic journey from antioxidant defence to diabetes management and beyond. *Phytochemistry Reviews* 1–19. DOI: 10.1007/s11101-024-10045-x.

Jurić, S., Jurić, M., Król-Kilińska, Ż., Vlahoviček-Kahlina, K., Vinceković, M. *et al*. (2022) Sources, stability, encapsulation and application of natural pigments in foods. *Food Reviews International* 38(8), 1735–1790. DOI: 10.1080/87559129.2020.1837862.

Kuddus, M., Singh, P., Raveendran, S. and Singh, R. (2024) *Microbial Pigments: Applications in Food and Beverage Industry*. CRC Press, Boca Raton, FL.

Kumar, P., Banik, S.P., Ohia, S.E., Moriyama, H., Chakraborty, S. *et al*. (2024) Current insights on the photoprotective mechanism of the macular carotenoids, lutein and zeaxanthin: Safety, efficacy and bio-delivery. *Journal of the American Nutrition Association* 43(6), 505–518. DOI: 10.1080/27697061.2024.2319090.

López-Cruz, R., Sandoval-Contreras, T. and Iñiguez-Moreno, M. (2023) Plant pigments: Classification, extraction, and challenge of their application in the food industry. *Food and Bioprocess Technology* 16(12), 2725–2741. DOI: 10.1007/s11947-023-03075-4.

López-Díaz, T.M., Alegría, Á., Rodríguez-Calleja, J.M., Combarros-Fuertes, P., Fresno, J.M. *et al.* (2023) Blue cheeses: Microbiology and its role in the sensory characteristics. *Dairy* 4(3), 410–422. DOI: 10.3390/dairy4030027.

Lu, W., Shi, Y., Wang, R., Su, D., Tang, M. *et al.* (2021) Antioxidant activity and healthy benefits of natural pigments in fruits: A review. *International Journal of Molecular Sciences* 22(9), 4945. DOI: 10.3390/ijms22094945.

Luzardo-Ocampo, I., Ramírez-Jiménez, A.K., Yañez, J., Mojica, L. and Luna-Vital, D.A. (2021) Technological applications of natural colorants in food systems: A review. *Foods* 10(3), 634. DOI: 10.3390/foods10030634.

Maoka, T. (2020) Carotenoids as natural functional pigments. *Journal of Natural Medicines* 74(1), 1–16. DOI: 10.1007/s11418-019-01364-x.

Martinez, R.M., Melo, C.P.B., Pinto, I.C., Mendes-Pierotti, S., Vignoli, J.A. *et al.* (2024) Betalains: A narrative review on pharmacological mechanisms supporting the nutraceutical potential towards health benefits. *Foods* 13(23), 3909. DOI: 10.3390/foods13233909.

Martins, T., Barros, A.N., Rosa, E. and Antunes, L. (2023) Enhancing health benefits through chlorophylls and chlorophyll-rich agro-food: A comprehensive review. *Molecules* 28(14), 5344. DOI: 10.3390/molecules28145344.

Meruvu, H. and Dos Santos, J.C. (2021) Colors of life: A review on fungal pigments. *Critical Reviews in Biotechnology* 41(8), 1153–1177. DOI: 10.1080/07388551.2021.1901647.

Miller, M.D., Steinmaus, C., Golub, M.S., Castorina, R., Thilakartne, R. *et al.* (2022) Potential impacts of synthetic food dyes on activity and attention in children: A review of the human and animal evidence. *Environmental Health* 21(1), 45. DOI: 10.1186/s12940-022-00849-9.

Miranda, P.H.S., dos Santos, A.C., Freitas, B.C.B., de Souza Martins, G.A., Boas, E.V. d. B.V. *et al.* (2021) A scientific approach to extraction methods and stability of pigments from amazonian fruits. *Trends in Food Science & Technology* 113, 335–345. DOI: 10.1016/j.tifs.2021.04.047.

Mitra, S., Rauf, A., Tareq, A.M., Jahan, S., Emran, T.B. *et al.* (2021) Potential health benefits of carotenoid lutein: An updated review. *Food and Chemical Toxicology* 154, 112328. DOI: 10.1016/j.fct.2021.112328.

Mohankumari, H.P., Naidu, K.A., Narasimhamurthy, K. and Vijayalakshmi, G. (2021) Bioactive pigments of *Monascus purpureus* attributed to antioxidant, HMG-coa reductase inhibition and anti-atherogenic functions. *Frontiers in Sustainable Food Systems* 5, 590427. DOI: 10.3389/fsufs.2021.590427.

Mohd Shukri, A., Alias, A.K., Murad, M., Yen, K.S. and Cheng, L.H. (2022) A review of natural cheese and imitation cheese. *Journal of Food Processing and Preservation* 46(1), e16112. DOI: 10.1111/jfpp.16112.

Molina, A.K., Corrêa, R.C.G., Prieto, M.A., Pereira, C. and Barros, L. (2023) Bioactive natural pigments' extraction, isolation, and stability in food applications. *Molecules* 28(3), 1200. DOI: 10.3390/molecules28031200.

Morón-Ortiz, Á., Mapelli-Brahm, P. and Meléndez-Martínez, A.J. (2024) Sustainable green extraction of carotenoid pigments: Innovative technologies and bio-based solvents. *Antioxidants* 13(2), 239. DOI: 10.3390/antiox13020239.

Mouritsen, O.G., Styrbæk, K. and Johansen, M. (2024) *Plant-Forward Cuisine: Basic Concepts and Practical Applications*. Taylor & Francis, Abingdon, UK.

Mummaleti, G., Udo, T., Mohan, A. and Kong, F. (2024) Synthesis, characterization and application of microbial pigments in foods as natural colors. *Critical Reviews in Food Science and Nutrition* 1–30. DOI: 10.1080/10408398.2024.2417802.

Nabi, B.G., Mukhtar, K., Ahmed, W., Manzoor, M.F., Ranjha, M.M.A.N. *et al.* (2023) Natural pigments: anthocyanins, carotenoids, chlorophylls, and betalains as colorants in food products. *Food Bioscience* 52, 102403. DOI: 10.1016/j.fbio.2023.102403.

Nirmal, N.P., Medhe, S., Dahal, M., Koirala, P., Nirmal, S. *et al.* (2024) Betalains protect various body organs through antioxidant and anti-inflammatory pathways. *Food Science and Human Wellness* 13(3), 1109–1117. DOI: 10.26599/FSHW.2022.9250093.

Oancea, S. (2021) A review of the current knowledge of thermal stability of anthocyanins and approaches to their stabilization to heat. *Antioxidants* 10(9), 1337. DOI: 10.3390/antiox10091337.

Pagels, F., Pereira, R.N., Vicente, A.A. and Guedes, A.C. (2021) Extraction of pigments from microalgae and cyanobacteria—A review on current methodologies. *Applied Sciences* 11(11), 5187. DOI: 10.3390/app11115187.

Patil, A.D., Kasabe, P.J. and Dandge, P.B. (2022) Pharmaceutical and nutraceutical potential of natural bioactive pigment: Astaxanthin. *Natural Products and Bioprospecting* 12(1), 25. DOI: 10.1007/s13659-022-00347-y.

Peter, A.P., Koyande, A.K., Chew, K.W., Ho, S.-H., Chen, W.-H. *et al.* (2022) Continuous cultivation of microalgae in photobioreactors as a source of renewable energy: Current status and future challenges. *Renewable and Sustainable Energy Reviews* 154, 111852. DOI: 10.1016/j.rser.2021.111852.

Polyakov, N.E., Focsan, A.L., Gao, Y. and Kispert, L.D. (2023) The endless world of carotenoids-structural, chemical and biological aspects of some rare carotenoids. *International Journal of Molecular Sciences* 24(12), 9885. DOI: 10.3390/ijms24129885.

Prieto-Santiago, V., Cavia, M.M., Alonso-Torre, S.R. and Carrillo, C. (2020) Relationship between color and betalain content in different thermally treated beetroot products. *Journal of Food Science and Technology* 57(9), 3305–3313. DOI: 10.1007/s13197-020-04363-z.

Qi, Q., Chu, M., Yu, X., Xie, Y., Li, Y. *et al.* (2023) Anthocyanins and proanthocyanidins: Chemical structures, food sources, bioactivities, and product development. *Food Reviews International* 39(7), 4581–4609. DOI: 10.1080/87559129.2022.2029479.

Ramadhani, E.P., Mustofa, A., Puyanda, I.R. and Jokopriyambodo, W. (2024) Effect of adding chlorophyll extract of suji leaves (*Dracaena angustifolia*) and fermentation time on yogurt characteristics. *Agric* 36(2), 213–230. DOI: 10.24246/agric.2024.v36.i2.p213-230.

Rao, A.S., Deka, S.P., More, S.S., Nair, A., More, V.S. *et al.* (2021) A comprehensive review on different microbial-derived pigments and their multipurpose activities. *Microbial Polymers: Applications and Ecological Perspectives* 479–519. DOI: 10.1007/978-981-16-0045-6_20.

Ravindran, P. (2023) Annatto, kaffir lime, licorice, star anise, sumac, and tarragon. In: *Handbook of Spices in India: 75 Years of Research and Development*. Springer, New Delhi, India, pp. 3369–3435.

Renita, A.A., Gajaria, T.K., Sathish, S., Kumar, J.A., Lakshmi, D.S. *et al.* (2023) Progress and prospective of the industrial development and applications of ECO-friendly colorants: An insight into environmental impact and sustainability issues. *Foods* 12(7), 1521. DOI: 10.3390/foods12071521.

Rodella, U., Honisch, C., Gatto, C., Ruzza, P. and D'Amato Tóthová, J. (2023) Antioxidant nutraceutical strategies in the prevention of oxidative stress related eye diseases. *Nutrients* 15(10), 2283. DOI: 10.3390/nu15102283.

Rodriguez-Amaya, D.B., Esquivel, P. and Meléndez-Martínez, A.J. (2023) Comprehensive update on carotenoid colorants from plants and microalgae: Challenges and advances from research laboratories to industry. *Foods* 12(22), 4080. DOI: 10.3390/foods12224080.

Roriz, C.M.L. d. F. (2021) New natural sources of betacyanins: Optimization of the extraction process, stability studies and incorporation as a colorant in different foods. Available at: https://docta.ucm.es/entities/publication/c96f5010-64da-44c4-867a-e9d56d2e91bf (accessed 21 August 2025).

Roy, D., Verma, S., Das, S., Chakdar, H. and Pabbi, S. (2024) Production and applications of cyanobacterial phycocyanin: Trends and prospects. *Indian Journal of Microbiology* 1–19. DOI: 10.1007/s12088-024-01332-4.

Sanlier, N., Yildiz, E. and Ozler, E. (2024) An overview on the effects of some carotenoids on health: Lutein and zeaxanthin. *Current Nutrition Reports* 13(4), 828–844. DOI: 10.1007/s13668-024-00579-z.

Shafqat, A. (2024) Insight into the anti-cancer and anti-viral therapeutic properties of biological active molecule prodigiosin. *Asia-Pacific Journal of Pharmacotherapy & Toxicology* 4, 122–136. DOI: 10.32948/ajpt.2024.12.19.

Sharifi-Rad, J., Rayess, Y.E., Rizk, A.A., Sadaka, C., Zgheib, R. *et al.* (2020) Turmeric and its major compound curcumin on health: Bioactive effects and safety profiles for food, pharmaceutical, biotechnological and medicinal applications. *Frontiers in Pharmacology* 11, 01021. DOI: 10.3389/fphar.2020.01021.

Singh, S., Aeri, V. and Sharma, V. (2023) Encapsulated natural pigments: Techniques and applications. *Journal of Food Process Engineering* 46(12), e14311. DOI: 10.1111/jfpe.14311.

Teleanu, D.M., Niculescu, A.-G., Lungu, I.I., Radu, C.I., Vladâcenco, O. *et al.* (2022) An overview of oxidative stress, neuroinflammation, and neurodegenerative diseases. *International Journal of Molecular Sciences* 23(11), 5938. DOI: 10.3390/ijms23115938.

Thakur, M. and Modi, V.K. (2024) Biocolorants in food: Sources, extraction, applications and future prospects. *Critical Reviews in Food Science and Nutrition* 64(14), 4674–4713. DOI: 10.1080/10408398.2022.2144997.

Tufail, T., Bader Ul Ain, H., Noreen, S., Ikram, A., Arshad, M.T. *et al*. (2024) Nutritional benefits of lycopene and beta-carotene: A comprehensive overview. *Food Science & Nutrition* 12(11), 8715–8741. DOI: 10.1002/fsn3.4502.

Villegas-Méndez, M.Á., Montañez, J., Contreras-Esquivel, J.C., Salmerón, I., Koutinas, A.A. *et al*. (2023) Scale-up and fed-batch cultivation strategy for the enhanced co-production of microbial lipids and carotenoids using renewable waste feedstock. *Journal of Environmental Management* 339, 117866. DOI: 10.1016/j.jenvman.2023.117866.

Wei, J., Li, Y., Ye, Z., Li, Y. and Zhou, Z. (2023) Citrus carotenoid extracts exert anticancer effects through anti-proliferation, oxidative stress, and mitochondrial-dependent apoptosis in MCF-7 cells. *Foods* 12(18), 3469. DOI: 10.3390/foods12183469.

Wu, Y., Yuan, Y., Jiang, W., Zhang, X., Ren, S. *et al*. (2022) Enrichment of health-promoting lutein and zeaxanthin in tomato fruit through metabolic engineering. *Synthetic and Systems Biotechnology* 7(4), 1159–1166. DOI: 10.1016/j.synbio.2022.08.005.

Xu, M., Fang, D., Kimatu, B.M., Lyu, L., Wu, W. *et al*. (2024) Recent advances in anthocyanin-based films and its application in sustainable intelligent food packaging: A review. *Food Control* 162, 110431. DOI: 10.1016/j.foodcont.2024.110431.

Zannou, O., Oussou, K.F., Chabi, I.B., Odouaro, O.B.O., Deli, M.G.E.P. *et al*. (2024) A comprehensive review of recent development in extraction and encapsulation techniques of betalains. *Critical Reviews in Food Science and Nutrition* 64(30), 11263–11280. DOI: 10.1080/10408398.2023.2235695.

Zhang, H., Ahima, J., Yang, Q., Zhao, L., Zhang, X. *et al*. (2021) A review on citrinin: Its occurrence, risk implications, analytical techniques, biosynthesis, physiochemical properties and control. *Food Research International* 141, 110075. DOI: 10.1016/j.foodres.2020.110075.

Zia-Ul-Haq, M. (2021) Historical and introductory aspects of carotenoids. In: *Carotenoids: Structure and Function in the Human Body*. Springer, New York, pp. 1–42. DOI: 10.1007/978-3-030-46459-2_1.

Ziyaei, K., Abdi, F., Mokhtari, M., Daneshmehr, M.A. and Ataie, Z. (2023) Phycocyanin as a nature-inspired antidiabetic agent: A systematic review. *Phytomedicine* 119, 154964. DOI: 10.1016/j.phymed.2023.154964.

11 Applications of Microbial Pigments in Pharmaceuticals and Cosmetics

Jatin A. Tekawade[1], Mayuresh R. Redkar[1,2*], Vinod R. Biradar[3], Pranita K. Pawar[1] and Sachinkumar V. Patil[4]

[1]Department of Pharmaceutics, Yashwantrao Bhonsale College of Pharmacy, Sawantwadi, Maharashtra, India; [2]Rajarambapu College of Pharmacy, Kasegaon, Maharashtra, India; [3]Department of Pharmaceutical Chemistry, Yashwantrao Bhonsale College of Pharmacy, Sawantwadi, Maharashtra, India; [4]Department of Pharmaceutics, Dr Ashok Gujar Institute of Pharmacy, Karad, Maharashtra, India

11.1 Introduction

As civilization grew, coloring became more and more common. Pigments are now an essential part of our daily life and are used in a variety of industries, such as agriculture, textiles, food, medicine, and cosmetics (Yusuf *et al.*, 2017; Paillié-Jiménez *et al.*, 2020; Venil *et al.*, 2020a). Prior to the widespread use of synthetic pigments, natural pigments were the only source of color. Since Perkin first created synthetic dyes in 1856, the availability of affordable and convenient synthetic pigments has led to a decrease in the use of natural colors. It is well known that the production of some synthetic dyes is forbidden due to the carcinogenic nature of the product and the environmental effects of disposing of its industrial waste (Zollinger, 2003; Venil *et al.*, 2013).

Colors are among the first, if not the first, characteristics that humans notice and assess when they view any kind of object, including food, clothing, flowers, plants, animals, and other people. As a result, the most diverse industries spend money on R&D to produce new colors and sources, which in turn inspires new market trends that drive innovation (Kumar *et al.*, 2015; Yadav *et al.*, 2023). Chemicals known as pigments or dyes preferentially absorb light and reflect it at wavelengths between 400 and 800 nm, allowing the human eye to perceive a particular color. These ensuing hues physiologically incorporate initial human sensations, which are a component of visual information processing (Cuthill *et al.*, 2017). Despite their common uses in giving items color, dyes and pigments differ from one another. Water-soluble dyes have a molecular degree of dispersion that gives them more vibrant colors, but they are not very stable in the presence of light. Pigments, on the other hand, are hardly soluble in water by nature and disperse into particles; they are unaffected by the inclusion vehicle and continue to absorb and mirror light (Lincke, 2003; Rothon, 2012).

Pigments are of natural and synthetic types. Synthetic pigments are produced artificially. They come in a variety of colors and are frequently utilized in industries such as ceramics,

*Corresponding author: mayuresh.redkar253@gmail.com

 Biopigments of Microbial Origin: Innovations and Applications Across Industries (eds Madan *et al.*)
DOI: 10.1079/ 9781836990505.0011

paints, coatings, and plastics. Cadmium red, bismuth vanadate, and iron oxide yellow and black are a few examples. Minerals are the source of several artificial inorganic colors, such as chromium oxide and titanium dioxide. The stability and longevity of these pigments are well known (Cserháti, 2006; Rothon, 2012). Synthetic pigment offers an economic benefit due to its low manufacturing costs and steady, well-known synthesis method. Both natural and synthetic dyes are widely employed in many aspects of daily life, including the manufacturing of food, pharmaceuticals, textiles, paper, and agricultural techniques (Dufossé, 2009).

Green technology claims that modern manufacturing lines benefit from using more natural source materials and fewer hazardous goods. It is commonly known that the production of some synthetic dyes is forbidden because of the carcinogenicity of the product or precursor and the impact that the disposal of their industrial wastes has on the environment. Natural pigments exhibit beneficial biological functions as antioxidants and anticancer agents in addition to their ability to boost a product's marketability. Investigating different natural sources of colorants and their uses is crucial. Some bacterial pigments for use in food, cosmetics, or textiles can now be produced by the industry (Soliev *et al.*, 2011).

The biological colors that come from microbial cells are called microbial pigments. A number of microbial pigments are used as colorants and to provide vital chemicals or nutrients that are beneficial to human health. Some also have biological benefits, such as antiinflammatory, immunosuppressive, anticancer, and antibacterial properties (Venil *et al.*, 2014). Depending on their color and biological activity, microbial pigments are used for a number of reasons. For instance, fluorescent microbial pigments are used in laboratories to mark antibodies. By altering their color, certain pigments can also be used to gauge pH variations or demonstrate the advancement of particular processes (Dufossé, 2019). Microbial pigments are in high demand because of their many uses in a variety of fields, including biomedical, agricultural, evolutionary, ecological, and industrial (Chatragadda and Dufossé, 2021). They are also easy to cultivate, inexpensive, stable, biodegradable, and environmentally friendly.

A large variety of pigments with multiple biological characteristics and a range of commercial applications are produced by several microorganisms. Furthermore, because they guard against solar radiation, act as a method of adaptation to a variety of harsh environments, and participate in functional activities like photosynthesis, microbial pigments are important in the molecular physiological processes of microorganisms (Sutthiwong and Caro, 2014). Because microbes are influenced by their surroundings, they generate a variety of pigments with distinct characteristics that are linked to the microorganism–ecosystem relationship. A wide range of microbial pigments, including melanin, carotenoids, flavins, phenazines, quinones, monascines, and violaceins, have been identified and found in various ecosystems (Gao and Garcia-Pichel, 2011). These natural pigments have been shown in numerous studies to possess an array of beneficial properties, including anticancer and pro-vitamin A properties, as well as desirable qualities including photostability, pH stability, and thermal stability (Malik *et al.*, 2012). Bacteria, yeasts, fungi, and microalgae are among the many microorganisms that create pigments (Celedón and Díaz, 2021).

11.2 Microbial Pigments and Historical Perspectives

Microbial pigments have gained attention as sustainable alternatives to synthetic dyes due to their biodegradable nature and potential health benefits. Unlike synthetic dyes, which are widely used for their affordability and consistent color quality, microbial pigments offer additional advantages such as antioxidant, antimicrobial, and antiinflammatory properties (Paillié-Jiménez *et al.*, 2020). Growing concerns over the environmental hazards and health risks associated with synthetic dyes, such as their toxicity, nonbiodegradability, and carcinogenic potential, have encouraged the search for natural substitutes (Wilson, 2016).

Historically, pigments have been integral to various industries, including textiles, cosmetics, pharmaceuticals, and food (Paillié-Jiménez *et al.*, 2020). While plant- and animal-based dyes were traditionally used, microorganisms have emerged as an efficient and renewable source of natural pigments. Bacteria, fungi, and microalgae have demonstrated remarkable pigment-producing capabilities, with species like *Monascus*, *Serratia*, *Streptomyces*, *Penicillium*, and *Talaromyces* being widely studied for their vibrant colorants (Wynn-Williams *et al.*, 2002). Advances in biotechnology, particularly in omics sciences such as proteomics and metabolomics, have provided deeper insights into microbial pigment production, allowing for improved yields and industrial applications (Ramesh *et al.*, 2019).

With the rising demand for natural and eco-friendly colorants, the pigment market is expanding rapidly, experiencing an annual growth of approximately 7%. Researchers have focused on enhancing microbial pigment production through fermentation technology while utilizing organic waste and agricultural byproducts as cost-effective and sustainable raw materials (Sumathy, n.d.). This approach not only reduces production costs but also aligns with global efforts to promote a circular bioeconomy (Dufossé, 2009).

Among microbial pigments, those derived from microalgae have proven to be highly valuable across multiple industries (Chatragadda and Dufossé, 2021). Beyond serving as natural colorants, they also provide protection against UV radiation, support immune function, and exhibit antiaging and cardiovascular benefits. Notable microalgae species such as *Spirulina*, *Chlorella*, *Haematococcus*, and *Phaeodactylum* are known for producing carotenoids, chlorophylls, and phycobiliproteins, which have applications in food, cosmetics, pharmaceuticals, and even biomedical research (Akilandeswari and Pradeep, 2017). Their strong antioxidant properties and fluorescence capabilities have also made them useful in molecular biology and medical imaging (Lazcano, 2010).

Despite their potential, large-scale microbial pigment production faces challenges, including the need for genetic and metabolic optimization to enhance yield and cost-effectiveness. Additionally, factors such as stability under varying environmental conditions and regulatory approvals must be addressed to ensure widespread industrial adoption (Hari *et al.*, 1994). Researchers are also exploring innovative extraction and purification techniques, such as enzyme-assisted and supercritical fluid extraction, to improve pigment quality and sustainability (Britton and Britton, 1983).

Over time, microorganisms have proven to be promising sources of natural pigments with diverse industrial applications. As biotechnological advancements continue, microbial pigments are expected to play a crucial role in replacing synthetic dyes, offering a safer and more environmentally responsible solution for multiple industries.

11.3 Natural Pigments and Artificial Colors

Color is a vital aspect of the pharmaceutical and cosmetic industries, serving both functional and aesthetic roles. In pharmaceutical dosage forms, color is crucial for identification during manufacturing and distribution (Tuli *et al.*, 2015). Many patients depend on color to recognize their prescribed medication and ensure the correct dosage. Although synthetic colors are commonly used due to their stability and cost-effectiveness, safety concerns have led to increased interest in natural pigments as safer alternatives (Kanekar and Khale, 2017).

Natural pigments are bioactive compounds responsible for the diverse coloration found in plants, including their leaves, fruits, flowers, and seeds. The primary plant pigments include chlorophyll (green), carotenoids (yellow, red, orange), flavonoids (anthocyanins, anthoxanthins), and betalains (red, yellow, purple). These pigments not only contribute to visual appeal but also provide health benefits, such as antioxidant, antimicrobial, and anticancer properties (Dave *et al.*, 2022). Some of the richest edible sources of these pigments include berries, grapes, purple cabbage, and eggplants. Microorganisms capable of producing pigments also hold significant commercial value, offering a sustainable source of natural colorants with added biological properties (Schoneker, 2013).

Artificial colorants, on the other hand, have long been utilized in various industries, including pharmaceuticals, textiles, cosmetics, and food, to enhance product appearance, maintain uniformity, and improve shelf stability. Synthetic dyes are preferred in pharmaceutical formulations because only small quantities are needed to achieve the desired effect. However, their excessive use has been associated with health concerns, including allergic reactions, carcinogenic effects, and genetic alterations (Abel, 2012). Due to these risks, there is a growing shift toward using natural pigments in pharmaceuticals and cosmetics, encouraged by increasing consumer awareness and stricter regulatory policies (Allam *et al.*, 2011). Beyond their coloring properties, natural pigments offer additional functional benefits such as antioxidant and antiinflammatory effects, making them valuable in medicinal and pharmacological applications.

Regulatory agencies such as the US FDA (21 CFR part), EU Commission, and Indian FDA (D&C Act 1940) have established stringent guidelines for the use of coloring agents in pharmaceuticals and cosmetics. Compliance with these regulations is crucial to prevent health hazards, especially for vulnerable groups like children and the elderly (Venil *et al.*, 2020b). Since coloring agents are essential in pharmaceutical formulations, manufacturers must adhere to permissible concentrations and provide justification for their inclusion in specific products. To ensure safety, preclinical and clinical trials are conducted to establish appropriate usage limits and minimize toxicity risks (Kumar *et al.*, 2021).

The increasing preference for natural pigments over synthetic alternatives underscores the significance of ongoing research into microbial and plant-derived colorants. Advances in biotechnology have improved the extraction and stabilization processes of these natural dyes, making them more commercially viable for use in pharmaceuticals and cosmetics (Sharma *et al.*, 2024). With the rising demand for sustainable, health-conscious products, natural pigments are emerging as promising alternatives to synthetic dyes, offering both safety and efficacy across various industries (Morocho-Jácome *et al.*, 2020).

11.4 Classification of Microbial Pigments

Microbial pigments are naturally occurring colorants derived from bacteria, fungi, and algae, widely recognized for their broad applications in pharmaceuticals, cosmetics, and the food industry (Kumar *et al.*, 2015). Unlike synthetic dyes, which may pose health hazards, these pigments provide a sustainable and biocompatible alternative with additional therapeutic benefits, including antioxidant, antimicrobial, anticancer, and antiinflammatory properties (Kiki, 2023). As secondary metabolites, microbial pigments are produced through specialized biosynthetic pathways and play a key role in microbial survival by offering protection against environmental stressors such as UV radiation and oxidative damage (Venil *et al.*, 2013). Their stability, biodegradability, and potential health benefits have made them an attractive option for commercial applications, particularly in natural food colorants, pharmaceutical formulations, and cosmetic products. With advancements in biotechnology, the large-scale production of microbial pigments has become more feasible, further boosting their industrial significance as mentioned in Table 11.1.

11.5 Applications of Microbial Pigments

11.5.1 Pharmaceutical applications

Microbial pigments, a diverse class of pigmented secondary metabolites, have garnered significant attention in pharmaceutical research due to their extensive therapeutic potential. These bioactive compounds, derived from various microorganisms, exhibit a broad spectrum of pharmacological properties. Their unique structural diversity and biochemical mechanisms render them promising candidates for drug development, particularly in the management of malignancies such as cancer and leukemia, as well as metabolic disorders like diabetes mellitus. Ongoing research continues to explore their molecular interactions, biosynthetic pathways, and potential clinical applications, paving the way for innovative therapeutic strategies. Several microbial pigments with notable bioactivity have been identified and investigated for their translational potential in modern medicine.

11.5.1.1 Anthocyanins

Anthocyanins, a subclass of flavonoid pigments, are hydrophilic bioactive compounds extensively studied for their multifaceted pharmacological properties. These secondary metabolites exhibit potent antioxidant activity, primarily attributed to their phenolic hydroxyl moieties, which facilitate the donation of hydrogen atoms or electrons to neutralize reactive oxygen species (ROS). Beyond their antioxidative potential,

Table 11.1. Different types of Microbial Pigments and their properties

Class of pigments	Color	Properties	Examples
Carotenoids	Yellow, orange, red	Antioxidant, photoprotection, immune system support	**Bacteria**: *Rhodotorula glutinis, Serratia marcescens*, *Micrococcus luteus* **Fungi**: *Blakeslea trispora*, *Neurospora crassa* **Algae**: *Dunaliella salina* (rich in β-carotene)
Anthraquinones	Red, orange, purple	Antimicrobial, anticancer, widely used in textile and pharmaceutical industries	**Bacteria**: *Streptomyces coelicolor* (produces actinorhodin) **Fungi**: *Monascus purpureus* (produces *Monascus* pigments)
Melanins	Black, brown	UV protection, antioxidant, used in cosmetics and medicine	**Bacteria**: *Pseudomonas aeruginosa* (pyomelanin), *Streptomyces* species **Fungi**: *Aspergillus niger*, *Cryptococcus neoformans*
Phenazines	Blue-green	Antimicrobial, anticancer, biofilm inhibition	**Bacteria**: *Pseudomonas aeruginosa* (produces pyocyanin), *Streptomyces* species
Violaceins	Purple, violet	Antibacterial, antiviral, anticancer	**Bacteria**: *Chromobacterium violaceum*, *Janthinobacterium lividum*
Prodigiosins	Red	Antimicrobial, anticancer, immunosuppressive	**Bacteria**: *Serratia marcescens*, *Streptomyces coelicolor*
Flavins	Yellow-green (fluorescent)	Function as cofactors in biological oxidation-reduction reactions	**Bacteria**: *Bacillus subtilis*, *Escherichia coli* (produces riboflavin)

anthocyanins play a pivotal role in modulating immune responses and mitigating the risk of oncogenesis. Their anticancer efficacy is predominantly mediated through two fundamental mechanisms: (i) modulation of redox homeostasis, thereby attenuating oxidative stress-induced carcinogenesis, and (ii) interference with critical cellular processes, including cell cycle regulation, apoptotic pathways, inflammatory cascades, angiogenic signaling, and metastatic progression. These mechanistic insights underscore the therapeutic promise of anthocyanins in the prevention and treatment of malignancies (Wang and Mazza, 2002; Katsube *et al.*, 2003; Martin *et al.*, 2003; Lazzè *et al.*, 2004).

11.5.1.2 Prodigiosin

Prodigiosin, a bioactive tripyrrole pigment, exhibits a diverse spectrum of pharmacological activities, making it a promising candidate for therapeutic applications. This cytotoxic secondary metabolite is synthesized by various eubacteria, including *Vibrio psychroerythrus* (D'Aoust and Gerber, 1974), *Serratia marcescens*, and *Pseudomonas magnesiorubra* (Lewis and Corpe, 1964). Initially identified in *S. marcescens*, a Gram-negative bacterium known for producing a nondiffusible red pigment (Kalesperis *et al.*, 1975; Gargallo-Viola, 1989), prodigiosin has also been biosynthesized using *Streptomyces* and *Serratia* species. Notably, prodigiosin exerts potent immunosuppressive effects and demonstrates significant antiproliferative and cytotoxic activity against a panel of 60 human tumor cell lines, encompassing malignancies of the lung, colon, liver, ovary, and brain, as well as melanoma and leukemia (Kalesperis *et al.*, 1975; Gargallo-Viola, 1989). Furthermore, it has been implicated in the prevention and management of diabetes mellitus, highlighting its potential as a multifaceted therapeutic agent (Pandey *et al.*, n.d.).

11.5.1.3 Violacein

Violacein, a bioactive indole-derived pigment, is predominantly synthesized by *Chromobacterium violaceum* and exhibits a broad spectrum of pharmacological properties, including antitumoral, antiparasitic, antiprotozoal, anticancer, antiviral, antibacterial, and antioxidant activities (Lichstein and Van De Sand, 1946; Konzen *et al.*, 2006; Nakamura *et al.*, 2024) . The therapeutic potential of violacein extends to antimycobacterial applications, as highlighted by Mojib *et al.* (2010), who reported its efficacy alongside flexirubin, a yellow-orange pigment derived from *Flavobacterium* sp. Ant 342, in targeting *Mycobacterium* species. These findings suggest that violacein and related bacterial pigments may be valuable candidates for the development of novel chemotherapeutic agents against tuberculosis. Given its diverse bioactivity, violacein represents a promising compound with significant translational potential in biomedical research and therapeutic interventions (Richard, n.d.). The exploration of bacterial pigments continues to unveil novel bioactive molecules with extensive clinical applicability.

11.5.1.4 Red yeast rice

Red yeast rice (RYR) is a traditionally fermented rice product derived from the metabolic activity of *Monascus* species, including *M. ruber*, *M. purpureus*, *M. ruber*, and *M. pilosus*. These fungi are known for their ability to biosynthesize secondary metabolites with polyketide structures, alongside the production of bioactive yellow, orange, and red pigments. Notably, *M. ruber* has historically been utilized in the preparation of angkak, a fermented rice product exhibiting potent hypocholesterolemic effects (Panda *et al.*, 2010) . RYR has been identified as a rich source of pharmacologically active constituents, including statin-like compounds, unsaturated fatty acids, sterols, and B-complex vitamins, which contribute to its lipid-lowering and cardioprotective properties. Furthermore, emerging evidence suggests that both RYR and statins play a crucial role in glycemic regulation, demonstrating the potential to reduce blood glucose levels in diabetic individuals (Chang *et al.*, 2006). These findings underscore the therapeutic relevance of RYR in metabolic disorder management and cardiovascular health.

11.5.2 Cosmeceutical applications

Historically, long before the advent of modern cosmetics, individuals utilized naturally derived substances to enhance their aesthetic appeal and promote dermatological health. Mineral-based compounds, botanical extracts, and lipid-rich oils were extensively employed for cosmetic formulations across diverse cultures (Pagels *et al.*, 2022). Contemporary cosmetics are meticulously designed to modify and enhance the skin's morphology, texture, and overall appearance through bioactive constituents tailored to various skin types (Mourelle *et al.*, n.d.).

In recent years, there has been a paradigm shift within the cosmetics industry toward environmentally sustainable practices, encompassing both raw material sourcing and eco-conscious packaging strategies. This transition has given rise to cosmeceuticals, an emerging sector at the intersection of dermatology and pharmacology, wherein bioactive compounds are integrated into cosmetic formulations to confer therapeutic benefits beyond aesthetic enhancement. While the term "natural" does not inherently equate to "safe" or "efficacious," natural cosmetic ingredients often exhibit bioactive properties that may contribute to dermatological wellbeing (Pagels *et al.*, 2022).

Microbial pigments have emerged as promising cosmeceutical agents due to their potent antioxidant and antiaging properties, positioning them as viable candidates for inclusion in next-generation cosmetic formulations. Additionally, their vibrant pigmentation offers an aesthetically appealing alternative to synthetic colorants. Although several cosmetic products incorporating microbial pigments have been commercialized, regulatory and safety considerations have limited their widespread adoption. The antioxidative, melanogenesis-inhibitory, and photoprotective properties of these pigments underscore their potential for future cosmetic applications that not only enhance appearance but also impart substantial dermatological benefits.

11.5.2.1 Photoprotection and antioxidant capacity

Endogenous antioxidant defense mechanisms within the skin play a crucial role in neutralizing ROS, thereby mitigating oxidative stress and preserving cellular homeostasis. However, excessive ROS generation, particularly due to ultraviolet (UV) radiation exposure, can overwhelm these protective systems, leading to oxidative damage at the molecular level. This oxidative burden contributes to lipid peroxidation, protein degradation, and DNA instability, ultimately compromising cellular integrity (Wang *et al.*, 2015). Moreover, ROS accumulation is implicated in apoptotic and necrotic cell death pathways, manifesting clinically as premature skin aging, characterized by the formation of wrinkles, xerosis, and loss of dermal elasticity. Persistent oxidative stress further exacerbates photoaging, serving as a key etiological factor in the pathogenesis of cutaneous malignancies, including melanoma and other UV-induced inflammatory dermatoses (Wang *et al.*, 2015).

Conventional chemical UV filters, despite their efficacy in absorbing ultraviolet radiation, may not confer comprehensive photoprotection as they primarily function by absorbing rather than reflecting incident UV rays. Additionally, concerns regarding their photostability, potential to induce cutaneous irritation, and risk of allergic hypersensitivity reactions have been well documented. Given these limitations, the integration of antioxidant molecules into photoprotective formulations is imperative. Incorporating bioactive antioxidants into topical sunscreen preparations can enhance their efficacy in attenuating UV-induced oxidative damage, thereby providing a multifaceted approach to skin photoprotection and minimizing the deleterious consequences of prolonged UV exposure (Mendes-Silva *et al.*, 2020).

The cosmetics industry is increasingly focusing on the development of innovative, sustainable, and UV-resistant bioactive compounds to meet the growing demand for eco-friendly formulations (Ding *et al.*, 2022). Microbial-derived pigments, particularly those synthesized by microalgae and cyanobacteria, have emerged as promising candidates due to their dual functionality as natural colorants and potent antioxidants. These biochromes, predominantly photosynthetic pigments such as carotenoids, exhibit remarkable antioxidative properties, making them highly valuable for cosmetic applications (Mourelle *et al.*, n.d.).

Carotenoids have garnered significant interest within the cosmetic sector owing to their robust free radical-scavenging capacity, which effectively mitigates oxidative stress and reduces UV-induced dermal photodamage (Terao *et al.*, 2010). Additionally, their intrinsic photoprotective attributes position them as viable alternatives or complementary agents to conventional chemical UV filters. The incorporation of carotenoids into photoprotective cosmetic formulations presents a strategic approach to enhancing skin defense against UV radiation while minimizing reliance on synthetic sunscreen agents, thereby promoting the development of safer and more biocompatible skincare products (Mendes-Silva *et al.*, 2020).

Lycopene, astaxanthin, and β-carotene are among the most extensively studied carotenoids, widely recognized for their potent antioxidative and photoprotective properties (Stahl and Sies, 2012). β-Carotene, a key precursor of provitamin A, serves as an effective free radical scavenger, with its cutaneous accumulation following dietary intake playing a pivotal role in mitigating oxidative stress. Upon metabolic conversion to vitamin A, β-carotene facilitates melanin biosynthesis, thereby contributing to photoprotection and skin homeostasis (Ding *et al.*, 2022; Mourelle *et al.*, n.d.).

Lycopene, another structurally significant carotenoid, exhibits superior antioxidative efficacy, rendering it a valuable bioactive ingredient in sunscreen formulations. Its capacity to neutralize oxygen-derived free radicals underscores its role in shielding the skin from UV-induced oxidative damage (Ding *et al.*, 2022; Mourelle *et al.*, n.d.). Additionally, lycopene has been shown to exert profound effects on skin texture, particularly in reducing dermal roughness. This suggests that augmenting the skin's endogenous antioxidant reservoir through lycopene-enriched formulations may substantially enhance cutaneous smoothness and resilience (Tosato *et al.*, 2016).

Astaxanthin is widely regarded as one of the most potent naturally occurring antioxidants, exhibiting remarkable efficacy in counteracting oxidative stress and inhibiting lipid peroxidation (Tosato *et al.*, 2016; Mendes-Silva *et al.*, 2020). This robust antioxidative capability has been closely associated with its ability to attenuate the formation of wrinkles and hyperpigmented lesions, thereby contributing to skin rejuvenation and photoprotection (Tosato *et al.*, 2016; Mendes-Silva *et al.*, 2020).

Beyond its dermatological benefits, astaxanthin plays a critical role in human metabolism, particularly in safeguarding the skin against UV-induced photooxidative damage. Its integration into natural sunscreen formulations has been proposed as a viable strategy for enhancing cutaneous defense against solar radiation. Notably, astaxanthin demonstrates a tenfold higher antioxidant capacity compared to other carotenoids such as zeaxanthin, lutein, and β-carotene, a property attributed to the presence of carbonyl functional groups in its ionone ring structure (Pan *et al.*, 2012). Owing to its exceptional free radical-scavenging potential, astaxanthin has been incorporated as a key bioactive ingredient in various cosmetic formulations worldwide, reinforcing its relevance in advanced cosmeceutical applications.

Emerging research on cosmetic formulations incorporating carotenoid-based bioactives has demonstrated promising efficacy, particularly in the context of oral supplementation, thereby reinforcing the potential of topical carotenoid application for mitigating UV-induced oxidative stress (Tosato *et al.*, 2016). Microbial-derived carotenoids have garnered significant attention in dermatological and photoprotective formulations due to their superior antioxidative potency, lipophilic nature, and enhanced stability. These attributes have facilitated their increasing utilization in sunscreen formulations and anti-UV therapies, positioning them as viable alternatives to conventional photoprotective agents (Mendes-Silva *et al.*, 2020).

A novel melanin-enriched formulation, designated as F3Cream, was developed by integrating bacterially derived melanin from *Halomonas venusta* with a bioactive seaweed extract obtained from *Gelidium spinosum*. This formulation exhibited significant photoprotective properties, characterized by high sun protection factor (SPF) values, along with notable wound-healing efficacy (Poulose *et al.*, n.d.; Nawaz *et al.*, 2020). Additionally, F3Cream demonstrated potent antimicrobial activity against common dermatological pathogens, including *Staphylococcus aureus* and *Streptococcus pyogenes*, highlighting its potential as a multifunctional cosmeceutical agent (Poulose *et al.*, n.d.; Nawaz *et al.*, 2020). Furthermore, numerous other investigations have explored the photoprotective efficacy of microbial pigments, as summarized in Table 11.2.

11.5.2.2 Antiaging compounds and skincare products

The rapid expansion of urbanization and globalization has intensified environmental stressors, including pollution and psychological stress, which collectively contribute to the premature aging of the skin (Brandt *et al.*, 2011). Cutaneous aging manifests through a spectrum of morphological and structural alterations, such as epidermal thinning, wrinkle formation, fine lines, enlarged pores, increased fragility, decreased elasticity, and xerosis. Skin aging can be categorized into intrinsic and extrinsic processes. Intrinsic aging, governed by genetic and metabolic factors, results in the progressive decline of dermal elasticity, heightened transparency, and enhanced vascular prominence. These physiological changes are reflected in the tonal variations observed across different age groups, from neonates to young adults and the elderly (Wang *et al.*, 2015). Conversely, extrinsic aging is primarily driven by chronic exposure to UV radiation and environmental pollutants, leading to keratinocyte dysplasia, degradation of the dermal extracellular matrix, hyperpigmentation, laxity, wrinkle formation, and textural coarseness (Pagels *et al.*, 2022).

Ultraviolet radiation is classified into three subtypes: UVA, UVB, and UVC. UVA penetrates deep into the dermis, inducing oxidative stress and structural degradation that accelerate wrinkle formation. UVB radiation, in contrast, is more cytotoxic, leading to erythema, epidermal burns, and an elevated risk of photocarcinogenesis. UVC radiation is largely attenuated by the stratospheric ozone layer, preventing its direct impact on human skin (Hussein, 2005; Wang *et al.*, 2015).

The growing demand for antioxidant-rich and antiaging formulations stems from the increasing consumer desire for enhanced aesthetic appeal and dermatological health. With heightened awareness regarding the potential toxicity and safety concerns of synthetic compounds, there has been a significant shift toward the utilization of naturally derived bioactive ingredients, which offer superior biocompatibility

Table 11.2. Some studies focused on the photoprotection activity of different microbial pigments.

Pigment	Source	Producer microbes
Carotenoids: yellow pigment	Bacteria	*Flavobacterium* sp.
Carotenoids: orange pigment	Bacteria	*Brevibacterium* sp.
Melanin	Bacteria	*Vibrio natriegens*
Carotenoids: lycopene	Cyanobacteria	*Anabena* sp.
Carotenoids: scytonemin	Cyanobacteria	*Nostoc* sp.
Biopterin	Cyanobacteria	*Oscillatoria* sp.
Phycocyanobilin	Cyanobacteria	*Spirulina* sp.
Carotenoids: astaxanthin	Chlorophyta	*Haematococcus lacustris*
Phycoerythrobilins	Rhodophyta	*Porphyridium* sp.
Phycobiliproteins	Rhodophyta	*Porphyridium aerogineum*
Pink-red	Bacteria	*Agrobacterium aurantiacum*
Pink-red	Bacteria	*Paracoccus carotinifaciens*
Dark red	Bacteria	*Bradyrhizobium* sp.
Yellow	Bacteria	*Flavobacterium* sp. *Paracoccus zeaxanthinifaciens*
Blue	Bacteria	*Corynebacterium insidiosum*
Red	Bacteria	*Rugamonas rubra*, *Streptoverticillium*
Pink-red	Bacteria	*Xanthophyllomyces dendrorhous*
Dark red	Bacteria	*Haloferax alexandrinus*
Golden yellow	Bacteria	*Staphylococcus aureus*
Purple	Bacteria	*Chromobacterium violaceum*
Red	Bacteria	*Serratia marcescens*, *Serratia rubidaea*
Blue-green	Bacteria	*Pseudomonas aeruginosa*
Yellow	Bacteria	*Xanthomonas oryzae*
Purple	Bacteria	*Janthinobacterium lividum*
Red	Algae	*Dunaliella salina*
Cream	Fungi	*Blakeslea trispora*
Red	Fungi	*Fusarium sporotrichioides*
Red	Fungi	*Haematococcus pluvialis*
Red-orange	Fungi	*Monascus* sp.
Red-yellow	Fungi	*Monascus purpureus*
Orange-pink	Fungi	*Monascus roseus*
Yellow	Fungi	*Monascus* sp.
Red	Fungi	*Penicillium oxalicum*
Red	Fungi	*Blakeslea trispora*
Deep blood red	Fungi	*Cordyceps unilateralis*
Yellow	Fungi	*Ashbya gossypi*

Continued

Table 11.2. Continued

Pigment	Source	Producer microbes
Yellow-orange	Fungi	*Mucor circinelloides*, *Neurospora crassa*
Red	Fungi	*Pacilomyces farinosus*
Pink-red	Yeast	*Phaffia rhodozyma*
Orange-red	Yeast	*Rhodotorula* sp., *Rhodotorula glutinis*
Red	Actinomycetes	*Streptoverticillium rubrireticuli*
Red	Actinomycetes	*Streptomyces echinoruber*

and reduced adverse effects (Brandt *et al.*, 2011). This paradigm shift has accelerated the pursuit of sustainable and efficacious cosmetic ingredients within the industry, fostering extensive research into novel bioactive compounds.

Among the most extensively studied natural pigments, carotenoids have emerged as pivotal microbial-derived constituents in cosmetic formulations due to their inherent safety profile, vibrant pigmentation, and multifunctional dermatological benefits (Novoveská *et al.*, 2019). The biological roles of carotenoids in cutaneous physiology are attributed to their function as precursors of vitamin A and their potent photoprotective properties against UV radiation. Their mechanism of action primarily involves the direct absorption of UV rays, mitigating photodamage and oxidative stress-induced aging. Clinical investigations have further substantiated the efficacy of carotenoids as nutricosmetic agents, demonstrating their potential in delaying premature skin aging and enhancing overall skin resilience (Yeager and Lim, 2019).

β-Cryptoxanthin, a bioactive carotenoid predominantly found in *Dunaliella salina* and various other microalgal species, has been reported to enhance the biosynthesis of hyaluronic acid, a critical glycosaminoglycan responsible for maintaining skin hydration and structural integrity (Stahl and Sies, 2012; Sayo *et al.*, 2013). Additionally, the cyanobacterium *Anabaena vaginicola* has demonstrated a remarkably higher lycopene yield compared to conventional natural sources, making it a promising candidate for commercial exploitation (Hashtroudi *et al.*, 2013). Lycopene, widely recognized for its antiaging properties, has already been integrated into personal care formulations, underscoring the potential of cyanobacteria and microalgae as sustainable sources of this potent antioxidant for cosmeceutical applications (Mourelle *et al.*, n.d.).

Astaxanthin derived from *Haematococcus pluvialis* has been shown to improve cutaneous parameters, including epidermal texture, wrinkle reduction, corneocyte cohesion, and overall moisture retention (Tominaga *et al.*, 2012). Numerous commercial formulations incorporate this bioactive pigment due to its extensive dermatological benefits. For instance, the Japanese cosmetics company Kose has developed AstaBlanc, a skincare product specifically designed to mitigate wrinkles and hyperpigmentation, leveraging the antioxidant potency of astaxanthin. Similarly, the Swedish corporation AstaReal AB has introduced AstaReal®, a natural astaxanthin formulation reputed for its ability to rejuvenate photodamaged skin, attenuate wrinkle formation, and enhance dermal elasticity (Arad and Yaron, 1992).

A prior investigation assessed the dermatological benefits of lutein and zeaxanthin supplementation in three distinct cohorts. The first cohort received both carotenoids via oral and topical administration, the second cohort was subjected exclusively to topical application, while the third group received only oral supplementation. The findings indicated that while all intervention strategies conferred notable enhancements in skin hydration and elasticity, the synergistic effect of the combined oral and topical regimen exhibited the most pronounced improvement. Additionally, the highest degree of antioxidant defense against oxidative stress-induced dermal damage was achieved through this combinatory approach (Palombo *et al.*, 2007).

In a separate study, the bioactive potential of phycobiliproteins and carotenoids derived from the cyanobacterium *Cyanobium* sp. was systematically evaluated. The results demonstrated that phycobiliprotein extracts exhibited potent antihyaluronidase and anticollagenase activities, while the carotenoid extract primarily exerted antihyaluronidase effects. Given their capacity to inhibit enzymatic degradation of hyaluronic acid and collagen, both extracts present significant promise as natural antiaging agents. Moreover, their intrinsic bioactivity, coupled with stability and ease of formulation, positions these cyanobacterial metabolites as viable, sustainable candidates for incorporation into cosmeceutical applications (Pagels *et al.*, 2022).

11.5.2.3 Whitening effect

The skin serves as a critical barrier, with its primary layers, the epidermis and dermis, offering protection against environmental insults and pathogenic invasion. The epidermis, the outermost layer, is predominantly composed of keratinocytes interspersed with melanocytes, which play a pivotal role in photoprotection. Melanocytes synthesize melanin and transfer it to keratinocytes, facilitating the formation of supranuclear melanin caps that shield cellular DNA from UV-induced genotoxicity (Costin and Hearing, 2007; Ding *et al.*, 2022). The pigmentation of the skin is primarily governed by the quantity, distribution, and biochemical nature of melanin (Dessinioti *et al.*, 2009).

Tyrosinase, a rate-limiting copper-containing enzyme, is integral to melanogenesis. It catalyzes the hydroxylation of L-tyrosine to L-dopa (L-3,4-dihydroxyphenylalanine) and subsequently oxidizes L-dopa to dopaquinone. Dopaquinone undergoes a series of complex oxidative polymerizations, culminating in the biosynthesis of eumelanin or pheomelanin, contingent upon the availability of cysteine and other modulatory factors (Agrawal *et al.*, 2018; Alves *et al.*, 2020; Ding *et al.*, 2022).

Skin depigmentation is a widely sought after cosmetic intervention globally, with a pronounced demand in Asian countries (Li *et al.*, n.d.). The preference for lighter skin tones in various Asian cultures has driven the substantial expansion of the skin-whitening sector within the cosmeceutical industry, with projections indicating continued market growth. Cutaneous exposure to UV radiation upregulates tyrosinase activity and enhances melanosome biogenesis. These melanosomes facilitate melanogenesis, transferring melanin to keratinocytes, thereby contributing to tanning and skin pigmentation. Over time, epidermal turnover and desquamation result in the gradual degradation and elimination of melanin, leading to the attenuation of hyperpigmentation (Wang *et al.*, 2011, 2015).

Among the most effective strategies for skin depigmentation is the inhibition of tyrosinase, a critical enzyme in melanogenesis that catalyzes the conversion of tyrosine to melanin. In the dermatological and pharmaceutical industries, tyrosinase inhibitors are of considerable therapeutic interest due to their efficacy in mitigating pigmentary disorders. These inhibitors are incorporated into cosmeceutical formulations to address hyperpigmentation concerns, including ephelides (freckles) and melasma, by downregulating melanin biosynthesis (Claus and Decker, 2006). Moreover, targeting tyrosinase and its regulatory pathways has been explored as a potential therapeutic approach for managing hypopigmentary conditions such as albinism and piebaldism, which stem from defective melanogenesis (Likhitwitayawuid, 2008; Manivasagan *et al.*, 2013).

Despite the extensive identification of *in vitro* tyrosinase inhibitors, only a limited number have demonstrated significant therapeutic efficacy in clinical applications. Contemporary efforts in the search for natural tyrosinase inhibitors have predominantly focused on marine-derived bioactives, particularly those derived from macroalgae. Notably, fucoxanthin, a xanthophyll carotenoid extracted from *Laminaria japonica*, has exhibited potent inhibitory effects on tyrosinase activity in UVB-irradiated guinea pigs and has been shown to suppress melanogenesis in UVB-exposed murine models (Thomas and Kim, 2013; Wang *et al.*, 2015). Furthermore, oral administration of fucoxanthin resulted in a downregulation of melanogenesis-associated mRNA expression in the skin, suggesting its capacity to modulate melanogenesis at the transcriptional level (Thomas and Kim, 2013).

Astaxanthin, the principal carotenoid produced by *Haematococcus pluvialis*, possesses enzymatic antioxidant properties akin to superoxide dismutase and catalase, thereby mitigating oxidative damage to human lymphocyte proteins and essential lipids. Its antioxidant potential surpasses that of conventional antioxidants such as vitamins C, E, and other carotenoids (Pan *et al.*, 2012). Empirical studies indicate that both topical and oral administration of astaxanthin contributes to the attenuation of skin hyperpigmentation, inhibition of melanin biosynthesis, and overall improvement in skin health (Tominaga *et al.*, 2012; Wang *et al.*, 2015). Additionally, astaxanthin derived from marine yeast has been demonstrated to mitigate age-related pigmentation disorders, including age spots (Costa and Santos, 2017).

Several marine-derived compounds with tyrosinase inhibitory activity have been successfully integrated into commercial cosmeceutical formulations; however, certain agents, such as hydroquinone, have been prohibited in all European nations due to their potential health hazards (Burger *et al.*, n.d.).

11.5.2.4 Pigment substances

The application of synthetic pigments in cosmetic formulations, including nail polishes and lipsticks, primarily involves derivatives of benzene, toluene, and other petrochemical intermediates. However, concerns regarding their potential toxicological impact and environmental persistence have led to increasing consumer apprehension. Consequently, natural pigments have gained prominence due to their biocompatibility, safety, and stability (Kalra *et al.*, 2020). Extensive research has been conducted to explore microorganisms as a sustainable and economically viable source of biopigments with novel chromophoric properties (Ding *et al.*, 2022). Notably, certain marine invertebrates, such as corals and sponges, exhibit vivid coloration, which is hypothesized to arise from the presence of symbiotic microorganisms harboring photosynthetic pigments (Mourelle *et al.*, n.d.). The utilization of marine microbial pigments in the cosmetic industry has expanded significantly due to their structural diversity and functional attributes (Ding *et al.*, 2022).

Among these, phycobiliproteins such as phycocyanins, synthesized by thermophilic cyanobacteria, have demonstrated potential as natural colorants in eye cosmetics, including eyeshadows (Bermejo *et al.*, 2003). A noteworthy industrial example is the incorporation of phycocyanin extracted from *Spirulina* spp. as a bio-derived pigment in eye make-up by the Japanese company Ink (Yarkent *et al.*, 2020). Additionally, carotenoids and phycobiliproteins obtained from red microalgae serve as valuable chromophores for formulating pink and purple hues in cosmetic products (Ding *et al.*, 2022). For instance, biopigment extracts from diverse red microalgal species have been successfully employed in the fabrication of lipsticks and eyeshadows, either as cream-based or powder formulations (Arad and Yaron, 1992).

Biotechnological advancements have facilitated the utilization of phycoerythrin as a naturally derived chromophore in the formulation of tinted creams and cosmetic products. Notably, phycoerythrobilins, derived from *Spirulina* (Cyanobacteria) and *Porphyridium* (Rhodophyta), have been explored for their potential application in lipsticks and eyeliners due to their vibrant pigmentation and bioactive properties (Bedoux *et al.*, 2014; Hamed, n.d.). Furthermore, phycobiliproteins extracted from *Porphyridium aerogineum* have been effectively employed as cosmetic colorants, exhibiting remarkable pH resilience within the range of 4–5 and exceptional photostability under exposure to light (Mourelle *et al.*, n.d.; Chandra *et al.*, n.d.).

Among these biopigments, phycocyanin has garnered significant attention due to its multifunctional properties, including antioxidant activity, immunomodulatory effects, and antiinflammatory potential. Its inherent stability further enhances its applicability in cosmetic formulations, where it serves as a key pigment in the production of eyeliners, eyeshadow powders, and lipsticks (Sekar and Chandramohan, 2008).

A patent (US6740313 B2) has documented the application of ankaflavin, a principal pigment biosynthesized by *Monascus* spp., as an active ingredient in dermatological formulations designed to impart a persistent skin pigmentation akin to natural tanning. The patented formulation demonstrated immediate and uniform coloration in an *in vitro* skin model, highlighting its efficacy as a bio-based tanning agent (Aruldass *et al.*, 2018).

In a separate study, a bio-lip balm formulated with crude pigment extracted from *Streptomyces bellus*, in conjunction with shredded beeswax, lanolin, and coconut oil, provided evidence supporting the potential of melanin pigments as primary constituents in a range of cosmetic and personal care applications (Srinivasan *et al.*, n.d.). Additionally, the pronounced green pigmentation of microalgae facilitates the extraction of chlorophyll, a pigment extensively employed in the cosmetic industry. Chlorophyll derived from chlorophytes is not only utilized as a natural colorant but also exhibits antiinflammatory properties, making it a valuable component in dermatological formulations (Yarkent *et al.*, 2020).

Recent comprehensive reviews have further emphasized the multifunctional attributes of chlorophyll, including its odor-neutralizing capability, which renders it an essential additive in deodorants, toothpastes, and various personal hygiene products (Mourelle *et al.*, n.d.).

Moreover, extensive ecological surveys have been conducted to identify novel microbial sources of biopigments with applications in the cosmetic industry. Several pigment-producing strains have been characterized, underscoring the biodiversity of marine-derived fungi and their biosynthetic potential. Notably, variecolorquinones, yellow-hued secondary metabolites, are synthesized by *Aspergillus glaucus* and isolated from mangrove root ecosystems. The same fungal species also produces aspergilolide, a red pigment obtained from the identical habitat. Similarly, bostrycin, a red pigment, is biosynthesized by *Aspergillus* spp. inhabiting coral reef environments. *Eurotium cristatum*, a marine sponge-associated fungus, synthesizes physcion, a yellow anthraquinone derivative. Another red pigment, NIOM-02, is produced by *Penicillium* spp. and sourced from marine sedimentary deposits, while 2,3-dihydrocitromycin, a yellow secondary metabolite, is generated by *Penicillium bilaii* and recovered from the Huon estuary. This diverse spectrum of fungal-derived bio-pigments highlights the ecological and biopharmaceutical significance of marine mycobiota in the development of novel cosmetic formulations (Mourelle *et al.*, n.d.).

11.6 Challenges in Pigment Compound Development

Microbial pigments have drawn increasing interest in pharmaceuticals and cosmetics due to their natural origin, potential health benefits, and environmentally friendly attributes. Despite these advantages, several obstacles impede their large-scale commercialization (Ahmad *et al.*, 2012). Key challenges include regulatory requirements, production limitations, extraction difficulties, stability issues, and consumer acceptance. This chapter explores these challenges in depth and presents possible strategies to promote the integration of microbial pigments into pharmaceutical and cosmetic applications (Nigam and Pandey, 2009).

Ensuring the safety of microbial pigments is a fundamental concern in their use for pharmaceuticals and cosmetics. Regulatory agencies such as the US Food and Drug Administration (FDA) and the European Medicines Agency (EMA) enforce strict guidelines to minimize risks related to toxicity and allergic reactions. Before receiving approval, microbial pigments must undergo rigorous safety assessments, including toxicological and dermatological evaluations (Dufossé, 2006). The potential presence of endotoxins and microbial contaminants further complicates regulatory compliance. Variations in pigment production due to differences in microbial strains and fermentation conditions necessitate stringent quality control measures. Despite ongoing research, standardized protocols for pigment characterization and purity assessment are still being developed.

The economic viability of microbial pigment production remains a significant concern, as yields are generally lower than those of synthetic dyes. Many microbial strains produce pigments in limited quantities, requiring optimization to enhance productivity (Joshi *et al.*, n.d.). Additionally, high substrate costs and slow microbial growth rates contribute to increased production expenses. Large-scale fermentation demands precise control over parameters such as temperature, pH, and oxygen levels, all of which raise operational costs (Kumar *et al.*, 2022). To achieve commercial feasibility, improvements in bioreactor designs and fermentation techniques are essential. Compared to synthetic dyes, microbial pigments often necessitate complex extraction and purification methods, further elevating production costs. Many microbial pigments are intracellular, necessitating cell disruption for extraction (Narsing Rao *et al.*, 2017). Traditional solvent-based extraction methods raise environmental and toxicity concerns. Furthermore, purification techniques such as chromatography and filtration require substantial investment and time. For pharmaceutical and cosmetic applications, achieving high purity levels is imperative, adding another layer of complexity to the processing of microbial pigments (Martins *et al.*, 2011).

Stability remains a major concern for microbial pigments, as they tend to degrade when exposed to environmental factors such as light, temperature, and pH fluctuations. Exposure to ultraviolet light, oxidation, and temperature changes can cause color degradation, impacting product quality and shelf-life (Soliev *et al.*, 2011). Additionally, variations in pH can influence pigment stability, complicating their formulation in pharmaceutical and cosmetic products. Research into stabilization techniques, such as encapsulation using biopolymers or liposomes, offers promising solutions. The incorporation of antioxidants and stabilizers can also enhance stability, though they must comply with safety regulations (Silva *et al.*, 2021).

Scaling up microbial pigment production presents additional challenges due to inconsistencies in yield, fermentation conditions, and reproducibility. Variations in microbial strain performance can lead to fluctuations in pigment output. To address these issues, optimization of bioprocesses and strain engineering is required to ensure consistent production. Moreover, integrating microbial pigments into existing pharmaceutical and cosmetic manufacturing processes often necessitates adjustments in formulation strategies (Mansour, 2018). Automation and continuous fermentation approaches are being explored to improve industrial scalability. Despite their natural and sustainable appeal, microbial pigments face stiff competition from synthetic dyes, which are widely used in pharmaceuticals and cosmetics due to their cost-effectiveness and stability. Consumer skepticism regarding the efficacy and safety of microbial-derived pigments poses another barrier to widespread adoption (Arad and Yaron, 1992). Increasing consumer awareness through educational initiatives and scientific validation of the benefits of microbial pigments can foster greater acceptance. Regulatory backing and transparent labeling practices can further enhance consumer confidence in these natural colorants.

Genetic and metabolic engineering have significantly contributed to improving microbial pigment production but ethical and regulatory concerns continue to limit progress. The use of genetically modified organisms (GMOs) in pigment production raises ethical and legal challenges, with stringent policies governing their commercialization (Bhatt, 2022). Advances in CRISPR technology and metabolic engineering have shown potential in boosting pigment production efficiency, but regulatory restrictions pose limitations. Synthetic biology techniques offer innovative approaches to refining microbial metabolic pathways for enhanced yield and stability, yet their industrial implementation remains a complex issue (Lee *et al.*, 2025).

Several strategies are being explored to overcome these challenges. Optimization of bioprocesses through improved fermentation techniques and cost-effective substrates can enhance pigment yield. Sustainable extraction methods, such as supercritical fluid extraction and enzymatic hydrolysis, offer eco-friendly alternatives to conventional processes. Encapsulation and stabilization techniques leveraging nanotechnology and bio-based carriers aim to extend the shelf-life of microbial pigments (Ishimoto *et al.*, 2024). Metabolic engineering and synthetic biology approaches continue to refine microbial strains for increased pigment production. Additionally, establishing clear regulatory guidelines and industry standards will facilitate the safe and effective integration of microbial pigments into pharmaceuticals and cosmetics (Vishnupriya *et al.*, 2021).

The use of microbial pigments in pharmaceutical and cosmetic products holds considerable promise for sustainability, safety, and bioactivity. However, overcoming challenges related to regulatory approval, production scalability, extraction efficiency, stability, and market acceptance is crucial for their successful commercialization. With advancements in bioprocessing, genetic engineering, and industrial-scale production, microbial pigments have the potential to serve as viable alternatives to synthetic dyes, transforming the pharmaceutical and cosmetic industries in the future.

11.7 Pathogenicity of Pigmented Microbes

Microbial pigmentation is a fascinating phenomenon that extends beyond mere identification markers in clinical microbiology. Many pathogenic microbes synthesize biological pigments that significantly contribute to their virulence (Nosanchuk and Casadevall, 2003). These pigments aid microbial survival by providing protection against environmental stressors, enabling evasion of host immune responses, and enhancing resistance to antimicrobial agents. Understanding the functional roles of these pigments is crucial in the development of innovative therapeutic strategies aimed at combating infections caused by pigmented microbes (Liu and Nizet, 2009).

Among the various microbial pigments, melanin plays a particularly vital role in microbial virulence. It is a key determinant of pathogenicity in several microorganisms, enhancing their ability to persist within host tissues. Studies on *Cryptococcus neoformans* reveal that melanin-deficient strains exhibit reduced virulence, suggesting that while melanin is not essential for microbial viability, it significantly aids in infection (Nosanchuk and Casadevall, 2006). The primary function of melanin in pathogenic microbes is to offer protection against oxidative stress and host immune responses, allowing these organisms to persist in adverse environments.

Melanin-coated microbial cells exhibit increased resistance to enzymatic degradation by hydrolytic enzymes produced by host immune cells. For instance, *C. neoformans* with a melanin-rich cell wall is more resistant to immune attack compared to its nonmelanized counterparts (Zhang *et al.*, 2018). This protective function extends to various fungal pathogens, including *Rhizoctonia* spp., *Sclerotinia* spp., and *Alternaria kikuchiana*, where melanin acts as both a physical and biochemical shield in host defenses. The protective role of melanin extends beyond host–pathogen interactions to environmental stressors (Spiga *et al.*, 2021). It provides defense against UV radiation, oxidative stress, and extreme temperature fluctuations. In *Bacillus subtilis*, for example, spores contain a melanin-like pigment in their protein coat, which enhances their survival under UV exposure and oxidative damage (Slots and Genco, 1984). Similarly, *Bacillus thuringiensis* and *Pseudomonas aeruginosa* utilize melanin to withstand UV radiation and toxic chemicals, further emphasizing its role in microbial persistence.

Another crucial aspect of melanin's function in microbial pathogenicity is immune evasion. Melanin-rich cells are less susceptible to phagocytosis, thereby enabling prolonged persistence within the host. In *C. neoformans*, melanized cells exhibit reduced uptake by macrophages, allowing them to establish chronic infections (Ni *et al.*, 2020). Beyond vertebrate immune defenses, melanin also plays a role in microbial survival against environmental predators such as amoebae, reinforcing its adaptive significance. Additionally, melanin contributes to microbial resistance against heavy metal toxicity. It has the ability to bind heavy metals, thereby reducing their toxic effects on microbial cells (Taborda *et al.*, 2008). *C. neoformans* demonstrates increased resistance to silver nitrate, a well-known antimicrobial compound, due to its melanin-rich cell wall. Similar metal-binding properties have been observed in other fungal melanins, suggesting that this pigment aids microbial survival in contaminated environments (Moura-Alves *et al.*, 2014).

Melanin also plays a role in microbial morphogenesis and adaptation. Melanotic fungi incorporate melanin into their cell walls and remodel it continuously during growth, allowing them to respond effectively to environmental pressures (Narsing Rao *et al.*, 2017). Although the regulatory mechanisms governing melanin synthesis and its structural modifications remain largely unexplored, ongoing research seeks to elucidate how this pigment influences microbial development and long-term persistence.

Beyond melanin, several other microbial pigments contribute significantly to virulence. These pigments enhance immune evasion, promote inflammation, and exert toxic effects on host cells. Understanding their functional mechanisms can help identify novel targets for antimicrobial therapies. One such pigment is staphyloxanthin, the golden pigment produced by *Staphylococcus aureus* (Rao *et al.*, 2021). Staphyloxanthin acts as an antioxidant, neutralizing ROS generated by the host immune system. This enhances bacterial resistance to immune defenses, facilitating the persistence of *S. aureus* in infected tissues. Similarly, pyocyanin, a blue-green pigment produced by *Pseudomonas aeruginosa*, disrupts host cell function by generating ROS. This leads to oxidative stress, immune suppression, and enhanced bacterial survival, particularly in chronic infections such as cystic fibrosis (Solano, 2014).

Melanin also plays a vital role in fungal pathogens such as *Cryptococcus neoformans* and *Aspergillus* spp. These dark pigments protect fungal cells from oxidative stress and increase resistance to antifungal treatments (Casadevall and Pirofski, 2001). This enhances their persistence within immunocompromised hosts, complicating treatment strategies and making infections more challenging to manage.

Given the crucial role of microbial pigments in virulence, targeting pigment biosynthesis presents a promising therapeutic avenue. By inhibiting pigment production, pathogens could become more susceptible to immune clearance and antimicrobial treatment. Research into pigment-targeting therapies is ongoing, and potential strategies include the development of inhibitors that block pigment synthesis or disrupt pigment function within microbial cells. Such approaches could pave the way for novel treatments for infections caused by pigmented microbes (Casadevall and Pirofski, 2001).

Microbial pigments play a multifaceted role in pathogenicity. They contribute to immune evasion, enhance environmental resilience, and increase resistance to antimicrobial agents (Dadachova and Casadevall, 2008). Among these pigments, melanin is a particularly critical virulence factor, supporting microbial survival and pathogenicity across multiple species (Plonka and Grabacka, 2006). Understanding the genetic and biochemical pathways underlying pigment production will be essential for developing targeted therapies to mitigate the impact of pigment-associated infections. Future research should focus on elucidating the molecular mechanisms governing pigment synthesis and function, ultimately leading to improved treatment strategies for infections caused by pigmented microbial pathogens (Plonka and Grabacka, 2006).

References

Abel, A. (2012) The history of dyes and pigments: From natural dyes to high performance pigments. In: Best, J. (ed.) *Colour Design*, 2nd edn. Woodhead Publishing, pp. 557–587. DOI: 10.1016/B978-0-08-101270-3.00024-2.

Agrawal, S., Adholeya, A., Barrow, C.J. and Deshmukh, S.K. (2018) Marine fungi: An untapped bioresource for future cosmeceuticals. *Phytochemistry Letters* 23, 15–20. DOI: 10.1016/j.phytol.2017.11.003.

Ahmad, W.A., Ahmad, W.Y.W., Zakaria, Z.A. and Yusof, N.Z. (2012) Application of bacterial pigments as colorant. In: *Application of Bacterial Pigments as Colorants*. Springer, Berlin, pp. 57–74. DOI: 10.1007/978-3-642-24520-6_4.

Akilandeswari, P. and Pradeep, B.V. (2017) Microbial pigments: Potential functions and prospects. DOI: 10.1002/9781119166191.ch11.

Allam, K. V. and Kumar, G. P. (2011). *Colorants-the cosmetics for the pharmaceutical dosage forms. Int J Pharm Pharm Sci*, 3(Suppl 3), 9

Alves, A., Sousa, E., Kijjoa, A. and Pinto, M. (2020) Marine-derived compounds with potential use as cosmeceuticals and nutricosmetics. *Molecules* 25(11), 2536. DOI: 10.3390/molecules25112536.

Arad, S. and Yaron, A. (1992) Natural pigments from red microalgae for use in foods and cosmetics. *Trends in Food Science & Technology* 3, 92–97. DOI: 10.1016/0924-2244(92)90145-M.

Aruldass, C.A., Dufossé, L. and Ahmad, W.A. (2018) Current perspective of yellowish-orange pigments from microorganisms – a review. *Journal of Cleaner Production* 180, 168–182. DOI: 10.1016/j.jclepro.2018.01.093.

Bedoux, G., Hardouin, K., Burlot, A.S. and Bourgougnon, N. (2014) Bioactive components from seaweeds: Cosmetic applications and future development. In: *Advances in Botanical Research*. Academic Press, New York, pp. 345–378. DOI: 10.1016/B978-0-12-408062-1.00012-3.

Bermejo, R., Acién, F.G., Ibáñez, M.J., Fernández, J.M., Molina, E. *et al.* (2003) Preparative purification of B-phycoerythrin from the microalga *Porphyridium cruentum* by expanded-bed adsorption chromatography. *Journal of Chromatography. B, Analytical Technologies in the Biomedical and Life Sciences* 790(1–2), 317–325. DOI: 10.1016/s1570-0232(03)00168-5.

Bhatt, P. (2022) *Industrial Applications of Microbial Enzymes*. CRC Press, Boca Raton, FL.

Brandt, F.S., Cazzaniga, A. and Hann, M. (2011) Cosmeceuticals: Current trends and market analysis. *Seminars in Cutaneous Medicine and Surgery* 30(3), 141–143. DOI: 10.1016/j.sder.2011.05.006.

Britton, G. and Britton, G. (1983) *The Biochemistry of Natural Pigments*. Cambridge University Press, Cambridge, UK.

Burger, P., Landreau, A., Azoulay, S., Michel, T. and Fernandez, X. (n.d.) Skin whitening cosmetics: Feedback and challenges in the development of natural skin lighteners. Available at: www.mdpi.com/2079-9284/3/4/36 (accessed 27 February 2025).

Casadevall, A. and Pirofski, L. (2001) Host–pathogen interactions: The attributes of virulence. *Journal of Infectious Diseases* 184(3), 337–344. DOI: 10.1086/322044.

Celedón, R.S. and Díaz, L.B. (2021) Natural pigments of bacterial origin and their possible biomedical applications. *Microorganisms* 9(4), 739. DOI: 10.3390/microorganisms9040739.

Chandra, R., Parra, R. and Iqbal, H.M.N. (n.d.) Phycobiliproteins: A novel green tool from marine origin blue-green algae and red algae. Available at: https://pubmed.ncbi.nlm.nih.gov/27491380/ (accessed 27 February 2025).

Chang, J.C., Wu, M.C., Liu, I.M. and Cheng, J.T. (2006) Plasma glucose-lowering action of Hon-Chi in streptozotocin-induced diabetic rats. *Hormone and Metabolic Research* 38(2), 76–81. DOI: 10.1055/s-2006-925116.

Chatragadda, R. and Dufossé, L. (2021) Ecological and biotechnological aspects of pigmented microbes: A way forward in development of food and pharmaceutical grade pigments. DOI: 10.3390/microorganisms9030637. (accessed 27 February 2025).

Claus, H. and Decker, H. (2006) Bacterial tyrosinases. *Systematic and Applied Microbiology* 29(1), 3–14. DOI: 10.1016/j.syapm.2005.07.012.

Costa, R. and Santos, L. (2017) Delivery systems for cosmetics – from manufacturing to the skin of natural antioxidants. *Powder Technology* 322, 402–416. DOI: 10.1016/j.powtec.2017.07.086.

Costin, G.-E. and Hearing, V.J. (2007) Human skin pigmentation: Melanocytes modulate skin color in response to stress. *FASEB Journal* 21(4), 976–994. DOI: 10.1096/fj.06-6649rev.

Cserháti, T. (2006) *Liquid Chromatography of Natural Pigments and Synthetic Dyes*. Elsevier, St Louis, MO.

Cuthill, I.C., Allen, W.L., Arbuckle, K., Caspers, B., Chaplin, G. *et al.* (2017) The biology of color. *Science* 357(6350), eaan0221. DOI: 10.1126/science.aan0221.

Dadachova, E. and Casadevall, A. (2008) Ionizing radiation: How fungi cope, adapt, and exploit with the help of melanin. *Current Opinion in Microbiology, Growth and Development: Eukaryotes/Prokaryotes* 11, 525–531. DOI: 10.1016/j.mib.2008.09.013.

D'Aoust, J.Y. and Gerber, N.N. (1974) Isolation and purification of prodigiosin from Vibrio psychroerythrus. *Journal of Bacteriology* 118, 756–757. DOI: 10.1128/jb.118.2.756-757.1974.

Dave, S., Das, J., Varshney, B. and Sharma, V.P. (2022) Dyes and pigments: Interventions and how safe and sustainable are colors of life. In: Dave, S. and Das, J. (eds) *Trends and Contemporary Technologies for Photocatalytic Degradation of Dyes*. Springer International Publishing, Cham, Switzerland, pp. 1–20. DOI: 10.1007/978-3-031-08991-6_1.

Dessinioti, C., Stratigos, A.J., Rigopoulos, D. and Katsambas, A.D. (2009) A review of genetic disorders of hypopigmentation: Lessons learned from the biology of melanocytes. *Experimental Dermatology* 18, 741–749. DOI: 10.1111/j.1600-0625.2009.00896.x.

Ding, J., Wu, B. and Chen, L. (2022) Application of marine microbial natural products in cosmetics. *Frontiers in Microbiology* 13. DOI: 10.3389/fmicb.2022.892505.

Dufossé, L. (2006) Microbial production of food grade pigments. *Food Technology and Biotechnology* 313–321. Available at: https://hrcak.srce.hr/109913

Dufossé, L. (2009) Pigments. *Microbial* 4, 457–471. DOI: 10.1016/B978-012373944-5.00155-3.

Dufossé, L. (2019) Research, Development, and Production of Microalgal and Microbial Biocolorants - Bioprocessing for Biomolecules Production. Available at: https://onlinelibrary.wiley.com/doi/10.1002/9781119434436.ch4 (accessed 3 April 2025).

Gao, Q. and Garcia-Pichel, F. (2011) Microbial ultraviolet sunscreens. *Nature Reviews Microbiology* 9(11), 791–802. DOI: 10.1038/nrmicro2649.

Gargallo-Viola, D. (1989) Enzyme polymorphism, prodigiosin production, and plasmid fingerprints in clinical and naturally occurring isolates of *Serratia marcescens*. *Journal of Clinical Microbiology* 27(5), 860–868. DOI: 10.1128/jcm.27.5.860-868.1989.

Hamed, I. (n.d.) The evolution and versatility of microalgal biotechnology: A review. Available at: https://pubmed.ncbi.nlm.nih.gov/33401835/ (accessed 27 February 2025).

Hari, R.K., Patel, T.R. and Martin, A.M. (1994) An overview of pigment production in biological systems: Functions, biosynthesis, and applications in food industry. *Food Reviews International* 10(1), 49–70. DOI: 10.1080/87559129409540985.

Hashtroudi, M.S., Shariatmadari, Z., Riahi, H. and Ghassempour, A. (2013) Analysis of anabaena vaginicola and nostoc calcicola from Northern Iran, as rich sources of major carotenoids. *Food Chemistry* 136(3–4), 1148–1153. DOI: 10.1016/j.foodchem.2012.09.055.

Hussein, M.R. (2005) Ultraviolet radiation and skin cancer: Molecular mechanisms. *Journal of Cutaneous Pathology* 32(3), 191–205. DOI: 10.1111/j.0303-6987.2005.00281.x.

Ishimoto, C.K., Paulino, B.N., Neri-Numa, I.A. and Bicas, J.L. (2024) The blue palette of life: A comprehensive review of natural bluish colorants with potential commercial applications. *Food Research International* 196, 115082. DOI: 10.1016/j.foodres.2024.115082.

Joshi, V.K., Attri, D., Bala, A. and Bhushan, S. (n.d.) NIScPR online periodical repository: Microbial pigments. Available at: https://nopr.niscpr.res.in/handle/123456789/11334 (accessed 3 May 2025).

Kalesperis, G.S., Prahlad, K.V. and Lynch, D.L. (1975) Toxigenic studies with the antibiotic pigments from *Serratia marcescens*. *Canadian Journal of Microbiology* 21(2), 213–220. DOI: 10.1139/m75-030.

Kalra, R., Conlan, X.A. and Goel, M. (2020) Fungi as a potential source of pigments: Harnessing filamentous fungi. *Frontiers in Chemistry* 8, 369. DOI: 10.3389/fchem.2020.00369.

Kanekar, H. and Khale, A. (2017) Coloring agents: Current regulatory perspective for coloring agents intended for pharmaceutical and cosmetic use. *International Journal of Pharmaceutical and Phytopharmacological Research*. Available at: www.iacmcolor.org/wp-content/uploads/2014/08/Kanekar-and-Khale-2014-colors-in-pharma-and-cosmetics.pdf (accessed 12 August 2025).

Katsube, N., Iwashita, K., Tsushida, T., Yamaki, K. and Kobori, M. (2003) Induction of apoptosis in cancer cells by Bilberry (*Vaccinium myrtillus*) and the anthocyanins. *Journal of Agricultural and Food Chemistry* 51(1), 68–75. DOI: 10.1021/jf025781x.

Kiki, M.J. (2023) Biopigments of microbial origin and their application in the cosmetic industry. *Cosmetics* 10(2), 47. DOI: 10.3390/cosmetics10020047.

Konzen, M., De Marco, D., Cordova, C.A.S., Vieira, T.O., Antônio, R.V. *et al.* (2006) Antioxidant properties of violacein: Possible relation on its biological function. *Bioorganic & Medicinal Chemistry* 14(24), 8307–8313. DOI: 10.1016/j.bmc.2006.09.013.

Kumar, A., Vishwakarma, H.S., Singh, J. and Kumar, M. (2015) Microbial pigments: Production and their applications in various industries. *International Journal of Pharmaceutical, Chemical and Biological Sciences* 5, 203–212.

Kumar, A., Dixit, U., Singh, K., Prakash Gupta, S. and Jamal Beg, M. (2021) Structure and properties of dyes and pigments. In: *Dyes and Pigments*. IntechOpen, London, pp. 1–19. DOI: 10.5772/intechopen.97104.

Kumar, S., Kumar, V., Ambika, A.A.A., Nag, D., Kumar, V. *et al.* (2022) Microbial pigments: Learning from himalayan perspective to industrial applications. *Journal of Industrial Microbiology & Biotechnology* 49(5), kuac017. DOI: 10.1093/jimb/kuac017.

Lazcano, A. (2010) Historical development of origins research. *Cold Spring Harbor Perspectives in Biology* 2(11), a002089. DOI: 10.1101/cshperspect.a002089.

Lazzè, M.C., Savio, M., Pizzala, R., Cazzalini, O., Perucca, P. *et al.* (2004) Anthocyanins induce cell cycle perturbations and apoptosis in different human cell lines. *Carcinogenesis* 25(8), 1427–1433. DOI: 10.1093/carcin/bgh138.

Lee, S.Y., Coutinho, J.A.P. and Weingarten, M. (2025) Sustainable recovery of microbial-derived natural pigments using deep eutectic solvents: Advances, potential, and challenges. *Separation and Purification Technology* 361, 131413. DOI: 10.1016/j.seppur.2025.131413.

Lewis, S.M. and Corpe, W.A. (1964) Prodigiosin-producing bacteria from marine sources. *Applied Microbiology* 12(1), 13–17. DOI: 10.1128/am.12.1.13-17.1964.

Li, E.P.H., Min, H.J., Belk, R.W., Kimura, J. and Bahl, S. (n.d.) Skin lightening and beauty in four Asian cultures. *Advances in Consumer Research* 35, 444–449.

Lichstein, H.C. and Van De Sand, V.F. (1946) The antibiotic activity of violacein, prodigiosin, and phthiocol. *Journal of Bacteriology* 52(1), 145–146. DOI: 10.1128/jb.52.1.145-146.1946.

Likhitwitayawuid, K. (2008) Stilbenes with tyrosinase inhibitory activity. Available at: https://openurl.ebsco.com/contentitem/gcd:31291373?sid=ebsco:plink:crawler&id=ebsco:gcd:31291373 (accessed 27 February 2025).

Lincke, G. (2003) Molecular stacks as a common characteristic in the crystal lattice of organic pigment dyes a contribution to the "soluble–insoluble" dichotomy of dyes and pigments from the technological point of view. *Dyes and Pigments* 59(1), 1–24. DOI: 10.1016/S0143-7208(03)00097-4.

Liu, G.Y. and Nizet, V. (2009) Color me bad: Microbial pigments as virulence factors. *Trends in Microbiology* 17(9), 406–413. DOI: 10.1016/j.tim.2009.06.006.

Malik, K., Tokkas, J. and Goyal, S. (2012) Microbial pigments: A review. *International Journal of Microbial Resource Technology* 1(4), 361–365.

Manivasagan, P., Venkatesan, J., Sivakumar, K. and Kim, S.-K. (2013) Actinobacterial melanins: Current status and perspective for the future. *World Journal of Microbiology & Biotechnology* 29(10), 1737–1750. DOI: 10.1007/s11274-013-1352-y.

Mansour, R. (2018) Natural dyes and pigments: Extraction and applications. In: *Handbook of Renewable Materials for Coloration and Finishing*. John Wiley & Sons, Chichester, UK, pp. 102–175. DOI: 10.1002/9781119407850.

Martin, S., Giannone, G., Andriantsitohaina, R. and Martinez, M.C. (2003) Delphinidin, an active compound of red wine, inhibits endothelial cell apoptosis via nitric oxide pathway and regulation of calcium homeostasis. *British Journal of Pharmacology* 139(6), 1095–1102. DOI: 10.1038/sj.bjp.0705347.

Martins, S., Mussatto, S.I., Martínez-Avila, G., Montañez-Saenz, J., Aguilar, C.N. *et al.* (2011) Bioactive phenolic compounds: Production and extraction by solid-state fermentation. A Review. *Biotechnology Advances* 29(3), 365–373. DOI: 10.1016/j.biotechadv.2011.01.008.

Mendes-Silva, T.C.D., da Andrade, R.F.S., Ootani, M.A., Mendes, P.V.D., da Silva, M.R.F. *et al.* (2020) Biotechnological potential of carotenoids produced by extremophilic microorganisms and application prospects for the cosmetics industry. *Advances in Microbiology* 10(8), 397–410. DOI: 10.4236/aim.2020.108029.

Mojib, N., Philpott, R., Huang, J.P., Niederweis, M. and Bej, A.K. (2010) Antimycobacterial activity *in vitro* of pigments isolated from Antarctic bacteria. *Antonie van Leeuwenhoek* 98(4), 531–540. DOI: 10.1007/s10482-010-9470-0.

Morocho-Jácome, A.L., Ruscinc, N., Martinez, R.M., de Carvalho, J.C.M., Santos de Almeida, T. *et al.* (2020) (Bio)Technological aspects of microalgae pigments for cosmetics. *Applied Microbiology and Biotechnology* 104(22), 9513–9522. DOI: 10.1007/s00253-020-10936-x.

Moura-Alves, P., Faé, K., Houthuys, E., Dorhoi, A., Kreuchwig, A. *et al.* (2014) AhR sensing of bacterial pigments regulates antibacterial defence. *Nature* 512(7515), 387–392. DOI: 10.1038/nature13684.

Mourelle, M.L., Gómez, C.P. and Legido, J.L. (n.d.) The potential use of marine microalgae and cyanobacteria in cosmetics and thalassotherapy. Available at: www.mdpi.com/2079-9284/4/4/46 (accessed 26 February 2025).

Nakamura, Y., Sawada, T., Morita, Y. and Tamiya, E. (2024) Isolation of a psychrotrophic bacterium from the organic residue of a water tank keeping rainbow trout and antibacterial effect of violet pigment produced from the strain. *Biochemical Engineering Journal* 12, 79–86. DOI: |10.1016/S1369-703X(02)00079-7.

Narsing Rao, M.P., Xiao, M. and Li, W.-J. (2017) Fungal and bacterial pigments: Secondary metabolites with wide applications. *Frontiers in Microbiology* 8, 1113. DOI: 10.3389/fmicb.2017.01113.

Nawaz, A., Chaudhary, R., Shah, Z., Dufossé, L., Fouillaud, M. *et al.* (2020) An overview on industrial and medical applications of bio-pigments synthesized by marine bacteria. *Microorganisms* 9(1), 11. DOI: 10.3390/microorganisms9010011.

Ni, S., Li, B., Xu, Y., Mao, F., Li, X. *et al.* (2020) Targeting virulence factors as an antimicrobial approach: Pigment inhibitors. *Medicinal Research Reviews* 40(1), 293–338. DOI: 10.1002/med.21621.

Nigam, P.S.-N. and Pandey, A. (2009) *Biotechnology for Agro-Industrial Residues Utilisation: Utilisation of Agro-Residues*. Springer Science & Business Media, Dordrecht, The Netherlands.

Nosanchuk, J.D. and Casadevall, A. (2003) The contribution of melanin to microbial pathogenesis. *Cellular Microbiology* 5(4), 203–223. DOI: 10.1046/j.1462-5814.2003.00268.x.

Nosanchuk, J.D. and Casadevall, A. (2006) Impact of melanin on microbial virulence and clinical resistance to antimicrobial compounds. *Antimicrobial Agents and Chemotherapy* 50(11), 3519–3528. DOI: 10.1128/AAC.00545-06.

Novoveská, L., Ross, M.E., Stanley, M.S., Pradelles, R., Wasiolek, V. *et al.* (2019) Microalgal carotenoids: A review of production, current markets, regulations, and future direction. *Marine Drugs* 17(11), 640. DOI: 10.3390/md17110640.

Pagels, F., Almeida, C., Vasconcelos, V. and Guedes, A.C. (2022) Cosmetic potential of pigments extracts from the marine cyanobacterium *Cyanobium* sp. *Marine Drugs* 20(8), 481. DOI: 10.3390/md20080481.

Pailliè-Jiménez, M.E., Stincone, P. and Brandelli, A. (2020) Natural pigments of microbial origin. *Frontiers in Sustainable Food Systemsi* 4. DOI: 10.3389/fsufs.2020.590439.

Palombo, P., Fabrizi, G., Ruocco, V., Ruocco, E., Fluhr, J. *et al.* (2007) Beneficial long-term effects of combined oral/topical antioxidant treatment with the carotenoids Lutein and Zeaxanthin on human skin: A double-blind, placebo-controlled study. *Skin Pharmacology and Physiology* 20(4), 199–210. DOI: 10.1159/000101807.

Pan, J.-L., Wang, H.-M., Chen, C.-Y. and Chang, J.-S. (2012) Extraction of astaxanthin from *Haematococcus pluvialis* by supercritical carbon dioxide fluid with ethanol modifier. *Engineering in Life Sciences* 12(6), 638–647. DOI: 10.1002/elsc.201100157.

Panda, B. P., Javed, S., & Ali, M. (2010). Production of Angkak Through Co-Culture of Monascus Purpureus and MONASCUS RUBER. Brazilian journal of microbiology : [publication of the Brazilian Society for Microbiology], 41(3), 757–764. https://doi.org/10.1590/S1517-83822010000300028"

Pandey, R., Chander, R. and Sainis, K.B. (n.d.) Prodigiosins: A novel family of immunosuppressants with anti-cancer activity. Available at: https://pubmed.ncbi.nlm.nih.gov/18341204/ (accessed 26 February 2025).

Plonka, P.M. and Grabacka, M. (2006) Melanin synthesis in microorganisms – biotechnological and medical aspects. *Acta Biochimica Polonica* 53(3), 429–443. DOI: 10.18388/abp.2006_3314.

Poulose, N., Sajayan, A., Ravindran, A., Sreechithra, T.V., Vardhan, V. *et al*. (n.d.) Photoprotective effect of nanomelanin-seaweed concentrate in formulated cosmetic cream: With improved antioxidant and wound healing properties. Available at: www.sciencedirect.com/science/article/abs/pii/S1011134419315842 (accessed 26 February 2025).

Ramesh, C., Vinithkumar, N.V., Kirubagaran, R., Venil, C.K. and Dufossé, L. (2019) Multifaceted applications of microbial pigments: Current knowledge, challenges and future directions for public health implications. *Microorganisms* 7(7), 186. DOI: 10.3390/microorganisms7070186.

Rao, A.S., Deka, S.P., More, S.S., Nair, A., More, V.S. *et al*. (2021) A comprehensive review on different microbial-derived pigments and their multipurpose activities. In: Vaishnav, A. and Choudhary, D.K. (eds) *Microbial Polymers: Applications and Ecological Perspectives*. Springer, Singapore, pp. 479–519. DOI: 10.1007/978-981-16-0045-6_20.

Richard, C. (n.d.) [Chromobacterium violaceum, opportunist pathogenic bacteria in tropical and subtropical regions]. Available at: https://pubmed.ncbi.nlm.nih.gov/8219788/ (accessed 26 February 2025).

Rothon, R. (2012) Pigment and Nanopigment Dispersion Technologies. Available at: www.ebooks.com/en-us/book/886748/pigment-and-nanopigment-dispersion-technologies/roger-rothon/ (accessed 3 April 2025).

Sayo, T., Sugiyama, Y. and Inoue, S. (2013) Lutein, a Nonprovitamin A, activates the retinoic acid receptor to induce HAS3-dependent hyaluronan synthesis in keratinocytes. *Bioscience, Biotechnology, and Biochemistry* 77(6), 1282–1286. DOI: 10.1271/bbb.130124.

Schoneker, D.R. (2013) Coloring agents for use in pharmaceuticals. In: *Encyclopedia of Pharmaceutical Science and Technology. CRC Press*. CRC Press, Boca Raton, FL.

Sekar, S. and Chandramohan, M. (2008) Phycobiliproteins as a commodity: Trends in applied research, patents and commercialization. *Journal of Applied Phycology* 20(2), 113–136. DOI: 10.1007/s10811-007-9188-1.

Sharma, N., Shekhar, P., Kumar, V., Kaur, H. and Jayasena, V. (2024) Microbial pigments: Sources, current status, future challenges in cosmetics and therapeutic applications. *Journal of Basic Microbiology* 64(1), 4–21. DOI: 10.1002/jobm.202300214.

Silva, T.R.E., Silva Jr, L.C.F., de Queiroz, A.C., Alexandre Moreira, M.S., de Carvalho Fraga, C.A. *et al*. (2021) Pigments from antarctic bacteria and their biotechnological applications. *Critical Reviews in Biotechnology* 41(6), 809–826. DOI: 10.1080/07388551.2021.1888068.

Slots, J. and Genco, R.J. (1984) Black-pigmented Bacteroides species, Capnocytophaga species, and Actinobacillus actinomycetemcomitans in human periodontal disease: Virulence factors in colonization, survival, and tissue destruction. *Journal of Dental Research* 63(3), 412–421. DOI: 10.1177/00220345840630031101.

Solano, F. (2014) Melanins: Skin pigments and much more—types, structural models, biological functions, and formation routes. *New Journal of Science* 498276, 1–28. DOI: 10.1155/2014/498276.

Soliev, A.B., Hosokawa, K. and Enomoto, K. (2011) Bioactive pigments from marine bacteria: Applications and physiological roles. *Evidence-Based Complementary and Alternative Medicine* 670349. DOI: 10.1155/2011/670349.

Spiga, L., Jimenez, A.G., Santos, R.L. and Winter, S.E. (2021) How microbiological tests reflect bacterial pathogenesis and host adaptation. *Brazilian Journal of Microbiology* 52(4), 1745–1753. DOI: 10.1007/s42770-021-00571-7.

Srinivasan, M., Merlyn Keziah, S., Hemalatha, M. and Subathra Devi, C. (n.d.) Pigment from Streptomyces bellus MSA1 isolated from marine sediments. Available at: https://iopscience.iop.org/article/10.1088/1757-899X/263/2/022049/meta (accessed 27 February 2025).

Stahl, W. and Sies, H. (2012) β-Carotene and other carotenoids in protection from sunlight. *American Journal of Clinical Nutrition* 96(5), 1179S–84S. DOI: 10.3945/ajcn.112.034819.

Sumathy, B. (n.d.) (PDF) microbial pigments. DOI: 10.1007/978-1-4020-9942-7_8.

Sutthiwong, N. and Caro, Y. (2014) Bacteria belonging to the extremely versatile genus Arthrobacter as novel source of natural pigments with extended hue range. *Food Research International* 65, 156–162.

Taborda, C.P., da Silva, M.B., Nosanchuk, J.D. and Travassos, L.R. (2008) Melanin as a virulence factor of *Paracoccidioides brasiliensis* and other dimorphic pathogenic fungi: A minireview. *Mycopathologia* 165(4–5), 331–339. DOI: 10.1007/s11046-007-9061-4.

Terao, J., Minami, Y. and Bando, N. (2010) Singlet molecular oxygen-quenching activity of carotenoids: Relevance to protection of the skin from photoaging. *Journal of Clinical Biochemistry and Nutrition* 48, 57–62. DOI: 10.3164/jcbn.11-008FR. (accessed 26 February 2025).

Thomas, N.V. and Kim, S.-K. (2013) Beneficial effects of marine algal compounds in cosmeceuticals. *Marine Drugs* 11(1), 146–164. DOI: 10.3390/md11010146.

Tominaga, K., Hongo, N., Karato, M. and Yamashita, E. (2012) Cosmetic benefits of astaxanthin on humans subjects. *Acta Biochimica Polonica* 59(1), 43–47.

Tosato, M.G., Orallo, D.E., Fangio, M.F., Diz, V., Dicelio, L.E. *et al.* (2016) Nanomaterials and natural products for UV-photoprotection. In: Grumezescu, A.M. (ed.) *Surface Chemistry of Nanobiomaterials*. William Andrew Publishing, New York, pp. 359–392. DOI: 10.1016/B978-0-323-42861-3.00012-1.

Tuli, H.S., Chaudhary, P., Beniwal, V. and Sharma, A.K. (2015) Microbial pigments as natural color sources: Current trends and future perspectives. *Journal of Food Science and Technology* 52(8), 4669–4678. DOI: 10.1007/s13197-014-1601-6.

Venil, C.K., Zakaria, Z.A. and Ahmad, W.A. (2013) Bacterial pigments and their applications. *Process Biochemistry* 48(7), 1065–1079. DOI: 10.1016/j.procbio.2013.06.006.

Venil, C.K., Aruldass, C.A., Dufossé, L., Zakaria, Z.A. and Ahmad, W.A. (2014) Current perspective on bacterial pigments: Emerging sustainable compounds with coloring and biological properties for the industry – an incisive evaluation. *RSC Advances* 4(74), 39523. DOI: 10.1039/C4RA06162D.

Venil, C.K., Devi, P.R. and Ahmad, W.A. (2020a) Agro-industrial waste as substrates for the production of bacterial pigment. In: Zakaria, Z.A., Boopathy, R. and Dib, J.R. (eds) *Valorisation of Agro-Industrial Residues – Volume I: Biological Approaches*. Springer International, Cham, Switzerland, pp. 149–162. DOI: 10.1007/978-3-030-39137-9_7.

Venil, C.K., Dufossé, L. and Renuka Devi, P. (2020b) Bacterial pigments: Sustainable compounds with market potential for pharma and food industry. *Frontiers in Sustainable Food Systems* 4. DOI: 10.3389/fsufs.2020.00100.

Vishnupriya, S., Bhavaniramya, S., Baskaran, D. and Karthiayani, A. (2021) Microbial pigments and their application. In: Vaishnav, A. and Choudhary, D.K. (eds) *Microbial Polymers: Applications and Ecological Perspectives*. Springer, Singapore, pp. 193–214. DOI: 10.1007/978-981-16-0045-6_9.

Wang, J. and Mazza, G. (2002) Effects of anthocyanins and other phenolic compounds on the production of tumor necrosis factor alpha in LPS/IFN-gamma-activated RAW 264.7 macrophages. *Journal of Agricultural and Food Chemistry* 50(15), 4183–4189. DOI: 10.1021/jf011613d.

Wang, H.-M., Chen, C.-Y. and Wen, Z.-H. (2011) Identifying melanogenesis inhibitors from *Cinnamomum subavenium* with *in vitro* and *in vivo* screening systems by targeting the human tyrosinase. *Experimental Dermatology* 20(3), 242–248. DOI: 10.1111/j.1600-0625.2010.01161.x.

Wang, H.M.D., Chen, C.C., Huynh, P. and Chang, J.S. (2015) Exploring the potential of using algae in cosmetics. *Bioresource Technology* 184, 355–362. DOI: 10.1016/j.biortech.2014.12.001.

Wilson, M. (2016) The human microbiota and chronic disease: Dysbiosis as a cause of human pathology. Available at: https://doi.org/10.10/9781118982907.ch1

Wynn-Williams, D.D., Edwards, H.G.M., Newton, E.M. and Holder, J.M. (2002) Pigmentation as a survival strategy for ancient and modern photosynthetic microbes under high ultraviolet stress on planetary surfaces. *International Journal of Astrobiology* 1(1), 39–49. DOI: 10.1017/S1473550402001039.

Yadav, S., Tiwari, K.S., Gupta, C., Tiwari, M.K., Khan, A. *et al.* (2023) A brief review on natural dyes, pigments: Recent advances and future perspectives. *Results in Chemistry* 5, 100733. DOI: 10.1016/j.rechem.2022.100733.

Yarkent, Ç., Gürlek, C. and Oncel, S.S. (2020) Potential of microalgal compounds in trending natural cosmetics: A review. *Sustainable Chemistry and Pharmacy* 17, 100304. DOI: 10.1016/j.scp.2020.100304.

Yeager, D.G. and Lim, H.W. (2019) What's new in photoprotection: A review of new concepts and controversies. *Dermatologic Clinics* 37, 149–157. DOI: 10.1016/j.det.2018.11.003.

Yusuf, M., Shabbir, M. and Mohammad, F. (2017) Natural colorants: Historical, processing and sustainable prospects. *Natural Products and Bioprospecting* 7(1), 123–145. DOI: 10.1007/s13659-017-0119-9.

Zhang, J., Suo, Y., Zhang, D., Jin, F., Zhao, H. *et al.* (2018) Genetic and virulent difference between pigmented and non-pigmented staphylococcus aureus. *Frontiers in Microbiology* 9, 598. DOI: 10.3389/fmicb.2018.00598.

Zollinger, H. (2003) *Color Chemistry: Syntheses, Properties, and Applications of Organic Dyes and Pigments*. John Wiley & Sons, New York.

12 Microbial Pigments in Textile and Dyeing Industries

Mridul Umesh[1], Nilina James[2*], Liya Merin Stanly[1], Sreehari Suresh[3] and Adhithya Sankar Santhosh[1]

[1]Department of Life Sciences, Christ University, Bangalore, Karnataka, India; [2]Department of Botany, Teresian College, Mysore, Karnataka, India; [3]Centre for Global Health Research, Saveetha Medical College and Hospital, Saveetha Institute of Medical and Technical Sciences (SIMATS), Chennai, Thandalam, India

12.1 Introduction

Dyes and coloring are an inevitable part of the textile industry as it is one of the major criteria that determines consumer preference and sales value of the fabricated textiles. Pigments or dyes are substances that absorb light of a specific wavelength and reflect the rest at a wavelength ranging between 400 and 800 nm, thereby making a specific wavelength visible to the naked eye. When integrated with photoreceptors in the human retina, these colors provide the visual perception of different colors and shades. Dyes are used to impart color to yarns, fabrics, and fibers (Kramar and Kostic, 2022).

The use of dyes in textile processing dates back to ancient civilizations and is assumed to be at least 4000 years old. Although the discovery and use of synthetic pigments have revolutionized the textile industry, accelerating its market value and export demand due to their low cost, durability, and stability, the issues related to their toxicity, nonbiodegradability, and environmental persistence have lead research to focus on more eco-friendly and sustainable alternatives to synthetic pigments in the textile industry (Barreto *et al.*, 2023). The entry of synthetic pigments into the food chain or their exposure to humans via any route have been reported to induce carcinogenesis and allergic reactions (Olas *et al.*, 2021). To date, more than 10,000 different synthetic dyes and their derivatives have been employed in the textile industry and their production globally is as high as 7×10^7 tons per year. The textile industry on an average releases about 10–45% of dyes into the wastewater coming from dyeing units and around 1,40,000 [Hr1]tons of textile dyes are released in wastewater per year (Islam *et al.*, 2023).

Microorganisms are often seen as cell factories that can be used to produce commercially important metabolites, including pigments. Microbial pigment production offers potential benefits as they can be cultivated in a bioreactor under precisely controlled physicochemical conditions (Aman Mohammadi *et al.*, 2022). Secondly, production can be achieved using easily available and cheap agroindustrial waste substrates, emphasizing the concept of waste valorization and making the production process economically viable and cost-effective. Microbial

*Corresponding author: nilina.james@res.christuniversity.in

DOI: 10.1079/ 9781836990505.0012

pigment production is not influenced by environmental factors so steady production can be achieved throughout the year under optimized conditions (Venil *et al.*, 2013). The emergence of synthetic biology and genetic engineering procedures has helped to accelerate and enhance microbial pigment production, thereby making its application feasible in the textile industry. Microbial pigments are produced from bacteria, fungi, algae, and actinomycetes and have potential application in the textile industry as a sustainable alternative to synthetic dyes (Rajendran *et al.*, 2023).

Pigment-producing bacteria can be found in diverse habitats, including marine and terrestrial environments. Previous studies reveal that pigment-producing microorganisms can withstand a wide variety of physiological conditions in a dynamic environment and are thought to have increased stress tolerance. Pigments in microbes help them to tolerate UV radiation and high light intensity (Muhammad *et al.*, 2024).

Commercial application of microbial pigments in the textile industry requires the isolation of potent pigment-producing microorganisms, designing of cost-effective media for commercial production of pigments, enhancing of pigment production through strain improvement methods, purification of microbial methods using various analytical methods, and application for textile dyeing (Orlandi *et al.*, 2022). These microbial pigments have other commercial applications due to their antibacterial, antioxidant, and other biological properties.

This chapter describes the diversity of microbial pigments, extraction and processing techniques, methods for textile dyeing using microbial pigments, and the advantages, limitations, and challenges associated with the use of microbial pigments in the textile industry (Huang *et al.*, 2024).

12.2 Types of Microbial Pigments Used in Textile Dyeing

12.2.1 Key microbial pigments suitable for textiles

Microbes produce pigments as secondary metabolites in response to external stimuli. These compounds also exhibit properties such as antimicrobial, anticancer, and antioxidative activity (Table 12.1) (Bano *et al.*, 2023). These properties can enhance textiles dyed with these dyes, providing additional benefits (Repon *et al.*, 2023). Key microbial pigments which are used for textile dyeing include carotenoids, indigo, violacein, prodigiosin, and melanin (Fig. 12.1).

Carotenoids are a class of isoprenoid-derived compounds which exhibit vibrant yellow, orange, red, and purple hues. These compounds are generally present in archaea, eubacteria, algae, fungi, and plants. The majority of the carotenoid-producing organisms synthesize carotenoids from phytoene (C40) (Barreto *et al.*, 2023). Carotenoids have a polyene chain with nine or more conjugated double bonds and terminal groups on both ends. The class is divided into carotenoids (nonoxygenated) and xanthophylls (oxygenated) (Kumar *et al.*, 2024). Some of the commonly used carotenoids are β-carotene, astaxanthin, canthaxanthin, and lutein. Organisms such as *Blakeslea trispora*, *Rhodotorula glutinis*, and *Paracoccus carotinifaciens* are employed for the large-scale production of these compounds for textile use (Devi *et al.*, 2024).

Indigo is another commonly used microbial pigment in traditional textile dyeing. Traditionally indigo, an azo dye, was produced from *Indigofera tinctoria* (Li *et al.*, 2022), but recently space constraints and lack of availability of the plants have forced the scientific community to produce indigo using organisms such as *Pseudomonas putida* and *Escherichia coli*. These organisms can oxidize indole into indigo, which reduces the chemical treatment processes and promotes environmental compatibility for the synthesis of indigo (Mindt *et al.*, 2023).

Microbes such as *Chromobacterium violaceum* and *Janthinobacterium lividum* can synthesize the unique compound violacein, through the enzymatic transformation of the amino acid tryptophan (Park *et al.*, 2021). Violacein exhibits a vivid violet hue, which makes it ideal for dyeing textiles. It possesses unique bioactive properties which are suitable for fabrics with protective properties. Violacein is commonly used to dye healthcare items, sportswear, and outdoor gear (Ghosh *et al.*, 2025).

Prodigiosines are red hydrophobic tripyrrole pigments synthesized by organisms such as certain marine bacteria, actinomycetes,

Table 12.1. Common pigment-producing microorganisms.

Microorganism type	Organism name	Pigment name	Color	Application in textile/ dyeing industries	Reference
Bacteria	*Janthinobacterium lividu*	Violacein	Deep purple	Dyeing polyamide fabrics	(Kanelli *et al.*, 2018)
	Serratia marcescens SB08	Prodigiosin	Red	Pure silk, China silk, cotton yarns	(Venil *et al.*, 2021)
	Rhodonellum psychrophilum GL8	Prodigiosin	Red	Cotton, glacé cotton, silk, and rayon fabric	(Bisht *et al.*, 2020)
	Pseudomonas aeruginosa	Pyocyanin	Yellow	Polyester fabric	(Govindaraj *et al.*, 2021)
	Streptomyces sp., NP4	Prodigiosin	Red/blue	Cotton and viscose fabric	(Kramar *et al.*, 2021)
Fungi	*Aspergillus strains* AN01	Asperyellone	Yellow	Silk, cotton, synthetic and wool fabrics	(Gnanamani *et al.*, 2019)
	Penicillium murcianum	Carotenoids	Yellow	Wool	(Hernández *et al.*, 2019)
	Talaromyces australis	Carotenoids	Red	Wool	(Hernández *et al.*, 2019)
	Fusarium oxysporum	Anthraquinone	Blue, violet	Cotton fabric	(Chadni *et al.*, 2017)
	Nigrospora aurantiaca	Bostrycin	Red	Cotton fabrics	(Suwannarach *et al.*, 2019)
Actinomycetes	*Micromonospora* sp. K7-06	–	Blue	Cotton and wool	(Abidin *et al.*, 2023)
	Streptomyces coacervatus	–	Dark violet	Silk	(Urtgam *et al.*, 2023)
	Streptomyces cyaneofuscatus	Actinomycins	Yellow	Silk	(Chen *et al.*, 2021)

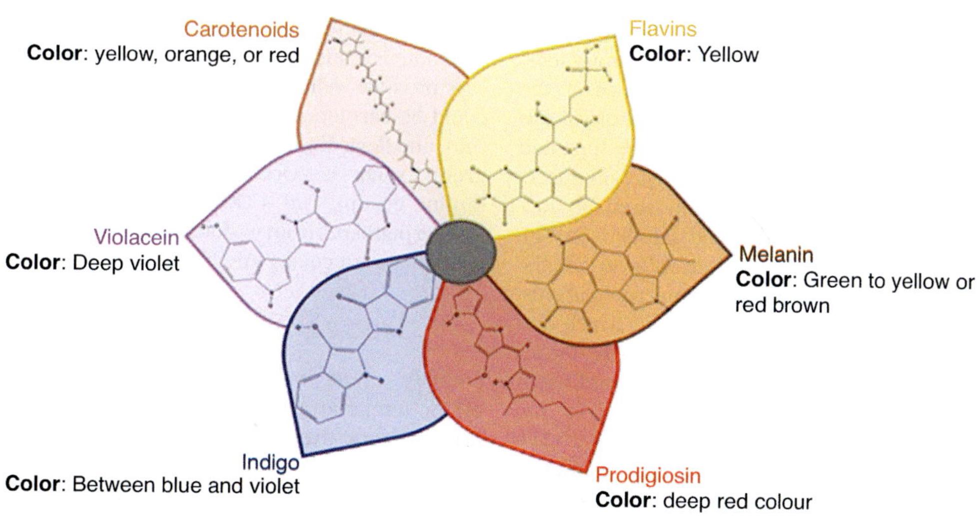

Fig. 12.1. Key microbial pigments used in textile dyeing.

and *Serratia* spp. (Barreto *et al.*, 2023). Prodigiosin is the most common prodigiosine used in textile industries. Prodigiosines can exist as straight-chain compounds (prodigiosin, undecyl prodigiosin) and cyclic derivatives (cyclononylprodigiosin and butyl-meta-cycloheptylprodiginine) (Parthasarathy and Manikkam, 2024). The microbes synthesize prodigiosines through enzymatic condensation of 4-methoxy-2–2′-bipyrrole-5-carbaldehyde (MBC) and a bipyrrole molecule with a monopyrrole unit (Srilekha *et al.*, 2024). Like prodigiosines, there exists a class of yellow compounds called tambjamines, which features a bipyrrole core condensed with a primary amine. This class of compounds are produced by bacteria such as *Pseudoalteromonas* spp. (Tang *et al.*, 2025).

Melanins are naturally occurring pigments with high molecular weight, which are produced by the oxidative polymerization of phenolic compounds or tyrosine in the microbes (Singh *et al.*, 2021). Melanin production is observed in species such as *Aspergillus niger*, *Streptomyces*, and *Cryptococcus neoformans* (Singh *et al.*, 2021). Melanins are renowned for their stability, insolubility, and gradient colorations from dark brown to black. There are two types of melanin: eumelanins and pheomelanins. Microbes generally produce the eumelanins through the enzymatic action (tyrosinase) on tyrosine (El-Naggar and Saber, 2022).

Flavins are another class of yellow pigments which are produced from pteridine compounds. They are characterized by the presence of the N-heterocyclic isoalloxazine ring. *Pseudomonas fluorescens*, *Escherichia coli*, *Streptomyces* spp., and certain archaea are the major flavin producers which are employed for the industrial production of flavins (Pavlovska and Cibulka, 2021). Riboflavin (vitamin B2) and deazaflavin are the major flavins synthesized by microorganisms. Riboflavin is a water-soluble pigment which acts as the precursor for all the biologically significant flavins. Flavin mononucleotide (FMN) and flavin adenine dinucleotide (FAD) are utilized for the synthesis of riboflavin (Cibulka and Fraaije, 2021).

In addition to these microbial pigments, several other key pigments produced by microbes can be utilized in the textile industry. These include chlorophylls, betalains, polyketide pigments, quinones, and azaphenols (Barreto *et al.*, 2023). All these dyes have remarkable properties such as antioxidant and antimicrobial activities as well as the ability to offer natural ultraviolet (UV) protection. These factors, along with the coloring aspect, enhance their applicability. Some of these pigments also have properties such as wound healing, which potentially extend the application of clothes dyed with such colors into medical fields. The high thermal stability of pigments makes them ideal for a wide range of industrial dyeing processes.

12.2.2 Pigment properties (color, stability, fastness) that make them suitable for textiles

12.2.2.1 Color intensity

Microbial pigments have high chromatic intensity due to their complex molecular structure which can absorb visible light across specific wavelengths. A specific group of atoms called chromophores which can absorb light between 400 nm and 700 nm wavelength, along with associated functional groups, allow the microbial pigment to absorb and reflect light to exhibit color. Microorganisms of the same genus can produce pigments with slight shade variation due to genetic, metabolic, and environmental conditioning. This natural variability is an advantage which makes microbes produce a unique palette of color. For instance, *Chryseobacterium* strain can produce yellowish-orange pigment which could be carotenoids, anthraquinones, zeaxanthin, and flexirubin. Differences in the functional group, such as -OH (hydroxyl group), $-NH_2$ (amino group), and -COOH (carboxylic group), among pigments from various strains make them exhibit vibrant colors (Aruldass *et al.*, 2018).

12.2.2.2 Chemical stability

Stability under varying conditions such as pH or temperature is desirable for a pigment used in fabric dyeing. Microbial pigments have been found to have high stability over varying bioprocess conditions which is attributed to the conditions of the microbial strain from which the pigment has been extracted. For example, a purple-blue pigment isolated from *Streptomyces* sp. A1013Y, which was identified as 4,8,13-trihydroxy-6,11-dione-trihydrogranaticin A

(TDTA), is stable from pH 3 to pH 11, between 20 and 100°C, and under UV light. In addition to this, TDTA exhibited free radical scavenging properties which give added benefit to dyeing (Aman Mohammadi *et al.*, 2022).

12.2.2.3 Binding affinity

The primary determinant for the suitability of microbial pigment as a dye is its efficiency in binding to fabric. Interactions including hydrogen bond, covalent bonding, and electrostatic interactions of the pigment's functional group with the fabric fibers make them adhere to the cloth. Binding efficiency can be improved by either changing pH or using auxiliaries like mordants (Kramar and Kostic, 2022).

12.2.2.4 Light and wash fastness

Microbial dyes exhibit significant resistance to light and laundering conditions compared to synthetic dyes which usually leach out when washed or fade due to photodegradation. For instance, fabric stained with pigments extracted from *Serratia marcescens* showed resistance to alkali, acid, and detergent treatment with 100% staining, with only a small amount of discoloration (Agha *et al.*, 2019). Pyocyanin pigment produced from *Pseudomonas aeruginosa* was used to stain cotton and polyester cloth and it was observed that the pigment resisted light to a greater extent with an excellent light fastness property according to International Organization for Standardization (ISO) greyscale standards (Sengupta and Bhowal, 2023).

These properties collectively make microbial dyes a highly suitable and sustainable choice for textile dyeing, offering durability, vibrant coloration, and environmental benefits over conventional synthetic dyes.

12.3 Extraction and Processing of Microbial Pigments for Textiles

12.3.1 Cultivation of pigment-producing microbes for industrial use

As discussed above, microorganisms, including bacteria, molds, yeasts, and fungi, have the ability to produce pigments based on the stimuli to which they are exposed. Fungal genera including *Aspergillus*, *Penicillium*, *Rhodotorula*, and *Monascus* are known producers of biopigments (Meruvu and Dos Santos, 2021). Common pigment-producing bacterial strains include *Streptoverticillium rubrireticuli*, *Pseudomonas magneslorubra*, *Vibrio gazogenes*, *Vibrio psychroerythrus*, *Serratia marcescens*, *Alteromonas rubra*, *Streptomyces longisporus*, *Rugamonas rubra*, and, *S. rubidaea* (Amorim *et al.*, 2022). Yeasts are also used for pigment production due to their unicellular structure and high growth rates. Yeast cells can produce carotenoids depending on the media and process parameters (Rocha Balbino *et al.*, 2023).

Maize steep liquor, which is a by-product of the maize wet-milling industry, is a common substrate for the production of pigments as it is rich in crude nitrogen sources, carbohydrates, salts, and vitamins which promote the growth of microbes (Tahmasebi *et al.*, 2024). Fruit and vegetable wastes (peel, pulp, seeds, and wastewater) can be a nutritious substrate for microbial fermentation (Sharma *et al.*, 2021). The byproducts generated from agricultural products such as bagasse, husks, molasses, bran, and cobs can provide essential nutrients and a surface to anchor for the microbes to produce pigments (Lopes and Ligabue-Braun, 2021). Waste dairy whey can be an excellent source of nitrogen and lactose, which can promote pigment production in microbes (Lopes and Ligabue-Braun, 2021).

For commercial pigment production, the chosen microbe should satisfy parameters such as nonpathogenicity and nontoxicity. These organisms should be able to utilize a wide variety of carbon and nitrogen sources, providing high pigment yields. Finally, organisms are selected based on ease of separation from the fermented media. This approach promotes sustainability and reduces overall waste production from agriculture and industries (Sandybayeva *et al.*, 2022). Coupling pigment production with waste management can reduce the cost of production of pigments, providing easy access to pigments worldwide.

12.3.2 Extraction methods and purification processes

Pigment extraction methods should be selected based on the type of pigment produced, microbial source and the components of the production media (Pagels *et al.*, 2021). Selection is further based on stability preservation of pigments, cost-effectiveness, scalability, and environmental sustainability. Extraction methods include solvent extraction, ultrasonication, mechanical disruption, enzymatic digestion methods, acid/base extraction, supercritical fluid extraction, and adsorption/desorption techniques (De Cesaris *et al.*, 2024). Filtration techniques and centrifugation techniques are often employed in the initial steps of pigment extraction to separate the biomass and microbial cells (Nemer *et al.*, 2021).

The most common pigment extraction method is solvent extraction, which is based on the solubility of microbial pigments in organic solvents such as methanol, acetone, ethanol, and hexane (Molina *et al.*, 2023). Pigments such as carotenoids and violacein are generally extracted through solvent extraction methods. Acid/base extraction relies on pH adjustments to solubilize the pigments (Huang *et al.*, 2025). Generally, pigments such as melanin are extracted via this method, followed by precipitation. Ultrasonication can disintegrate the microbial cells and release the intracellular pigments, which aids in pigment extraction from tough cell walls (Šovljanski *et al.*, 2022). Bead beating, grinding, and high-pressure homogenization are common mechanical disruption methods for tough microbial structures such as fungal mycelia (Jawed *et al.*, 2023). Enzyme digestion can also be used to lyze the cell walls of microbes to release intracellular pigments. Enzymes such as lysozyme are commonly used for microbial pigment extraction. While this is an efficient method of extraction, the enzymes may increase the cost of production (Rajendran *et al.*, 2023). Liquid-liquid extraction is a handy technique to separate out pigments based on their solubility in organic and inorganic liquids. This method is effective in extraction of pigments such as carotenoids (Mukhopadhyay and Banerjee, 2023).

Supercritical fluid extraction can be used to extract high-quality pigments while preserving their stability. This approach utilizes supercritical CO_2 combined with other solvents for efficient extraction. It requires a minimum amount of solvents and causes minimal environmental concerns, but the high cost of equipment poses a challenge for large-scale usage (Abou Elmaaty *et al.*, 2022). The chromatographic techniques such as thin-layer chromatography (TLC) and high-performance liquid chromatography (HPLC) can extract highly purified pigments, but they have the same challenges as supercritical fluid extraction (Wang *et al.*, 2021).

The adsorption/desorption technique is a hybrid method which employs adsorbent materials such as activated charcoal or resins to remove the pigment from the fermentation media. Later these materials are subjected to solvent-based desorption to obtain pure, concentrated pigments. This method allows reuse of the adsorbent materials, which makes it cost-effective (Rápó and Tonk, 2021). Precipitation techniques can also be used for effective pigment extraction. This method involves altering the pH or changing the polarity of the solution by adding salts such as ammonium sulfate (Singh *et al.*, 2021).

12.3.3 Techniques to enhance pigment yield and stability for textile applications

Enhancing microbial pigment yield and stability is crucial for enabling their large-scale applications in industries like textiles. This can be achieved through effective genetic modifications, process and product optimization, and optimized stabilization techniques (Fig. 12.2) (Lyu *et al.*, 2022). Genetic modifications can be achieved by traditional strain improvement methods such as UV and chemical mutagenesis, as well as modern genome editing tools such as CRISPR-Cas9. These techniques can precisely modify the biosynthesis pathway of pigment production, thereby facilitating stable pigment production with higher yields. Genome editing can minimize the production of unwanted secondary metabolites that are otherwise synthesized alongside the desired compound (Trovão *et al.*, 2022). The transfer and expression of pigment production genes into common microbial hosts like *Escherichia coli* or *Saccharomyces cerevisiae* have been successfully demonstrated. These organisms require minimal resources for growth and gene expression, enabling efficient pigment production with reduced contamination (Lyu *et al.*, 2022).

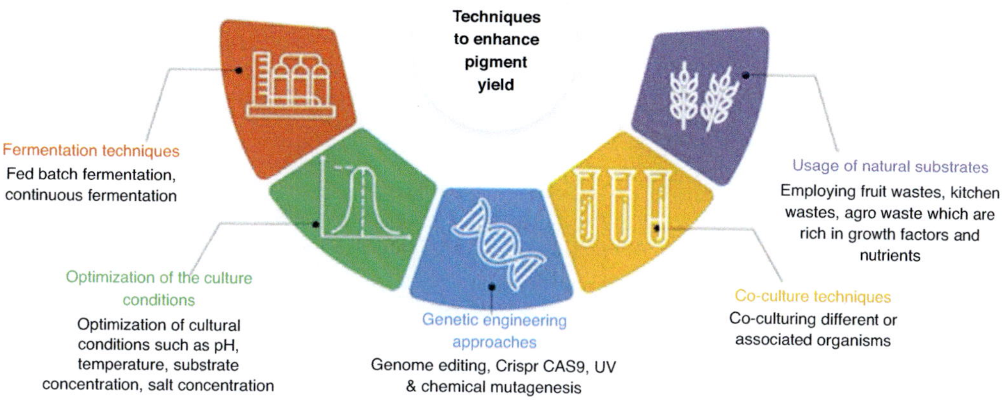

Fig. 12.2. Techniques used to enhance pigment yield.

Optimization of fermentation conditions such as pH, temperature, substrate concentration, aeration, and agitation can achieve the maximum pigment production from the organisms (Yazdian and Mirshafiei, 2024). Studies have shown that altering the carbon-to-nitrogen ratio and supplementing media with precursors of the pigment can significantly increase production (Chaudhuri and Balasubramanian, 2025). The use of naturally occurring substrate tends to produce stable pigments (Umesh *et al.*, 2023). The utilization of agroindustrial and other industrial waste not only promotes sustainability but also enhances pigment production (Davis *et al.*, 2024). Approaches such as solid-state fermentation often tend to increase pigment production in organisms such as fungi (Ismail *et al.*, 2021) and also promotes cost-effective pigment production. Fed-batch fermentation and continuous fermentation can enhance pigment production as they add the nutrients periodically in the media, thereby mitigating the inhibitory effects of excessive nutrient concentrations (Paul *et al.*, 2023). Strategies such as cocultivation of complementary microorganisms increase pigment generation in the target organisms, while coculturing certain microbes tends to produce enzymes and growth factors which can improve the metabolic activity of the pigment producers (Tong *et al.*, 2023).

Environmental factors such as heat, light, and oxygen can affect the stability of microbial pigments. Hence stabilization methods such as encapsulation and blending with stabilizers need to be employed to produce a longer shelf-life of these pigments (Villa *et al.*, 2022). The encapsulation of pigments in polymer matrices produced from materials such as alginate, chitosan, or gelatin can shield the pigments from exposure to environmental factors. Blending the pigments with antioxidants and UV blockers further stabilizes the pigment (Wang *et al.*, 2024a).

These strategies can improve the yield and quality of microbial pigments, additionally promoting cost-effectiveness and stability which are necessary for scaled-up production. These advancements have made microbial pigments a promising alternative to synthetic dyes, aligning with the growing demand for eco-friendly solutions.

12.4 Methods of Dyeing with Microbial Pigments

12.4.1 Methods for applying microbial pigments to fabrics

Several techniques are used to dye textiles with microbial pigments which can be differentiated based on the type of fabric, pigment used, and nature of the final product (Fig. 12.3).

Direct dyeing is the simplest and most widely used method. The fabric is given a dye bath by immersing it in a container with microbial

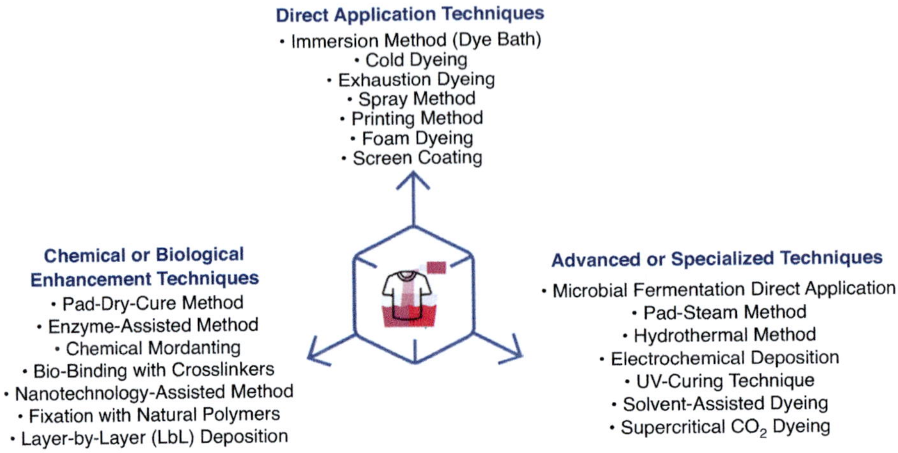

Fig. 12.3. Techniques for applying microbial pigments to fabrics.

pigment. Absorbance of the dye by the fabric is direct, without the use of mordants or fixatives (Aman Mohammadi *et al.*, 2022). This method is suitable for fabrics that possess natural ability to absorb pigments well, such as cotton, silk or wool. The advantages of this method are that it is simple and cost-effective, suitable for natural fibers, and does not require any additional chemicals or mordants. The disadvantages are its lower effectiveness on synthetic fibers and the fixation of the pigment may not be as strong as other methods in which mordants are involved (Liu *et al.*, 2021).

Another popular method is mordanting which uses a chemical agent, generally metal salt, that can improve bonding between the pigment and the fabric. Mordants are chemicals such as alum, iron or copper which are applied to the fabric before dyeing. Biomordanting is the use of natural substances like tannins, alums, and plant extracts as mordants to avoid the use of heavy metals, which reduces health risks for the workers. Some mordants can be obtained from agricultural waste, promoting sustainability. The mordant binds pigments to the fabric by forming complexes that improve its fixation (Table 12.2) (Rather *et al.*, 2023).

Potassium aluminum sulfate (alum) is commonly used to achieve bright colors, especially for anthocyanin pigments. Iron helps in darkening the shades and improving the color fastness of the pigment. Tannins are plant-based mordants that give additional eco-friendly benefits. The limitations of this method include the addition of chemicals for pretreatment, thus increasing the complexity of the process, and some mordants can also be toxic and need to be disposed of carefully. Restriction in color range and time consumption are limitations when using biomordants (Ren *et al.*, 2021).

Enzymatic pretreatment is another method used to enhance uptake of microbial pigments by fabric. Cellulases or pectinases help break down substances in fabric that hinder the absorption of dye, allowing the fabric to better absorb the pigment. One advantage of this method is that it increases dye absorption by opening up the structure of fabric. This method is considered more eco-friendly as it avoids the use of harsh chemicals. It can also be applied to both natural and synthetic fabrics. The limitations of this method lie in careful control of enzyme activity as excess can damage the fabric. Enzymes can be costly and require specific conditions for optimal activity (Sengupta and Bhowal, 2023).

Waterless dyeing is another biological method which uses supercritical carbon dioxide as a medium to attach and fix colors onto fabric. The pressurized carbon dioxide (supercritical state) exerts properties of both liquid and gas to allow the efficient dissolution and penetration of pigments into the fabrics. Advantages of this method include avoidance of water, as the overuse of water is one of the most problematic environmental issues in textile dyeing. Reusability of carbon dioxide is another

Table 12.2. Comparative analysis of eco-friendly dyeing techniques using microbial pigments and traditional synthetic dyeing methods.

Aspect	Eco-friendly dyeing method	Traditional dyeing method	References
Pigment source	Microorganisms	Petroleum-derived chemical	(Ramesh *et al.*, 2022)
Environmental impact	Biodegradable, minimal water pollution, negligible harmful byproducts	Toxic effluents, cause water pollution and are nonbiodegradable	(Saravanan *et al.*, 2021)
Toxicity	Nontoxic and safe for workers and consumers	Highly toxic to workers, consumers and ecosystem	(Dutta *et al.*, 2024)
Energy use	Less energy required	High energy intensive synthesis process	(Singha *et al.*, 2021)
Water use	Eco-friendly mordant and processes reduce water usage	Excessive water consumption in dyeing and printing	(Mondal *et al.*, 2022)
Fixation process	Natural and mild chemical mordants are required	Harsh chemical mordants such as heavy metals and formaldehydes are required	(Pranta and Rahaman, 2024)
Color variability	Limited to natural use	Wide range of vibrant and consistent colors	(Alegbe and Uthman, 2024)
Cost	High initial cost as microbial cultivation and extraction is a tedious process	Low initial cost and widely available materials	(Behera *et al.*, 2021b)
Durability	Good after optimization, requires research for fastness	Color fastness is superior to light, water and washing	(Zhang *et al.*, 2022)
Scalability	Emerging technology as large-scale production is still challenging	Well-established industrial production	(Sharma *et al.*, 2022)
Health hazards	Safe in handling and usage	Risk of allergies, carcinogenicity, and respiratory issues	(Al-Tohamy *et al.*, 2022)
Sustainability	Highly sustainable and renewable	Not sustainable as it relies on finite fossil fuel resources	(Pranta and Rahaman, 2024)
Waste management	Minimal waste and byproducts are often nontoxic and biodegradable	Toxic waste requires costly treatment to manage pollution	(Saravanan *et al.*, 2021)

advantage that helps to promote sustainability (Rajendran *et al.*, 2023). This method is limited because supercritical carbon dioxide is expensive and specialized training is required to manage the technology efficiently. Wastewater recycling using advanced filtration methods such as reverse osmosis and nanofiltration can promote reuse of water to reduce environmental pollution (Kramar and Kostic, 2022).

12.4.2 Factors affecting dye uptake and filtration in textile dyeing

Major factors that influence dye uptake and fixation in fabrics are pH of the dye bath, temperature, fabric type, dye concentration, time of dyeing, mordant type and concentration, ionic strength of the dye bath, and agitation or stirring (Marszałek and Żyłła, 2021).

pH is crucial in ionizing both dye and fabric and has a greater impact on how well the dye molecules attach to the fibers. pH can also affect the stability of the dye molecules as some dyes bind better in acidic conditions whereas others have better binding in alkaline conditions. In mordanting, the pH has a major role in the formation of metal bonds in complexes which improve fixation at optimal pH range (Sharma and Malaviya, 2022).

Another important factor is temperature which can impact the movement of dye molecules into the fabric. Higher temperature usually increases the diffusion rate of dye into the fabric, although extremely high temperatures can degrade heat-sensitive microbial pigments. Thus, maintaining optimal temperature is crucial as moderate heating can provide better dye fixation by facilitating chemical bonding between dye molecules and fabric fibers. The type of fabric is another major factor that can influence the dyeing process (Mcyotto *et al.*, 2021). Natural fabrics like wool, silk, and cotton usually have higher affinity for microbial pigment compared to synthetic fibers like polyester. The presence of hydroxyl groups on natural fibers allows stronger hydrogen bonding with microbial pigments. Therefore, it is essential for synthetic fibers to undergo pretreatment with mordants to improve dye uptake.

The concentration of dye plays a major role as higher concentration can give deep coloration but also increases the risk of uneven dye distribution, while at low concentration, the saturation of dye on fabric may not be adequate, leading to lighter shades. Hence, it is important to maintain a concentration suitable for more dye to interact with fabric, resulting in stronger fixation, and to avoid oversaturation leading to poor fixation and color bleeding (Bai *et al.*, 2023). Duration of dyeing also plays a very important role as longer exposure can produce more uniform and deeper coloration. However, excessively long dyeing periods lead to degradation of sensitive microbial pigments.

Mordant type and concentration also play key roles in dye fixation and uptake. The metal-based mordants such as alum form complexes with microbial pigments, allowing the fabric to absorb the pigment readily; the type and concentration of the mordants vary the intensity of the color absorbed by the fabric. Therefore, controlling the concentration of mordants is crucial to yield the desired fixation.

The ionic strength of the dye bath is the concentration of ions in the solution which can interact between dye and fabric. Excessive ionic strength can destroy the fabric structure, leading to poor fixation and weakening of dye bonds. Agitation is another factor that ensures even uptake of dye throughout the fabric. Gentle agitation is preferred as excessive agitation may damage delicate fabric and cause uneven dyeing (Srivastava *et al.*, 2022).

12.5 Color Fastness and Performance of Microbial Pigments

12.5.1 Assessment of wash, light, and rub fastness of microbial dyes on textiles

The suitability of pigments to be used for dyeing fabrics depends on color fastness, resistance to washing and fading, vibrant coloration over the textile fabric ensuring long-lasting performance, and aesthetic appeal (Table 12.3) (Khattab *et al.*, 2020).

Microbial pigments offer a promising alternative to synthetic textile dyes due to their sustainability and eco-friendliness. Building on this potential, six different microbial pigments were extracted from actinomycetes and bacterial strains, which effectively stained silk, jute, and synthetic fibers, emphasizing their applicability for use in fabric dyeing (Nair *et al.*, 2017). Notably, prodigiosin extracted from *Serratia rubidaea* RAM_Alex showed resistance to color fading when it was exposed to visible and UV light in fabrics like satin and chiffon, while fabrics like linen, baft, gabardine, Dacron, and jersey showed color fading. The reason for this is the breaking of bonds of microbial pigment in the presence of light which was enhanced by the presence of oxygen and humidity in the surroundings. Fading of fabrics dyed with microbial or natural pigment is a commonly encountered problem which could be resolved simply by using a reliable mordant of either synthetic or natural source. When the above fabrics were subjected to a washing test to check color fastness, they were resistant to wash out with little or no residue in the washed water sample (Forster *et al.*, 2017; Metwally *et al.*, 2021).

Out of the many ways to improve color fastness and washing resistance, linseed oil was used as a carrier for a fungal pigment extracted from *Scytalidium cuboideum*. Linseed oil improved the color binding efficiency of the pigment to the sample, especially natural fibers. SEM analysis of the colored fabric showed a layer of linseed oil mixed with pigment over the fabric. Tensile strength and tear strength tests revealed that the dyed fabric showed no change in mechanical properties over stress of any kind. Laundry and bleaching tests showed that once the dyed fabric was bleached or washed, there was a change of shade of color in terms of its intensity

Table 12.3. Comparison of microbial dye with synthetic dye.

Dye source	Color	Name	Durability	Color retention	Resistance to wash	Environmental impact	References
Microbial	Red	Prodigiosin	Maximum dyeing was for wool and least was for cotton	Stable in sunlight. Maximum color fastness was for polyester and minimum for cotton	Resistance to wash in water and detergent	Nontoxic	(Shalinimol, 2019)
Synthetic		Reactive red	Durability of dye in cotton fabric was good	Good color retention	Not much resistance to washing	The effluent released from the washing test was less polluted. However, it consumes salt and alkalis for maintaining color strength	(Irfan *et al.*, 2020)
Microbial	Blue	Pyocyanin	The stain could impart stable coloration to different fabrics like cotton, silk, and wool	Dye absorption to fabric ranged from 50 to 65%	Resistance to wash in water	Nontoxic	(Sengupta and Bhowal, 2023)
Synthetic		Anthraquinone dye derivatives	All compounds of dye derivative exhibited dye coloration even at low dye concentration	Polyester fabric dyed with the dye showed deep coloration	Resistant to washing	Dyeing process involves boiling, pressurization, and depressurization. Comparatively safer for the environment	(Penthala *et al.*, 2019)
Microbial	Yellow	Carotenoids	The stain could impart yellow to orange color to fabrics like cotton and polyester	Cotton and polyester fabric retained color in alkaline solution while cotton fabric alone retained color in acid solution	Samples dyed in alkaline solution demonstrated greater resistance to water	Nontoxic	(Daramola and Awojobi, 2023) (Zohari *et al.*, 2021)
Synthetic		Remazol yellow RR	Maximum durability at pH 12	Maximum color fastness at pH 12	Samples dyed at acidic pH demonstrated greater resistance to water	Toxic to the environment as the procedure involves use of chemicals (acids and alkalis)	(Moula *et al.*, 2022)

or the shade of color itself but no loss of the whole of the color from the fabric was observed. This means that the dyed fabric retained color but change of shade of color was observed. In contrast to this, heating of the fabric resulted in the transformation of color to bright color, creating a waterproof layer over the fabric, maybe because of the linseed oil itself (Palomino Agurto *et al.*, 2020). Hence, the resistance to washing, bleaching, and light fastness underscores the potential of microbial pigments as textile dyes.

12.5.2 Techniques for enhancing fastness of microbial pigments

The wide range of available colors, stability, and versatility across fabrics make microbial pigments adaptable to various needs of textile dyeing. Antimicrobial activity, antioxidant properties, and their sustainable nature are additional properties that enhance their application, but several techniques are proven to be effective in increasing the fastness of microbial pigments (Fig. 12.4).

The use of sustainable biomordants is a technique that enhances microbial pigment adherence to fabrics. L-cysteine (L-Cys), a common biomordant, showed enhanced color fastness when mixed with prodigiosin, extracted from *Serratia plymuthica* (Mouro *et al.*, 2023). Another study reported improved adherence of microbial pigment from *Citricoccus* sp. to wool and cotton by using copper sulfate ($CuSO_4$) as mordant (Ezhil *et al.*, 2022). Optimizing the fermentation process and dyeing parameters can effectively improve dying performance and microbial yield. A study conducted by Naaz *et al.* (2024) showed optimal pigment production by *Penicillium* sp. for silk dyeing, by adjusting fermentation conditions such as pH (pH 5). In addition, immobilization of enzymes in textile processing leads to substantial improvement in fastness properties of microbial pigments. Stabilizing microbial pigments by immobilizing lignin peroxidase in calcium-alginate beads improves wash fastness (Shaheen *et al.*, 2017).

Developments in nanotechnology and biotechnological techniques have provided novel approaches to improving fastness properties of microbial pigments. These techniques focus on resisting microbial degradation to increase the durability of textiles. Chatragadda and Dufossé (2021) explored the dual functionality of prodigiosin and carotenoids to resist microbial degradation, indirectly improving color fastness and color retention. Implementing nanotechnology to form nanomicelles and nanosuspensions opens up the possibility of deeper pigment penetration into fibers. A study conducted by Ren *et al.* (2017) showed that nanomicelles of prodigiosin exhibited improved color penetration, UV resistance and resistance to washing, when dispersed in nanosuspensions.

Surface modification and crosslinking is another technique that showed substantial enhancement in microbial pigment fastness. Crosslinking fluorocarbons and fabric precationization created stronger bonding between microbial pigments and textile fibers (Hassabo *et al.*, 2023). Microbial pigments have been shown to have improved fastness properties when biomordants, fermentation optimization, enzyme immobilization, nanotechnology, and crosslinkers are employed. This provides environmentally friendly and sustainable solutions for the dyeing process of the textile industry.

12.6 Environmental Benefits of Microbial Pigments in Textile Dyeing

12.6.1 Reduction in water pollution and waste production

The textile dyeing industry is one of the most polluting sectors globally, resulting in water pollution, high waste generation, and ecological degradation. Microbial pigments are a promising replacement to synthetic dyes as they are environmentally friendly and produced using sustainable processes (Lara *et al.*, 2022).

The traditional textile dyeing process generates huge quantities of effluents containing toxic chemicals, heavy metals, and nonbiodegradable substances. The contamination of water bodies with this effluent affects human health and aquatic ecosystems. Microbial pigments in textile dyeing have many advantages including reduced effluent toxicity, efficient utilization of resources, and minimized chemical dependency.

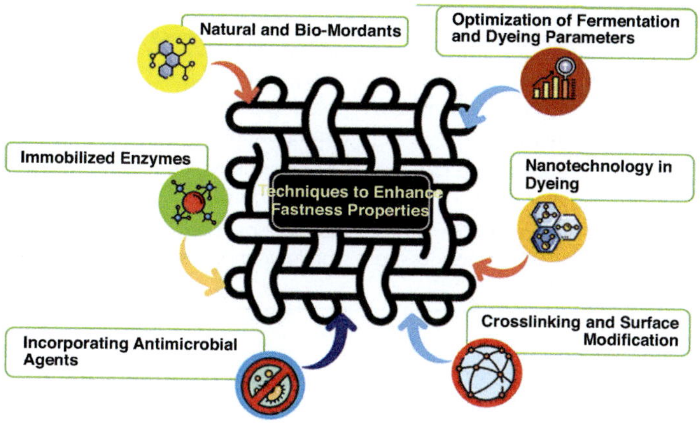

Fig. 12.4. Strategies for improving the color fastness of microbial pigments.

As microbial pigments occur naturally, they are biodegradable (Table 12.4). The effluent produced will be significantly less toxic compared to synthetic dye, thus reducing the release of harmful chemicals into the environment. The synthesis of microbial pigment includes use of organic waste substances such as feedstock, agricultural waste, and byproducts of the food industry (Mouro *et al.*, 2023).

12.6.2 Biodegradability and lower impact of microbial dyes

Microbial pigments are sourced from natural sources like bacteria, fungi, and algae which has a significant impact on sustainability, seamlessly aligning with the principles of green chemistry to promote renewable and eco-friendly practices. Biodegradation is the natural decomposition in the environment due to microbial and enzymatic activity. The biodegradable nature of microbial pigments is an alternative to synthetic dyes which are a persistent cause of pollution by remaining in the ecosystem for extended periods (Tacas *et al.*, 2021). Synthetic dyes accumulate and disrupt the ecosystem while microbial dyes leave no residue in the environment.

One of the major advantages of microbial pigments is their method of degradation which does not pose any risk of bioaccumulation or toxicity. The nontoxic intermediates formed during the degradation of microbial pigments cause no harm to plants, animals or aquatic organisms. Synthetic dyes accumulate in organisms by entering their food chain. The production of microbial dyes is more stable compared to synthetic dyes as it involves fermentation techniques with factors such as low temperature and low energy input, thereby decreasing greenhouse gas emission (Senthil Rathi and Senthil Kumar, 2022). The use of agricultural waste or organic byproducts for the production of microbial pigments reduces reliance on petrochemical-based feedstocks. Thus, the utilization of organic waste and production of biodegradable microbial pigments help promote a circular economy by minimizing waste and enhancing resource efficiency.

12.6.3 Energy efficiency in the microbial dyeing process

The microbial dyeing process has significant advantages in saving energy compared to traditional dyeing methods. The benefits arise from the properties of microbial pigments and the advancements in biotechnological technique which make microbial dye a more suitable and energy-efficient alternative to synthetic dye. Low-temperature dyeing is one of the major advantages of microbial pigment as excellent dyeing outcomes can be achieved at lower temperature, unlike synthetic dyes. Microbial dyeing also decreases carbon footprint as less

Table 12.4. Comparison of synthetic dyes and microbial pigments: key parameters.

Parameter	Synthetic dyes	Microbial pigments	References
Toxicity of effluent	Highly toxic due to the presence of heavy metals and other toxins	Less toxic as biodegradable components are present	(Kramar and Kostic, 2022)
Resource utilization	Petroleum-based raw materials	Feedstock derived from organic waste	(Lara *et al.*, 2022)
Dependency on chemicals	Highly dependable as fixative and mordants are required	Less dependent as it is a biological process	(Mouro *et al.*, 2023)

heating is required and is cost-effective even for large-scale industrial application (Kapoor *et al.*, 2021). The development of efficient fermentation techniques such as use of optimized microbial strains and controlled environmental conditions have enhanced pigment production with minimal energy input. Use of organic waste further reduces the energy expenditure associated with raw material procurement and processing. Innovative reactors and bioprocess automation have ensured energy-efficient production of microbial pigment even on a large scale.

The use of solar energy to power the fermentation heating system improves the dyeing process. Dyeing processes have reduced their reliance on fossil fuels significantly (Rafaqat *et al.*, 2022). Organic waste used in the production of microbial pigment is utilized to generate biogas as an energy source for fermentation and auxiliary operations. By combining multiple renewable energy sources like wind and solar, the microbial dyeing process has enhanced its overall sustainability and reduced energy costs. These techniques not only reduce the environmental impact but also pave the way for greener practices in dyeing industries (Khan *et al.*, 2023).

12.7 Challenges in Using Microbial Pigments in the Textile Industry

In spite of having excellent potential to replace synthetic dyes in the textile industry, commercialization of microbial pigments for textile application still faces major hurdles. Microbial pigments are more expensive than their synthetic counterparts, limiting their availability to small and medium textile industries. Their expected market value is predicted to increase from USD 1.5 billion to USD 2.5 billion by 2025 (Downham and Collins, 2000).

Irrespective of the nature of microbial pigments or their source of extraction, their commercial viability is based on the cost-effectiveness of the pigments extracted. This in turn is affected by multiple factors including the cost of media used for cultivation of pigment-producing microorganisms, type of solvents used for extraction of microbial pigments, requirement of bioreactor design for commercial production of microbial pigments and designing of optimized production processes for commercial textile dyeing (Sen *et al.*, 2019).

The cost of the production media (synthetic media) is a major factor that affects the cost of the final product derived from the pigmented microorganisms. As an alternative to this, cheap agroindustrial residues are used as substrates for commercial production of microbial pigments (Grewal *et al.*, 2022). This strategy in turn has another issue associated with the heterogeneity of composition of the agroindustrial waste and also the changes contributed by seasonal variations. This poses challenges for the optimization of the agro-based fermentation process for microbial pigment production (Rajendran *et al.*, 2023). The yield of microbial pigments produced from a carefully controlled fermentation process is significantly less compared to the synthetic dyes. Dyeing of fabrics using natural pigments demands their use in high quantities to ensure longevity of color and to provide resistance to fading (De Carvalho *et al.*, 2014). The color range of microbial pigments is often limited and their compatibility for mixing, which is important for textile applications, is often restricted. The sensitivity of microbial pigments to temperature, pH, detergents, and other factors that can have a direct effect on their stability also serves as a major obstacle for commercial application (Rather *et al.*, 2023).

Extraction of microbial pigments using a solvent extraction process, although successful at laboratory scale, cannot be applied at commercial scale due to the high volume of solvent required. The release of organic solvents into the environment post extraction of microbial pigments was found to be toxic to the environment and other organisms. Thus this process needs modification to be eco-friendly and economically viable (Rajendran *et al.*, 2023).

Another major concern hindering the commercialization of microbial pigments is associated with the pathogenicity of the producer strain or immunogenicity of certain microbial pigments that makes them unsuitable for the textile industry. Pyomelanin produced from *Vibrio campbellii*, pyocyanin produced from *Pseudomonas aeruginosa*, and violacein produced from *Chromobacterium violaceum*, although found to be highly stable dyes in laboratory conditions, cannot be successful at commercial scale because of the pathogenicity associated with the production strains (Liu and Nizet, 2009).

Another challenge associated with the use of microbial pigments is their limited color fastness. color fastness refers to the ability of the dyed fabric to maintain its color irrespective of the environmental conditions and types of washing employed. Optimal color fastness using microbial pigments depends on the type of the pigment used, nature of the fabrics to be dyed, and the dyeing method employed (Khambhaty and Venugopal, 2024). The commercial value of the fabrics dyed using microbial pigments is determined by their color fastness to light, washing, perspiration, and rubbing. The restrictions associated with the use of chemical mordants due to ecotoxicity further reduce the spectrum of naturally occurring microbial pigments in textile dyeing. Absence of a standard global protocol for dyeing using microbial pigments can also adversely affect the color fastness of the fabrics (Behera *et al.*, 2021a).

The commercial viability and sustainability of the microbial pigment textile dyeing process depend on designing and developing products and processes that ensure a stable color along with zero toxicity/allergic reactions and high durability. This can be achieved by using the right mordants, enzymatic methods, and irradiation methods to enhance the uptake of dye by the fabrics. Furthermore, genetic engineering and strain improvement approaches can be used to produce more pH-tolerant and detergent-tolerant microbial pigments, thereby improving their application in the textile industry (Rather *et al.*, 2023).

12.8 Innovations and Future Prospects in Microbial Textile Dyes

Environmental concerns and approaches by nations to achieve sustainable development have encouraged the textile industry to reduce its reliance on chemical dyes that are hazardous to nature. The application of microbial textile dyes is promising, but this paradigm shift requires further improvements for fully fledged application in the future. Biotechnology and genetic engineering are the spearhead of sustainable development, providing novel approaches and innovations in microbial textile dyes.

Production of novel pigments using genetic engineering and synthetic biology can transform microbial textile dyes in different ways (Fig. 12.5). Engineering synthetic pathways is a method that holds huge potential for creating novel enhanced pigments. Advanced genetic engineering tools like CRISPR-Cas9 are being explored to create these pathways (Alam *et al.*, 2021). Genetic engineering is used to increase production of carotenoids like deinoxanthin in *Deinococcus* bacteria to allow the application of these radiation-resistant pigments in industry (Wang *et al.*, 2024b). Heterologous expression is an effective way to produce pigments. A study by Wehrs *et al.* (2018) reported production of indigoidine in *Saccharomyces cerevisiae* by integrating specific genes from *Streptomyces lavendulae*. Metabolic engineering is another promising method in which the metabolic networks in a microorganism are modified to increase the availability of precursors. The production of violacein, a natural violet pigment in *Chromobacterium violaceum*, is increased by enhancing the availability of tryptophan, which acts as a precursor for the pigment (Hoshino, 2011).

Combinatorial biosynthesis is a robust strategy where the genes from two different species are combined to create new pigments with enhanced properties. Most of the time, the intricacy and duration of the recombinant pathway cause issues since one strain is unable to meet the demands of the system, a condition known as metabolic burden (Jawed *et al.*, 2019). Combinatorial biosynthesis holds huge potential as it can outgrow the limitations posed by metabolic burden by using multiple engineered microbial strains to distribute metabolic burdens. Benefiting from high-throughput transcriptomics, proteomics, and metabolomics, omics-guided engineering can identify and target key pathways for pigment production. To fully comprehend the convoluted metabolic network, a number of systems biology techniques have been created, such as the development and simulation of *in silico* genome-scale metabolic models (GEMs) and multi-omics analysis (Ko *et al.*, 2020). Utilizing omics data in omics-guided engineering is beneficial for optimizing microbial production pathways. The development in omics also paves the way for utilization of biobricks for modular pathway assembly and fine-tuning of gene expression. Biobricks and modular designing is a unique strategy for developing novel synthetic biology tools customized for pathway optimization and combinatorial assembly (Xu *et al.*, 2012).

The role of biotechnology in providing robust, sustainable, and eco-friendly dyeing solutions for the textile industry is huge. These efforts ensure that future developments for widespread adoption of microbial dyes as a key component of the eco-friendly textile industry align with the concept of a circular economy. Enzyme-based innovations in the textile industry can provide economical and energy-efficient solutions. In textile manufacturing processes, enzymes such as amylase, cellulase, catalase, protease, pectinase, laccase, and lipase are frequently employed as eco-friendly substitutes for toxic and dangerous chemicals (Kumar *et al.*, 2023). They also yield high-quality products while using less energy, water, and time. Eco-friendly coloration of textiles using polydopamine (PDA) structural color nanofilm with high color fastness can replace chemical pigments and dyes used in the conventional textile dyeing method, which results in significant environmental contamination due to the massive volume of colored wastewater produced (Fang *et al.*, 2021).

Collaboration among organizations, industry, and academia is imperative for the sustainable transformation of one of the most problematic industries by fostering sustainability-focused innovations, advancing collaborative research, and promoting effective knowledge sharing (Marsh *et al.*, 2022).

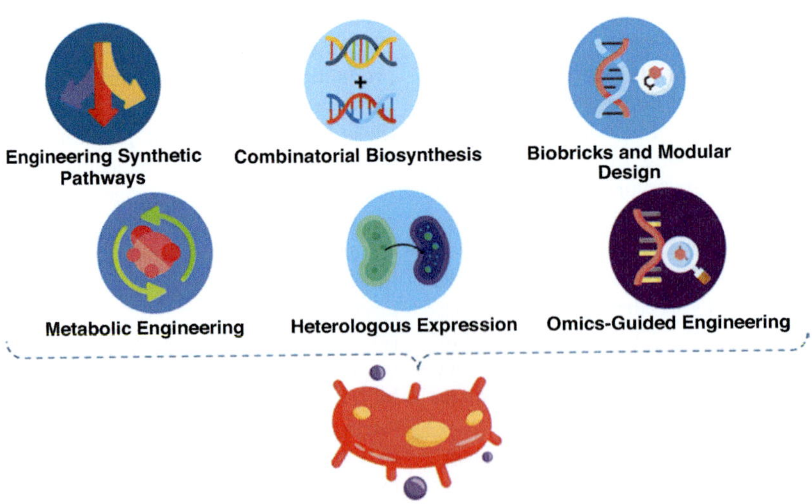

Fig. 12.5. Genetic engineering and synthetic biology for novel pigment production.

References

Abidin, Z.A.Z., Azemi, S.E.F. and Zainuddin, Z. (2023) Diffusible pigment of actinomycete as a source of textile dye. *Revelation and Science* 13(2). DOI: 10.31436/revival.v13i2.370.

Abou Elmaaty, T., Sayed-Ahmed, K., Elsisi, H. and Magdi, M. (2022) Optimization of extraction of natural antimicrobial pigments using supercritical fluids: A review. *Processes* 10(10), 2111. DOI: 10.3390/pr10102111.

Agha, Y.Y., Bahjat, S.A. and Thanoon, M.F. (2019) Assessment of bacterial pigments as textile colorants. *Indian Journal of Public Health Research & Development* 10(12), 1565. DOI: 10.37506/v10/i12/2019/ijphrd/192431.

Alam, K., Hao, J., Zhang, Y. and Li, A. (2021) Synthetic biology-inspired strategies and tools for engineering of microbial natural product biosynthetic pathways. *Biotechnology Advances* 49, 107759. DOI: 10.1016/j.biotechadv.2021.107759.

Alegbe, E.O. and Uthman, T.O. (2024) A review of history, properties, classification, applications and challenges of natural and synthetic dyes. *Heliyon* 10(13), e33646. DOI: 10.1016/j.heliyon.2024.e33646.

Al-Tohamy, R., Ali, S.S., Li, F., Okasha, K.M., Mahmoud, Y.A.-G. *et al.* (2022) A critical review on the treatment of dye-containing wastewater: Ecotoxicological and health concerns of textile dyes and possible remediation approaches for environmental safety. *Ecotoxicology and Environmental Safety* 231, 113160. DOI: 10.1016/j.ecoenv.2021.113160.

Aman Mohammadi, M., Ahangari, H., Mousazadeh, S., Hosseini, S.M. and Dufossé, L. (2022) Microbial pigments as an alternative to synthetic dyes and food additives: A brief review of recent studies. *Bioprocess and Biosystems Engineering* 45(1), 1–12. DOI: 10.1007/s00449-021-02621-8.

Amorim, L.F.A., Fangueiro, R. and Gouveia, I.C. (2022) Characterization of bioactive colored materials produced from bacterial cellulose and bacterial pigments. *Materials* 15(6), 2069. DOI: 10.3390/ma15062069.

Aruldass, C.A., Dufossé, L. and Ahmad, W.A. (2018) Current perspective of yellowish-orange pigments from microorganisms – a review. *Journal of Cleaner Production* 180, 168–182. DOI: 10.1016/j.jclepro.2018.01.093.

Bai, X., Shi, J., Zhang, Z., Hu, M., Chai, J. *et al.* (2023) Efficient persulfate activation by photo-excited organic dyes: Mechanism and application for actual dyeing wastewater self-purification. *Journal of Cleaner Production* 421, 138375. DOI: 10.1016/j.jclepro.2023.138375.

Bano, A., Qadri, T.A. and Khan, N. (2023) Bioactive metabolites of plants and microbes and their role in agricultural sustainability and mitigation of plant stress. *South African Journal of Botany* 159, 98–109. DOI: 10.1016/j.sajb.2023.05.049.

Barreto, J.V. de O., Casanova, L.M., Junior, A.N., Reis-Mansur, M.C.P.P. and Vermelho, A.B. (2023) Microbial pigments: Major groups and industrial applications. *Microorganisms* 11(12), 2920. DOI: 10.3390/microorganisms11122920.

Behera, H.T., Mojumdar, A., Nivedita, S. and Ray, L. (2021a) Microbial pigments: Secondary metabolites with multifaceted roles. In: *Microbial Polymers*. Springer Singapore, Singapore, pp. 631–654.

Behera, M., Nayak, J., Banerjee, S., Chakrabortty, S. and Tripathy, S.K. (2021b) A review on the treatment of textile industry waste effluents towards the development of efficient mitigation strategy: An integrated system design approach. *Journal of Environmental Chemical Engineering* 9(4), 105277. DOI: 10.1016/j.jece.2021.105277.

Bisht, G., Srivastava, S., Kulshreshtha, R., Sourirajan, A., Baumler, D.J. *et al.* (2020) Applications of red pigments from psychrophilic *Rhodonellum psychrophilum* GL8 in health, food and antimicrobial finishes on textiles. *Process Biochemistry* 94, 15–29. DOI: 10.1016/j.procbio.2020.03.021.

Chadni, Z., Rahaman, M.H., Jerin, I., Hoque, K.M.F. and Reza, M.A. (2017) Extraction and optimisation of red pigment production as secondary metabolites from *Talaromyces verruculosus* and its potential use in textile industries. *Mycology* 8(1), 48–57. DOI: 10.1080/21501203.2017.1302013.

Chatragadda, R. and Dufossé, L. (2021) Ecological and biotechnological aspects of pigmented microbes: A way forward in development of food and pharmaceutical grade pigments. *Microorganisms* 9(3). DOI: 10.3390/microorganisms9030637.

Chaudhuri, R. and Balasubramanian, P. (2025) Enhancing nutrient removal from hydroponic effluent with simultaneous production of lipid-rich biomass through mixotrophic cultivation of microalgae. *Journal of Applied Phycology* 37(2), 957–971. DOI: 10.1007/s10811-024-03425-5.

Chen, W., Ye, K., Zhu, X., Zhang, H., Si, R. *et al.* (2021) Actinomycin X2, an antimicrobial *depsipeptide* from marine-derived *Streptomyces cyaneofuscatus* applied as a good natural dye for silk fabric. *Marine Drugs* 20(1), 16. DOI: 10.3390/md20010016.

Cibulka, R. and Fraaije, M.W. (2021) Modes of flavin-based catalysis. In: *Flavin-Based Catalysis*. Wiley, New York, pp. 97–124. DOI: 10.1002/9783527830138.

Daramola, I.V. and Awojobi, K.O. (2023) Bioproduction and application of orange and yellow pigments obtained from *Chryseomicrobium palamuruense* and micrococcus luteus. *Journal of Advances in Microbiology* 23(8), 36–50. DOI: 10.9734/jamb/2023/v23i8742.

Davis, D., Umesh, M., Santhosh, A.S., Suresh, S., Shanmugam, S. *et al.* (2024) Extraction of fungal chitosan by leveraging pineapple peel substrate for sustainable biopolymer production. *Polymers* 16(17), 2455. DOI: 10.3390/polym16172455.

De Carvalho, J.C., Cardoso, L.C., Ghiggi, V., Woiciechowski, A.L., Vandenberghe, L.P.S. *et al.* (2014) Microbial pigments. In: *Biotransformation of Waste Biomass into High Value Biochemicals*. Springer, New York, pp. 73–97.

De Cesaris, M.G., Antonelli, L., Lucci, E., Felli, N., Dal Bosco, C. *et al.* (2024) Current trends to green food sample preparation. A review. *Journal of Chromatography Open* 6(100170), 100170. DOI: 10.1016/j.jcoa.2024.100170.

Devi, M.P., Joshi, V.K., Hemanta, L. and Aresh, J. (2024) Biocolor. In: *Bioreactor Technology in Food Processing*. CRC Press, Boca Raton, FL, pp. 533–566.

Downham, A. and Collins, P. (2000) Colouring our foods in the last and next millennium. *International Journal of Food Science and Technology* 35(1), 5–22. DOI: 10.1046/j.1365-2621.2000.00373.x.

Dutta, S., Adhikary, S., Bhattacharya, S., Roy, D., Chatterjee, S. *et al.* (2024) Contamination of textile dyes in aquatic environment: Adverse impacts on aquatic ecosystem and human health, and its management using bioremediation. *Journal of Environmental Management* 353, 120103. DOI: 10.1016/j.jenvman.2024.120103.

El-Naggar, N.E.-A. and Saber, W.I.A. (2022) Natural melanin: Current trends, and future approaches, with especial reference to microbial source. *Polymers* 14(7), 1339. DOI: 10.3390/polym14071339.

Ezhil, D., Meignanalakshmi, S. and Haritha, B. (2022) Yellow pigment from fast-growing marine soil bacteria *Citricoccus* sp. for dyeing cotton fabrics. *Current Journal of Applied Science and Technology* 24–34. DOI: 10.9734/cjast/2022/v41i413994.

Fang, Y., Liu, X., Zheng, H. and Liu, H. (2021) Eco-friendly colorization of textile originating from polydopamine nanofilm structural color with high colorfastness. *Journal of Cleaner Production* 295, 126523. DOI: 10.1016/j.jclepro.2021.126523.

Forster, A.L., Bitter, J.L., Rosenthal, S., Brooks, S. and Watson, S.S. (2017) Photofading in cotton fibers dyed using red, yellow, and blue direct dyes during examination with microspectrophotometry (MSP). *Forensic Chemistry* 5, 72–78. DOI: 10.1016/j.forc.2017.06.006.

Ghosh, J., Repon, Md., Pranta, A.D., Rupanty, N.S., Khan, F. *et al.* (2025) Bioactive component integrated textiles: A promising source of medicine and healthcare. *Journal of Engineered Fibers and Fabrics* 20. DOI: 10.1177/15589250241308561.

Gnanamani, A., Shanuja, S.K., Giri Dev, V.R. and Gnanamani., A. (2019) Asperyellone – a suitable coloring agent for protein based textile fabrics: An approach on production, characterization and application. *Journal of Textile Engineering & Fashion Technology* 5(2). DOI: 10.15406/jteft.2019.05.00184.

Govindaraj, S., Ugamoorthi, R. and Ramarethinam, S. (2021) Isolation of pseudomonas aeruginosa for bacterial pigment production and its application on synthetic knitted fabric. *Indian Journal of Fibre & Textile Research* 46(2). DOI: 10.56042/ijftr.v46i2.31716.

Grewal, J., Woła Cewicz, M., Pyter, W., Joshi, N., Drewniak, L. *et al.* (2022) Colorful treasure from agro-industrial wastes: A sustainable chassis for microbial pigment production. *Frontiers in Microbiology* 13, 832918. DOI: 10.3389/fmicb.2022.832918.

Hassabo, A., Shaker, S., Abd El-Salam, N., Mohamed, N., Gouda, N. *et al.* (2023) Latexes and their applications in textile industry. *Egyptian Journal of Chemistry* 67, 251–268. DOI: 10.21608/ejchem.2023.216967.8120.

Hernández, V.A., Galleguillos, F., Thibaut, R. and Müller, A. (2019) Fungal dyes for textile applications: Testing of industrial conditions for wool fabrics dyeing. *Journal of the Textile Institute* 110(1), 61–66. DOI: 10.1080/00405000.2018.1460037.

Hoshino, T. (2011) Violacein and related tryptophan metabolites produced by *Chromobacterium violaceum*: Biosynthetic mechanism and pathway for construction of violacein core. *Applied Microbiology and Biotechnology* 91(6), 1463–1475. DOI: 10.1007/s00253-011-3468-z.

Huang, R., Xia, S., Gong, S., Wang, J., Zhang, W. *et al.* (2025) Enhancing sensitivity and stability of natural pigments in pH-responsive freshness indicators: A review. *Food Chemistry* 463(Pt 3), 141357. DOI: 10.1016/j.foodchem.2024.141357.

Huang, X., Gan, L., He, Z., Jiang, G. and He, T. (2024) Bacterial pigments as a promising alternative to synthetic colorants: From fundamentals to applications. *Journal of Microbiology and Biotechnology* 34(11), 2153–2165. DOI: 10.4014/jmb.2404.04018.

Irfan, M., Xie, K. and Hou, A. (2020) Effect of reactive dye structures and substituents on cellulose fabric dyeing. *Fibers and Polymers* 21(9), 2018–2023. DOI: 10.1007/s12221-020-9472-0.

Islam, T., Repon, M.R., Islam, T., Sarwar, Z. and Rahman, M.M. (2023) Impact of textile dyes on health and ecosystem: A review of structure, causes, and potential solutions. *Environmental Science and Pollution Research International* 30(4), 9207–9242. DOI: 10.1007/s11356-022-24398-3.

Jawed, K., Yazdani, S.S. and Koffas, M.A. (2019) Advances in the development and application of microbial consortia for metabolic engineering. *Metabolic Engineering Communications* 9, e00095. DOI: 10.1016/j.mec.2019.e00095.

Ismail, G.A., Fitriana, A.D. and Sukandar, U. (2021) Medium optimization for production of *Monascus* purpureus pigment through solid-state fermentation. *IOP Conference Series* 1143(1), 012019. DOI: 10.1088/1757-899X/1143/1/012019.

Jawed, A., Bhatt, A.K., Bhatia, R.K. and Bhalla, T.C. (eds) (2023) Biomolecules: Types, homogenization, bead beater, sonication. In: *Basic Biotechniques for Bioprocess and Bioentrepreneurship*. Elsevier, pp. 125–131. DOI: 10.1016/B978-0-12-816109-8.00006-4.

Kanelli, M., Mandic, M., Kalakona, M., Vasilakos, S., Kekos, D. *et al.* (2018) Microbial production of violacein and process optimization for dyeing polyamide fabrics with acquired antimicrobial properties. *Frontiers in Microbiology* 9, 1495. DOI: 10.3389/fmicb.2018.01495.

Kapoor, R.T., Danish, M., Singh, R.S., Rafatullah, M. and Abdul Khalil, H.P.S. (2021) Exploiting microbial biomass in treating azo dyes contaminated wastewater: Mechanism of degradation and factors affecting microbial efficiency. *Journal of Water Process Engineering* 43, 102255. DOI: 10.1016/j.jwpe.2021.102255.

Khambhaty, Y. and Venugopal, S. (2024) Embracing bacterial pigments for eco-friendly leather coloring. *Clean Technologies and Environmental Policy* 1–12. DOI: 10.1007/s10098-024-03046-z.

Khan, M.D., Singh, A., Khan, M.Z., Tabraiz, S. and Sheikh, J. (2023) Current perspectives, recent advancements, and efficiencies of various dye-containing wastewater treatment technologies. *Journal of Water Process Engineering* 53, 103579. DOI: 10.1016/j.jwpe.2023.103579.

Khattab, T.A., Abdelrahman, M.S. and Rehan, M. (2020) Textile dyeing industry: Environmental impacts and remediation. *Environmental Science and Pollution Research International* 27(4), 3803–3818. DOI: 10.1007/s11356-019-07137-z.

Ko, Y.-S., Kim, J.W., Lee, J.A., Han, T., Kim, G.B. *et al.* (2020) Tools and strategies of systems metabolic engineering for the development of microbial cell factories for chemical production. *Chemical Society Reviews* 49(14), 4615–4636. DOI: 10.1039/d0cs00155d.

Kramar, A. and Kostic, M.M. (2022) Bacterial secondary metabolites as biopigments for textile dyeing. *Textiles* 2(2), 252–264. DOI: 10.3390/textiles2020013.

Kramar, A.D., Ilic-Tomic, T.R., Lađarević, J.M., Nikodinovic-Runic, J.B. and Kostic, M.M. (2021) Halochromic cellulose textile obtained via dyeing with biocolorant isolated from *Streptomyces* sp. strain NP4. *Cellulose* 28(13), 8771–8784. DOI: 10.1007/s10570-021-04071-7.

Kumar, D., Bhardwaj, R., Jassal, S., Goyal, T., Khullar, A. *et al.* (2023) Application of enzymes for an eco-friendly approach to textile processing. *Environmental Science and Pollution Research* 30(28), 71838–71848. DOI: 10.1007/s11356-021-16764-4.

Kumar, P., Banik, S.P., Ohia, S.E., Moriyama, H., Chakraborty, S. *et al.* (2024) Current insights on the photoprotective mechanism of the macular carotenoids, lutein and zeaxanthin: Safety, efficacy and bio-delivery. *Journal of the American Nutrition Association* 43(6), 505–518. DOI: 10.1080/27697061.2024.2319090.

Lara, L., Cabral, I. and Cunha, J. (2022) Ecological approaches to textile dyeing: A review. *Sustainability* 14, 8353. DOI: 10.3390/su14148353.

Li, S., Cunningham, A.B., Shi, Y., Qiu, Z., Hartl, A. *et al.* (2022) Blue to black: Hypotheses on plant use complexity in traditional dyeing processes in southeast Asia and China. *Industrial Crops and Products* 188, 115706. DOI: 10.1016/j.indcrop.2022.115706.

Liu, G.Y. and Nizet, V. (2009) Color me bad: Microbial pigments as virulence factors. *Trends in Microbiology* 17(9), 406–413. DOI: 10.1016/j.tim.2009.06.006.

Liu, Y., Yu, Z., Zhang, Y. and Wang, H. (2021) Microbial dyeing for inoculation and pigment used in wood processing: Opportunities and challenges. *Dyes and Pigments* 186, 109021. DOI: 10.1016/j.dyepig.2020.109021.

Lopes, F.C. and Ligabue-Braun, R. (2021) Agro-industrial residues: Eco-friendly and inexpensive substrates for microbial pigments production. *Frontiers in Sustainable Food Systems* 5. DOI: 10.3389/fsufs.2021.589414.

Lyu, X., Lyu, Y., Yu, H., Chen, W., Ye, L. *et al.* (2022) Biotechnological advances for improving natural pigment production: A state-of-the-art review. *Bioresources and Bioprocessing* 9(1), 8. DOI: 10.1186/s40643-022-00497-4.

Marsh, J., Boszhard, I., Contargyris, A., Cullen, J., Junge, K. *et al.* (2022) A value-driven business ecosystem for industrial transformation: The case of the EU's H2020 "Textile and clothing business labs. *Sustainability: Science Practice and Policy* 18(1), 263–277. DOI: 10.1080/15487733.2022.2039491.

Marszałek, J. and Żyłła, R. (2021) Recovery of water from textile dyeing using membrane filtration processes. *Processes* 9(10), 1833. DOI: 10.3390/pr9101833.

Mcyotto, F., Wei, Q., Macharia, D.K., Huang, M., Shen, C. *et al.* (2021) Effect of dye structure on color removal efficiency by coagulation. *Chemical Engineering Journal* 405, 126674. DOI: 10.1016/j.cej.2020.126674.

Meruvu, H. and Dos Santos, J.C. (2021) Colors of life: A review on fungal pigments. *Critical Reviews in Biotechnology* 41(8), 1153–1177. DOI: 10.1080/07388551.2021.1901647.

Metwally, R.A., El Sikaily, A., El-Sersy, N.A., Ghozlan, H.A. and Sabry, S.A. (2021) Antimicrobial activity of textile fabrics dyed with prodigiosin pigment extracted from marine *Serratia rubidaea* ramalex bacteria. *Egyptian Journal of Aquatic Research* 47(3), 301–305. DOI: 10.1016/j.ejar.2021.05.004.

Mindt, M., Ferrer, L., Bosch, D., Cankar, K. and Wendisch, V.F. (2023) De novo tryptophanase-based indole production by metabolically engineered *Corynebacterium glutamicum*. *Applied Microbiology and Biotechnology* 107(5–6), 1621–1634. DOI: 10.1007/s00253-023-12397-4.

Molina, A.K., Corrêa, R.C.G., Prieto, M.A., Pereira, C. and Barros, L. (2023) Bioactive natural pigments' extraction, isolation, and stability in food applications. *Molecules* 28(3), 1200. DOI: 10.3390/molecules28031200.

Mondal, B.V., Kabir, S.M.F., Ali, A. and Hannan, Md. (2022) Sustainable natural printing of cotton fabric without metal-based mordant. *Journal of Natural Fibers* 19(17), 15327–15342. DOI: 10.1080/15440478.2022.2124215.

Moula, A.T.M.G., Hosen, M.D., Siddiquee, M.A.B., Momin, M.A., Kaisar, Z. *et al.* (2022) Effect of dye bath ph in dyeing of cotton knitted fabric with reactive dye (Remazol yellow RR) in exhaust method: Impact on color strength, chromatic values and fastness properties. *Heliyon* 8(11), e11246. DOI: 10.1016/j.heliyon.2022.e11246.

Mouro, C., Gomes, A.P., Costa, R.V., Moghtader, F. and Gouveia, I.C. (2023) The sustainable bioactive dyeing of textiles: A novel strategy using bacterial pigments, natural antibacterial ingredients, and deep eutectic solvents. *Gels* 9(10), 800. DOI: 10.3390/gels9100800.

Muhammad, M., Basit, A., Majeed, M.Z., Shah, A.A., Ullah, I. *et al.* (2024) Bacterial pigments and their applications. Elsevier, St Louis, MO, pp. 277–298. DOI: 10.1016/B978-0-323-95251-4.00019-3.

Mukhopadhyay, C. and Banerjee, B. (eds) (2023) *Organic Synthesis, Natural Products Isolation, Drug Design, Industry and the Environment*. De Gruyter, Berlin. DOI: 10.1515/9783111243993.

Naaz, S., Gupta, C. and Aggarwal, S. (2024) Optimizing pigment production from Penicillium sp. for sustainable silk dyeing. *ShodhKosh: Journal of Visual and Performing Arts* 5(4). DOI: 10.29121/shodhkosh.v5.i4.2024.2093.

Nair, A.S., Kumar, B.P. and Geo, J.A. (2017) Microbial production of textile grade pigments. *African Journal of Microbiology Research* 11(42), 1532–1537. DOI: 10.5897/AJMR2017.8205.

Nemer, G., Louka, N., Vorobiev, E., Salameh, D., Nicaud, J.-M. *et al.* (2021) Mechanical cell disruption technologies for the extraction of dyes and pigments from microorganisms: A review. *Fermentation* 7(1), 36. DOI: 10.3390/fermentation7010036.

Olas, B., Białecki, J., Urbańska, K. and Bryś, M. (2021) The effects of natural and synthetic blue dyes on human health: A review of current knowledge and therapeutic perspectives. *Advances in Nutrition* 12(6), 2301–2311. DOI: 10.1093/advances/nmab081.

Orlandi, V.T., Martegani, E., Giaroni, C., Baj, A. and Bolognese, F. (2022) Bacterial pigments: A colorful palette reservoir for biotechnological applications. *Biotechnology and Applied Biochemistry* 69(3), 981–1001. DOI: 10.1002/bab.2170.

Pagels, F., Pereira, R.N., Vicente, A.A. and Guedes, A.C. (2021) Extraction of pigments from microalgae and cyanobacteria—A review on current methodologies. *Applied Sciences* 11(11), 5187. DOI: 10.3390/app11115187.

Palomino Agurto, M.E., Vega Gutierrez, S.M., Court, R.C., Chen, H.-L. and Robinson, S.C. (2020) Oil-based fungal pigment from *Scytalidium cuboideum* as a textile dye. *Journal of Fungi* 6(2), 53. DOI: 10.3390/jof6020053.

Park, H., Park, S., Yang, Y.-H. and Choi, K.-Y. (2021) Microbial synthesis of violacein pigment and its potential applications. *Critical Reviews in Biotechnology* 41(6), 879–901. DOI: 10.1080/07388551.2021.1892579.

Parthasarathy, K. and Manikkam, R. (eds) (2024) *Translational Research in Biomedical Sciences: Recent Progress and Future Prospects*. Springer, Singapore. DOI: 10.1007/978-981-97-1777-4.

Paul, D., Kumari, P.K. and Siddiqui, N. (2023) Yeast carotenoids: Cost-effective fermentation strategies for health care applications. *Fermentation* 9(2), 147. DOI: 10.3390/fermentation9020147.

Pavlovska, T. and Cibulka, R. (2021) Structure and properties of flavins. In: *Flavin-based Catalysis*. Wiley, Chichester, UK, pp. 1–27. DOI: 10.1002/9783527830138.

Penthala, R., Kumar, R.S., Heo, G., Kim, H., Lee, I.Y. *et al.* (2019) Synthesis and efficient dyeing of anthraquinone derivatives on polyester fabric with supercritical carbon dioxide. *Dyes and Pigments* 166, 330–339. DOI: 10.1016/j.dyepig.2019.03.027.

Pranta, A.D. and Rahaman, M.T. (2024) Extraction of eco-friendly natural dyes and biomordants for textile coloration: A critical review. *Nano-Structures & Nano-Objects* 39, 101243. DOI: 10.1016/j.nanoso.2024.101243.

Rafaqat, S., Ali, N., Torres, C. and Rittmann, B. (2022) Recent progress in treatment of dyes wastewater using microbial-electro-fenton technology. *RSC Advances* 12(27), 17104–17137. DOI: 10.1039/d2ra01831d.

Rajendran, P., Somasundaram, P. and Dufossé, L. (2023) Microbial pigments: Eco-friendly extraction techniques and some industrial applications. *Journal of Molecular Structure* 1290, 135958. DOI: 10.1016/j.molstruc.2023.135958.

Ramesh, C., Prasastha, V.R., Venkatachalam, M. and Dufossé, L. (2022) Natural substrates and culture conditions to produce pigments from potential microbes in submerged fermentation. *Fermentation* 8(9), 460. DOI: 10.3390/fermentation8090460.

Rápó, E. and Tonk, S. (2021) Factors affecting synthetic dye adsorption; Desorption studies: A review of results from the last five years (2017–2021). *Molecules* 26(17), 5419. DOI: 10.3390/molecules26175419.

Rather, L.J., Mir, S.S., Ali, S. and Salam, S. (2023) Research progress, challenges, and perspectives in microbial pigment production for industrial applications – A review. *Dyes and Pigments* 210, 110989. DOI: 10.1016/j.dyepig.2022.110989.

Ren, Y., Gong, J., Fu, R., Li, Z., Yu, Z. *et al.* (2017) Dyeing and functional properties of polyester fabric dyed with *Prodigiosins nanomicelles* produced by microbial fermentation. *Journal of Cleaner Production* 148, 375–385. DOI: 10.1016/j.jclepro.2017.01.168.

Ren, Y., Fu, R., Fang, K., Xie, R., Hao, L. *et al.* (2021) Clean dyeing of acrylic fabric by sustainable red bacterial pigment based on nano-suspension system. *Journal of Cleaner Production* 281, 125295. DOI: 10.1016/j.jclepro.2020.125295.

Repon, M.R., Islam, T., Islam, T., Ghorab, A.E. and Rahman, M.M. (2023) Cleaner pathway for developing bioactive textile materials using natural dyes: A review. *Environmental Science and Pollution Research International* 30(17), 48793–48823. DOI: 10.1007/s11356-023-26131-0.

Rocha Balbino, T., Sánchez-Muñoz, S., Díaz-Ruíz, E., Moura Rocha, T., Mier-Alba, E. *et al.* (2023) Lignocellulosic biorefineries as a platform for the production of high-value yeast derived pigments – A review. *Bioresource Technology* 386, 129549. DOI: 10.1016/j.biortech.2023.129549.

Sandybayeva, S.K., Kossalbayev, B.D., Zayadan, B.K., Sadvakasova, A.K., Bolatkhan, K. *et al.* (2022) Prospects of cyanobacterial pigment production: Biotechnological potential and optimization strategies. *Biochemical Engineering Journal* 187, 108640. DOI: 10.1016/j.bej.2022.108640.

Saravanan, A., Senthil Kumar, P., Jeevanantham, S., Karishma, S., Tajsabreen, B. *et al.* (2021) Effective water/wastewater treatment methodologies for toxic pollutants removal: Processes and applications towards sustainable development. *Chemosphere* 280, 130595. DOI: 10.1016/j.chemosphere.2021.130595.

Sen, T., Barrow, C.J. and Deshmukh, S.K. (2019) Microbial pigments in the food industry-challenges and the way forward. *Frontiers in Nutrition* 6, 7. DOI: 10.3389/fnut.2019.00007.

Sengupta, S. and Bhowal, J. (2023) Characterization of a blue-green pigment extracted from pseudomonas aeruginosa and its application in textile and paper dyeing. *Environmental Science and Pollution Research* 30(11), 30343–30357. DOI: 10.1007/s11356-022-24241-9.

Senthil Rathi, B. and Senthil Kumar, P. (2022) Sustainable approach on the biodegradation of azo dyes: A short review. *Current Opinion in Green and Sustainable Chemistry* 33, 100578. DOI: 10.1016/j.cogsc.2021.100578.

Shaheen, R., Asgher, M., Hussain, F. and Bhatti, H.N. (2017) Immobilized lignin peroxidase from *Ganoderma lucidum* IBL-05 with improved dye decolorization and cytotoxicity reduction properties. *International Journal of Biological Macromolecules* 103, 57–64. DOI: 10.1016/j.ijbiomac.2017.04.040.

Shalinimol, C.R. (2019) Extraction of pink pigment (Prodigiosin) from *Serratia marcescens* MTCC 10774 bacteria and their application in industries. *Think India Journal* 22(4), 4272–4282.

Sharma, R. and Malaviya, P. (2022) Constructed wetlands for textile wastewater remediation: A review on concept, pollutant removal mechanisms, and integrated technologies for efficiency enhancement. *Chemosphere* 290, 133358. DOI: 10.1016/j.chemosphere.2021.133358.

Sharma, B., Kumari, N., Mathur, S. and Sharma, V. (2022) A systematic review on iron-based nanoparticle-mediated clean-up of textile dyes: Challenges and prospects of scale-up technologies. *Environmental Science and Pollution Research International* 29(1), 312–331. DOI: 10.1007/s11356-021-16846-3.

Sharma, M., Usmani, Z., Gupta, V.K. and Bhat, R. (2021) Valorization of fruits and vegetable wastes and by-products to produce natural pigments. *Critical Reviews in Biotechnology* 41(4), 535–563. DOI: 10.1080/07388551.2021.1873240.

Singh, S., Nimse, S.B., Mathew, D.E., Dhimmar, A., Sahastrabudhe, H. *et al.* (2021) Microbial melanin: Recent advances in biosynthesis, extraction, characterization, and applications. *Biotechnology Advances* 53, 107773. DOI: 10.1016/j.biotechadv.2021.107773.

Singha, K., Pandit, P., Maity, S. and Ahmed, S. (2021) Harmful environmental effects for textile chemical dyeing practice. In: *Green Chemistry for Sustainable Textiles*. Elsevier, pp. 153–164.

Šovljanski, O., Saveljić, A., Tomić, A., Šeregelj, V., Lončar, B. *et al.* (2022) Carotenoid-producing yeasts: Selection of the best-performing strain and the total carotenoid extraction procedure. *Processes* 10(9), 1699. DOI: 10.3390/pr10091699.

Srilekha, V., Krishna, G., Sreelatha, B., Jagadeesh Kumar, E. and Rajeshwari, K.V.N. (2024) Prodigiosin: A fascinating and the most versatile bioactive pigment with diverse applications. *Systems Microbiology and Biomanufacturing* 4(1), 66–76. DOI: 10.1007/s43393-023-00192-1.

Srivastava, A., Rani, R.M., Patle, D.S. and Kumar, S. (2022) Emerging bioremediation technologies for the treatment of textile wastewater containing synthetic dyes: A comprehensive review. *Journal of Chemical Technology & Biotechnology* 97(1), 26–41. DOI: 10.1002/jctb.6891.

Suwannarach, N., Kumla, J., Nishizaki, Y., Sugimoto, N., Meerak, J. *et al.* (2019) Optimization and characterization of red pigment production from an endophytic fungus, *Nigrospora aurantiaca* CMU-ZY2045, and its potential source of natural dye for use in textile dyeing. *Applied Microbiology and Biotechnology* 103(17), 6973–6987. DOI: 10.1007/s00253-019-09926-5.

Tacas, A.C.J., Tsai, P.-W., Tayo, L.L., Hsueh, C.-C., Sun, S.-Y. *et al.* (2021) Degradation and biotoxicity of azo dyes using indigenous bacteria-acclimated microbial fuel cells (MFCs). *Process Biochemistry* 102, 59–71. DOI: 10.1016/j.procbio.2020.12.003.

Tahmasebi, Z., Zilouei, H. and Kot, A.M. (2024) Investigating the concomitant production of carotenoids and lipids by the yeast *Rhodosporidium babjevae* using sugarcane bagasse hydrolysate or corn steep liquor. *Renewable Energy* 228, 120618. DOI: 10.1016/j.renene.2024.120618.

Tang, Q., Li, Z., Chen, N., Luo, X., Zhao, Q. *et al.* (2025) Natural pigments derived from plants and microorganisms: Classification, biosynthesis, and applications. *Plant Biotechnology Journal* 23(2), 592–614. DOI: 10.1111/pbi.14522.

Tong, C.Y., Honda, K. and Derek, C.J.C. (2023) A review on microalgal-bacterial co-culture: The multifaceted role of beneficial bacteria towards enhancement of microalgal metabolite production. *Environmental Research* 228, 115872. DOI: 10.1016/j.envres.2023.115872.

Trovão, M., Schüler, L.M., Machado, A., Bombo, G., Navalho, S. *et al.* (2022) Random mutagenesis as a promising tool for microalgal strain improvement towards industrial production. *Marine Drugs* 20(7), 440. DOI: 10.3390/md20070440.

Umesh, M., Suresh, S., Santosh, A.S., Prasad, S., Chinnathambi, A. *et al.* (2023) Valorization of pineapple peel waste for fungal pigment production using *Talaromyces albobiverticillius*: Insights into antibacterial, antioxidant and textile dyeing properties. *Environmental Research* 229, 115973. DOI: 10.1016/j.envres.2023.115973.

Urtgam, S., Sumpradit, T. and Thurnkul, N. (2023) The antibacterial activity and silk dyeing of the crude pigment extract from J4 actinobacteria strain. *Journal of Current Science and Technology* 13(2), 455–464. DOI: 10.59796/jcst.V13N2.2023.262.

Venil, C.K., Zakaria, Z.A. and Ahmad, W.A. (2013) Bacterial pigments and their applications. *Process Biochemistry* 48(7), 1065–1079. DOI: 10.1016/j.procbio.2013.06.006.

Venil, C.K., Dufossé, L., Velmurugan, P., Malathi, M. and Lakshmanaperumalsamy, P. (2021) Extraction and application of pigment from serratia marcescens SB08, an insect enteric gut bacterium, for textile dyeing. *Textiles* 1(1), 21–36. DOI: 10.3390/textiles1010003.

Villa, F., Wu, Y.-L., Zerboni, A. and Cappitelli, F. (2022) In living color: Pigment-based microbial ecology at the mineral-air interface. *Bioscience* 72(12), 1156–1175. DOI: 10.1093/biosci/biac091.

Wang, M., Zhang, Y., Wang, R., Wang, Z., Yang, B. *et al*. (2021) An evolving technology that integrates classical methods with continuous technological developments: Thin-layer chromatography bioautography 26(15), 4647. DOI: 10.3390/molecules26154647.

Wang, Y., Liu, J., Yi, Y., Zhu, L., Liu, M. *et al*. (2024a) Insights into the synthesis, engineering, and functions of microbial pigments in *Deinococcus bacteria*. *Frontiers in Microbiology* 15, 1447785. DOI: 10.3389/fmicb.2024.1447785.

Wang, Y., Zhong, J., McClements, D.J., Zhang, Z., Zhang, R. *et al*. (2024b) Stability enhancement methods for natural pigments in intelligent packaging: A review. *Critical Reviews in Food Science and Nutrition* 1–16. DOI: 10.1080/10408398.2024.2437570.

Wehrs, M., Prahl, J.-P., Moon, J., Li, Y., Tanjore, D. *et al*. (2018) Production efficiency of the bacterial non-ribosomal peptide *indigoidine* relies on the respiratory metabolic state in s. cerevisiae. *Microbial Cell Factories* 17(1), 193. DOI: 10.1186/s12934-018-1045-1.

Xu, P., Vansiri, A., Bhan, N. and Koffas, M.A.G. (2012) EPathBrick: A synthetic biology platform for engineering metabolic pathways in E. coli. *ACS Synthetic Biology* 1(7), 256–266. DOI: 10.1021/sb300016b.

Yazdian, F. and Mirshafiei, M. (2024) *Bioreactor Technology, Fermentation Strategy, and Statistical Approaches for Pigment Production by Microorganisms*. CRC Press, Boca Raton, FL, pp. 444–464.

Zhang, Y., ul-Islam, S., Rather, L.J. and Li, Q. (2022) Recent advances in the surface modification strategies to improve functional finishing of cotton with natural colourants – a review. *Journal of Cleaner Production* 335, 130313. DOI: 10.1016/j.jclepro.2021.130313.

Zohari, M., Sepahy, A.A. and Amini, K. (2021) Molecular cloning and anti-cancer activity of carotenoid pigments isolated from micrococcus spp. and rhodotorula spp. *Journal of Sciences* 32(1), 29–37. DOI: 10.22059/jsciences.2020.297577.1007499.

13 Antioxidant and Antimicrobial Properties of Microbial Pigments

Hemanta Kumar Mondal[1*#], Mohd Tariq[2,3], Apeksha Sawant[1#], Ramandeep Saini[4#] and Ayush Madan[5]

[1]Department of Life Sciences, Parul Institute of Applied Sciences, Parul University, Vadodara 391760, Gujarat, India; [2]Department of Academics, Sumandeep Vidyapeeth, Deemed to be University, Vadodara-391760, Gujarat, India; [3]Department of Biotechnology, Graphic Era (Deemed to be University), Dehradun, Uttarakhand-248002, India; [4]Department of Biotechnology, Chandigarh College of Technology, Chandigarh Group of Colleges, Landran (140307), Greater Mohali, Punjab, India; [5]Department of Biotechnology, School of Research & Technology (SORT), People's University, Bhopal, Madhya Pradesh, India

13.1 Introduction

Microbial pigments constitute a structurally diverse group of naturally occurring biochromes synthesized by bacteria, fungi, and microalgae (Rather *et al.*, 2025). These pigments encompass a wide spectrum of chemical classes, including carotenoids, phenazines, melanins, quinones, and tetrapyrroles, each possessing distinct chromophoric groups responsible for their characteristic coloration and bioactivity profiles. Their structural diversity is closely linked to their biosynthetic origins, with pigments arising from various primary and secondary metabolic pathways such as polyketide, terpenoid, and shikimate-derived routes (Bills and Gloer, 2016). Beyond imparting color, microbial pigments are increasingly recognized for their potent biological functions, notably antioxidant and antimicrobial activities, positioning them as valuable bioactive agents in multiple industrial sectors.

13.1.1 Ecological significance and natural roles

In natural ecosystems, microbial pigments serve critical ecological functions, often acting as protective molecules that enable microorganisms to survive under adverse environmental conditions (Villa *et al.*, 2022). Many pigments provide photoprotection by absorbing harmful ultraviolet (UV) radiation and mitigating photooxidative damage. Others function as redox mediators, facilitating electron transport or neutralizing reactive oxygen species (ROS) generated during metabolic stress (Singh *et al.*, 2023). Certain pigments, such as violacein and phenazines, exhibit antimicrobial activity that contributes to microbial competition, thereby influencing community structure and ecological balance. Furthermore, pigment production can be triggered as adaptive metabolites in microbial survival strategies (Zandi and Schnug, 2022).

*Corresponding author: hemanta.mondal36265@paruluniversity.ac.in

#These authors have contributed equally to this work and share first authorship.

 Biopigments of Microbial Origin: Innovations and Applications Across Industries (eds Madan *et al.*)
DOI: 10.1079/ 9781836990505.0013

13.1.2 Industrial and economic relevance

The global pigment market is witnessing an increasing shift towards naturally derived colorants due to concerns over the environmental toxicity and health risks associated with synthetic dyes (Dave *et al.*, 2022). Microbial pigments present significant advantages over plant- or animal-derived pigments, including rapid growth rates, independence from seasonal variation, and the ability to utilize inexpensive substrates, including agro-industrial wastes, as culture media (Nath, 2025). These features translate into high scalability potential for industrial production under controlled fermentation conditions, ensuring consistent pigment yield and quality.

Industries such as food, pharmaceuticals, cosmetics, and textiles are integrating microbial pigments not only for their coloration properties but also for their functional bioactivities. Carotenoids such as astaxanthin are incorporated in nutraceuticals for their antioxidant potential, while prodigiosin and violacein are investigated for their antimicrobial and anticancer properties (Negi *et al.*, 2025). In cosmetics, pigments with photoprotective and free radical-scavenging properties are used in antiaging formulations, and in food processing, pigments with antimicrobial potential serve as natural preservatives.

13.1.3 Bioactive importance in healthcare, food, and cosmetics

The antioxidant activities of microbial pigments stem from their ability to neutralize free radicals, chelate prooxidant metals, and inhibit lipid peroxidation, thereby mitigating oxidative stress implicated in chronic diseases such as cancer, neurodegenerative disorders, and cardiovascular pathologies (Rani *et al.*, 2021). For example, β-carotene exhibits singlet oxygen quenching, while melanin scavenges hydroxyl radicals and peroxynitrite. Similarly, pigments such as phycocyanin from cyanobacteria demonstrate both antioxidant and antiinflammatory activities, making them promising candidates for therapeutic applications.

From an antimicrobial perspective, pigments like prodigiosin disrupt bacterial membrane integrity, violacein induces apoptosis-like death in protozoa and certain fungi, and phenazine derivatives inhibit biofilm formation in pathogenic bacteria (Guryanov and Naumenko, 2024). The dual antioxidant–antimicrobial nature of some pigments offers the possibility of multifunctional therapeutic agents, especially relevant in addressing oxidative stress-associated infections or conditions involving immune compromise.

In the food industry, antioxidant pigments contribute to oxidative stability, extending product shelf-life while also imparting natural coloration. Pigments with antimicrobial properties, when incorporated into edible coatings or packaging materials, can inhibit microbial spoilage and food-borne pathogens, aligning with consumer preferences for clean-label products (Chauhan and Rao, 2024). In cosmetics, antioxidant pigments reduce UV-induced skin damage, enhance photoprotection, and prevent premature aging, while antimicrobial pigments contribute to product preservation without synthetic preservatives (Flieger *et al.*, 2024).

13.1.4 Sustainability and eco-friendly advantages

The production of microbial pigments aligns with the principles of green chemistry and sustainable biotechnology. Microbial cultivation requires less land, water, and energy compared to conventional pigment sources such as crops or livestock-derived materials (Grewal *et al.*, 2022). Additionally, microbial fermentation can be conducted on renewable feedstocks, including lignocellulosic biomass, glycerol, and food processing residues, contributing to waste valorization and circular bioeconomy practices. The use of extremophiles capable of growing under saline, acidic, or thermophilic conditions further reduces contamination risks and allows process optimization in nonsterile environments, potentially lowering production costs.

Moreover, microbial pigment production generates fewer hazardous byproducts compared to synthetic pigment manufacturing, reducing the environmental burden of chemical waste disposal (Lopes and Ligabue-Braun, 2021). This eco-friendly advantage has led to increasing regulatory and commercial interest, with microbial pigments positioned as sustainable alternatives in markets where environmental compliance is a growing priority.

13.1.5 Scope and objectives of the chapter

This chapter critically examines the antioxidant and antimicrobial properties of microbial pigments, emphasizing their structural diversity, mechanisms of bioactivity, and application potential across multiple industries. The scope includes a systematic classification of microbial pigments based on chemical structure and biosynthetic pathways, followed by a detailed exploration of their bioactive mechanisms, supported by specific examples. Industrial applications are discussed with an emphasis on the integration of these bioactivities into functional foods, pharmaceuticals, cosmetics, and materials science.

Furthermore, the chapter addresses the synergistic roles of pigments possessing both antioxidant and antimicrobial properties, considering their potential in novel therapeutic strategies and in tackling emerging issues such as antibiotic resistance. The safety, toxicity, and regulatory aspects of pigment utilization are evaluated to provide a framework for industrial adoption and compliance with global standards. Finally, emerging trends in genetic engineering, nanoformulation, and sustainable production processes are analyzed to forecast the future trajectory of microbial pigment research and commercialization.

The comprehensive synthesis of ecological, biochemical, and industrial perspectives presented herein is intended to bridge fundamental research with applied innovation, thereby supporting the development of microbial pigments as next-generation bioactive agents in the context of health, industry, and environmental sustainability.

13.2 Sources and Classification of Microbial Pigments

Microbial pigments are produced by a variety of bacteria, fungi, and algae, each group contributing chemically distinct compounds with diverse ecological roles and industrial relevance. Their origin and biosynthetic pathways determine both their structural characteristics and bioactivity potential (Barreto *et al.*, 2023).

13.2.1 Sources of microbial pigments

13.2.1.1 Bacterial pigments

Bacteria produce a wide spectrum of pigments with notable functional properties. *Serratia marcescens* synthesizes prodigiosin, a red tripyrrole pigment with antibacterial, antifungal, and anticancer activities, as shown in Fig. 13.1. *Chromobacterium violaceum* generates violacein, a bisindole violet pigment with antimicrobial, antiparasitic, and antioxidant functions (Choi *et al.*, 2021). *Pseudomonas* spp. produce phenazine derivatives such as pyocyanin and phenazine-1-carboxamide, which act as redox mediators and disrupt competing microbial populations.

Carotenoids, including β-carotene, zeaxanthin, and astaxanthin, are synthesized by bacteria such as *Paracoccus* and *Rhodobacter* spp., functioning as photoprotective and antioxidant molecules (Mapelli-Brahm *et al.*, 2023). Extremophilic halophiles like *Halobacterium salinarum* produce bacterioruberin, a C50 carotenoid that stabilizes membranes under hypersaline conditions and protects against UV damage.

13.2.1.2 Fungal pigments

Fungi produce pigments through specialized secondary metabolism. *Monascus* spp. are prolific producers of polyketide-derived pigments, ranging from yellow to red hues, widely used as natural food colorants with additional antioxidant and antimicrobial activities (Deshmukh *et al.*, 2022). Melanins, phenolic polymers present in *Aspergillus*, *Penicillium*, and *Alternaria*, contribute to stress resistance and are implicated in fungal pathogenicity.

Fungal anthraquinones, such as emodin and dermocybin from*Fusarium* and*Trichoderma* spp., display intense coloration and bioactivities, including antimicrobial and cytotoxic effects (Fouillaud *et al.*, 2016). Lichens, symbiotic associations of fungi with algae or cyanobacteria, also produce unique pigment metabolites with ecological and pharmacological significance.

13.2.1.3 Algal and cyanobacterial pigments

Microalgae and cyanobacteria produce pigments with high biotechnological value. Phycobiliproteins such as phycocyanin from *Spirulina platensis* and

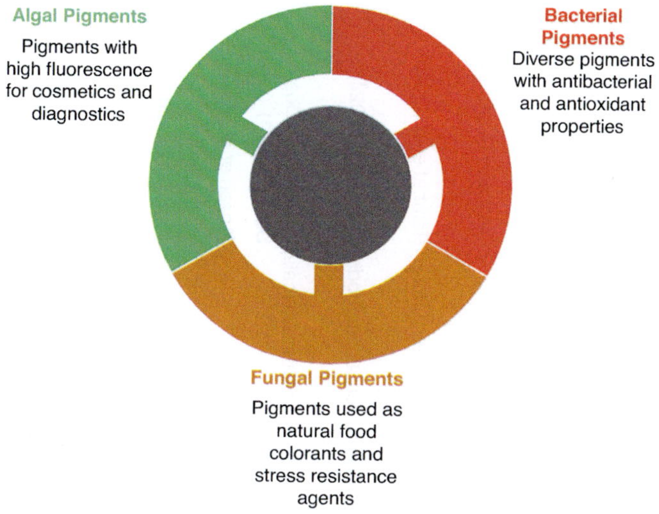

Fig. 13.1. Sources and functional roles of algal, bacterial, and fungal pigments.

phycoerythrin from *Porphyridium cruentum* display vibrant coloration and high fluorescence, making them suitable for cosmetics and diagnostic uses (Tounsi *et al.*, 2023). Chlorophyll derivatives from *Chlorella vulgaris* show antioxidant and antiinflammatory effects. Carotenoids like astaxanthin from *Haematococcus pluvialis* and fucoxanthin from brown algae (*Undaria pinnatifida*) have potent radical-scavenging abilities.

Cyanobacteria produce scytonemin, a lipid-soluble pigment with UV-screening capability, enhancing survival in high-radiation environments (Nandagopal *et al.*, 2021). A summary of representative pigments, their sources, colors, bioactivities, and applications is presented in Table 13.1, providing a comparative overview for selection in industrial and biomedical contexts.

13.2.2 Classification of microbial pigments

13.2.2.1 Classification based on chemical structure

Pigments may be categorized according to their chromophore type and functional groups. Carotenoids are isoprenoid compounds divided into carotenes (e.g. β-carotene, lycopene) and xanthophylls (e.g. zeaxanthin, astaxanthin), generally hydrophobic and lipid soluble (Perveen *et al.*, 2021). Phenolics, including melanins and phenolic quinones, display potent radical-scavenging and metal-chelating activities. Quinones, such as anthraquinones, are conjugated compounds with antimicrobial and cytotoxic potential. Tetrapyrroles, including chlorophylls and phycobilins, are macrocyclic pigments essential in photosynthetic light harvesting. Indole derivatives such as violacein originate from tryptophan metabolism and exhibit antimicrobial and anticancer properties (Ramesh *et al.*, 2025).

13.2.2.2 Classification based on biosynthetic pathways

Pigments can also be classified by their metabolic origin. Polyketide-derived pigments include *Monascus* pigments and anthraquinones, synthesized via polyketide synthases (Tang *et al.*, 2025). Terpenoid-derived pigments, such as carotenoids, are produced through the mevalonate or methylerythritol phosphate pathways. Shikimate-derived pigments, including violacein and phenazines, originate from the shikimate pathway. Mixed-origin pigments, such as scytonemin, integrate precursors from multiple biosynthetic routes, reflecting complex enzymatic regulation (Sen and Mallick, 2022).

Table 13.1. Representative microbial pigments, natural sources, visible coloration, major bioactivities, and industrial applications.

Pigment	Source organism(s)	Color	Major bioactivities	Industrial applications	References
Prodigiosin	*Serratia marcescens*, *Hahella chejuensis*	Red	Antibacterial, antifungal, anticancer	Pharmaceuticals, antimicrobial coatings	(Sahoo and Pradhan, 2024)
Violacein	*Chromobacterium violaceum*, *Janthinobacterium* spp.	Violet	Antibacterial, antiparasitic, antioxidant	Pharmaceuticals, biocontrol agents	(Park *et al*., 2021)
Phenazine derivatives	*Pseudomonas aeruginosa*	Blue-green	Antibacterial, antibiofilm, redox mediator	Antifouling materials, agricultural biocontrol	(Abdelaziz *et al*., 2023)
β-Carotene	*Paracoccus marcusii*, *Dunaliella salina*	Orange	Antioxidant, provitamin A	Nutraceuticals, food colorant	(Naik and Gupte, 2024)
Astaxanthin	*Haematococcus pluvialis*, *Paracoccus carotinifaciens*	Red-orange	Antioxidant, antiinflammatory	Nutraceuticals, aquaculture feed, cosmetics	(Tambat *et al*., 2025)
Bacterioruberin	Halophilic archaea (*Halobacterium salinarum*)	Pink-red	UV protection, antioxidant	Cosmetics, functional foods	(Giani *et al*., 2024)
Melanin	*Aspergillus niger*, *Cryptococcus neoformans*	Brown-black	UV protection, antioxidant, virulence factor	Cosmetics, pharmaceuticals	(Meruvu and Dos Santos, 2021)
Anthraquinones	*Fusarium oxysporum*, *Trichoderma* spp.	Yellow-red	Antimicrobial, cytotoxic	Natural dyes, pharmaceuticals	(Ansari *et al*., 2025)
Phycocyanin	*Spirulina platensis*	Blue	Antioxidant, antiinflammatory, fluorescence marker	Nutraceuticals, cosmetics, diagnostic assays	(Omar *et al*., 2022)
Phycoerythrin	*Porphyridium cruentum*	Pink-red	Antioxidant, fluorescent label	Cosmetics, biomedical imaging	(Tan *et al*., 2022)
Fucoxanthin	*Undaria pinnatifida*, brown algae	Brown-orange	Antioxidant, antiobesity	Nutraceuticals, cosmetics	(Tiwari *et al*., 2022)
Scytonemin	Cyanobacteria (*Nostoc commune*)	Yellow-brown	UV screening, antioxidant	Sunscreens, cosmetics	(Ručová *et al*., 2023)

13.2.2.3 Industrial implications of source and classification knowledge

Understanding the source and classification of microbial pigments facilitates targeted strain selection and industrial application. Structural classification aids in predicting pigment solubility, stability, and compatibility with formulations, whereas biosynthetic classification supports metabolic engineering strategies for yield improvement and functional modification. The integration of this knowledge enables efficient production and maximizes the bioactive and economic potential of microbial pigments.

13.3 Antioxidant Properties of Microbial Pigments

Microbial pigments are increasingly recognized as potent natural antioxidants capable of mitigating oxidative stress in biological systems and preserving the quality of perishable products (Rani *et al.*, 2021). Their antioxidant potential is linked to structural features such as extended conjugated double-bond systems, aromatic moieties, and redox-active functional groups that enable electron transfer, hydrogen atom donation, and metal ion chelation. The antioxidant action of microbial pigments is not merely an incidental property but often plays an ecological role, protecting producer organisms from photooxidative damage, reactive oxygen species (ROS), and environmental stressors (Rather *et al.*, 2025).

These antioxidant properties have positioned microbial pigments as promising candidates for use in food preservation, cosmetic photoprotection, functional nutraceuticals, and the prevention of oxidative stress-associated pathologies in humans.

13.3.1 Mechanisms of antioxidant action

The antioxidant capacity of microbial pigments is mediated through multiple mechanisms, often acting synergistically to provide comprehensive protection against oxidative damage.

13.3.1.1 Radical scavenging activity

Many pigments act as direct scavengers of free radicals such as hydroxyl radicals (•OH), superoxide anions (O_2•–), and peroxyl radicals (ROO•). In the DPPH (2,2-diphenyl-1-picrylhydrazyl) assay, pigments donate electrons or hydrogen atoms to the DPPH radical, reducing it to a stable form, while in the ABTS (2,2′-azinobis-(3-ethylbenzothiazoline-6-sulfonic acid)) assay, pigments quench the ABTS•+ cation radical, leading to decolorization proportional to antioxidant strength (Wołosiak *et al.*, 2021). Carotenoids such as β-carotene, lycopene, and astaxanthin are particularly effective radical quenchers due to their conjugated polyene chains, which delocalize unpaired electrons, stabilizing the radical species.

13.3.1.2 Metal chelation

Transition metals such as Fe^{2+} and Cu^{2+} catalyze ROS production via Fenton and Haber–Weiss reactions, generating highly reactive hydroxyl radicals. Phenolic pigments, including melanins and anthraquinones, chelate these metals, reducing their catalytic activity and thereby limiting ROS formation. This chelation capacity is particularly relevant in food preservation and in mitigating metal-induced oxidative stress in biological tissues (Timoshnikov *et al.*, 2022).

13.3.1.3 Inhibition of lipid peroxidation

Polyunsaturated fatty acids in cellular membranes and food lipids are susceptible to peroxidation, producing cytotoxic and proinflammatory aldehydes such as malondialdehyde (MDA) and 4-hydroxynonenal (HNE). Pigments like phycocyanin from *Spirulina platensis* and bacterioruberin from halophiles can intercept lipid peroxyl radicals, breaking the chain reaction of lipid peroxidation and preserving membrane integrity.

13.3.1.4 Modulation of endogenous antioxidant systems

Beyond direct scavenging, some microbial pigments upregulate endogenous antioxidant defenses. Astaxanthin and phycocyanin have been shown to induce nuclear factor erythroid 2-related factor 2 (Nrf2), which activates the transcription of antioxidant enzymes such as superoxide dismutase (SOD), catalase (CAT), and glutathione peroxidase (GPx). This indirect mechanism enhances long-term oxidative resilience in cells and tissues (Ashrafizadeh *et al.*, 2022).

13.3.2 Representative pigments with antioxidant activity

Carotenoids are the most widely studied antioxidant pigments. β-Carotene, from *Paracoccus marcusii* and *Dunaliella salina*, quenches singlet oxygen with an efficiency that protects both photosynthetic organisms and human tissues from photooxidative stress (Fong, 2023).

Astaxanthin, produced by *Haematococcus pluvialis* and some bacteria, exhibits superior singlet oxygen quenching compared to β-carotene and α-tocopherol due to its polar ionone rings that anchor in membrane surfaces while the polyene chain spans the bilayer. Lycopene, derived from *Blakeslea trispora* and certain marine bacteria, is notable for its strong peroxyl radical scavenging ability.

Phenolic pigments like melanin, synthesized by *Aspergillus niger* and *Cryptococcus neoformans*, display multifunctional antioxidant effects, including hydroxyl radical scavenging, UV absorption, and transition metal chelation. These properties contribute to both microbial survival and potential therapeutic applications (Suthar *et al.*, 2023).

Phycobiliproteins, such as C-phycocyanin from *Spirulina platensis* and phycoerythrin from *Porphyridium cruentum*, are water-soluble pigments with antioxidant capacity derived from their linear tetrapyrrole chromophores. These pigments scavenge alkoxyl and hydroxyl radicals and inhibit lipid peroxidation while exhibiting high photostability.

Quinone pigments, including anthraquinones like emodin from *Fusarium* spp., have multiple hydroxyl substitutions that enhance electron-donating capacity and radical stabilization.

C50 carotenoids such as bacterioruberin from *Halobacterium salinarum* are particularly effective in protecting lipid membranes against peroxidation and quenching singlet oxygen in extreme environments (Ben Hamad Bouhamed *et al.*, 2024).

13.3.3 Applications of antioxidant properties

The application of microbial pigments as antioxidants spans food, cosmetics, nutraceuticals, and pharmaceuticals, often combining functionality with aesthetic value.

In food systems, oxidative degradation of lipids and pigments leads to rancidity and discoloration, reducing shelf-life and sensory quality. Incorporating β-carotene or astaxanthin as natural antioxidants in dairy, meat, and confectionery products enhances oxidative stability while providing color. Pigments with antioxidant properties are also used in active food packaging, where they are embedded in biopolymer films to prevent oxidation during storage (Kumar *et al.*, 2024).

In cosmetic products, antioxidant pigments protect against photoaging by neutralizing ROS generated by UV exposure. Astaxanthin formulations have been shown to improve skin elasticity, reduce wrinkle formation, and decrease erythema following UV exposure. Phycocyanin-containing products provide both photoprotection and antiinflammatory benefits for skin health (Ramos *et al.*, 2025).

In pharmaceutical and nutraceutical contexts, antioxidant pigments are investigated for roles in preventing or mitigating oxidative stress-related disorders. For example, supplementation with astaxanthin reduces oxidative biomarkers in cardiovascular disease patients, while phycocyanin has been explored as an adjuvant therapy in neurodegenerative conditions due to its neuroprotective effects.

Table 13.2 provides a detailed summary of key antioxidant pigments, their microbial sources, underlying mechanisms, and industrial applications, serving as a quick reference for targeted selection in product development.

13.4 Antimicrobial Properties of Microbial Pigments

Microbial pigments possess significant antimicrobial potential, functioning not only as secondary metabolites with ecological importance but also as bioactive agents of industrial and

Table 13.2. Representative microbial pigments with antioxidant properties, sources, mechanisms of action, and industrial applications.

Pigment	Source organism(s)	Mechanism(s) of action	Industrial applications	References
β-Carotene	*Paracoccus marcusii*, *Dunaliella salina*	Singlet oxygen quenching, peroxyl radical scavenging	Nutraceuticals, food antioxidants, cosmetics	(Naik and Gupte, 2024)
Astaxanthin	*Haematococcus pluvialis*, *Paracoccus carotinifaciens*	Singlet oxygen quenching, membrane protection, and Nrf2 activation	Nutraceuticals, cosmetics, functional foods	(Tambat *et al*., 2025)
Lycopene	*Blakeslea trispora*, a marine bacterium	Radical scavenging, membrane lipid stabilization	Food additives, nutraceuticals	(Han *et al*., 2025)
Melanin	*Aspergillus niger*, *Cryptococcus neoformans*	Hydroxyl radical scavenging, UV absorption, metal ion chelation	Pharmaceuticals, cosmetics	(Meruvu and Dos Santos, 2021)
Phycocyanin	*Spirulina platensis*	Hydroxyl and alkoxyl radical scavenging, lipid peroxidation inhibition, and antiinflammatory effects	Nutraceuticals, cosmetics, pharmaceuticals	(Omar *et al*., 2022)
Phycoerythrin	*Porphyridium cruentum*	Radical scavenging, lipid peroxidation inhibition	Cosmetics, biomedical imaging	(Tan *et al*., 2022)
Emodin	*Fusarium oxysporum*	Metal chelation, radical scavenging, modulation of inflammatory pathways	Pharmaceuticals, dietary supplements	(Shalapy *et al*., 2024)
Bacterioruberin	*Halobacterium salinarum*	Lipid peroxidation inhibition, singlet oxygen quenching	Functional foods, cosmetics	(Giani *et al*., 2024)

clinical interest. The antimicrobial activity of these pigments stems from diverse structural and physicochemical attributes, enabling them to target multiple microbial systems simultaneously. Such properties make them valuable in food preservation, pharmaceutical development, and biofouling prevention (Kazi *et al.*, 2022).

13.4.1 Mechanisms of antimicrobial action

The antimicrobial activity of microbial pigments is multifactorial, involving both direct biochemical interactions with target organisms and indirect modulation of microbial environments.

13.4.1.1 Disruption of cell membrane integrity

Many hydrophobic pigments, such as prodigiosin and bacterioruberin, integrate into lipid bilayers, causing increased membrane fluidity and permeability. This leads to leakage of intracellular contents and loss of essential metabolites. In Gram-negative bacteria, pigments can destabilize the outer membrane by interacting with lipopolysaccharide components, whereas in fungi, they disrupt ergosterol-rich membranes (Zhao *et al.*, 2025).

13.4.1.2 Inhibition of macromolecular synthesis

Certain pigments interfere with DNA, RNA, or protein synthesis. Prodigiosin, for example, can intercalate into DNA, thereby blocking replication and transcription. Phenazine derivatives from *Pseudomonas aeruginosa* inhibit respiratory enzymes and nucleic acid synthesis, impairing bacterial metabolism and proliferation. Violacein exhibits similar DNA-binding properties and can induce apoptosis-like cell death in protozoa and yeast (Aman Mohammadi *et al.*, 2022).

13.4.1.3 ROS-mediated microbial killing

Pigments with redox activity can catalyze ROS generation in microbial cells. Phenazines, due to their ability to undergo redox cycling, produce superoxide radicals that damage cell membranes, nucleic acids, and proteins. This mechanism is particularly effective against biofilm-forming pathogens, where ROS penetration into biofilm matrices enhances microbial eradication.

13.4.1.4 Disruption of biofilm formation

Microbial pigments such as pyocyanin and violacein inhibit biofilm formation by interfering with quorum-sensing systems. This not only prevents microbial adhesion to surfaces but also enhances susceptibility to antimicrobial agents (Marín-Sanhueza *et al.*, 2022).

13.4.2 Examples of pigments with antimicrobial potential

Prodigiosin, produced by *Serratia marcescens* and *Hahella chejuensis*, exhibits broad-spectrum antibacterial activity against Gram-positive and Gram-negative bacteria, antifungal activity against *Candida albicans*, and antiparasitic effects against *Plasmodium falciparum*. Its activity is attributed to membrane disruption, DNA intercalation, and pH-dependent protonophoric action.

Violacein, from *Chromobacterium violaceum* and *Janthinobacterium* spp., is active against *Staphylococcus aureus*, *Escherichia coli*, and certain protozoan parasites. Its purple bisindole structure facilitates DNA binding, ROS generation, and inhibition of quorum sensing (Park *et al.*, 2021).

Phenazine derivatives, such as pyocyanin and phenazine-1-carboxamide, produced by *Pseudomonas aeruginosa*, are effective against bacterial and fungal pathogens, especially in polymicrobial environments. They disrupt respiratory electron transport chains and induce oxidative stress.

Monascus pigments, while primarily recognized for their coloring properties, possess antibacterial activity against *Listeria monocytogenes* and antifungal effects against *Aspergillus* spp., mediated by membrane disruption and interference with spore germination.

Bacterioruberin, from halophilic archaea, inhibits Gram-positive bacteria and certain molds, combining hydrophobic membrane disruption with ROS generation under photoactivation (Giani *et al.*, 2024).

Anthraquinones, such as emodin, have demonstrated inhibitory activity against *Helicobacter pylori* and dermatophytes through disruption of enzymatic activity and interference with energy metabolism.

13.4.3 Applications of antimicrobial activity

The antimicrobial activity of microbial pigments supports a wide range of applications in food safety, pharmaceutical therapy, and material protection.

13.4.3.1 Food preservation and packaging

Natural pigments with antimicrobial properties are increasingly incorporated into edible films and coatings to inhibit food-borne pathogens and spoilage organisms. For example, phycocyanin-containing coatings have been applied to seafood to delay microbial growth, while prodigiosin-enriched packaging films have been tested for meat preservation. Such applications align with consumer demand for clean-label, additive-free foods (Rana *et al.*, 2021).

13.4.3.2 Pharmaceutical and therapeutic applications

Pigments such as violacein and prodigiosin have attracted attention as leads for novel antibiotics, particularly in the context of multidrug-resistant infections. Phenazine derivatives are being explored as adjunct therapies for chronic biofilm-associated infections, including cystic fibrosis lung disease caused by *Pseudomonas aeruginosa*. The dual antimicrobial and anticancer activity of some pigments also presents opportunities for multifunctional drug development (Kantak and Bharate, 2022).

13.4.3.3 Antifouling and material protection

Marine biofouling, the accumulation of microorganisms and larger organisms on submerged surfaces, can be mitigated by pigment-based coatings. Violacein and phenazines incorporated into antifouling paints inhibit microbial colonization without relying on environmentally persistent heavy metals. Similar strategies are being evaluated for hospital surfaces and medical devices to reduce healthcare-associated infections (Nwuzor *et al.*, 2021).

13.5 Synergistic Bioactivities and Multifunctional Roles

Microbial pigments that simultaneously exhibit antioxidant and antimicrobial activities are of particular interest in industrial biotechnology, food preservation, healthcare, and cosmetics. This dual bioactivity enables them to address oxidative damage while suppressing microbial growth, offering multifunctional value that extends beyond the capabilities of single-function additives. In natural ecosystems, such synergistic bioactivities confer adaptive advantages to pigment-producing microorganisms, enhancing survival under oxidative and microbial stress conditions.

From an application standpoint, the combination of antioxidant and antimicrobial properties reduces the need for multiple additives, thereby simplifying formulations, improving product stability, and appealing to consumer preferences for multifunctional, natural bioactives.

13.5.1 Pigments with dual antioxidant–antimicrobial properties

Several microbial pigments identified in sections 13.3 and 13.4 have demonstrated both antioxidant and antimicrobial properties in experimental studies.

Astaxanthin, derived from *Haematococcus pluvialis* and certain bacteria, is a potent singlet oxygen quencher that also disrupts microbial membranes, particularly under oxidative stress conditions. Its amphipathic structure allows integration into lipid bilayers, where it protects host cells while destabilizing pathogen membranes.

Prodigiosin, from *Serratia marcescens*, combines membrane-disruptive antimicrobial action with ROS-scavenging ability. Interestingly, its antioxidant capacity may protect host cells during infection while its antimicrobial activity targets pathogens directly (Negi *et al.*, 2025).

Violacein, from *Chromobacterium violaceum*, has been shown to inhibit bacterial quorum sensing and biofilm formation while also neutralizing free radicals. Its bisindole chromophore is responsible for both DNA-binding antimicrobial effects and antioxidant radical scavenging.

Phycocyanin, from *Spirulina platensis*, provides ROS scavenging and lipid peroxidation inhibition alongside mild bactericidal effects against Gram-positive and Gram-negative bacteria.

Bacterioruberin, a C50 carotenoid from halophilic archaea, combines exceptional photoprotective antioxidant activity with antimicrobial action via photoactivated ROS generation, making it relevant in photodynamic antimicrobial therapy.

13.5.2 Mechanistic basis for synergistic effects

Dual bioactivity arises from the inherent chemical properties of these pigments.

- *Shared reactive interfaces*: conjugated double bonds and aromatic structures allow both ROS quenching and oxidative stress-based microbial killing.
- *Membrane affinity*: amphipathic pigments localize at membrane interfaces, enabling them to protect host membranes from oxidation while simultaneously disrupting pathogen membranes (Struyfs *et al.*, 2021).
- *Redox cycling*: certain pigments, such as phenazines, can function as antioxidants under low redox potential but as prooxidants in the presence of pathogens, generating ROS to kill them.
- *Metal ion interaction*: pigments like melanins can chelate metal ions, preventing ROS generation in host tissues while also depriving pathogens of essential cofactors.

The balance between antioxidant and antimicrobial roles can depend on environmental factors such as pH, light exposure, and oxygen availability. For example, bacterioruberin demonstrates strong antioxidant effects under dark conditions but shifts toward antimicrobial photooxidation under high-intensity light (Lee *et al.*, 2024).

13.5.3 Industrial and clinical implications

In food systems, dual-function pigments act as natural preservatives that both delay oxidative rancidity and inhibit microbial spoilage. Phycocyanin-enriched edible films have extended the shelf-life of refrigerated seafood by inhibiting *Listeria monocytogenes* growth while maintaining product color and freshness.

In cosmetics, pigments such as astaxanthin and bacterioruberin serve as UV-protective antioxidants that also prevent microbial contamination in topical formulations, reducing reliance on synthetic preservatives (Tounsi *et al.*, 2023).

In pharmaceuticals, dual-function pigments hold promise for managing oxidative stress-associated infections, such as chronic wounds, where biofilm-forming bacteria create an oxidatively stressed environment. Here, pigments like violacein could reduce both microbial load and inflammatory oxidative damage (Kantak and Bharate, 2022).

In medical devices, coatings containing dual-active pigments could provide prolonged antimicrobial protection while limiting oxidative degradation of the device materials, particularly in catheters, stents, or wound dressings.

13.6 Safety, Toxicity, and Regulatory Aspects

The utilization of microbial pigments in food, pharmaceuticals, cosmetics, and other industrial products necessitates a comprehensive safety evaluation to ensure that these compounds are nontoxic, nonallergenic, and compliant with established regulatory frameworks. Although many of these pigments are of natural origin, the assumption that "natural" implies "safe" is unfounded, making a detailed toxicological assessment indispensable (Ghosh *et al.*, 2025).

Safety considerations involve evaluating potential cytotoxicity, mutagenicity, allergenicity, and acceptable daily intake (ADI), and ensuring compliance with international standards.

13.6.1 Toxicological considerations

Toxicological assessment of microbial pigments reveals that their biological effects can vary widely depending on concentration, molecular structure, and route of exposure. Certain pigments, such as prodigiosin, display selective cytotoxicity toward cancer cells, which is advantageous in oncology applications but may become harmful to normal cells if administered at higher concentrations. Anthraquinones like emodin have shown hepatotoxicity in animal models at elevated doses, underscoring the importance of defining safe exposure limits through preclinical studies (Chatragadda and Dufossé, 2021).

Mutagenicity and genotoxicity are critical parameters in pigment safety evaluation, particularly for compounds with DNA-binding capabilities. Violacein and phenazines, while effective antimicrobial agents, can interact directly with genetic material, necessitating careful dose optimization to prevent genotoxic effects in host tissues. Additionally, pigments produced by certain fungi, such as *Monascus* pigments, may be contaminated with citrinin, a nephrotoxic mycotoxin, if fermentation conditions are not meticulously controlled. This emphasizes the need for stringent quality control during manufacturing to ensure the absence of such contaminants.

Allergenicity is another concern, especially for proteinaceous pigments. Phycocyanin, derived from *Spirulina platensis*, is generally well tolerated but may provoke allergic responses in individuals sensitized to cyanobacterial proteins. Allergenicity testing, including *in vitro* and *in vivo* assays, is therefore essential before these pigments are incorporated into cosmetic or topical formulations (Sharma *et al.*, 2024).

13.6.2 Acceptable daily intake and safety testing

Determining the acceptable daily intake for a pigment is a crucial step in establishing its safety profile for human use. This process involves identifying the no-observed-adverse-effect level (NOAEL) through animal studies and applying appropriate safety factors to extrapolate a safe level for human consumption. For example, the Joint FAO/WHO Expert Committee on Food Additives (JECFA) has set the ADI for naturally derived β-carotene at 5–6 mg/kg body weight per day. Similarly, astaxanthin from *Haematococcus pluvialis* has an ADI of 0.034 mg/kg/day, derived from human supplementation trials demonstrating both efficacy and safety (Bjørklund *et al.*, 2022).

Safety testing follows a structured progression, beginning with *in vitro* assays such as cytotoxicity (MTT) and genotoxicity (Ames and comet assays) to screen for potential hazards. Promising pigments that pass these initial evaluations undergo *in vivo* studies to assess acute, subchronic, and chronic toxicity in relevant animal models. If these are successful, human clinical trials are conducted, particularly for pigments intended for use in nutraceuticals and pharmaceuticals, to confirm tolerability and identify any adverse effects under conditions of intended use.

13.6.3 Regulatory frameworks

The regulatory approval process for microbial pigments depends on the intended application and the jurisdiction in which the product will be marketed. In food applications within the United States, the Food and Drug Administration (FDA) classifies pigments as "color additives" and further designates them as either "exempt from certification" if naturally derived or "subject to certification" if synthetic or chemically modified. Many microbial pigments may qualify for Generally Recognized as Safe (GRAS) status, provided that comprehensive safety documentation is available. In the European Union, the European Food Safety Authority (EFSA) assigns E-numbers to approved colorants, such as E160a for β-carotene, and sets maximum permitted levels for their inclusion in various food products (Öztekin *et al.*, 2022).

Pharmaceutical applications are regulated according to International Council for Harmonisation (ICH) guidelines, requiring compliance with Good Manufacturing Practice (GMP), impurity profiling, and, when relevant, pharmacokinetic and pharmacodynamic evaluations. Pigments used as excipients in drug formulations must meet both safety and functional requirements, while those acting as active pharmaceutical ingredients undergo even more stringent evaluation.

13.6.4 Case examples of regulatory challenges

Several microbial pigments illustrate the complexities of safety and regulatory approval. *Monascus* pigments, while widely used in Asian countries, face restrictions in the European Union and North America due to the risk of citrinin contamination. This challenge underscores the importance of strain selection and rigorous fermentation monitoring. Phenazine derivatives, despite their potent antimicrobial properties, have limited approval for large-scale antifouling use because of their environmental toxicity to nontarget marine organisms (Neves *et al.*, 2024). Astaxanthin is another example; although it is widely approved for use in aquaculture feed, its inclusion in human food products is permitted only when derived from authorized sources such as *H. pluvialis*.

13.7 Emerging Trends and Future Perspectives

The rapidly advancing fields of biotechnology, synthetic biology, and nanotechnology are transforming the research, production, and application of microbial pigments. Beyond their traditional use as natural colorants, these compounds are increasingly recognized as multifunctional bioactive agents with antioxidant, antimicrobial, anticancer, and photoprotective properties. The growing consumer preference for natural, sustainable products, coupled with regulatory restrictions on synthetic additives, has created a favorable environment for microbial pigments to emerge as strategic resources in multiple industrial sectors. However, realizing their full potential requires addressing scientific, technological, regulatory, and economic challenges.

13.7.1 Technological advancements in production

One of the most significant trends is the shift toward precision fermentation and metabolic engineering to enhance pigment yields and bioactivity. Advances in genome editing tools, such as CRISPR-Cas systems, have enabled targeted modifications to biosynthetic gene clusters in bacteria, fungi, and algae, resulting in increased pigment production, improved stability, and tailored bioactivity profiles. Heterologous expression systems, where pigment biosynthetic pathways are transferred into well-characterized microbial hosts like *Escherichia coli* or *Saccharomyces cerevisiae*, offer the advantage of faster growth rates, reduced contamination risk, and easier regulatory approval (Acharya *et al.*, 2024).

Bioprocess optimization is also gaining attention, particularly the use of fed-batch and continuous cultivation systems, low-cost feedstocks, and bioreactor designs that facilitate pigment recovery. Downstream processing innovations, including aqueous two-phase extraction and membrane-based separation, are reducing solvent usage and improving environmental sustainability. Moreover, the use of immobilized cell systems and cocultivation strategies is being explored to exploit microbial symbiosis for higher pigment yields and diversified metabolite profiles.

13.7.2 Novel applications beyond coloration

The next generation of microbial pigment applications is moving far beyond coloration. In food systems, dual-function pigments with antioxidant and antimicrobial properties are being developed as active packaging components, where they serve as both visual indicators of product freshness and preservatives that extend shelf-life. In the pharmaceutical sector, pigments are being investigated for targeted drug delivery using nanocarrier systems, leveraging their bioactive properties alongside colorimetric monitoring capabilities (Maghsoudi *et al.*, 2022).

In the cosmetics industry, pigments such as astaxanthin and bacterioruberin are gaining traction for their photoprotective and antiaging effects. The integration of these pigments into sunscreen formulations provides both UV absorption and free radical scavenging, delivering synergistic skin protection. Similarly, pigment-based photodynamic therapy is emerging as a minimally invasive approach for treating skin infections and certain cancers, wherein pigments act as photosensitizers that generate reactive oxygen species upon light activation.

The agricultural sector is exploring microbial pigments as biocontrol agents to suppress plant pathogens while reducing reliance on synthetic pesticides. Pigments with antimicrobial activity can be incorporated into seed coatings or foliar sprays, offering environmentally friendly crop protection strategies.

13.7.3 Integration with nanotechnology

Nanotechnology offers transformative possibilities for microbial pigment applications. Encapsulation in nanocarriers such as liposomes, polymeric nanoparticles, and nanoemulsions can enhance pigment stability against light, heat, and pH fluctuations. Such encapsulation also improves bioavailability in nutraceutical and pharmaceutical products by facilitating controlled release and targeted delivery to specific tissues (Bhandari *et al.*, 2024).

Metallic nanoparticle synthesis using microbial pigments as reducing and stabilizing agents is another emerging trend. Pigments such as prodigiosin and violacein can mediate the biosynthesis of silver and gold nanoparticles, yielding hybrid materials with enhanced antimicrobial and catalytic properties. These pigment-mediated nanoparticles have potential applications in wound healing, water purification, and biosensing technologies.

13.7.4 Addressing safety, scalability, and regulatory gaps

Despite the growing interest, large-scale commercialization of microbial pigments still faces barriers. The safety profile of some pigments remains incompletely characterized, particularly concerning long-term exposure and potential bioaccumulation. This is especially relevant for pigments intended for food and pharmaceutical applications. Comprehensive toxicological studies and standardized safety testing protocols are needed to facilitate regulatory approval.

Scalability is another challenge. While laboratory-scale yields can be high, translating these processes to industrial scales often leads to reduced productivity due to shear stress, oxygen transfer limitations, and metabolic bottlenecks. Advances in bioreactor engineering, process monitoring via artificial intelligence, and predictive metabolic modeling are likely to address these constraints (Elazzazy *et al.*, 2025).

From a regulatory perspective, the lack of harmonized international standards for microbial pigments creates market entry barriers. The development of global safety databases, pigment-specific monographs, and consensus-based ADI values would streamline approvals and encourage cross-border trade.

13.8 Conclusion

Microbial pigments represent a versatile and sustainable class of bioactive compounds with applications extending far beyond their traditional role as natural colorants. As demonstrated in this chapter, many pigments possess intrinsic antioxidant and antimicrobial properties, offering multifunctional benefits for the food, pharmaceutical, cosmetic, and agricultural sectors. Advances in microbial strain engineering, precision fermentation, and bioprocess optimization are enabling higher yields, improved stability, and tailored functional profiles, while integration with nanotechnology is expanding their potential in targeted delivery, active packaging, and photodynamic therapy. The dual bioactivity of several pigments not only enhances product performance but also aligns with global trends favoring natural, multifunctional additives over synthetic counterparts. Nevertheless, large-scale adoption is contingent upon addressing safety and regulatory requirements through comprehensive toxicological evaluation, establishing harmonized international standards, and ensuring consistent product quality.

Future development will depend on interdisciplinary collaboration, combining synthetic biology, nanotechnology, materials science, and process engineering to unlock novel pigment functionalities and economically viable production systems. With continued innovation and rigorous safety validation, microbial pigments are poised to become strategic components in next-generation products that meet consumer demands for efficacy, safety, and sustainability, while contributing to the global shift toward cleaner and more environmentally responsible industrial practices.

References

Abdelaziz, A.A., Kamer, A.M.A., Al-Monofy, K.B. and Al-Madboly, L.A. (2023) Pseudomonas aeruginosa's greenish-blue pigment pyocyanin: Its production and biological activities. *Microbial Cell Factories* 22(1), 110. DOI: 10.1186/s12934-023-02122-1.

Acharya, K., Shaw, S., Bhattacharya, S.P., Biswas, S., Bhandary, S, *et al*. (2024) Pigments from pathogenic bacteria: A comprehensive update on recent advances. *World Journal of Microbiology & Biotechnology* 40(9), 270. DOI: 10.1007/s11274-024-04076-x.

Aman Mohammadi, M., Ahangari, H., Mousazadeh, S., Hosseini, S.M. and Dufossé, L. (2022) Microbial pigments as an alternative to synthetic dyes and food additives: A brief review of recent studies. *Bioprocess and Biosystems Engineering* 45(1), 1–12. DOI: 10.1007/s00449-021-02621-8.

Ansari, T.N., Iqbal, S. and Shrivastava, A. (2025) Fungal pigments: A sustainable alternative to synthetic colors. *Microbial Colorants: Chemistry, Biosynthesis and Applications* 51–80. DOI: 10.1002/9781394287888.

Ashrafizadeh, M., Ahmadi, Z., Yaribeygi, H., Sathyapalan, T. and Sahebkar, A. (2022) Astaxanthin and Nrf2 signaling pathway: A novel target for new therapeutic approaches. *Mini Reviews in Medicinal Chemistry* 22(2), 312–321. DOI: 10.2174/1389557521666210505112834.

Barreto, J.V.D.O., Casanova, L.M., Junior, A.N., Reis-Mansur, M.C.P.P. and Vermelho, A.B. (2023) Microbial pigments: Major groups and industrial applications. *Microorganisms* 11(12), 2920. DOI: 10.3390/microorganisms11122920.

Ben Hamad Bouhamed, S., Chaari, M., Baati, H., Zouari, S. and Ammar, E. (2024) Extreme halophilic *Archaea*: *Halobacterium salinarum* carotenoids characterization and antioxidant properties. *Heliyon* 10(17), e36832. DOI: 10.1016/j.heliyon.2024.e36832.

Bhandari, M., Sharma, R., Sharma, S., Bobade, H. and Singh, B. (2024) Recent advances in nanoencapsulation of natural pigments: Emerging technologies, stability, therapeutic properties and potential food applications. *Pigment & Resin Technology* 53(1), 53–61. DOI: 10.1108/PRT-04-2022-0050.

Bills, G.F. and Gloer, J.B. (2016) Biologically active secondary metabolites from the fungi. *Microbiology Spectrum* 4(6), 10–1128. DOI: 10.1128/microbiolspec.FUNK-0009-2016.

Bjørklund, G., Gasmi, A., Lenchyk, L., Shanaida, M., Zafar, S, *et al*. (2022) The role of astaxanthin as a nutraceutical in health and age-related conditions. *Molecules* 27(21), 7167. DOI: 10.3390/molecules27217167.

Chatragadda, R. and Dufossé, L. (2021) Ecological and biotechnological aspects of pigmented microbes: A way forward in development of food and pharmaceutical grade pigments. *Microorganisms* 9(3), 637. DOI: 10.3390/microorganisms9030637.

Chauhan, K. and Rao, A. (2024) Clean-label alternatives for food preservation: An emerging trend. *Heliyon* 10(16), e35815. DOI: 10.1016/j.heliyon.2024.e35815.

Choi, S.Y., Lim, S., Yoon, K.H., Lee, J.I. and Mitchell, R.J. (2021) Biotechnological activities and applications of bacterial pigments violacein and prodigiosin. *Journal of Biological Engineering* 15(1), 10. DOI: 10.1186/s13036-021-00262-9.

Dave, S., Das, J., Varshney, B. and Sharma, V.P. (2022) Dyes and pigments: Interventions and how safe and sustainable are colors of life!!! In: *Trends and Contemporary Technologies for Photocatalytic Degradation of Dyes*. Springer, Cham, Switzerland, pp. 1–20. DOI: 10.1007/978-3-031-08991-6_1.

Deshmukh, S.K., Lagashetti, A., Singh, S.K., Badgujar, H.F. and Kumar, U. (2022) Fungal pigment research in India: An overview. *Progress in Mycology: Biology and Biotechnological Applications* 519–544. DOI: 10.1007/978-981-16-3307-2_17.

Elazzazy, A.M., Baeshen, M.N., Alasmi, K.M., Alqurashi, S.I., Desouky, S.E. *et al*. (2025) Where biology meets engineering: Scaling up microbial nutraceuticals to bridge nutrition, therapeutics, and global impact. *Microorganisms* 13(3), 566. DOI: 10.3390/microorganisms13030566.

Flieger, J., Raszewska-Famielec, M., Radzikowska-Büchner, E. and Flieger, W. (2024) Skin protection by carotenoid pigments. *International Journal of Molecular Sciences* 25(3), 1431. DOI: 10.3390/ijms25031431.

Fong, C.W. (2023) The origin of the antioxidant capacity against oxidative and photooxidative stress, singlet oxygen reactivity, and the excited states in skin and eye-related diseases of astaxanthin and other carotenoids, isomers, and esters in the cell membrane and cytosol. Doctoral dissertation, Eigenenergy Adelaide, Australia.

Fouillaud, M., Venkatachalam, M., Girard-Valenciennes, E., Caro, Y. and Dufossé, L. (2016) Anthraquinones and derivatives from marine-derived fungi: Structural diversity and selected biological activities. *Marine Drugs* 14(4), 64. DOI: 10.3390/md14040064.

Ghosh, J., Repon, M.R., Pranta, A.D., Rupanty, N.S., Khan, F, *et al.* (2025) Bioactive component integrated textiles: A promising source of medicine and healthcare. *Journal of Engineered Fibers and Fabrics* 20, 1–44. DOI: 10.1177/15589250241308561.

Giani, M., Pire, C. and Martínez-Espinosa, R.M. (2024) Bacterioruberin: Biosynthesis, antioxidant activity, and therapeutic applications in cancer and immune pathologies. *Marine Drugs* 22(4), 167. DOI: 10.3390/md22040167.

Grewal, J., Woła Cewicz, M., Pyter, W., Joshi, N., Drewniak, L. *et al.* (2022) Colorful treasure from agro-industrial wastes: A sustainable chassis for microbial pigment production. *Frontiers in Microbiology* 13, 832918. DOI: 10.3389/fmicb.2022.832918.

Guryanov, I. and Naumenko, E. (2024) Bacterial pigment prodigiosin as multifaceted compound for medical and industrial application. *Applied Microbiology* 4(4), 1702–1728. DOI: 10.3390/applmicrobiol4040115.

Han, A., Liu, Y., Gao, Z., Wang, J., Zhang, T, *et al.* (2025) Analysis of the mechanism of tripropylamine-enhanced lycopene accumulation in *Blakeslea trispora* through transcriptome. *International Microbiology* 1–13. DOI: 10.1007/s10123-025-00681-4.

Kantak, M.N. and Bharate, S.S. (2022) Analysis of clinical trials on biomaterial and therapeutic applications of Chitosan: A review. *Carbohydrate Polymers* 278, 118999. DOI: 10.1016/j.carbpol.2021.118999.

Kazi, Z., Hungund, B.S., Yaradoddi, J.S., Banapurmath, N.R., Yusuf, A.A. *et al.* (2022) Production, characterization, and antimicrobial activity of pigment from *Streptomyces* species. *Journal of Nanomaterials* 2022(1), 3962301. DOI: 10.1155/2022/3962301.

Kumar, Y., Bist, Y., Thakur, D., Nagar, M. and Saxena, D.C. (2024) A review on the role of pH-sensitive natural pigments in biopolymers based intelligent food packaging films. *International Journal of Biological Macromolecules* 276(Pt 1), 133869. DOI: 10.1016/j.ijbiomac.2024.133869.

Lee, H., Cho, E.S., Hwang, C.Y., Cao, L., Kim, M.B, *et al.* (2024) Bacterioruberin extract from Haloarchaea *Haloferax marinum*: Component identification, antioxidant activity and anti-atrophy effect in LPS-treated C2C12 myotubes. *Microbial Biotechnology* 17(9), e70009. DOI: 10.1111/1751-7915.70009.

Lopes, F.C. and Ligabue-Braun, R. (2021) Agro-industrial residues: Eco-friendly and inexpensive substrates for microbial pigments production. *Frontiers in Sustainable Food Systems* 5, 589414. DOI: 10.3389/fsufs.2021.589414.

Maghsoudi, S., Taghavi Shahraki, B., Rabiee, N., Fatahi, Y., Bagherzadeh, M. *et al.* (2022) The colorful world of carotenoids: A profound insight on therapeutics and recent trends in nano delivery systems. *Critical Reviews in Food Science and Nutrition* 62(13), 3658–3697. DOI: 10.1080/10408398.2020.1867958.

Mapelli-Brahm, P., Gómez-Villegas, P., Gonda, M.L., León-Vaz, A., León, R. *et al.* (2023) Microalgae, seaweeds and aquatic bacteria, archaea, and yeasts: Sources of carotenoids with potential antioxidant and anti-inflammatory health-promoting actions in the sustainability era. *Marine Drugs* 21(6), 340. DOI: 10.3390/md21060340.

Marín-Sanhueza, C., Echeverría-Vega, A., Gómez, A., Cabrera-Barjas, G., Romero, R. *et al.* (2022) Stress dependent biofilm formation and bioactive melanin pigment production by a thermophilic *Bacillus* species from Chilean hot spring. *Polymers* 14(4), 680. DOI: 10.3390/polym14040680.

Meruvu, H. and Dos Santos, J.C. (2021) Colors of life: A review on fungal pigments. *Critical Reviews in Biotechnology* 41(8), 1153–1177. DOI: 10.1080/07388551.2021.1901647.

Naik, R. and Gupte, S. (2024) Characterization of pigment produced by high carotenoid yielding bacteria *Paracoccus marcusii* RSPO1 and evaluation of its anti-diabetic, anti-microbial and antioxidant properties. *Natural Product Research* 38(6), 968–977. DOI: 10.1080/14786419.2023.2208358.

Nandagopal, P., Steven, A.N., Chan, L.W., Rahmat, Z., Jamaluddin, H. *et al.* (2021) Bioactive metabolites produced by cyanobacteria for growth adaptation and their pharmacological properties. *Biology* 10(10), 1061. DOI: 10.3390/biology10101061.

Nath, S. (2025) Integration of microbial proteins into traditional food systems: Innovations, challenges, and future perspectives. *Food Reviews International* 1–26. DOI: 10.1080/87559129.2025.2520453.

Negi, A., Aggarwal, S., Kathuria, D., Barthwal, R. and Singh, N. (2025) An overview on natural marine pigments: Extraction, potential health benefits and stability in food applications. *Food Reviews International* 1–34. DOI: 10.1080/87559129.2025.2528956.

Neves, A.R., Godinho, S., Gonçalves, C., Gomes, A.S., Almeida, J.R. *et al.* (2024) A chemical toolbox to unveil synthetic nature-inspired antifouling (NIAF) compounds. *Marine Drugs* 22(9), 416. DOI: 10.3390/md22090416.

Nwuzor, I.C., Idumah, C.I., Nwanonenyi, S.C. and Ezeani, O.E. (2021) Emerging trends in self-polishing anti-fouling coatings for marine environment. *Safety in Extreme Environments* 3(1), 9–25. DOI: 10.1007/s42797-021-00031-3.

Omar, A.E., Al-Khalaifah, H.S., Osman, A., Gouda, A., Shalaby, S.I. *et al.* (2022) Modulating the growth, antioxidant activity, and immunoexpression of proinflammatory cytokines and apoptotic proteins in broiler chickens by adding dietary *Spirulina platensis* phycocyanin. *Antioxidants* 11(5), 991. DOI: 10.3390/antiox11050991.

Öztekin, S., Anaya, K. and Yurdunuseven-Yıldız, A. (2022) Regulation of natural food additives. In: *Natural Additives in Foods*. Springer, Cham, Switzerland, pp. 343–372. DOI: 10.1007/978-3-031-17346-2_13.

Park, H., Park, S., Yang, Y.H. and Choi, K.Y. (2021) Microbial synthesis of violacein pigment and its potential applications. *Critical Reviews in Biotechnology* 41(6), 879–901. DOI: 10.1080/07388551.2021.1892579.

Perveen, S., Zafar, S., Iqbal, N. and Riaz, M. (2021) Provitamin a carotenoids. In: *Carotenoids: Structure and Function in the Human Body*. Springer, Cham, Switzerland, pp. 775–797. DOI: 10.1007/978-3-030-46459-2_26.

Ramesh, P.V., Sangeetha Gowda, K.R. and More, S.S. (2025) Introduction to chemistry of microbial colorants: Structures, properties, and biosynthesis. *Microbial Colorants: Chemistry, Biosynthesis and Applications* 129–175. DOI: 10.1002/9781394287888.

Ramos, S.D.P., Bürck, M., Costa, S.F.F. da, Assis, M. and Braga, A.R.C. (2025) Spirulina as a key ingredient in the evolution of eco-friendly cosmetics. *Biotech* 14(2), 41. DOI: 10.3390/biotech14020041.

Rana, B., Bhattacharyya, M., Patni, B., Arya, M. and Joshi, G.K. (2021) The realm of microbial pigments in the food color market. *Frontiers in Sustainable Food Systems* 5, 603892. DOI: 10.3389/fsufs.2021.603892.

Rani, A., Saini, K.C., Bast, F., Mehariya, S., Bhatia, S.K. *et al.* (2021) Microorganisms: A potential source of bioactive molecules for antioxidant applications. *Molecules* 26(4), 1142. DOI: 10.3390/molecules26041142.

Rather, L.J., Mir, S.S. and Islam, Z. (2025) Introduction to microbial colorants. *Microbial Colorants: Chemistry, Biosynthesis and Applications* 1–32. DOI: 10.1002/9781394287888.ch1.

Ručová, D., Vilková, M., Sovová, S., Vargová, Z., Kostecká, Z. *et al.* (2023) Photoprotective and antioxidant properties of scytonemin isolated from Antarctic cyanobacterium Nostoc commune Vaucher ex Bornet & Flahault and its potential as sunscreen ingredient. *Journal of Applied Phycology* 35(6), 2839–2850. DOI: 10.1007/s10811-023-03109-6.

Sahoo, S.R. and Pradhan, A.K. (2024) Prodigiosin: An in-depth exploration of a bioactive compound from *Serratia* sp. *Current Bioactive Compounds* 20(3), 1–13. DOI: 10.2174/011573407227541623092707 4446.

Sen, S. and Mallick, N. (2022) Scytonemin: Unravelling major progress and prospects. *Algal Research* 64, 102678. DOI: 10.1016/j.algal.2022.102678.

Shalapy, N.M., Liu, M. and Kang, W. (2024) Protective effects of hepatic diseases by bioactive phytochemicals in *Fusarium oxysporum* – A review. *Heliyon* 10(5), e26562. DOI: 10.1016/j.heliyon.2024.e26562.

Sharma, N., Shekhar, P., Kumar, V., Kaur, H. and Jayasena, V. (2024) Microbial pigments: Sources, current status, future challenges in cosmetics and therapeutic applications. *Journal of Basic Microbiology* 64(1), 4–21. DOI: 10.1002/jobm.202300214.

Singh, V.K., Jha, S., Rana, P., Mishra, S., Kumari, N. *et al.* (2023) Resilience and mitigation strategies of cyanobacteria under ultraviolet radiation stress. *International Journal of Molecular Sciences* 24(15), 12381. DOI: 10.3390/ijms241512381.

Struyfs, C., Cammue, B.P.A. and Thevissen, K. (2021) Membrane-interacting antifungal peptides. *Frontiers in Cell and Developmental Biology* 9, 649875. DOI: 10.3389/fcell.2021.649875.

Suthar, M., Dufossé, L. and Singh, S.K. (2023) The enigmatic world of fungal melanin: A comprehensive review. *Journal of Fungi* 9(9), 891. DOI: 10.3390/jof9090891.

Tambat, V.S., Singhania, R.R., Sumathi, Y., Chen, C.W., Dong, C.D. *et al.* (2025) Astaxanthin: Nature's multifunctional molecule, natural sources, health benefits, and process advancements. *Critical Reviews in Biotechnology* 1–19. DOI: 10.1080/07388551.2025.2537816.

Tan, H.T., Yusoff, F.M., Khaw, Y.S., Noor Mazli, N.A.I., Nazarudin, M.F. *et al.* (2022) A review on a hidden gem: Phycoerythrin from blue-green algae. *Marine Drugs* 21(1), 28. DOI: 10.3390/md21010028.

Tang, Q., Li, Z., Chen, N., Luo, X. and Zhao, Q. (2025) Natural pigments derived from plants and microorganisms: Classification, biosynthesis, and applications. *Plant Biotechnology Journal* 23(2), 592–614. DOI: 10.1111/pbi.14522.

Timoshnikov, V.A., Selyutina, O.Y., Polyakov, N.E., Didichenko, V. and Kontoghiorghes, G.J. (2022) Mechanistic insights of chelator complexes with essential transition metals: Antioxidant/pro-oxidant activity and applications in medicine. *International Journal of Molecular Sciences* 23(3), 1247. DOI: 10.3390/ijms23031247.

Tiwari, O.N., Ghosh, D., Debnath, S., Sahu, M. and Vanitha, K. (2022) Multifaceted utilization of microalgal biomass towards industrial applications. In: *Biotechnology for Waste Biomass Utilization*. Apple Academic Press, Palm Bay, FL, pp. 79–127.

Tounsi, L., Ben Hlima, H., Hentati, F., Hentati, O., Derbel, H, *et al*. (2023) Microalgae: A promising source of bioactive phycybiliproteins. *Marine Drugs* 21, 440. DOI: 10.3390/md21080440.

Villa, F., Wu, Y.L., Zerboni, A. and Cappitelli, F. (2022) In living color: Pigment-based microbial ecology at the mineral-air interface. *Bioscience* 72(12), 1156–1175. DOI: 10.1093/biosci/biac091.

Wołosiak, R., Drużyńska, B., Derewiaka, D., Piecyk, M., Majewska, E. *et al*. (2021) Verification of the conditions for determination of antioxidant activity by ABTS and DPPH assays-Apractical approach. *Molecules* 27(1), 50. DOI: 10.3390/molecules27010050.

Zandi, P. and Schnug, E. (2022) Reactive oxygen species, antioxidant responses and implications from a microbial modulation perspective. *Biology* 11(2), 155. DOI: 10.3390/biology11020155.

Zhao, L., Qu, Y., Shang, G., Jiang, W., Shen, Q. *et al*. (2025) Natural pigments for microbial control in food systems: Mechanisms, preservation applications, and formulation strategies. *Food Control* 180, 111629. DOI: 10.1016/j.foodcont.2025.111629.

14 Biopigments in Environmental and Agricultural Applications

Jinesh Kaneriya[1], Vivek Pattani[1], Krishna Joshi[1] and Gaurav Sanghvi[2*]
[1]Department of Microbiology, Atmiya University, Rajkot, Gujarat, India; [2]Department of Microbiology, Marwadi University, Rajkot, Gujarat, India

14.1 Introduction

Biopigments are natural colorants produced by various organisms including plants, fungi, bacteria, algae, and marine organisms. Biopigments are naturally occurring compounds that provide color to various organisms, and have gained attention for their ecological and health benefits. These pigments are derived from diverse sources, including plants, microalgae, and microorganisms, each of which offers unique properties and applications. These pigments have gained significant attention owing to their eco-friendly nature and potential health benefits compared to those of synthetic colorants. Microorganisms, particularly fungi and bacteria, are noteworthy sources of biopigment because of their all-season production capability, stability, and cost-effectiveness (Sánchez-Muñoz *et al.*, 2020; Meruvu and Dos Santos, 2021). Interestingly, although fungi and bacteria are prominent sources, microalgae and cyanobacteria are emerging as potential candidates for biopigment production. These organisms have fast-growing abilities and high pigment content and are generally recognized as safe (GRAS) (Saini *et al.*, 2020). Marine sources, especially algae and marine microorganisms, have been identified as rich sources of active natural pigments since the early 20th century (Ye *et al.*, 2019).

Chlorophylls, which are essential for photosynthesis, are abundant in green plants and microalgae, particularly in edible leaves (Sousa, 2022). Carotenoids, responsible for yellow, orange, and red hues, are commonly found in fruits and vegetables, such as carrots and tomatoes, as well as in microalgae, such as *Dunaliella salina*, a notable source of β-carotene (Silva *et al.*, 2020). Anthocyanins, known for their antioxidant properties, are present in fruits, such as blueberries and red pitaya (Sousa, 2022). Melanin, a dark pigment found in various organisms, has attracted interest for its diverse applications, with microbial melanin being particularly promising (El-Naggar and Saber, 2022). Microalgae such as *Spirulina platensis* and *Haematococcus pluvialis* are prominent sources of phycocyanin and astaxanthin, respectively (Silva *et al.*, 2020). Fungi and bacteria, including genetically engineered strains, are being explored for sustainable pigment production, often utilizing agro-industrial waste as a feedstock (Grewal *et al.*, 2022; Kumar *et al.*, 2022). Plants, especially edible leaves, are

*Corresponding author: gv.sanghvi@gmail.com

 Biopigments of Microbial Origin: Innovations and Applications Across Industries (eds Madan *et al.*)
DOI: 10.1079/ 9781836990505.0014

rich in chlorophylls and carotenoids, which contribute to their nutritional value and health benefits.

The diversity of biopigment sources allows for a wide range of applications in various industries including food, cosmetics, pharmaceuticals, and textiles. The growing consumer demand for natural alternatives to synthetic colorants has driven research into sustainable and efficient methods for the production of these pigments. However, challenges such as low yield, high production costs, and downstream recovery must be addressed for large-scale commercial production (Sánchez-Muñoz *et al.*, 2020; Sandybayeva *et al.*, 2022). Despite the numerous advantages of biopigments, such as sustainability and health benefits, commercial viability of microbial pigments remains a challenge. Further research and technological advancements are crucial to improve production efficiency and achieve wider market acceptance.

14.1.1 Significance of biopigments as eco-friendly alternatives to synthetic dyes

The use of biopigments as eco-friendly alternatives to synthetic dyes has attracted significant attention in recent years. Microalgal pigments have been identified as a sustainable approach to textile dyeing, offering a natural and environmentally friendly option for the industry (Mutaf-Kılıc *et al.*, 2023). Similarly, bacterial pigments have emerged as promising alternatives to synthetic colorants, with research focusing on their safety and eco-friendliness (Huang *et al.*, 2024). Fungal-based dyes have also been explored as viable eco-friendly options for synthetic dyes, highlighting the environmental and health challenges associated with traditional dyes (Alegbe and Uthman, 2024). Microbial pigments, including those from bacteria, have been recognized for their therapeutic and industrial potential, further emphasizing their importance compared to synthetic dyes (Anshi *et al.*, 2024). Plant pigments have also been identified as attractive alternatives to synthetic dyes owing to their natural origin and multifunctional benefits (Di Salvo *et al.*, 2023).

Microbial pigments are rapidly replacing synthetic dyes in the food and health industries because of their eco-friendly and nontoxic nature. The cosmetic industry has also seen a shift towards organic pigments, offering a renewable and biodegradable alternative to synthetic dyes. In the textile industry, biopigments have the potential to reduce global water pollution by providing an eco-friendly alternative to traditional inorganic dyes. Bacterial pigments not only have been found to be environmentally friendly and safe for human health but also possess significant pharmacological activities, making them a valuable alternative to synthetic colorants (Huang *et al.*, 2024).

Overall, the significance of biopigments as eco-friendly alternatives to synthetic dyes is evident in various industries including textiles, cosmetics, food, and pharmaceuticals. The use of natural pigments offers a sustainable and environmentally friendly option that addresses the challenges associated with synthetic dyes, highlighting the importance of transitioning towards more eco-friendly practices.

The use of biopigment as an alternative to synthetic dyes in the textile industry offers significant environmental benefits by reducing pollution and enhancing sustainability. Biopigments derived from natural sources are generally biodegradable and less toxic, contributing to an eco-friendly dyeing process. This transition not only mitigates the harmful effects associated with synthetic dyes but also promotes the utilization of renewable resources. Natural dyes are typically biodegradable, unlike synthetic dyes which persist in the environment and contribute to pollution (Grande *et al.*, 2023). Additionally, biopigments exhibit lower toxicity levels, thereby reducing the risk of harmful effects on aquatic life and ecosystems when released into wastewater (Rassabina *et al.*, 2022). Biopigments also support sustainable resource utilization by being sourced from renewable materials, such as agricultural waste, including red onion skins and date palm fibers, thereby promoting waste valorization and resource efficiency (Baaka *et al.*, 2023). Innovations, such as waterless dyeing techniques using supercritical CO_2, further enhance sustainability by minimizing water usage and addressing water scarcity issues associated with traditional dyeing methods (Herrala *et al.*, 2022). Economically, the adoption of biomordants and natural dyes can reduce overall dyeing costs, as demonstrated in studies comparing sonochemical methods to conventional techniques (Periyasamy, 2022).

While the environmental advantages of biopigments are substantial, challenges remain in achieving dyeing efficiency and color fastness comparable to those of synthetic dyes. Further research is essential to optimize these properties and ensure the widespread adoption of biopigments in the textile industry.

14.1.2 Scope of environmental and agricultural applications

14.1.2.1 Environmental applications

BIOREMEDIATION. Pigment-producing microorganisms can help degrade pollutants and toxic compounds in contaminated soil and water. These pigment-producing microorganisms can also act as biosensors, helping to detect and monitor environmental pollutants. Additionally, some microbial pigments have been found to possess antimicrobial properties, which can be utilized in the development of eco-friendly alternatives to chemical pesticides and disinfectants. These pigments can enhance the efficiency of bioremediation processes by improving the survival and activity of microorganisms in polluted environments. Additionally, some microbial pigments have been found to directly bind to and sequester heavy metals, making them valuable tools for removing toxic elements from contaminated sites. Furthermore, the use of pigment-producing microorganisms in bioremediation offers a more sustainable and eco-friendlier alternative to traditional chemical-based remediation methods.

WASTEWATER TREATMENT. Some microbial pigments can act as biosorbents, removing heavy metals and other pollutants from industrial effluents. The application of microbial pigments in wastewater treatment extends beyond heavy metal removal, as they can also aid in the degradation of organic pollutants and the reduction of chemical oxygen demand. These pigments may also serve as indicators of water quality, changing color in response to specific contaminants or environmental conditions. Moreover, the integration of pigment-producing microorganisms into existing wastewater treatment systems could potentially enhance overall treatment efficiency while reducing operational costs.

BIOINDICATORS. Certain pigmented microorganisms can serve as indicators of environmental quality, detecting changes in pH, temperature, or pollutant levels. The use of pigmented microorganisms as bioindicators can provide real-time monitoring of water quality parameters, enabling rapid response to environmental changes in treatment systems. These color-changing microorganisms could be incorporated into sensor technologies for continuous, noninvasive monitoring of wastewater treatment processes. Furthermore, the multifunctional nature of microbial pigments in wastewater treatment presents opportunities for developing integrated, sustainable solutions that address multiple aspects of water purification simultaneously.

ECO-FRIENDLY DYES. Microbial pigments offer sustainable alternatives to synthetic dyes in various industries, reducing environmental impact. These natural pigments, produced by microorganisms such as bacteria, fungi, and algae, can be used in textiles, food coloring, and cosmetics. Their production requires fewer resources and generates less waste compared to traditional synthetic dye manufacturing processes. Moreover, microbial pigments often exhibit biodegradability and lower toxicity, making them a promising solution for addressing environmental concerns associated with conventional dye production and usage.

14.1.2.2 Agricultural applications

BIOPESTICIDES. Some microbial pigments exhibit antimicrobial properties, potentially serving as natural pesticides against plant pathogens. These biopesticides could offer an environmentally friendly alternative to synthetic chemical pesticides, reducing the ecological impact of crop protection. Additionally, microbial pigments may enhance plant resistance to pests and diseases by stimulating the plant's natural defense mechanisms. Further research is needed to explore the efficacy and safety of pigment-based biopesticides in various agricultural settings and crop types.

BIOFERTILIZERS. Pigment-producing microorganisms can enhance soil fertility and promote plant growth through various mechanisms. Some pigment-producing microorganisms can also fix atmospheric nitrogen, solubilize phosphorus, and produce plant growth-promoting hormones, contributing to improved soil health and crop yields. These beneficial microbes may form symbiotic relationships with plant roots, enhancing nutrient uptake and overall plant performance. Furthermore, the use of biofertilizers based on pigment-producing microorganisms could reduce reliance on chemical fertilizers, leading to more sustainable agricultural practices and reduced environmental impact.

CROP PROTECTION. Certain microbial pigments may induce systemic resistance in plants, improving their defense against pathogens and pests. The pigments produced by these microorganisms may also have antifungal or antibacterial properties, directly protecting plants from harmful pathogens. Additionally, some microbial pigments can act as natural UV protectants for plants, shielding them from excessive sun damage and improving their resilience to environmental stressors. These multifaceted benefits of pigment-producing microorganisms highlight their potential as a sustainable and eco-friendly solution for enhancing crop health and productivity in modern agriculture.

SEED COATING. Microbial pigments can be used to develop protective coatings for seeds, thereby enhancing their germination and early plant growth. These coatings can provide a barrier against pathogens and pests, reducing the need for chemical pesticides. Additionally, microbial pigments may offer UV protection to seeds, thus improving their viability during storage and transportation. The use of such coatings aligns with sustainable agricultural practices, potentially increasing crop yields while minimizing the environmental impact.

LIVESTOCK FEED ADDITIVES. Some microbial pigments have potential as natural feed colorants and nutritional supplements for animals, and the incorporation of microbial pigments into livestock feed can enhance the visual appeal of animal products, such as egg yolks and fish flesh, making them more attractive to consumers. Furthermore, certain microbial pigments possess antioxidant properties that may contribute to improved animal health and longevity. These natural additives could potentially reduce reliance on synthetic colorants and supplements in the livestock industry, aligning with growing consumer demand for more natural and sustainable animal products.

POSTHARVEST PRESERVATION. The antimicrobial properties of certain pigments may help extend the shelf-life of agricultural products, and the use of microbial pigments in postharvest preservation can lead to a significant reduction in food waste and economic losses in the agricultural sector. Additionally, these natural preservatives may offer safer alternatives to synthetic chemical treatments, potentially reducing the risk of harmful residues in food products. Further research on the efficacy and safety of microbial pigments as preservatives could pave the way for their widespread adoption in various food industries, from fresh produce to processed goods.

SOIL HEALTH INDICATORS. Pigmented microorganisms can serve as indicators of soil quality and health in agricultural ecosystems. These applications highlight the potential of microbial pigments to contribute to sustainable practices in both environmental management and agriculture. The use of microbial pigments as soil health indicators could lead to the development of innovative diagnostic tools for farmers and agricultural scientists. These tools may enable more precise and timely interventions in soil management, potentially improving crop yields and reducing the need for chemical fertilizers. Moreover, the multifaceted applications of microbial pigments in both food preservation and soil health assessments underscore their potential as versatile and eco-friendly solutions in the broader context of sustainable agriculture and food production.

14.2 Biopigments in Environmental Applications

Biopigments are compounds that are synthesized spontaneously by several organisms, including algae, bacteria, fungus, and plants. Apart from their physiological and biological purposes in natural systems, these pigments might be applied in other ways in the environment. By enhancing the condition of the soil and accelerating the breakdown of contaminants, biopigments can support long-term environmental management. The primary applications for biopigments in bioremediation, soil nutrition, and UV protection are discussed below.

14.2.1 Bioremediation: pigments aiding in pollutant degradation

Biopigments can actively break down or store pollutants as their molecule structure is arranged in distinctive ways. Biopolymer melanin may bind and store heavy metals like cadmium, lead, and mercury. Studies reveal that chelating metal ions requires functional groups of melanin like carboxyl, hydroxyl, and indole. However, the living entities find it more difficult to obtain these detrimental components in contaminated sites/surroundings. Pigments effective in destroying organic contaminants include violacein, produced by *Chromobacterium violaceum*, as they also possess antioxidative and antibacterial properties (Orlandi *et al.*, 2022). By facilitating reactive breakdown, violacein has been shown to degrade aromatic hydrocarbons, a main class of environmental toxins. Important for bioremediation efforts is this pigment's capacity to both reduce toxins and increase microbial activity (Nawaz *et al.*, 2021). Another example is pyomelanin, which is a pigment produced by several species of *Pseudomonas* that has demonstrated efficacy in decomposing phenolic compounds present in commercial waste (Krastanov *et al.*, 2013). Pyomelanin works as an electron mediator, increasing the activity of the enzymes of bacteria and therefore facilitating the breakdown of chronic toxins. Including pyomelanin-producing bacteria in contaminated sites seems to help break down contaminants in a more effective manner (Gadd, 2007).

Carotenoids can ais in oil spill clean-up. Many kinds of algae produce carotenoids, which protect microorganisms that break down oil from reactive stress (Chokshi *et al.*, 2017). Moreover, the slippery character of carotenoids facilitates their interaction with oil droplets. Microbes can thus easily access hydrocarbons and break them down faster. This use demonstrates how several types of pollutants may be cleaned by biopigments.

14.2.2 Biopigment application in wastewater treatment

Biopigments can absorb contaminants from wastewater, break them down, and neutralize them. In particular, chlorophyll compounds derived from algae are quite effective in eliminating nutrients from wastewater. Increasing oxygen content in water strengthens the photosynthesis of chlorophyll, therefore increasing aerobic microbe activity and facilitating the breakdown of pollutants (Mandal *et al.*, 2021).

Melanin pigment also performs quite effectively in biofiltration systems, since it can easily chelate heavy metals like mercury and arsenic (Muñoz-Torres *et al.*, 2024). Melanin is also an antioxidant, so it may neutralize reactive oxygen species (ROS), thereby preventing the deterioration of water quality. Melanin-based biofilters, according to Cordero *et al.* (2017), significantly reduced the heavy metal content in industrial wastewater; hence, they are a useful choice for many stakeholders including researchers and industries. Phycobiliproteins are particularly effective in removing pigments from textile industry wastewater. They are derived from cyanobacteria along with other pigments like phycocyanin and phycoerythrin. These pigments bind to molecules as biosorbents, therefore facilitating color extraction. They are both quantifiable and helpful due to their unique properties for facilitating the degradation process in real time (Kini *et al.*, 2020).

The red pigment prodigiosin produced by *Serratia marcescens* is antimicrobial and plays a crucial role in cleaning wastewater. Prodigiosin pigment stops the growth of harmful bacteria in wastewater, reducing the biological oxygen demand (BOD) of wastewater and therefore improving water quality (Tejasvini *et al.*, 2016). Prodigiosin pigment has gained attention due to its ability to break down phenolic compounds and other harmful substances. Additionally, carotenoids and other biopigments are used in bioreactors to assist the breakdown of hydrocarbons and other organic contaminants. Carotenoids protect microbial populations from oxidative damage resulting from generated reactive intermediates arising from the breakdown of pollutants. This step increases the lifetime and efficiency of the bioreactor. Adding bacteria that produce carotenoids to wastewater treatment facilities has improved their ability to eliminate petroleum products and other industry pollutants (Verma *et al.*, 2013).

14.2.3 Soil health and enrichment

14.2.3.1 *Enhancing soil microbial activity and fertility*

Soil fertility, condition, and utility depend heavily on microbial activity in soil. For breaking down organic matter, cycling nutrients, and creating soil structure, microorganisms are absolutely vital (Srivastava *et al.*, 2023). By means of microbial activity, soil becomes far richer and crop yields are greatly increased. Several approaches exist to enhance microbial populations and their functions.

One approach is to provide additional organic matter by means of compost, manure, and cover crops. Organic material provides nutrition for bacteria, enabling their growth and operation. Furthermore, applying many crop cycles can provide the soil with various carbon sources, therefore supporting a greater spectrum of microorganisms.

Reducing tilling and other techniques that disturb the soil also helps preserve its structure and provide more stable microbial niches. Excessive tillage can also alter the types of bacteria present and reduce the content of organic matter in the soil (Angon *et al.*, 2023). Reduced tillage or no-till conservation techniques help to maintain soil structure and increase microbial activity.

14.2.3.2 *Role of pigments in organic composting and soil structure*

Pigments play an essential role in the degradation of organic molecules and the enhancement of soil structure. The majority of these substances occurring in nature are derived from microorganisms and plants. They influence temperature regulation, microbial activity, and nutrient cycling, among various composting processes (Panesar *et al.*, 2015).

Carotenoids and chlorophyll are examples of pigments that operate as photosensitizers by collecting light energy and starting photochemical reactions as molecules break down. The breakdown of organic compounds is accelerated by the ROS produced by these reactions (Chokshi *et al.*, 2017). Furthermore, the species selection process can change the microbial communities' make-up.

The process of species selection can modify the composition of the microbial communities. The addition of pigments to compost prolongs the longevity of the soil. The humic acids present in compost interact with soil particles, increasing the adhesive properties of the soil and enhancing its capacity to retain moisture. The stabilized soil particles promote improved air circulation and drainage, thereby fostering optimal conditions for bacterial activity and root development.

14.2.4 UV protection and photoprotection

14.2.4.1 *Protective role of pigments in soil and plant health*

Ultraviolet (UV) light is one of the main natural stressors that affects plants and soil health by causing oxidative damage. Microorganisms contain two pigments that protect them from the harmful effects of UV rays: melanin and carotenoids. Melanin, which is resistant to a wide range of ultraviolet radiation, is produced by bacteria and fungi. This protects the microorganisms that live in the soil from oxidative stress. According to Gao and Garcia-Pichel (2011), this pigment protects bacteria that live in the soil from the DNA damage caused by ultraviolet light. This helps to keep microbial populations stable.

Carotenoids act as a shield for cells against the ROS that are created by ultraviolet radiation. The presence of these hues helps plants to become more robust by encouraging beneficial interactions between plants and bacteria, which in turn protects microbial cells. Singh *et al.* (2005) discovered that carotenoid-producing bacteria allow plants to survive and thrive in the presence of rising UV light. This is accomplished by reducing the amount of oxidative damage that occurs (Gao and Garcia-Pichel, 2011).

14.2.4.2 *Applications in reducing environmental UV impact*

Researchers are investigating the pigmentation of microorganisms with the intention of protecting the environment from the harmful effects of ultraviolet radiation. Plants can be protected against an excessive amount of UV light by employing biofertilizers that include melanin and insects that produce carotenoids. According to Verma *et al.* (2013), these biofertilizers assist plants in better coping with stress, which contributes to improved crop performance when exposed to UV radiation.

Although they are applied in agriculture, microbial pigments also find utility in the creation of UV light-immune objects. Melanin derived from microorganisms is being included in coatings and films in order to protect surfaces from UV damage. This treatment helps plastics and other synthetic materials withstand UV light degradation. Melanin-based treatments have been shown in studies by Muñoz-Torres *et al.* (2024) to extend the lifetime of outdoor objects.

Microbial hues are also being investigated by researchers to determine whether they may prevent UV rays from penetrating soil. Although UV radiation can damage plant roots and soil microorganisms, adding bacteria that produce colors to soil cleaners helps to reduce these effects. This approach helps to improve agricultural systems in areas where UV radiation is greater and enhances the health of the soil (Gadd, 2007).

14.3 Biopigments in Agricultural Applications

14.3.1 Plant growth promotion

14.3.1.1 Biopigments as biofertilizers and growth enhancers

Improvement of agricultural production on a daily basis is determined by a combination of global challenges, demands, and limitations of current agricultural systems. The world's population is projected to reach around 9.7 billion by 2050 (Ehrlich and Harte, 2015). This rapid population growth translates into an increased demand for food. The current scenario to improve agriculture production uses large-scale chemical-based fertilizers and pesticides (Fess *et al.*, 2011). To overcome this problem and improve agricultural production, alternative methods can be provided by biopigment-based fertilizers. These fertilizers can boost plant growth, improve soil health, increase resilience to pests and diseases, and reduce the environmental impact of agriculture (Arfa *et al.*, 2024). Biopigments derived from certain bacteria, fungi and algae can promote the growth of nitrogen-fixing bacteria, which increase soil nitrogen and reduce the need for chemical fertilizers. Additionally, the presence of biopigments can enhance the microbial diversity of the rhizosphere, improving soil structure and promoting sustainable agricultural practices. These biopigments enhance soil health and promote plant growth while minimizing environmental impact (Michalak *et al.*, 2021).

Biopigments have shown significant effects on plant growth and development. These pigments, such as carotenoids, chlorophylls, anthocyanins, and phycobilins, phycocyanin, melanin, flavanoids, etc., contribute various physiological processes to plant health, including light absorption, nitrogen fixing, antioxidant activity, oxidative stress, stress mitigation, nutrient cycling, biofortification and interaction with beneficial soil microorganisms (Barreto *et al.*, 2023).

Biopigments are beneficial for plants as they produce secondary metabolites that act as growth regulators or antioxidants. β-Carotene possesses antioxidant properties that can mitigate oxidative stress in plants (Dewanjee *et al.*, 2021), promoting better root development, improved photosynthesis, and enhanced resistance to environmental stress such as drought or salinity. Anthocyanin pigments are responsible for red, blue, and purple colors in plants and improve plant tolerance to UV radiation and heavy metal toxicity, enhancing crop resilience (Landi *et al.*, 2015). *Microcystis* is a genus of cynobacteria whose pigments include phycocyanin, chlorophyll a, and phycobilins (Saini *et al.*, 2018). *Microcystis* has been studied for its potential use in agriculture, especially as a source of natural pesticides. Phycocyanin has demonstrated antimicrobial properties that make it useful as a natural pesticide (Abd El-Baky and El-Baroty, 2012). It inhibits the growth of certain pathogenic bacteria, fungi, and algae, which could help in controlling pathogens and pests. Phycobilin is a promising pigment in the development of biopesticides and other bioactive compounds for sustainable farming (Hassan *et al.*, 2022).

Phycocyanin and phycobilins can inhibit plant pathogenic bacteria such as *Ralstonia solanacearum* (Orlandi *et al.*, 2022), which causes disease in solanaceous crops like tomatoes and potatoes (Ababa, 2024), and play a vital role in controlling *Xanthomonas* species, responsible for diseases like leaf spot on pepper and citrus plants (Joshi *et al.*, 2020). The antimicrobial effects of phycocyanin and phycobilins interact with microbial cells and disrupt cell membranes, inhibit cellular respiration, or interfere with protein synthesis in harmful microorganisms.

Serratia marcescens produces prodigosin (El-Bialy and Abou El-Nour, 2015), a red pigment with antibacterial, antifungal, and antiviral properties, which suppresses bacteria like *Pseudomonas syringae* (Kshetri *et al.*, 2019). Prodigiosin has antifungal activity against *Fusarium oxysporum*, *Rhizoctonia solani* (which causes root rot) (John Jimtha *et al.*, 2017), and *Botrytis cinerea* gray mold (Someya *et al.*, 2001). This makes *Serratia marcescens* a valuable tool in combating soil-borne and foliar fungal diseases. *Bacillus subtilis* is known for its beneficial role in agriculture, by producing the bioactive pigment melanin (Ghadge *et al.*, 2020). Melanin is a complex biopolymer associated with UV radiation resistance and oxidative stress tolerance and helps to protect plant tissues from UV exposure and prevent DNA mutations or protein denaturation (Glagoleva *et al.*, 2020). Melanin stimulates the antioxidant enzymes superoxide dismutase (SOD), catalase (CAT), and peroxidase (POD), which play important roles in detoxifying ROS (Cordero and Casadevall, 2020).

Rhodobacter sphaeroides, a bacterium that produces astaxanthin, exhibits both antifungal and antibacterial activities (Agarwal *et al.*, 2023), effectively inhibiting the growth of *Fusarium* and *Botrytis*. *Brevibacterium* sp. produces lutein and zeaxanthin, pigments which protect against pathogenic *Pseudomonas* spp. and *Fusarium* spp. (Orlandi *et al.*, 2022).

Several fungi are responsible for producing biopigments. *Monascus* spp. produce the red pigment monascorubrin which exhibits antifungal and antibacterial properties, inhibiting the growth of *Rhizopus* and *Penicillium*. *Monascus* is an effective biocontrol agent for preventing fungal infections in rice and vegetables (Ajithkumar *et al.*, 2024). *Aspergillus niger* produces anthocyanins, which have antifungal properties that are effective against the pathogens *Rhizopus* sp. and *Alternaria*. These anthocyanins help reduce fungal diseases in fruits and vegetables. The microalga *Haematococcus pluvialis* produces astaxanthin and has antifungal activity against *Rhizopus* (Badiale-Furlong *et al.*, 2023). It is useful in preventing fungal infections in crops post harvest. *Chlorella* sp. produce lutein and zeaxanthin, which exhibit antibacterial and antifungal properties against *Xanthomonas* and *Fusarium* (Gonçalves, 2021). The yeast species *Rhodotorula glutinis*, which produces β-carotene and torulene, is particularly effective against *Fusarium* and *Penicillium*. Similarly, *Phaffia rhodozyma* produces astaxanthin, which inhibits the growth of *Fusarium* and *Botrytis* (Georgescu *et al.*, 2024). Using these biopigments, farmers can reduce their chemical pesticide application, fostering more sustainable agricultural practices for maintaining crop health.

14.3.1.2 Role of carotenoids and anthocyanins in boosting plant health

Carotenoids and anthocyanins, known for their roles as pigments and antioxidants in plants, can also influence plant growth promotion. These compounds are involved in signaling, stress response, and protection mechanisms which can directly or indirectly enhance plant health and growth. Carotenoids are red, orange, and yellow pigments with antioxidant properties. *Rhodobacter sphaeroides*, *Brevibacterium* sp., and *Xanthobacter* sp. produce carotenoids including astaxanthin, lutein, zeaxanthin, lycopene, and β-carotene. *Fusarium* sp. and *Mucor* sp. produce β-carotene (Avalos and Carmen Limón, 2015). *Chlorella* sp. and *Haematococcus pluvialis* are notable for producing lutein, zeaxanthin, and astaxanthin (Cerón *et al.*, 2007; Bazarnova *et al.*, 2022). *Rhodotorula glutinis* and *Phaffia rhodozyma* are known for producing β-carotene and astaxanthin, respectively.

Carotenoids are crucial photosynthetic pigments. They act as accessory pigments, capturing light energy and transferring it to chlorophyll for efficient photosynthesis. Carotenoids such as β-carotene, lutein, and zeaxanthin function as antioxidants, protecting plants from oxidative stress induced by environmental factors such as UV radiation and drought by neutralizing ROS (Sandmann, 2019). Anthocyanins, water-soluble pigments that appear red, purple, or blue, are produced by *Pseudomonas* sp. under specific conditions (Garcia-Seco *et al.*, 2015). Microorganisms producing carotenoids and anthocyanins contribute to plant growth by enhancing resistance to pathogens, improving stress tolerance, and boosting nutrient uptake through improved soil microbiome interactions (Rahman *et al.*, 2018). These microbial activities help plants grow more effectively, even under challenging conditions.

14.3.2 Pest and disease management

14.3.2.1 Natural antimicrobial and antifungal properties of biopigments

Pest and disease management is a critical aspect of sustainable agriculture. Chemical pesticides and fungicides have been used to control pests and pathogens but increasing concerns about environmental damage and health risks suggest that natural biopigments are the alternative and eco-friendly solutions for pest and disease management (Singh *et al.*, 2023). Biopigments have antimicrobial and antifungal properties to combat plant diseases (Shah *et al.*, 2021).

14.3.2.2 Application of biopigments in integrated pest management strategies

Integrated pest management (IPM) has potential applications for different pest control scenarios such as repelling, attracting, or inhibiting pests. It holds promise for more integrated and nature-based pest management systems (Way and van Emden, 2000). Certain plant-derived anthocyanins and carotenoids possess insect repellent properties (Nigam and Luke, 2016). Pigments applied to plant surfaces reduce the feeding behavior of aphids, whiteflies, and caterpillars (Verma *et al.*, 2021). The presence of anthocyanins in plants can attract natural predators, such as ladybugs (Al-khayri *et al.*, 2023), by enhancing the visual appeal of plants. Biopigments stimulate defense mechanisms by phytoalexins to reduce pest attacks (Wink and Lehmann, 1996). Flavonoids in citrus fruits have been shown to enhance the plants' resistance to pests such as aphids and whiteflies (Ramaroson *et al.*, 2022). The flavonoid compounds trigger the plant's defense system, making it more difficult for the pests to infest the crop. Future developments could enhance their use in agro-ecosystems, promoting biodiversity and resilience against pest infestations.

14.3.3 Seed treatment and yield improvement

Coating seeds with biopigments has emerged as a promising strategy to enhance seed germination and crop yield. Biopigments can improve light absorption and protect seeds from environmental stress, thereby promoting healthier plant growth. Seed coatings containing biopigments can increase germination rates significantly (Zhang *et al.*, 2022). Food quality, nutritional value, and financial gain are all significantly reduced by postharvest crop loss and spoiling. Microbial growth, oxidation, and enzymatic degradation make it difficult to preserve harvested crops. Using biopigments with antioxidant properties as natural preservatives can improve the shelf-life of harvested crops.

Naturally occurring biopigments can protect against microbial and fungal contamination in postharvest crops as they have strong antioxidant, antimicrobial, and antifungal properties. Carotenoids such as β-carotene, lutein, and zeaxanthin are produced by organisms including *Bacillus* spp., *Streptomyces* spp., and *Xanthophyllomyces dendrorhous*. These pigments are used in seed coatings to extend shelf-life through antioxidant action, protect crops from pests, and enhance the color and nutritional content of fruits and vegetables (Gebregziabher *et al.*, 2022). Flavonoids, such as quercetin, kaempferol, and apigenin which are produced by *Bacillus* spp., *Pseudomonas* spp., and *Penicillium* spp., exhibit natural pest control properties due to their antimicrobial effects (Šovljanski *et al.*, 2023). They also protect seeds from oxidative stress, enhancing seed viability and promoting better germination rates. Additionally, flavonoids can influence seedling vigor by acting as natural growth regulators (De Oliveira *et al.*, 2024).

Chlorophyll a, b, and chlorophyllin pigments help improve seedling growth by protecting against oxidative stress and act as natural plant growth regulators, stimulating growth under environmental stress (Nazir *et al.*, 2020). They are also effective in postharvest preservation, preventing oxidative spoilage of crops. These pigments serve as natural colorants for fruits and vegetables, and provide protection against microbial spoilage due to their antifungal and antibacterial properties (Novais *et al.*, 2022). Xanthophylls, including lutein and zeaxanthin, are used in seed coatings to protect seeds from environmental stress and fungal pathogens and improve postharvest quality by preventing oxidative spoilage, and enhance crop nutritional value by increasing antioxidant content (Ghosh *et al.*, 2024). Violacein provides strong antimicrobial protection in seed coatings, improving germination rates (Anshi *et al.*, 2024). It provides antioxidant protection to seeds, enhancing their resistance to biotic and abiotic stresses. Phaeophytins increase postharvest preservation by reducing enzymatic degradation of crops (Lemes and Catão, 2024). Astaxanthin, a potent natural antioxidant, protect crops from oxidative stress during preservation and storage. It is particularly beneficial for fatty crops like avocados to protect against lipid oxidation and enhance growth stages (Ambati *et al.*, 2014) (Table 14.1).

Biopigments have diverse applications in agriculture, from improving seed viability and growth to protecting crops from microbial infections and oxidative damage. These pigments are recommended as sustainable, eco-friendly alternatives to enhance yield and extend the shelf-life of crops (Fig. 14.1).

14.4 Challenges and Limitations of Biopigments

Biopigments have gained increasing attention as alternatives to synthetic dyes in various industries, including agriculture, food production, and textiles. However, despite their promising environmental benefits and potential for sustainable production, biopigments still face several challenges and limitations that hinder their widespread use. These challenges span across production, regulatory concerns, environmental impact, and stability issues.

14.4.1 Production challenges

One of the primary challenges in biopigment production lies in the complexities related to cost and scalability. While biopigments are often touted for their environmental advantages over synthetic dyes, their production costs remain high due to several factors. The need for specialized equipment, time-consuming fermentation processes, and reliance on specific growing conditions for certain plants or microorganisms often result in higher production costs. Furthermore, scaling up biopigment production to meet industrial demands is difficult due to the relatively small-scale operations currently in place. The production of certain pigments may also require complex nutrient inputs or specialized substrates, adding to the overall expense and making large-scale production economically unfeasible in some cases (Di Salvo *et al.*, 2023).

Additionally, biopigments sourced from plants or microorganisms can face seasonal limitations and supply chain issues. For plant-based pigments, the availability of the raw material is often tied to specific harvest seasons, which can lead to shortages during off-peak times. Similarly, microbial production may suffer from variability in yields due to environmental conditions, such as temperature fluctuations or nutrient availability, further complicating the scalability of biopigment production.

14.4.2 Regulatory and safety concerns

In addition to production challenges, regulatory and safety concerns also play a crucial role in the limitations of biopigments. As biopigments are used in a wide range of applications, including food, cosmetics, and textiles, they are subject to strict regulations set by governmental bodies such as the FDA (Food and Drug Administration) or the EFSA (European Food Safety Authority) (Kumar *et al.*, 2015). These regulations aim to ensure that biopigments are safe for consumption and use, and their environmental impact is minimal.

One key regulatory challenge is the need for comprehensive environmental impact assessments (EIAs) for biopigment production. As biopigments are derived from living organisms, the cultivation and extraction processes

Table 14.1. Biopigments in agriculture.

Biopigment type	Producing organism	Biopigment name	Applications in agriculture	Reference
Carotenoids	*Bacillus* spp., *Streptomyces* spp., *Xanthophyllomyces dendrorhous*	β-Carotene, lutein, zeaxanthin	• Seed coating. Antioxidant properties extend shelf-life. Pest control. • Carotenoids possess antimicrobial effects. • Nutritional enrichment. Enhance color and nutrition in fruits and vegetables	(González-Peña *et al.*, 2023)
Phycocyanin/ phycobilins	*Cyanobacteria*	Phycocyanin, phycobilins,	• Natural pesticides, • antifungal activity	(Do Amaral *et al.*, 2023)
Anthocyanins	*Chromobacterium violaceum*, *Monascus* spp., *Aspergillus* spp.	Cyanidin, delphinidin, pelargonidin	• Antioxidant treatments. • Protect crops from oxidative damage. • Biological pest resistance. Inhibit fungal growth. • Seed coating. Reduce microbial infections and enhance germination	(Oliveira *et al.*, 2020)
Flavonoids	*Bacillus* spp., *Pseudomonas* spp., *Penicillium* spp.	Quercetin, kaempferol, apigenin	• Natural pest control. • Flavonoids show antimicrobial properties. • Protection against oxidative stress. • Enhances seed viability and germination. • Plant growth regulation. • Influence seedling vigor	(Khan *et al.*, 2024)
Chlorophylls	*Chlorobium* spp., *Anabaena* spp., *Chlorella* spp. *Microcystis*	Chlorophyll a, b, chlorophyllin	• Improve seedling growth. • Protect against oxidative stress. • Natural plant growth regulators. • Stimulate growth under stress. • Postharvest preservation. • Prevent postharvest oxidative spoilage	(Qiu *et al.*, 2019)

Continued

Table 14.1. Continued

Biopigment type	Producing organism	Biopigment name	Applications in agriculture	Reference
Betalains	*Beta vulgaris* (beetroot), *Cactaceae* family (prickly pear)	Betacyanins, betaxanthins	• Natural colorants for fruits and vegetables. • Protection against microbial spoilage. Betalains have antifungal and antibacterial properties. • Stress resistance. • Enhance plant tolerance to heat and drought stress	(Zannou *et al.*, 2024)
Xanthophylls	*Chlorobium* spp., *Corynebacterium* spp.	Lutein, zeaxanthin	• Seed coating. Protect seeds from environmental stress and fungal pathogens. • Improve postharvest quality • Prevent oxidative spoilage. • Nutrient enhancement of • crops through antioxidant enrichment	(Demmig-Adams *et al.*, 2020)
Violacein	*Chromobacterium violaceum*, *Pseudomonas* spp.	Violacein	• Antimicrobial seed coating. Reduces seed-borne diseases and enhances germination rates. • Antioxidant protection. • Protects seeds and seedlings from oxidative damage. • Stress tolerance. • Improves resistance to biotic and abiotic stresses	(Khaksar and Rigi, 2022)
Pheophytins	Chlorophyceae algae, *Corynebacterium* spp.	Pheophytin Chlorophyll a, b	• Postharvest preservation. • Reduce enzymatic degradation in crops. • Protection against oxidative damage, • Enhance shelf-life	(Dissanayake *et al.*, 2012)

Continued

Table 14.1. Continued

Biopigment type	Producing organism	Biopigment name	Applications in agriculture	Reference
Astaxanthin	*Haematococcus pluvialis*, *Chlorella* spp., *Phaffia rhodozyma*	Astaxanthin	• Natural antioxidant, protects crops against oxidative stress during storage. • Protection of fatty crops: prevents lipid oxidation in seeds like avocados. • Enhancement of seedling vigor. Improves early-stage growth by protecting against environmental stressors	(Oslan *et al.*, 2021)
Aspergillin	*Aspergillus* (fungus)	Citrinin Melanin Secondary metabolites	• Fungal biocontrol and antimicrobial properties	(Ismail *et al.*, 2024)
Penicillins	*Penicillium* (fungus)	Citrinin, melanin Secondary metabolites	• Used in antibiotic production (penicillin). • Food spoilage prevention (antimicrobial properties)	(Kumar *et al.*, 2018)
Isoflavones, carotenoids	*Glycine max* (soybean)	Isoflavonoids, carotenoids	• Used in biofortification of crops to improve nutritional content	(Summra *et al.*, 2024)
Chlorophyll	*Chlorella* (green algae)	Photosynthetic pigment	• Photosynthesis enhancement in plants. • Used in biofertilizers to promote growth and reduce the need for chemical fertilizers	(Dudina *et al.*, 2024)
Phycocyanin	*Spirulina*(blue-green algae)	Phycobilin, photosynthetic pigment	• Antioxidant properties (phycocyanin). • Enhances plant resistance to stress	(Castro-Gerónimo *et al.*, 2023)
Astaxanthin	*Haematococcus pluvialis* (green algae)	Carotenoid	• May help enhance plant growth through better stress tolerance	(Xing *et al.*, 2022)
Phycoerythrin	*Rhodomonas* (red algae)	Phycobilin, photosynthetic pigment	• Enhances plant photosynthesis in low light conditions	(Latsos *et al.*, 2021)

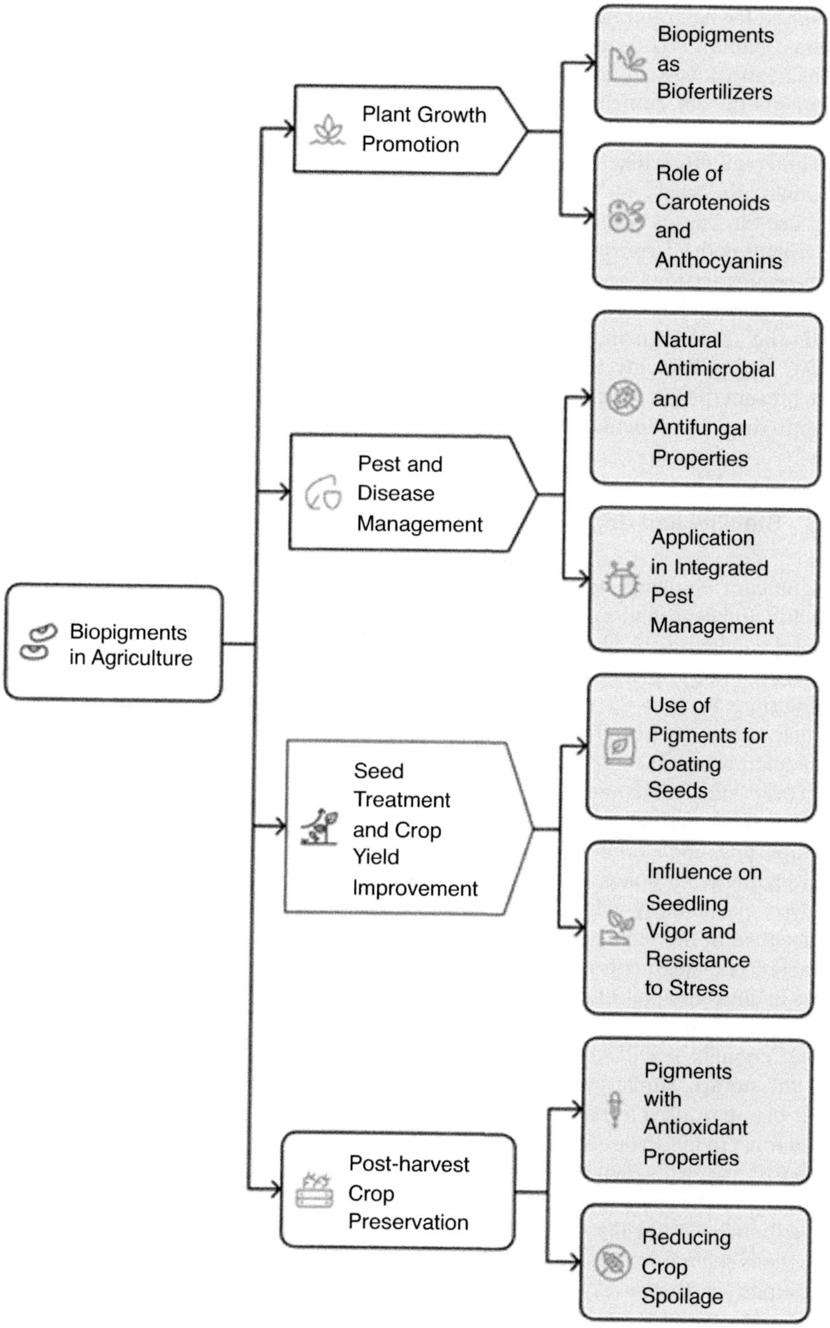

Fig. 14.1. Role of biopigments in agriculture.

must be assessed for potential negative impacts on ecosystems, such as soil degradation, water use, and the release of waste products. Ensuring that biopigments do not contribute to habitat destruction or biodiversity loss is a priority for researchers and regulators alike.

Additionally, the safety of biopigments in agricultural use must be closely monitored. For instance, pigments derived from certain microorganisms may require testing for potential toxicity, allergenicity, or other adverse effects on human health and the environment (Charalampia *et al.*, 2017). Stringent safety regulations are necessary to prevent the introduction of harmful substances into the environment or food chain.

14.4.3 Stability and degradation

Another significant challenge for biopigments is their stability and degradation under varying environmental conditions. Unlike synthetic pigments, which are often engineered to withstand extreme conditions such as UV radiation, temperature fluctuations, and exposure to air and moisture, biopigments are more susceptible to degradation over time. This can result in a reduction in color intensity, fading, or loss of efficacy during storage and application. For instance, plant-based pigments may degrade faster than synthetic dyes, particularly when exposed to sunlight, moisture, or high temperatures.

The stability of biopigments also affects their performance in different applications. In the food industry, for example, the shelf-life of biopigments must be long enough to withstand processing, packaging, and storage. Similarly, in agricultural applications, pigments used as natural colorants or fungicides must retain their efficacy over time to ensure consistent results (Lyu *et al.*, 2022). To overcome these challenges, further research is required to enhance the stability of biopigments, through methods such as encapsulation, chemical modifications, or the development of new biopigment sources with greater resilience.

14.5 Future Perspectives

The future of biopigments holds exciting possibilities, driven by advances in research and development (R&D), sustainable agricultural practices, and growing awareness of environmental concerns. Although challenges remain, continued efforts in these areas may lead to new solutions for scaling up biopigment production and improving their applications (Tuli *et al.*, 2015).

Future prospects for finding new biopigments include a variety of exotic fruits and vegetables with a potential spectrum of unique pigment coloring capabilities. These properties are characterized by additional use as functional food ingredients and novel pigment sources. The novel technique necessitates thorough food safety evaluation. Also a detailed survey was published on market-available colorants, highlighting legal restrictions, key chemical characteristics, color attributes, and the stability profiles of novel pigment alternatives (Brauch, 2016).

14.5.1 Research and development

One of the key areas of future growth for biopigments lies in the discovery of new pigment sources. While many biopigments have been derived from common plants and microorganisms, there is still significant untapped potential in exploring novel species, particularly those in underexplored or remote environments (Yadav *et al.*, 2023). This could lead to the identification of biopigments with unique colors, improved stability, and enhanced functionality, meeting the diverse needs of industries that require sustainable alternatives to synthetic dyes.

Genetic and metabolic engineering advancements also hold promise for improving the yield of biopigments. By modifying the genes responsible for pigment production in microorganisms or plants, researchers can potentially increase the efficiency of pigment biosynthesis, reduce production costs, and enhance the overall quality of the pigment (Rather *et al.*, 2023). For example, the introduction of specific enzymes or genetic pathways in microbial strains could result in higher pigment concentrations, thus making large-scale biopigment production more viable. Additionally, genetic engineering could enable the creation of novel biopigments with tailored properties, such as enhanced stability or improved solubility, to meet specific industrial requirements. These advancements in biotechnology could significantly expand the range of applications for biopigments, potentially leading to their adoption in sectors where synthetic dyes currently dominate. Furthermore, the development of more efficient extraction and purification techniques for biopigments is crucial for their widespread commercial use.

14.5.2 Sustainable agricultural practices

Integrating biopigments into sustainable farming models represents another promising avenue for future developments. Biopigments can be used in agriculture not only for their direct benefits in colorants and natural pesticides but also as a tool for enhancing soil health and supporting biodiversity (Mohammad Azmin *et al.*, 2022). Biopigments derived from microorganisms or plants can be employed to reduce the use of chemical pesticides and fertilizers. For example, certain biopigments possess antimicrobial and antifungal properties, which could help control plant diseases and reduce reliance on chemical agents.

The integration of biopigments in sustainable farming models could contribute to soil conservation by enhancing organic matter content and supporting the growth of beneficial microorganisms. By utilizing waste materials from biopigment production in composting or as fertilizers, agricultural systems can promote nutrient recycling and reduce the carbon footprint of farming practices.

Valorization of fruit and vegetable wastes and their byproducts can meet the demands of natural pigment production. These waste products are a rich source of natural pigments and can also contribute significantly to the development of functional foods. With a sustainability approach, extraction of natural pigments from vegetal wastes, extraction and processing technologies, encapsulation techniques and potential bioactivities can all be designed with an eco-friendly approach (Sharma *et al.*, 2021). The use of agro-industrial wastes as feedstocks may allow sustainable and cost-effective production of these valuable secondary metabolites with simultaneous waste management (Grewal *et al.*, 2022).

Future trends in eco-friendly pigment applications may also include the use of biopigments in biodegradable packaging materials, coatings, or textiles (Panesar *et al.*, 2015). As consumer demand for environmentally conscious products increases, the use of biopigments in these industries will likely rise, contributing to a circular economy where materials are renewable, reusable, and nontoxic. This approach not only addresses waste management challenges but also aligns with the principles of a circular economy by transforming waste into valuable products. The potential applications of biopigments extend beyond traditional industries, with emerging opportunities in fields such as smart packaging, biomedical devices, and sustainable construction materials. As research in this area progresses, we can expect to see novel production methods and applications that further enhance the environmental and economic benefits of biopigments.

14.5.3 Environmental benefits

The environmental benefits of biopigments are significant. Their use can reduce dependency on synthetic dyes, which are often derived from petrochemicals and may contain toxic heavy metals or other harmful substances. Biopigments, on the other hand, are generally biodegradable and nontoxic, reducing their environmental footprint. Over time, the widespread adoption of biopigments could lead to a reduction in water pollution, soil contamination, and the overall environmental degradation caused by synthetic dyes (Swami *et al.*, 2020). In agricultural applications, biopigments can also help improve water and soil quality. Certain biopigments derived from algae or other microorganisms can be cultivated in wastewater or other nonarable environments, reducing the need for fresh water and preventing contamination from agricultural run-off. Moreover, biopigments can potentially play a role in mitigating climate change by enhancing soil carbon sequestration and supporting more resilient agricultural ecosystems.

The widespread adoption of biopigments has the potential to revolutionize various industries while significantly reducing environmental pollution and enhancing agricultural sustainability. By leveraging natural sources and eco-friendly production methods, biopigments could contribute to improved water and soil quality, reduced contamination from agricultural run-off, and increased carbon sequestration in soils. This shift towards sustainable alternatives could have far-reaching positive impacts on ecosystem health and climate change mitigation efforts.

14.6 Conclusion

Biopigments offer a sustainable and environmentally friendly alternative to synthetic dyes, presenting significant benefits in ecological and agricultural domains. Their roles in bioremediation, wastewater treatment, and bioindication underscore their significance in improving ecological health. In agriculture, biopigments serve as biofertilizers, biopesticides, and growth enhancers, resulting in enhanced crop output, reduced dependence on chemical inputs, and improved postharvest preservation. Considering the production costs, stability, and scalability limitations, advancements in genetic engineering, microbial technology, and sustainable agriculture practices hold considerable promise for mitigating these issues.

Future research could focus on enhancing production techniques, exploring new pigment sources, and increasing the longevity of biopigments to ensure their effectiveness and broader acceptance. The increasing global need for sustainable practices positions biopigments as an innovative solution for industrial applications, capable of reducing environmental harm and promoting a circular economy. Biopigments, due to their many roles, can substantially aid in achieving long-term sustainability goals in environmental management and agricultural practices.

References

Ababa, G. (2024) Pathogenic diversity, ecology, epidemiology, and management practices of the potato bacterial wilt (*Ralstonia solanacearum*) disease. *Cogent Food & Agriculture* 10(1), 2407953. DOI: 10.1080/23311932.2024.2407953.

Abd El-Baky, H.H. and El-Baroty, G.S. (2012) Characterization and bioactivity of phycocyanin isolated from spirulina maxima grown under salt stress. *Food & Function* 3(4), 381. DOI: 10.1039/c2fo10194g.

Agarwal, H., Bajpai, S., Mishra, A., Kohli, I., Varma, A. *et al.* (2023) Bacterial pigments and their multifaceted roles in contemporary biotechnology and pharmacological applications. *Microorganisms* 11(3), 11. DOI: 10.3390/microorganisms11030614.

Ajithkumar, K., Savitha, A.S., Renuka, M. and Naik, M.K. (2024) Citrinin. In: *Anti-Mycotoxin Strategies for Food and Feed*. John Wiley & Sons, Chichester, UK. DOI: 10.1002/9781394160839.

Alegbe, E.O. and Uthman, T.O. (2024) A review of history, properties, classification, applications and challenges of natural and synthetic dyes. *Heliyon* 10, E33646. DOI: 10.1016/j.heliyon.2024.e33646.

Al-khayri, J.M., Rashmi, R., Toppo, V., Chole, P.B., Banadka, A. *et al.* (2023) Plant secondary metabolites. *Metabolites* 13. DOI: 10.3390/metabo13060716.

Ambati, R.R., Phang, S.-M., Ravi, S. and Aswathanarayana, R.G. (2014) Astaxanthin: Sources, extraction, stability. Biological activities and its commercial applications—a *review*. *Marine Drugs* 12. DOI: 10.3390/md12010128.

Angon, P.B., Anjum, N., Akter, M.M., Shreejana, K.C., Suma, R.P. *et al.* (2023) An overview of the impact of tillage and cropping systems on soil health in agricultural practices. *Advances in Agriculture* 1–14. DOI: 10.1155/2023/8861216.

Anshi, K.S., Kapil, S., Goswami, L. and Sharma, V. (2024) Unveiling the intricacies of microbial pigments as sustainable alternatives to synthetic colorants: Recent trends and advancements. *Micro* 4(4), 621–640. DOI: 10.3390/micro4040038.

Arfa, A., Akhtar, K., Yousaf, R., Misbah, M., Shahzadi, N. *et al.* (2024) Exploring diverse biological fertilizers (micronutrients) for sustainable soil fertility: A review. DOI: 10.20944/preprints202410.1140.v1.

Avalos, J. and Carmen Limón, M. (2015) Biological roles of fungal carotenoids. *Current Genetics* 61(3), 309–324. DOI: 10.1007/s00294-014-0454-x.

Baaka, N., Khiari, R. and Haji, A. (2023) Ecofriendly dyeing of textile materials with natural colorants from date palm fiber *fibrillium*. *Sustainability* 15(2), 1688. DOI: 10.3390/su15021688.

Badiale-Furlong, E., Pagnussatt, F.A. and Scaglioni, P.T. (2023) Antioxidant and antifungal compounds of microalgae. In: Jacob-Lopes, E., Queiroz, M.I., Maroneze, M.M. and Zepka, L.Q. (eds) *Handbook of Food and Feed from Microalgae*. Academic Press, New York.

Barreto, J.V., Casanova, L.M., Junior, A.N., Reis-Mansur, M.C. and Vermelho, A.B. (2023) Microbial pigments: Major groups and industrial applications. *Microorganisms* 11(12), 2920. DOI: 10.3390/microorganisms11122920.

Bazarnova, J., Smyatskaya, Y., Shlykova, A., Balabaev, A. and Đurović, S. (2022) Obtaining fat-soluble pigments—carotenoids from the biomass of chlorella microalgae. *Applied Sciences* 12(7), 3246. DOI: 10.3390/app12073246.

Brauch, J.E. (2016) Underutilized fruits and vegetables as potential novel pigment sources. In: Carle, R. and Schweiggert, R.M. (eds) *Handbook on Natural Pigments in Food and Beverages*. Woodhead Publishing, New York. DOI: 10.1016/B978-0-08-100371-8.00015-4.

Castro-Gerónimo, V.D., García-Rodríguez, R.V., Sánchez-Medina, A., Chamorro-Cevallos, G.A., Sánchez-González, D.J. *et al*. (2023) C-phycocyanin: A phycobiliprotein from *spirulina* with metabolic syndrome and oxidative stress effects. *Journal of Medicinal Food* 27(9), 807–813. DOI: 10.1089/jmf.2022.0113.

Cerón, M.C., García-Malea, M.C., Rivas, J., Acien, F.G., Fernandez, J.M. *et al*. (2007) Antioxidant activity of *Haematococcus pluvialis* cells grown in continuous culture as a function of their carotenoid and fatty acid content. *Applied Microbiology and Biotechnology* 74(5), 1112–1119. DOI: 10.1007/s00253-006-0743-5.

Charalampia, D., Antonios, K.E., Constantina, N. and Haralabos, K.C. (2017) Current trends and emerging technologies in biopigment production processes: Industrial food and health applications. *International Journal of Horticulture, Agriculture and Food Science* 1, 33–46.

Chokshi, K., Pancha, I., Ghosh, A. and Mishra, S. (2017) Oxidative stress-induced bioprospecting of microalgae. In: Kumar, M. and Ralph, P. (eds) *Systems Biology of Marine Ecosystems*. Springer International Publishing, Cham, Switzerland.

Cordero, R.J.B. and Casadevall, A. (2020) Melanin. *Current Biology* 30(4), R142–R143. DOI: 10.1016/j.cub.2019.12.042.

Cordero, R.J.B., VIJ, R. and Casadevall, A. (2017) Microbial melanins for radioprotection and bioremediation. *Microbial Biotechnology* 10(5), 1186–1190. DOI: 10.1111/1751-7915.12807.

De Oliveira, I.C., Santana, D.C., De Oliveira, J.L.G., Silva, E.V.M., Da Silva Candido Seron, A.C. *et al*. (2024) Flavonoids and their relationship with the physiological quality of seeds from different soybean genotypes. *Scientific Reports* 14(1), 17008. DOI: 10.1038/s41598-024-68117-z.

Demmig-Adams, B., López-Pozo, M., Stewart, J.J. and Adams, W.W. (2020) Zeaxanthin and lutein: Photoprotectors, anti-inflammatories, and brain food. *Molecules* 25(16), 3607. DOI: 10.3390/molecules25163607.

Dewanjee, S., Bhattacharjee, N., Chakraborty, P. and Bhattacharjee, S. (2021) Carotenoids as antioxidants. In: Zia-ul-haq, M., Dewanjee, S. and Riaz, M. (eds) *Carotenoids: Structure and Function in the Human Body*. Springer International Publishing, Cham, Switzerland.

Di Salvo, E., Lo Vecchio, G., De Pasquale, R., De Maria, L., Tardugno, R. *et al*. (2023) Natural pigments production and their application in food, health and other industries. *Nutrients* 15(8). DOI: 10.3390/nu15081923.

Dissanayake, P., Yamauchi, N. and Shigyo, M. (2012) Presence of pheophytin and its formation as a chlorophyll derivative in selected crop species. *Journal of Agricultural Sciences* 7(3), 127–134. DOI: 10.4038/jas.v7i3.4876.

Do Amaral, S.C., Xavier, L.P., Vasconcelos, V. and Santos, A.V. (2023) Cyanobacteria: A promising source of antifungal metabolites. *Marine Drugs* 21(6), 359. DOI: 10.3390/md21060359.

Dudina, I., Kalashnikova, E. and Kirakosyan, R. (2024) Green microalgae chlorella in the study of the biosynthetic potential of higher plants *in vitro*. *BIO Web of Conferences* 113, 01010. DOI: 10.1051/bioconf/202411301010.

Ehrlich, P.R. and Harte, J. (2015) To feed the world in 2050 will require a global revolution. *Proceedings of the National Academy of Sciences USA* 112(48), 14743–14744. DOI: 10.1073/pnas.1519841112.

El-Bialy, H.A. and Abou El-Nour, S.A. (2015) Physical and chemical stress on *Serratia marcescens* and studies on prodigiosin pigment production. *Annals of Microbiology* 65(1), 59–68. DOI: 10.1007/s13213-014-0837-8.

El-Naggar, N.E.-A. and Saber, W.I.A. (2022) Natural melanin: Current trends, and future approaches, with especial reference to microbial source. *Polymers* 14(7), 14. DOI: 10.3390/polym14071339.

Fess, T.L., Kotcon, J.B. and Benedito, V.A. (2011) Crop breeding for low input agriculture: A sustainable response to feed a growing world population. *Sustainability* 3(10), 1742–1772. DOI: 10.3390/su3101742.

Gadd, G.M. (2007) Geomycology: Biogeochemical transformations of rocks, minerals, metals and radionuclides by fungi, bioweathering and bioremediation. *Mycological Research* 111(Pt 1), 3–49. DOI: 10.1016/j.mycres.2006.12.001.

Gao, Q. and Garcia-Pichel, F. (2011) Microbial ultraviolet sunscreens. *Nature Reviews. Microbiology* 9(11), 791–802. DOI: 10.1038/nrmicro2649.

Garcia-Seco, D., Zhang, Y., Gutierrez-Mañero, F.J., Martin, C. and Ramos-Solano, B. (2015) Application of Pseudomonas fluorescens to Blackberry under field conditions improves fruit quality by modifying flavonoid metabolism. *PLoS ONE* 10(11), e0142639. DOI: 10.1371/journal.pone.0142639.

Gebregziabher, B.S., Zhang, S., Ghosh, S., Shaibu, A.S., Azam, M. *et al.* (2022) Origin, maturity group and seed coat color influence carotenoid and chlorophyll concentrations in soybean seeds. *Plants* 11(7), 848. DOI: 10.3390/plants11070848.

Georgescu, A.-M., Corbu, V.M. and Csutak, O. (2024) Molecular basis of yeasts antimicrobial activity-developing innovative strategies for biomedicine and biocontrol. *Current Issues in Molecular Biology* 46(5), 4721–4750. DOI: 10.3390/cimb46050285.

Ghadge, V., Kumar, P., Singh, S., Mathew, D.E., Bhattacharya, S. *et al.* (2020) Natural melanin produced by the endophytic *bacillus subtilis* 4NP-BL associated with the halophyte *Salicornia brachiata*. *Journal of Agricultural and Food Chemistry* 68(25), 6854–6863. DOI: 10.1021/acs.jafc.0c01997.

Ghosh, N., Dey, S., Guha, T., Paul, S. and Kundu, R. (2024) Recent advances in nano-priming induced plant growth promotion and environmental stress tolerance. *The Nucleus* 67(3), 653–674. DOI: 10.1007/s13237-024-00506-1.

Glagoleva, A.Y., Shoeva, O.Y. and Khlestkina, E.K. (2020) Melanin pigment in plants: Current knowledge and future perspectives. *Frontiers in Plant Science* 11, 770. DOI: 10.3389/fpls.2020.00770.

Gonçalves, A.L. (2021) The use of microalgae and cyanobacteria in the improvement of agricultural practices: A review on their biofertilising, biostimulating and biopesticide roles. *Applied Sciences* 11(2), 871. DOI: 10.3390/app11020871.

González-Peña, M.A., Ortega-Regules, A.E., Anaya De Parrodi, C. and Lozada-Ramírez, J.D. (2023) Chemistry, occurrence, properties, applications, and encapsulation of carotenoids-a review. *Plants* 12(2), 313. DOI: 10.3390/plants12020313.

Grande, R., Räisänen, R., Dou, J., Rajala, S., Malinen, K. *et al.* (2023) *In situ* adsorption of red onion (*Allium cepa*) natural dye on cellulose model films and fabrics exploiting chitosan as a natural mordant. *ACS Omega* 8(6), 5451–5463. DOI: 10.1021/acsomega.2c06650.

Grewal, J., Woła Cewicz, M., Pyter, W., Joshi, N., Drewniak, L. *et al.* (2022) Colorful treasure from agro-industrial wastes: A sustainable chassis for microbial pigment production. *Frontiers in Microbiology* 13, 832918. DOI: 10.3389/fmicb.2022.832918.

Hassan, S., Meenatchi, R., Pachillu, K., Bansal, S., Brindangnanam, P. *et al.* (2022) Identification and characterization of the novel bioactive compounds from microalgae and cyanobacteria for pharmaceutical and nutraceutical applications. *Journal of Basic Microbiology* 62, 999–1029. DOI: 10.1002/jobm.202200170.

Herrala, M., Yli-Öyrä, J., De Albuquerque, A.F., De Farias, N.O., Morales, D.A. *et al.* (2022) Waterless dyeing and *in vitro* toxicological properties of biocolorants from *Cortinarius sanguineus*. *Journal of Fungi* 8(11), 1129. DOI: 10.3390/jof8111129.

Huang, X., Gan, L., He, Z., Jiang, G. and He, T. (2024) Bacterial pigments as a promising alternative to synthetic colorants: From fundamentals to applications. *Journal of Microbiology and Biotechnology* 34(11), 2153–2165. DOI: 10.4014/jmb.2404.04018.

Ismail, A.E., Elissawy, A.M., Gad, H.A. and Singab, A.-N.B. (2024) Unearthing the potential secondary metabolites isolated from the fungus aspergillus *flocculosus*: A comprehensive review. *Archives of Pharmaceutical Sciences Ain Shams University* 8(1), 42–58. DOI: 10.21608/aps.2024.264592.1156.

John Jimtha, C., Jishma, P., Sreelekha, S., Chithra, S. and Radhakrishnan, E.K. (2017) Antifungal properties of prodigiosin producing *Rhizospheric serratia* sp. *Rhizosphere* 3, 105–108. DOI: 10.1016/j.rhisph.2017.02.003.

Joshi, T., Sharma, P., Joshi, T., Pandey, S.C., Pande, V. *et al.* (2020) A spotlight on the recent advances in bacterial plant diseases and their footprint on crop production. In: De Mandal, S. and Bhatt, P. (eds) *Recent Advancements in Microbial Diversity*. Academic Press, New Delhi, India. DOI: 10.1016/B978-0-12-821265-3.00003-7.

Khaksar, F. and Rigi, G. (2022) Optimization of production conditions of natural antibacterial *Violacin* pigment from *Janthinobacterium lividum*. *Journal of Ilam University of Medical Sciences* 30(3), 29–43. DOI: 10.52547/sjimu.30.3.29.

Khan, M.S., Khan, S., Khan, N. and Khan, A.S. (2024) Dietary sources, classification, biosynthesis, and mechanism of action of flavonoids in combating oxidative stress. In: *Role of Flavonoids in Chronic Metabolic Diseases*. John Wiley & Sons, New York. DOI: 10.1002/9781394238071.

Kini, S., Divyashree, M., Mani, M.K. and Mamatha, B.S. (2020) Algae and cyanobacteria as a source of novel bioactive compounds for biomedical applications. In: Singh, P.K., Kumar, A., Singh, V.K. and Shrivastava, A.K. (eds) *Advances in Cyanobacterial Biology*. Academic Press, New York. DOI: 10.1016/B978-0-12-819311-2.00012-7.

Krastanov, A., Alexieva, Z. and Yemendzhiev, H. (2013) Microbial degradation of phenol and phenolic derivatives. *Engineering in Life Sciences* 13(1), 76–87. DOI: 10.1002/elsc.201100227.

Kshetri, L., Naseem, F. and Pandey, P. (2019) Role of Serratia sp. as biocontrol agent and plant growth stimulator, with prospects of biotic stress management in plant. In: Sayyed, R.Z. (ed.) *Plant Growth Promoting Rhizobacteria for Sustainable Stress Management: Volume 2: Rhizobacteria in Biotic Stress Management*. Springer Singapore, Singapore.

Kumar, A., Vishwakarma, H.S., Singh, J., Dwivedi, S. and Kumar, M. (2015) Microbial pigments: Production and their applications in various industries. *International Journal of Pharmaceutical, Chemical and Biological Sciences* 5, 203–212.

Kumar, A., Asthana, M., Gupta, A., Nigam, D. and Mahajan, S. (2018) Secondary metabolism and antimicrobial metabolites of penicillium. In: Gupta, V.K. and Rodriguez-Couto, S. (eds) *New and Future Developments in Microbial Biotechnology and Bioengineering*. Elsevier, Amsterdam.

Kumar, S., Kumar, V., Ambika, A.A.A., Nag, D., Kumar, V. *et al.* (2022) Microbial pigments: Learning from himalayan perspective to industrial applications. *Journal of Industrial Microbiology and Biotechnology* 49(5), kuac017. DOI: 10.1093/jimb/kuac017.

Landi, M., Tattini, M. and Gould, K.S. (2015) Multiple functional roles of anthocyanins in plant-environment interactions. *Environmental and Experimental Botany* 119, 4–17. DOI: 10.1016/j.envexpbot.2015.05.012.

Latsos, C., van Houcke, J., Blommaert, L., Verbeeke, G.P., Kromkamp, J. *et al.* (2021) Effect of light quality and quantity on productivity and phycoerythrin concentration in the cryptophyte *Rhodomonas* sp. *Journal of Applied Phycology* 33(2), 729–741. DOI: 10.1007/s10811-020-02338-3.

Lemes, E.M. and Catão, H.C.R.M. (2024) Soybean seed coat cracks and green seeds—predisposing conditions, identification and management. *Seeds* 3(1), 133–148. DOI: 10.3390/seeds3010011.

Lyu, X., Lyu, Y., Yu, H., Chen, W., Ye, L. *et al.* (2022) Biotechnological advances for improving natural pigment production: A state-of-the-art review. *Bioresources and Bioprocessing* 9(1), 8. DOI: 10.1186/s40643-022-00497-4.

Mandal, A., Dutta, A., Das, R. and Mukherjee, J. (2021) Role of intertidal microbial communities in carbon dioxide sequestration and pollutant removal: A review. *Marine Pollution Bulletin* 170, 112626. DOI: 10.1016/j.marpolbul.2021.112626.

Meruvu, H. and Dos Santos, J.C. (2021) Colors of life: A review on fungal pigments. *Critical Reviews in Biotechnology* 41(8), 1153–1177. DOI: 10.1080/07388551.2021.1901647.

Michalak, I., Norrie, J. and Critchley, A.T. (2021) Enhancing plant pigments using natural biostimulants. In: Gupta, S. and Staden, J. (eds) *Biostimulants for Crops from Seed Germination to Plant Development*. Academic Press, New York.

Mohammad Azmin, S.N.H., Sulaiman, N.S., Mat Nor, M.S., Abdullah, P.S., Abdul Kari, Z, *et al.* (2022) A review on recent advances on natural plant pigments in foods: Functions, extraction, importance and challenges. *Applied Biochemistry and Biotechnology* 194, 4655–4672. DOI: 10.1007/s12010-022-04050-z.

Mutaf-Kılıc, T., Demir, A., Elibol, M. and Oncel, S.S. (2023) Microalgae pigments as a sustainable approach to textile dyeing: A critical review. *Algal Research* 76, 103291. DOI: 10.1016/j.algal.2023.103291.

Muñoz-Torres, P., Cárdenas-Ninasivincha, S. and Aguilar, Y. (2024) Exploring the agricultural applications of microbial melanin. *Microorganisms* 12(7), 1352. DOI: 10.3390/microorganisms12071352.

Nawaz, A., Chaudhary, R., Shah, Z., Dufossé, L., Fouillaud, M. *et al.* (2021) An overview on industrial and medical applications of bio-pigments synthesized by Marine Bacteria. *Microorganisms* 7, 93–100. DOI: 10.3390/microorganisms9010011.

Nazir, F., Fariduddin, Q. and Khan, T.A. (2020) Hydrogen peroxide as a signalling molecule in plants and its crosstalk with other plant growth regulators under heavy metal stress. *Chemosphere* 252, 126486. DOI: 10.1016/j.chemosphere.2020.126486.

Nigam, P.S. and Luke, J.S. (2016) Food additives: Production of microbial pigments and their antioxidant properties. *Current Opinion in Food Science* 7, 93–100. DOI: 10.1016/j.cofs.2016.02.004.

Novais, C., Molina, A.K., Abreu, R.M.V., Santo-Buelga, C., Ferreira, I.C.F.R. *et al.* (2022) Natural food colorants and preservatives: A review, a demand, and a challenge. *Journal of Agricultural and Food Chemistry* 70(9), 2789–2805. DOI: 10.1021/acs.jafc.1c07533.

Oliveira, H., Correia, P., Pereira, A.R., Araújo, P., Mateus, N. *et al.* (2020) Exploring the applications of the photoprotective properties of anthocyanins in biological systems. *International Journal of Molecular Sciences* 21(20), 7464. DOI: 10.3390/ijms21207464.

Orlandi, V.T., Martegani, E., Giaroni, C., Baj, A. and Bolognese, F. (2022) Bacterial pigments: A colorful palette reservoir for biotechnological applications. *Biotechnology and Applied Biochemistry* 69(3), 981–1001. DOI: 10.1002/bab.2170.

Oslan, S.N.H., Tan, J.S., Oslan, S.N., Matanjun, P., Mokhtar, R.A.M. *et al.* (2021) *Haematococcus pluvialis* as a potential source of astaxanthin with diverse applications in industrial sectors: Current research and future directions. *Molecules* 26(21), 6470. DOI: 10.3390/molecules26216470.

Panesar, R., Kaur, S. and Panesar, P.S. (2015) Production of microbial pigments utilizing agro-industrial waste: A review. *Current Opinion in Food Science* 1, 70–76. DOI: 10.1016/j.cofs.2014.12.002.

Periyasamy, A.P. (2022) Natural dyeing of cellulose fibers using *Syzygium cumini* fruit extracts and a bio-mordant: A step toward sustainable dyeing. *Sustainable Materials and Technologies* 33, e00472. DOI: 10.1016/j.susmat.2022.e00472.

Qiu, N.W., Jiang, D.C., Wang, X.S., Wang, B.S. and Zhou, F. (2019) Advances in the members and biosynthesis of chlorophyll family. *Photosynthetica* 57(4), 974–984. DOI: 10.32615/ps.2019.116.

Rahman, M., Sabir, A.A., Mukta, J.A., Khan, M.M.A., Mohi-Ud-Din, M. *et al.* (2018) Plant probiotic bacteria Bacillus and Paraburkholderia improve growth, yield and content of antioxidants in strawberry fruit. *Scientific Reports* 8(1), 2504. DOI: 10.1038/s41598-018-20235-1.

Ramaroson, M.-L., Koutouan, C., Helesbeux, J.-J., Le Clerc, V., Hamama, L. *et al.* (2022) Role of phenylpropanoids and flavonoids in plant resistance to pests and diseases. *Molecules* 27(23), 27. DOI: 10.3390/molecules27238371.

Rassabina, A., Khabibrakhmanova, V., Babaev, V., Daminova, A. and Minibayeva, F. (2022) Melanins from the *Lichens Lobaria pulmonaria* and *Lobaria retigera* as eco-friendly adsorbents of synthetic dyes. *International Journal of Molecular Sciences* 23(24), 15605. DOI: 10.3390/ijms232415605.

Rather, L.J., Mir, S.S., Ganie, S.A., Shahid, U.L. and Q., L.I. (2023) Research progress, challenges, and perspectives in microbial pigment production for industrial applications – a review. *Dyes and Pigments* 210, 110989. DOI: 10.1016/j.dyepig.2022.110989.

Saini, D.K., Pabbi, S. and Shukla, P. (2018) Cyanobacterial pigments: Perspectives and biotechnological approaches. *Food and Chemical Toxicology* 120, 616–624. DOI: 10.1016/j.fct.2018.08.002.

Saini, D.K., Chakdar, H., Pabbi, S. and Shukla, P. (2020) Enhancing production of microalgal biopigments through metabolic and genetic engineering. *Critical Reviews in Food Science and Nutrition* 60(3), 391–405. DOI: 10.1080/10408398.2018.1533518.

Sánchez-Muñoz, S., Mariano-Silva, G., Leite, M.O., Mura, F.B., Verma, M.L. *et al.* (2020) Production of fungal and bacterial pigments and their applications. In: Verma, M.L. and Chandel, A.K. (eds) *Biotechnological Production Of Bioactive Compounds*. Elsevier, St Louis, MO.

Sandmann, G. (2019) Antioxidant protection from UV- and light-stress related to carotenoid structures. *Antioxidants* 8(7), 219. DOI: 10.3390/antiox8070219.

Sandybayeva, S.K., Kossalbayev, B.D., Zayadan, B.K., Sadvakasova, A.K., Bolatkhan, K. *et al.* (2022) Prospects of cyanobacterial pigment production. *Biotechnological Potential and Optimization Strategies. Biochemical Engineering Journal* 187, 108640. DOI: 10.1016/j.bej.2022.108640.

Shah, S.T., Basit, A., Ullah, I. and Mohamed, H.I. (2021) Cyanobacteria and algae as biocontrol agents against fungal and bacterial plant pathogens. In: Mohamed, H.I., El-Beltagi, H.E.-D.S. and Abd-Elsalam, K.A. (eds) *Plant Growth-Promoting Microbes for Sustainable Biotic and Abiotic Stress Management*. Springer International Publishing, Cham, Switzerland.

Sharma, M., Usmani, Z., Gupta, V.K. and Bhat, R. (2021) Valorization of fruits and vegetable wastes and by-products to produce natural pigments. *Critical Reviews in Biotechnology* 41(4), 535–563. DOI: 10.1080/07388551.2021.1873240.

Silva, S.C., Ferreira, I.C.F.R., Dias, M.M. and Barreiro, M.F. (2020) Microalgae-derived pigments: A 10-year bibliometric review and industry and market trend analysis. *Molecules* 25(15), 3406. DOI: 10.3390/molecules25153406.

Singh, S., Kate, B.N. and Banerjee, U.C. (2005) Bioactive compounds from cyanobacteria and microalgae: An overview. *Critical Reviews in Biotechnology* 25(3), 73–95. DOI: 10.1080/07388550500248498.

Singh, T., Pandey, V.K., Dash, K.K., Zanwar, S. and Singh, R. (2023) Natural bio-colorant and pigments: Sources and applications in food processing. *Journal of Agriculture and Food Research* 12, 100628. DOI: 10.1016/j.jafr.2023.100628.

Someya, N., Nakajima, M., Hirayae, K., Hibi, T. and Akutsu, K. (2001) Synergistic antifungal activity of chitinolytic enzymes and prodigiosin produced by biocontrol bacterium, *Serratia marcescens* strain B2 against gray mold pathogen, *Botrytis cinerea*. *Journal of General Plant Pathology* 67(4), 312–317. DOI: 10.1007/PL00013038.

Sousa, C. (2022) Anthocyanins, carotenoids and chlorophylls in edible plant leaves unveiled by tandem mass spectrometry. *Foods* 11, 13. DOI: 10.3390/foods11131924.

Šovljanski, O., Kljakić, A.C. and Tomić, A. (2023) Antibacterial and antifungal potential of plant secondary metabolites. In: Mérillon, J.-M. and Ramawat, K.G. (eds) *Plant Specialized Metabolites: Phytochemistry, Ecology and Biotechnology*. Springer Nature, Cham, Switzerland.

Srivastava, P., Kushal Sachan, K., Baskar, P., Saikanth, D.R.K., Lytand, W. *et al.* (2023) Soil microbes expertly balancing nutrient demands and environmental preservation and ensuring the delicate stability of our ecosystems – a review. *International Journal of Plant & Soil Science* 35(18), 989–1000. DOI: 10.9734/ijpss/2023/v35i183363.

Summra, S., Sana, E.M., Ali Aslam, S. and Memoona, A. (2024) Physiological and nutraceutical properties of soybean (*Glycine max* L). In: Jose, C.J.-L. and Julia, E.F. (eds) *Soybean Crop*. Intechopen, Rijeka, Croatia.

Swami, S.B., Ghgare, S.N., Swami, S.S., Shinde, K.J., Kalse, S.B. *et al.* (2020) Natural pigments from plant sources: A review. *Pharma Innovation Journal* 9, 56.

Tejasvini, S.P., Khanolkar, A.B. and Nadaf, N.H. (2016) Production, purification, identification of prodigiosin from *Serratia* sp. and its antimicrobial activity. *Research Journal of Life Sciences, Bioinformatics, Pharmaceutical and Chemical Sciences* 1, 326–337.

Tuli, H.S., Chaudhary, P., Beniwal, V. and Sharma, A.K. (2015) Microbial pigments as natural color sources: Current trends and future perspectives. *Journal of Food Science and Technology* 52(8), 4669–4678. DOI: 10.1007/s13197-014-1601-6.

Verma, D.K., Guzmán, K.N.R., Mohapatra, B., Talukdar, D., Chávez-González, M.L. *et al.* (2021) Recent trends in plant- and microbe-based biopesticide for sustainable crop production and environmental security. In: Prasad, R., Kumar, V., Singh, J. and Upadhyaya, C.P. (eds) *Recent Developments in Microbial Technologies*. Springer Nature Singapore, Singapore. DOI: 10.1007/978-981-15-4439-2_1.

Verma, J.P., Yadav, J., Tiwari, K.N. and Kumar, A. (2013) Effect of indigenous *Mesorhizobium* spp. and plant growth promoting rhizobacteria on yields and nutrients uptake of chickpea (*Cicer arietinum* L.) under sustainable agriculture. *Ecological Engineering* 51, 282–286. DOI: 10.1016/j.ecoleng.2012.12.022.

Way, M.J. and van Emden, H.F. (2000) Integrated pest management in practice — pathways towards successful application. *Crop Protection* 19(2), 81–103. DOI: 10.1016/S0261-2194(99)00098-8.

Wink, M. and Lehmann, P. (1996) Wounding- and elicitor-induced formation of coloured chalcones and flavans (as phytoalexins) in hippeastrum x hortorum. *Botanica Acta* 109(5), 412–421. DOI: 10.1111/j.1438-8677.1996.tb00591.x.

Xing, H., Zhao, Y., Li, T., Han, B., Zhao, P. *et al.* (2022) Enhancing astaxanthin and lipid coproduction in *Haematococcus pluvialis* by the combined induction of plant growth regulators and multiple stresses. *Bioresource Technology* 344(Pt A), 126225. DOI: 10.1016/j.biortech.2021.126225.

Yadav, S., Tiwari, K.S., Gupta, C., Tiwari, M.K., Khan, A. *et al.* (2023) A brief review on natural dyes, pigments: Recent advances and future perspectives. *Results in Chemistry* 5, 100733. DOI: 10.1016/j.rechem.2022.100733.

Ye, K.-X., Fan, T.-T., Keen, L.J. and Han, B.-N. (2019) A review of pigments derived from marine natural products. *Israel Journal of Chemistry* 59(5), 327–338. DOI: 10.1002/ijch.201800154.

Zannou, O., Oussou, K.F., Chabi, I.B., Odouaro, O.B.O., Deli, M.G.E.P. *et al.* (2024) A comprehensive review of recent development in extraction and encapsulation techniques of betalains. *Critical Reviews in Food Science and Nutrition* 64(30), 11263–11280. DOI: 10.1080/10408398.2023.2235695.

Zhang, Y., Li, Y., Liang, S., Zheng, W., Chen, X. *et al.* (2022) Study on the preparation and effect of tomato seedling disease biocontrol compound seed-coating agent. *Life* 12(6), 849. DOI: 10.3390/life12060849.

15 Economic and Commercial Viability of Microbial Pigments

Rashid Abbas Khan[1], Tauqeer Ahmed Qadri[2], Amjid Khan[3], Muhammad Anas[4]* and Divya Jain[5]

[1]Department of Biology, Faculty of Sciences, Prince Sonkla University, Hat Yai, Thailand; [2]Department of Biosciences, COMSATS University, Islamabad, Pakistan; [3]UNESCO UNISA Africa Chair in Nanoscience and Nanotechnology, College of Graduate Studies, University of South Africa, Pretoria, South Africa; [4]Department of Soil Science & Plant Nutrition, Selçuk University, Konya, Türkiye; [5]Department of Microbiology, School of Applied and Life Sciences, Uttaranchal University, Dehradun, Uttarakhand, India

15.1 Introduction

The pigment landscape is changing in light of the increasing need for eco-friendly, multipurpose coloring agents in the field of industrial biotechnology where cheap but damaging synthetic dyes were previously used. Modern scientific indications of environmental degradation, health risk, and government regulation associated with the use of petrochemical-based pigments have heightened calls to develop sustainable natural pigment alternatives worldwide (Renita *et al.*, 2023). Microbial pigments, bioactive compounds produced by bacteria, fungi, yeast, and microalgae, are one of the most promising candidates in the next generation of colorants and have an increasingly wide range of applications beyond simple color (Rather *et al.*, 2025).

Essentially, microbial pigments form a synthetic grouping of biomaterials that combine high color performance with biodegradability, biocompatibility, and often significant bioactivity, especially in antioxidant, antimicrobial and anticancer applications (Gomes *et al.*, 2025). These colorants are produced in fixed fermentation conditions that facilitate an easily reproducible, high-volume product that is environmentally independent of seasonality or geographical location (Gänzle, 2022). In addition, they can be made on renewable substrates, such as agro-industrial wastes, thus aligning with principles of a circular economy, as well as sustainability goals. This versatility has positioned microbial pigments as a valuable feedstock for the food, pharmaceuticals, cosmetics, textiles, and numerous other industries (Silva *et al.*, 2025).

Microbial pigments demonstrate high chromaticity and biodegradability, biocompatibility, and frequently strong bioactivity, such as antioxidant, antimicrobial, and anticancer effects. They are produced using regulated fermentation strategies and provide predictable quality, scalability,

*Corresponding author: muhammed.anas@selcuk.edu.tr

DOI: 10.1079/ 9781836990505.0015

and resilience to seasonality or geographical influences (Barbinta-Patrascu *et al.*, 2024). They can also be obtained using renewable materials, particularly agro-industrial waste streams, thus aligning the circular economy and sustainability goals (Trujillo-Cayado *et al.*, 2025).

In the last two centuries, the world pigment market has been shaped by the economical and technological benefits of synthetic dyes. However, there has been a steadily increasing unsustainable aggregate cost of these synthetic products measured in the form of environmental degradations and health risks (Seetharaman *et al.*, 2024). At the same time, current regulatory requirements governing the use of pigments in Europe, including REACH, as well as Food and Drug Administration (FDA) regulations regarding safety of colorants in the US are rapidly accelerating the switch to biologically produced pigments. At the same time, the rapid development of metabolic engineering, synthetic biology, and process intensification has increased the yields, stability, and industrial feasibility of pigments from microbes and made microbial pigments a realistic commercial prospect (Benucci *et al.*, 2022).

However, microbial pigments face huge challenges even with all these technological advances. Beyond technological innovation, defeating these barriers requires an in-depth knowledge of cost breakthroughs, market dynamics, and business strategies that can make microbial pigments cost-competitive in the world market. This chapter provides an assessment of the economic and commercial viability aspects of microbial production of pigments, by integrating the technical and commercial perspectives of microbial pigment production, and critically appraising the costing plan, market parameters, and business frameworks that will be required to advance their competitive position and enable their move to mainstream commercial colorants in a sustainable global economy.

15.2 Sources and Types of Microbial Pigments

Microbial pigments form a wide range of chemically complex pool of bioactive molecules synthetized by fungi, bacteria, and microalgae (Table 15.1). Due to the diverse molecular structure of such compounds and their multipurpose characteristics, they offer a strong alternative to synthetic colorants (Rao *et al.*, 2021). Microbial strains provide year-round production, consistent quality, and adjustable yields due to the ability to regulate pigment production under controlled fermentation conditions, qualities that are often lacking in plant-derived or mineral equivalent options. In addition to their chromogenic properties, many microbial pigments exhibit antioxidant, antimicrobial, anticancer and antiinflammatory effects, extending their roles to many industrial applications (Rodriguez-Amaya *et al.*, 2023).

15.2.1 Bacterial pigments

Biopigments produced by bacteria include several significant chemical clusters, each with quite different structures and commercial uses. Examples include the carotenoids, which form a huge family of terpenoid pigments produced by genera including *Serratia*, *Paracoccus*, and *Rhodobacter* (Agarwal *et al.*, 2023). They are defined as conjugated double-bond systems, which impart bright yellow, orange, or red color, and significant antioxidant behavior. The carotenoid nutraceuticals, astaxanthin, zeaxanthin, and canthaxanthin, find uses in the nutraceutical industry, aquaculture industry, and even the cosmetics industry due to their health benefits and skin protective effects (Adamantidi *et al.*, 2025). Prodigiosin is a red tripyrrole pigment produced by *Serratia marcescens* which shows antimicrobial, antimalarial, and anticancer effects, although it presents a formulation challenge due to its instability under light exposure and in alkaline environments (Paul *et al.*, 2022). Violacein is a purple bisindole pigment produced by *Chromobacterium violaceum* and some species of *Actinobacterium* that has been shown to have very potent antimicrobial and antiparasitic activity, as well as promising anticancer properties (Park *et al.*, 2021).

15.2.2 Fungal pigments

Of the fungal pigments, the *Monascus* species pigments (used for centuries in East Asia as natural food colorants and preservatives) are the

Table 15.1. Classification of microbial pigments with their sources, chemical nature, industrial uses.

Pigment name	Microbial source	Chemical nature	Industrial and commercial uses	Reference
Carotenoids	*Serratia marcescens*, *Rhodotorula glutinis*, *Dunaliella salina*	Terpenoid hydrocarbons and xanthophylls (C40 tetraterpenes)	Food colorants, nutraceuticals (provitamin A), cosmetics, antioxidant supplements	(Gupta *et al*., 2022; do Socorro do Souza Vilhena *et al*., 2024)
Prodigiosin	*Serratia marcescens*, *Hahella chejuensis*	Tripyrrole red pigment	Antimicrobial agent, anticancer research, textile dyes	(Rapson, 2004)
Violacein	*Chromobacterium violaceum*, *Janthinobacterium lividum*	Indole-derived violet pigment	Antibacterial, antiparasitic, anticancer, cosmetics	(Delhoyo, 2007)
Monascus pigments	*Monascus purpureus*, *M. ruber*	Azaphilone polyketides	Natural food colorants (rice, meat, dairy), functional foods, antioxidant applications	(Afroz Toma *et al*., 2023a)
Anthraquinones	*Aspergillus* spp., *Penicillium* spp.	Aromatic polyketide compounds	Textile dyes, antimicrobial agents, herbal medicine formulations	(do Socorro do Souza Vilhena *et al*., 2024)
Azaphilones	*Monascus* spp., *Talaromyces* spp.	Fungal polyketides with pyrone–quinone structure	Food and beverage coloring, cosmetics	(Lin and Xu, 2022)
Phycocyanin	*Spirulina platensis*, *Anabaena* spp.	Water-soluble phycobiliprotein	Natural blue food colorant, nutraceuticals, fluorescent markers in diagnostics	(Fabre *et al*., 2022)
Phycoerythrin	*Porphyridium cruentum*, *Gracilaria* spp. (microalgal symbionts)	Water-soluble phycobiliprotein	Fluorescent probes, cosmetics, high-value food products	(García-Gómez *et al*., 2025)
Chlorophyll derivatives	*Chlorella vulgaris*, *Scenedesmus obliquus*	Magnesium-porphyrin complexes	Food additives, pharmaceuticals (wound healing), cosmetics	(Hosikian *et al*., 2010)
Scytonemin	Cyanobacteria (*Nostoc* spp.)	Lipid-soluble indolic–phenolic pigment	UV-protective agent, sunscreen formulations	(Pathak *et al*., 2020)
Melanin	*Cryptococcus neoformans*, *Streptomyces* spp.	High-molecular weight polyphenols	Antioxidants, cosmetics, protective coatings	(Salian *et al*., 2021)

most widely used. *Monascus* pigments are red, orange, and yellow varieties of azaphilone compounds with a valued stability profile in various processing conditions and a potential cardiovascular effect as a result of their reduction of blood cholesterol (Chaudhary *et al.*, 2022).

Polycyclic aromatic pigments with yellow to deep red coloration, the anthraquinones are produced by *Aspergillus*, *Penicillium*, *Trichoderma*, and related genera. Those pigments tend to have antimicrobial, antiviral and antiinflammatory effects, even though some derivatives can cause toxicity issues, so specific strain selection and purification are highly important (Poorniammal *et al.*, 2025).

Another structurally divergent family found in filamentous fungi are the azaphilones, which exhibit a broad range of colors including yellow to crimson and are attracting interest regarding their antioxidant and antiinflammatory bioactivity (Gomes, 2024).

15.2.3 Algal pigments

Microalgae and cyanobacteria are the sources of algal pigments which have high-value applications in the nutraceuticals, cosmetics, and biomedical markets. Phycobiliproteins are water-soluble protein pigment complexes produced in red algae (*Rhodophyta*) and cyanobacteria, including phycoerythrin and phycocyanin, which yield bright red and blue colors respectively. They are of interest as food colorants and biomarkers in diagnostic assays because of their intense fluorescence and natural abundance (Uma *et al.*, 2023).

Extracts of microalgae are a source of natural green colorant from chlorophyll derivatives with antioxidant effect and photoprotective activities. However, their high sensitivity towards acidic media and light requires the use of stabilization processes in industry (Pucci *et al.*, 2021).

15.2.4 Chemical structures, stability, and functional properties

The diversity of chemistry of microbial pigments, from the isoprenoid chains in carotenoids to the polyketide backbones in anthraquinones, has a direct impact on their optical properties, stability, and bioactivity (Ramesh *et al.*, 2025). The stability of pigments is a fundamental factor that determines their commercial viability, because pigments used in food should perform in a variety of pH, temperature, and light conditions during processing and storage. Modern stabilization strategies such as encapsulation, microemulsion systems, and and genetic engineering are used to create stronger structures (Jurić *et al.*, 2022).

15.2.5 Comparative advantages over synthetic pigments

Compared to synthetic pigments, microbial pigments have evident advantages: they are biodegradable, with less impact on the environment; they can be manufactured using renewable feedstocks, including agro-industrial byproducts; and in many cases, they have biofunctional benefits, including antioxidant or antimicrobial effects (Marcelli *et al.*, 2025). These properties offer possibilities of product differentiation in markets that demand clean-label, sustainable, and health-promoting products. To achieve the full potential of these benefits, however, further advances in strain improvement, fermentation efficiencies, and cost reduction will be needed in order to compete effectively with existing synthetic products (Mirsalami and Mirsalami, 2025).

15.3 Production Technologies

Production of microbial pigments must be economical, efficient, and scalable to be commercially successful and this relates not only to the functioning property and market demand but also to the underlying production system (Putnik *et al.*, 2013). Advances in technology have changed pigment biosynthesis, once conducted as artisanal fermentations, into a highly optimized bioprocess (Fig. 15.1). This advancement includes classical fermentation methods, innovative molecular engineering procedures, use of cost-efficient substances, and strict downstream processing regimes, all of which define industrial competitiveness (Lyu *et al.*, 2022).

Fig. 15.1. Flowchart of microbial pigment production and processing.

15.3.1 Traditional fermentation methods

15.3.1.1 Submerged fermentation (SMF)

Probably the most common method of microbial pigment production is submerged fermentation, used mostly for bacteria and yeast. In this process, microorganisms are grown in aerated liquid nutrient media under aerobic and agitated conditions, with controlled temperatures and pH (Ramesh *et al.*, 2022). SMF has the benefit of process control, ease of scale-up, and the possibility of continuous monitoring of biomass development and pigment deposition. Nevertheless, the process is also very resource intensive, using substantial amounts of water and energy, in addition to producing considerable quantities of wastewater. This can affect overall economic feasibility unless they are incorporated with recycling or energy efficient planning (Rengifo *et al.*, 2022).

15.3.1.2 Solid-state fermentation (SSF)

Solid-state fermentation uses solid nutrient substrates with low water content, such as rice, wheat bran, or agro-industrial wastes, as a source of nutrients and structural support for microbial growth. SSF is specifically applicable to fungi, mostly to filamentous ones such as *Monascus* and *Penicillium* species that exhibit natural growth on solid media (Šelo *et al.*, 2021). The merits of SSF include the use of less water, less liquid waste in downstream systems, and the possibility of more pigment production. Economically, SSF may be attractive since it utilizes cheap agricultural byproducts, but scaling up may be a challenge since it is difficult to control temperature gradients, aeration, and homogeneity of the substrate in large reactors (Chilakamarry *et al.*, 2022).

15.3.2 Advanced approaches

Intracellular construction of microbial metabolic pathways has become an effective method to increase pigment production, stability, and diversity. Microorganisms can be reconfigured to produce increased amounts of desired pigments by overexpressing rate-limiting enzymes, deleting nontargeting pathways, and the addition of heterologous genes (Lyu *et al.*, 2022). As an example, engineered *Escherichia coli* and *Corynebacterium glutamicum* strains have surpassed natural carotenoid producers and their yields. Synthetic biology builds on metabolic engineering because it allows the design of new biosynthetic pathways that are not found in nature (Heider and Wendisch, 2015). Pigment production can now be controlled at both transcriptional and translational level; transferrable modular genetic toolkits, synthetic promoters, and CRISPR-based editing systems permit control over the biosynthesis of pigments. The method is also applicable in developing new pigment analogues that can be more stable or bioactive, thus extending the market to new molecules that do not occur in nature (Lyu *et al.*, 2022).

Traditional mutagenesis methodologies, such as UV irradiation and chemical mutagenesis, are economical in enhancing pigment-producing strains, especially in small- to medium-scale applications. Such techniques, together with adaptive lab evolution, can produce robust strains that can grow under industrial fermentation conditions (Han *et al.*, 2024).

15.3.3 Utilization of low-cost and waste substrates

Economic production factors are based on the selection of substrate. Agro-industrial byproducts, such as molasses, whey, rice bran, corn steep liquor, and fruit processing residues, can significantly lower the cost of raw materials input and, at the same time, help to valorize wastes and encourage the principles of circular bioeconomy (Mohamed *et al.*, 2021). As an example, *Monascus* fermentation on rice bran or wheat bran not only produces value-added pigments, but also utilizes cheap agricultural waste. Microalgae grown in wastewater streams can synthesize phycobiliproteins and at the same time bioremediate the waste streams. This may possibly require substrate pretreatment to provide nutrient bioavailability and limit introduction of inhibitory compounds (Kaur *et al.*, 2022).

15.4 Quality Control, Purification, and Standardization Protocols

The nature of downstream processing influences the eventual cost and quality of microbial pigments. Depending on pigment polarity and stability, extraction techniques differ, commonly used processes being solvent extraction, supercritical fluid extraction, and aqueous two-phase systems. The method employed influences not only pigment purity but also safety and compliance, especially where food and pharmaceutical applications are concerned (Ngamwonglumlert *et al.*, 2017).

Quality control is carried out by analytical testing of the composition, concentration, and stability of pigments, commonly through high-performance liquid chromatography (HPLC), spectrophotometry, and mass spectrometry. The standardization procedures allow pigment batches to comply with regulations regarding purity, freedom from contamination, and colorimetric properties (Pauk *et al.*, 2014).

The use of Good Manufacturing Practices (GMP) and hazard analysis critical control point (HACCP) models is critical in ensuring product safety and marketability, in particular when dealing with industries where product quality is of the utmost importance (Owusu-Apenten and Vieira, 2022).

15.5 Economic Analysis

15.5.1 Cost structure and profitability indicators

The competitiveness of the microbial pigment industry is an economic exercise, in which the net effect of the cost of production, attainable market price, and long-term profitability are considered. Analysis of expenditure shows that raw materials represent one of the biggest expense items, especially in cases where refined substrates are expensive (Bi *et al.*, 2024). This burden could be greatly mitigated by switching to low-cost agro-industrial residues like molasses, corn steep liquor, or fruit-processing waste. Other costly aspects include energy to produce aeration, agitation, temperature control, and sterilization mainly in submerged fermentation systems (Behl *et al.*, 2023). Labor costs are also variable based on the level of automation; the more automated the bioreactor system, the fewer labor resources are needed, although the latter requires larger capital investment (Mitra and Murthy, 2022).

Extraction, purification, and formulation of downstream processing may represent a 30–50% share of total costs, especially for high-purity pigments used in the food or pharmaceutical markets. Cost efficiency can be improved through optimization of extraction methods as well as recovery systems (Rodríguez-Sifuentes *et al.*, 2021).

Analysis of pricing trends demonstrates that natural microbial pigments normally attract a premium compared to synthetic pigments, based on their safety profile and sustainability credentials, as well as other biofunctional advantages (Randhawa, 2024). Nevertheless, in cost-influenced sectors like textiles and mass-market foods, sensitivity to the market is critical. The measures of profitability, including the return on investment (ROI), payback period, and gross margin, tend to change considerably depending on the scale of production and target market. Elite niche applications offer ROI within 3–5 years and commodity-level application may encounter even longer payback times without significantly reducing production cost (Thilmany, 2012).

15.5.2 Market trends and commercial success stories

The global market for natural pigments is expected to grow steadily because of the demand for clean-label products and stricter chemical regulations (Yaqoob *et al.*, 2021). Market projections estimate billions of dollars in revenues by the end of this decade, with 5–7% compound annual growth rates (CAGR) in key locales. The Asia-Pacific markets are emerging at pace with increased food processing products and cosmetics industries, alongside Europe and North America that still lead in regulatory-based adoption and consumer awareness (Yaqoob *et al.*, 2021).

A number of case studies demonstrate the commercial potential of microbial pigments. Astaxanthin manufacture with *Haematococcus pluvialis* and *Paracoccus* organisms has penetrated the nutraceutical market and the fish farming industry, helping to maintain profit at high production costs (Ariyadasa *et al.*, 2024). The stable performance of *Monascus*-based pigments in Asian markets is attributable to categorization into complex cuisines or drinks. Firms that utilize modern biotechnological practices, strategic alliances, and vertical integration have demonstrated that coordinating process optimization with deliberate marketing positioning can achieve stable profitability (Lin and Lekhawipat, 2023).

In short, cost-effective production, targeted marketing, and differentiation based on quality and biofunctionality will be key to the long-term economic viability of microbial pigments. The ability of producers to effectively combine technological innovation and market intelligence will allow them to obtain a competitive niche in the emerging global colorant market (Agarwal *et al.*, 2023).

15.6 Commercial Applications Across Industries

15.6.1 Food and beverage sector

Microbial pigments have been used extensively in food and beverage manufacture, mainly as natural coloring agents and functional food components. There is growing consumer demand for clean-label products without synthetic additives, enhancing the high market pull on pigments sourced through bacteria, fungi, and microalgae (Sen *et al.*, 2019). Astaxanthin and β-carotene offer robust, yellow-red colors and antioxidant effects, making them appropriate in bakery products, dairy products, beverages, and candy (Delgado-Vargas *et al.*, 2000). *Spirulina*-derived phycocyanin can provide a deep blue color with a high concentration and is accepted in many food categories in many countries (Thevarajah *et al.*, 2024).

In addition to being aesthetically beneficial, most microbial pigments have an added functional value that can justify their incorporation into a nutraceutical blend. As an example, astaxanthin-enriched formulation has the health benefits of eye and cardiovascular protection (Alugoju *et al.*, 2023), and *Monascus* pigments have a cholesterol-reducing effect known to be found in traditional fermented foods. Color enhancement and health promotion in the context of advertising create commercial value, which is especially significant in high-end products (Zhu *et al.*, 2023). Consumer awareness of the use of plant- and microbe-based ingredients, as well as regulatory acceptance in key markets, are drivers in this market segment.

15.6.2 Pharmaceuticals and nutraceuticals

Microbial pigments have bioactivity that makes them useful far beyond their aesthetic properties. Prodigiosin, violacein, and some anthraquinones exhibit antimicrobial, anticancer, and antiinflammatory capabilities which make them targets of therapeutical development. Most of these applications are still in preclinical or experimental phases, but there is increasing pharmaceutical interest in pigment-derived bioactives (Delhoyo, 2007; Araújo *et al.*, 2022).

Nutraceutical pigments, including lutein, zeaxanthin, and astaxanthin, are added to dietary supplements due to their antioxidant and photoprotective properties (Sajilata *et al.*, 2008; Patil *et al.*, 2022). The health advantages of these compounds in clinical studies include their use as photoprotection for the skin and maintenance of eye health, as well as immunotropic effects (Abdel-Aal *et al.*, 2013). This dual functionality as colorant and health benefit creates a strong value offering for product developers interested in health-conscious consumers. Economically, aggregate output of pharmaceutical-grade pigments is relatively low, and this is counteracted by higher unit prices commanding excellent margins in niche markets.

15.6.3 Cosmetics and personal care

Microbial pigments have been adopted by the cosmetics industry as it makes the transition towards more natural, sustainable, and ethically sourced ingredients. Pigments from *Monascus*, carotenoid bacteria, and microalgae are used in lipsticks, eye shadows, foundations, and skincare products. Besides imparting stable and vivid colors, these pigments have the capacity to improve product function owing to their antioxidant and antiinflammatory capabilities, which contribute to antiaging and skin-soothing effects (Basu *et al.*, 2022; Dussably *et al.*, 2022).

Clean-label marketing tactics enhance the marketing potential of such formulations since customers are increasingly relating natural colorants with safety, quality, and environmental sustainability. Premium cosmetic brands especially take advantage of microbial pigments as one way to distinguish their products in the highly competitive market, with transparency of ingredients being one of the elements of competition (Nabeshima *et al.*, 2024).

15.6.4 Textile industry

The textile industry is finding renewed interest in microbial pigments as a replacement for synthetic dyes, a key contributor to water pollution. Examples of pigments that have been tested to dye cotton, silk, and wool include violacein, prodigiosin, and carotenoids, with their use resulting in vivid color with adequate fastness (Gomes *et al.*, 2025). *Monascus* and *Talaromyces* species are examples of filamentous fungi that demonstrate promise as industrial producers of solid-state natural dyes in inexpensive agricultural byproducts (Afroz Toma *et al.*, 2023b).

The sustainability merit of microbial pigments in textiles is that they are biodegradable, have lower toxicities, and can be made using renewable resources (Fig. 15.2). Although commercial use is still at low levels because of scale-up issues and cost factors, niche markets, e.g. organic apparel and eco-fashion, are already making use of these advantages. With time, process optimization and market demand for sustainable textiles could help to optimize usage in mainstream manufacturing (Jamee and Siddique, 2019; Gomes *et al.*, 2025).

15.7 Challenges and Limitations

15.7.1 Cost-efficiency and yield inconsistency

The high cost of producing microbial pigments compared to synthetic pigments is one of the greatest challenges facing large-scale commercialization. Although there have been some successes in fermentation technology that have increased yield, the costs in terms of substrates, utilities, and downstream processing are still very high (Venil *et al.*, 2020). Submerged fermentation may demand a lot of energy in the form of gases and mixing whereas solid-state fermentation encounters scale-up problems which can hamper efficient production (Umrao *et al.*, 2024). Another major

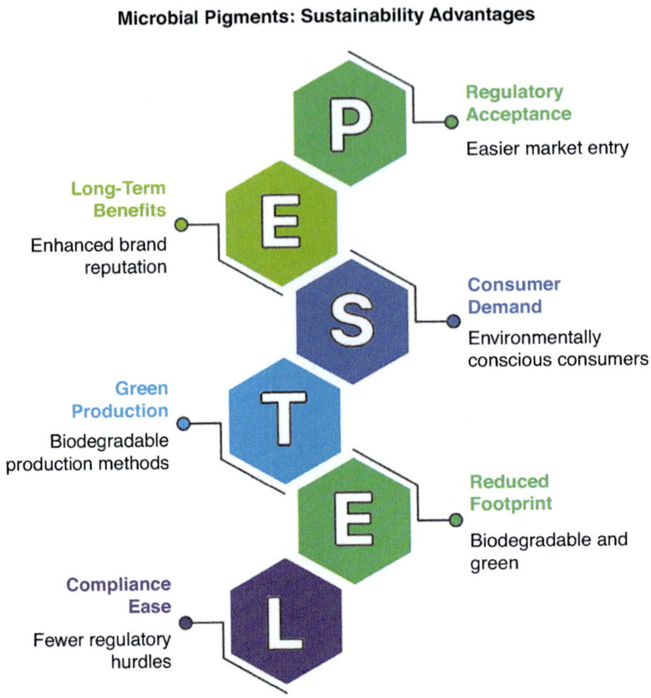

Fig. 15.2. Sustainability advantages of microbial pigments compared to synthetic pigments.

problem is yield variability; strain performance may fail in a manufacturing environment, and environmental shifts can affect pigment output. The absence of stable and predictable yields makes long-term contracts and pricing commitments risky for industrial buyers, which influences market penetration (Mulders *et al.*, 2014).

15.7.2 Stability, storage, and shelf-life constraints

The stability of pigments is a key determinant of product acceptance. Many microbial pigments are sensitive to environmental aspects including light, heat, pH, and exposure to oxygen (Chandra *et al.*, 2021). As an example, carotenoids are susceptible to oxidative breakdown and phycocyanin can fade under exposure to light or in low pH conditions. These fluctuations may restrict their application in goods that have a prolonged shelf-life or experience different processing conditions. Synthesis with stabilizers like microencapsulation, copigmentation, and antioxidant concentration formulation creates complexity and adds to the cost aspect of the manufacturing process, thereby making production inefficient and competitive (Malik *et al.*, 2012; Aman Mohammadi *et al.*, 2022).

15.7.3 Regulatory approval complexities

Another big challenge is the process of navigating food, pharmaceutical, and cosmetics regulations. Approval of new natural colorants takes time and money as studies involve investment in toxicological examinations and compositional investigation as well as observing regulations from various bodies including the US FDA, European Food Safety Administration (EFSA), and Japan Ministry of Health, Labour and Welfare. Pigments that may have a long history of use in some parts of the market might still need new safety analysis when presented to new markets or even to new applications. Such delays in regulation may hinder commercialization and raise the financial burden on producers (Aman Mohammadi *et al.*, 2022; Barreto *et al.*, 2023).

15.7.4 Consumer perception and market readiness

Although consumer demand is moving toward natural products, the levels of awareness and use of microbial pigments have been low. The word "microbial" might also have negative associations for consumers, which necessitates clear education and marketing focus on safety, natural origin, and sustainability in food and cosmetics (Tuli *et al.*, 2015). Moreover, markets differ with location and industry, in that a premium market can easily shift to microbial pigments even when the goods are expensive, but the price-sensitive marketplace may be slow to change from synthetics until the costs are equal.

15.8 Future Prospects and Innovations

In the coming years, microbial pigment development is likely to be characterized by rapid advances in genetic and metabolic engineering, as techniques such as CRISPR-Cas genome editing, synthetic biology toolboxes, and systems biology modeling allow researchers to develop pigment strains with greater productivity, more stable production, and larger color palettes. Through rationalization of metabolic routes, production systems can redirect capacity toward generating pigments by limiting the formation of unwanted byproducts and increasing efficiency. It is also possible to optimize engineered microbes to be tolerant to high pigment concentrations so that recovery efficiency increases and downstream processing is less costly. These developments not only make the strain development process more rapid, they also enable the production of new biofunctional pigments with designed properties, creating a new field of specialized products in areas like biomedical imaging, targeted drug delivery, and high-end cosmetics.

In line with such technological advances, there have been efforts to explore microbial diversity in extreme habitats, such as deep sea vents, salt flats, and geothermal springs, in the hope of finding pigment-producing species with distinct structural and functional properties. Extremophiles tend to develop pigments to act as protectants against strong UV radiation, salinity, or thermal stress, leading to an abundance of compounds with extraordinary stability that can help circumvent one of the key commercial inhibitors of existing microbial pigments. The speedy discovery of such candidates can be improved with the help of metagenomic and culture-independent screening technologies instead of using pigments from known sources.

Further cost-effective measures are achieved by producing microbial pigments using a circular bioeconomy approach because colocation with food processing facilities, breweries, or farms is possible to utilize nutrient-rich end-products as direct microbial growth substrates, further reducing the cost of raw materials and limiting waste. Bioprocesses intensification such as continuous fermentation and integrated biorefineries will increase the competitiveness of microbial pigments over synthetic ones in terms of costs and sustainability.

As the market in natural colorants is forecast to have a compound yearly growth rate of 5–7%, microbial pigments will gain an increasing share volume, assisted by regulatory forces and demand. In addition to traditional markets in food, beverages, and cosmetics, newer opportunities exist in areas such as natural inks and biodegradable plastics, smart textiles, and functional packaging, where pigments with multiple functions and bioactive properties can create new opportunities of high value, such as the combination of color and antimicrobial, antioxidant or UV-protective efficacy.

15.9 Conclusion

The pigment industry is undergoing a radical transformation as a result of the interplay between the environmental imperative, government regulation, and altered consumer demand, forming an ideal basis on which sustainable solutions like microbial pigments may develop. Bio-based colorants, biosynthesized by bacteria, fungi, yeasts, and microalgae, offer eye-catching and colorful colors together with intrinsic biodegradability, biocompatibility, and biofunctional effects, such as antimicrobial, antioxidant, and anticancer potential. They provide distinct benefits over synthetic dyes, such as nonseasonal production under controlled conditions, no dependence on agricultural and climatic constraints, and the potential to utilize low-cost renewable substrates.

In this chapter, the technological basis, cost schemes, market states and regulations that determine the economic and commercial feasibility of microbial pigments have been critically assessed. It has been found that these pigments are already commercially competitive within premium markets in nutraceuticals, cosmetics, and functional foods where the branding and bioactivity enable premium pricing to be applied. However, market acceptance into lower price-sensitive domains like mass-market foods and textiles will be contingent upon resolving issues that constrain adoption such as the high cost of production, lack of consistently high yields, stability considerations, and regulatory complexity. To overcome these bottlenecks, a multi-pronged approach is needed in which advanced metabolic engineering and synthetic biology tools are used to enhance yields or generate new pigments; sustainable feedstock use and process intensification enhance cost-effectiveness; stabilization technologies are applied to enhance shelf-life and further expand potential application. It is important to align scientific discoveries with business approaches, such as educating the markets to build trust in products achieved through microbial processes and offering positioning in high-value markets. Researchers are interested in finding new pigment-producing species, especially extremophilic ones, in which the optimum levels of stability characteristics might be expected, and in selection of strains with increased resistance to adverse conditions in industrial fermentation.

To industry stakeholders, investing in vertically integrated production systems, strategic partnerships, and compliant quality systems will be important to fast-track entry to the markets and expand operations. When properly integrated, such scientific, technical, and commercial avenues will not only supplant synthetic colorants in various industries, but will redefine the importance of colorants in a sustainable global economy, turning them into multifunctional agents that add profitability as well as health benefits for the global economy.

References

Abdel-Aal, E.-S.M., Akhtar, H., Zaheer, K. and Ali, R. (2013) Dietary sources of lutein and zeaxanthin carotenoids and their role in eye health. *Nutrients* 5(4), 1169–1185. DOI: 10.3390/nu5041169.

Adamantidi, T., Lafara, M.-P., Venetikidou, M., Likartsi, E., Toganidou, I. *et al.* (2025) Utilization and bio-efficacy of carotenoids, vitamin A and its vitaminoids in nutricosmetics, cosmeceuticals, and cosmetics' applications with skin-health promoting properties. *Applied Sciences* 15(3), 1657. DOI: 10.3390/app15031657.

Afroz Toma, M., Rahman, M.H., Rahman, M.S., Arif, M., Nazir, K.H.M.N.H. *et al.* (2023a) Fungal pigments: Carotenoids, riboflavin, and polyketides with diverse applications. *Journal of Fungi* 9(4), 454. DOI: 10.3390/jof9040454.

Afroz Toma, M., Rahman, M.H., Rahman, M.S., Arif, M., Nazir, K.N.M. *et al.* (2023b) Fungal pigments: Carotenoids, riboflavin, and polyketides with diverse applications. *Journal of Fungi* 9(4), 454. DOI: 10.3390/jof9040454.

Agarwal, H., Bajpai, S., Mishra, A., Kohli, I., Varma, A. *et al.* (2023) Bacterial pigments and their multifaceted roles in contemporary biotechnology and pharmacological applications. *Microorganisms* 11(3), 614. DOI: 10.3390/microorganisms11030614.

Alugoju, P., Krishna Swamy, V., Anthikapalli, N.V.A. and Tencomnao, T. (2023) Health benefits of astaxanthin against age-related diseases of multiple organs: A comprehensive review. *Critical Reviews in Food Science and Nutrition* 63(31), 10709–10774. DOI: 10.1080/10408398.2022.2084600.

Aman Mohammadi, M., Ahangari, H., Mousazadeh, S., Hosseini, S.M. and Dufossé, L. (2022) Microbial pigments as an alternative to synthetic dyes and food additives: A brief review of recent studies. *Bioprocess and Biosystems Engineering* 45(1), 1–12. DOI: 10.1007/s00449-021-02621-8.

Araújo, R.G., Zavala, N.R., Castillo-Zacarías, C., Barocio, M.E., Hidalgo-Vázquez, E. *et al.* (2022) Recent advances in prodigiosin as a bioactive compound in nanocomposite applications. *Molecules* 27(15), 4982. DOI: 10.3390/molecules27154982.

Ariyadasa, T.U., Thevarajah, B., Anthonio, R.A.D.P., Nimarshana, P. and Wasath, W.A.J. (2024) From present to prosperity: Assessing the current status and envisioning opportunities in the industrial-scale cultivation of haematococcus pluvialis for astaxanthin production. *Phytochemistry Reviews* 23(3), 749–779. DOI: 10.1007/s11101-023-09906-8.

Barbinta-Patrascu, M.-E., Bita, B. and Negut, I. (2024) From nature to technology: Exploring the potential of plant-based materials and modified plants in biomimetics, bionics, and green innovations. *Biomimetics* 9(7), 390. DOI: 10.3390/biomimetics9070390.

Barreto, J.V. d. O., Casanova, L.M., Junior, A.N., Reis-Mansur, M.C.P.P. and Vermelho, A.B. (2023) Microbial pigments: Major groups and industrial applications. *Microorganisms* 11(12), 2920. DOI: 10.3390/microorganisms11122920.

Basu, R., Ghosh, S., Dhara, B., Roy, I., Basumatary, E. *et al.* (2022) Extremophilic pigments in eukaryotic microbes: A tool for survivability. In: *Extremophiles: A Paradox of Nature with Biotechnological Implications*. De Gruyter Brill, New York, p. 149. DOI: 10.1515/9783110788488-008.

Behl, M., Thakar, S., Ghai, H., Sakhuja, D. and Bhatt, A.K. (2023) Fundamentals of fermentation technology. In: *Basic Biotechniques for Bioprocess and Bioentrepreneurship*. Elsevier, pp. 313–328.

Benucci, I., Lombardelli, C., Mazzocchi, C. and Esti, M. (2022) Natural colorants from vegetable food waste: Recovery, regulatory aspects, and stability – a review. *Comprehensive Reviews in Food Science and Food Safety* 21(3), 2715–2737. DOI: 10.1111/1541-4337.12951.

Bi, A., Umesh, K., Abdul, B.M., Sivakuma, D. and Srikanth, P. (2024) Economic and environmental sustainability of agriculture production at the crop level. *Global Journal of Environmental Science & Management* 10(3). DOI: 10.22034/gjesm.2024.03.24.

Chandra, R.D., Prihastyanti, M.N.U. and Lukitasari, D.M. (2021) Effects of pH, high pressure processing, and ultraviolet light on carotenoids, chlorophylls, and anthocyanins of fresh fruit and vegetable juices. *EFood* 2(3), 113–124. DOI: 10.2991/efood.k.210630.001.

Chaudhary, V., Katyal, P., Poonia, A.K., Kaur, J., Puniya, A.K. *et al.* (2022) Natural pigment from monascus: The production and therapeutic significance. *Journal of Applied Microbiology* 133(1), 18–38. DOI: 10.1111/jam.15308.

Chilakamarry, C.R., Sakinah, A.M., Zularisam, A.W., Sirohi, R., Khilji, I.A. *et al.* (2022) Advances in solid-state fermentation for bioconversion of agricultural wastes to value-added products: Opportunities and challenges. *Bioresource Technology* 343, 126065. DOI: 10.1016/j.biortech.2021.126065.

Delgado-Vargas, F., Jiménez, A. and Paredes-López, O. (2000) Natural pigments: Carotenoids, anthocyanins, and betalains – characteristics, biosynthesis, processing, and stability. *Critical Reviews in Food Science and Nutrition* 40(3), 173–289. DOI: 10.1080/10408690091189257.

Delhoyo, C. (2007) Layered double hydroxides and human health: An overview. *Applied Clay Science* 36(1–3), 103–121. DOI: 10.1016/j.clay.2006.06.010.

do Socorro do Souza Vilhena, K., da Silva, M.P., de Santana Botelho, A., Cardoso, F.J.B., Ferreira, O.O. *et al.* (2024) Biosynthesis of carotenoids in plants and microorganisms. In: *Carotenoids: Trends and Advances*. Springer, Dordrecht, The Netherlands, pp. 37–63.

Dussably, J., Mshvildadze, V., Pichette, A. and Ripoll, L. (2022) Microalgae and diatom – potential pharmaceutical and cosmetic resources – review. *Journal of Biomedical Research & Environmental Sciences* 3(9), 1082–1092. DOI: 10.37871/jbres1559.

Fabre, J.-F., Niangoran, N. g. U.F., Gaignard, C., Buso, D., Mouloungui, Z. *et al.* (2022) Extraction, purification and stability of C-phycocyanin from Arthrospira platensis. *European Food Research and Technology* 248(6), 1583–1599. DOI: 10.1007/s00217-022-03987-z.

Gänzle, M. (2022) The periodic table of fermented foods: Limitations and opportunities. *Applied Microbiology and Biotechnology* 106(8), 2815–2826. DOI: 10.1007/s00253-022-11909-y.

García-Gómez, C., Aguirre-Cavazos, D.E., Chávez-Montes, A., Ballesteros-Torres, J.M., Orozco-Flores, A.A. *et al.* (2025) Phycobilins versatile pigments with wide-ranging applications: Exploring their uses. *Biological Activities, Extraction Methods and Future Perspectives. Marine Drugs* 23(5), 201. DOI: 10.3390/md23050201.

Gomes, D.C. (2024) Fungal pigments: Applications and their medicinal potential. In: *Fungi Bioactive Metabolites: Integration of Pharmaceutical Applications*. Springer, Cham, Switzerland, pp. 651–681.

Gomes, M., Felgueiras, H.P., Leite, B.R. and Soares, G.M.B. (2025) Colourful protection: Challenges and perspectives of antibacterial pigments extracted from bacteria for textile applications. *Antibiotics* 14(5), 520. DOI: 10.3390/antibiotics14050520.

Gupta, I., Adin, S.N., Panda, B.P. and Mujeeb, M. (2022) β-carotene-production methods, biosynthesis from Phaffia rhodozyma, factors affecting its production during fermentation, pharmacological properties: A review. *Biotechnology and Applied Biochemistry* 69(6), 2517–2529. DOI: 10.1002/bab.2301.

Han, J., Ullah, M., Andoh, V., Khan, M.N., Feng, Y. *et al.* (2024) Engineering bacterial chitinases for industrial application: From protein engineering to bacterial strains mutation. A comprehensive review of physical, molecular, and computational approaches. *Journal of Agricultural and Food Chemistry* 72(42), 23082–23096. DOI: 10.1021/acs.jafc.4c06856.

Heider, S.A. and Wendisch, V.F. (2015) Engineering microbial cell factories: metabolic engineering of *Corynebacterium glutamicum* with a focus on non-natural products. *Biotechnology Journal* 10(8), 1170–1184. DOI: 10.1002/biot.201400590.

Hosikian, A., Lim, S., Halim, R. and Danquah, M.K. (2010) Chlorophyll extraction from microalgae: A review on the process engineering aspects. *International Journal of Chemical Engineering* 2010(1), 1–11. DOI: 10.1155/2010/391632.

Jamee, R. and Siddique, R. (2019) Biodegradation of synthetic dyes of textile effluent by microorganisms: An environmentally and economically sustainable approach. *European Journal of Microbiology & Immunology* 9(4), 114–118. DOI: 10.1556/1886.2019.00018.

Jurić, S., Jurić, M., Król-Kilińska, Ż., Vlahoviček-Kahlina, K., Vinceković, M. *et al.* (2022) Sources, stability, encapsulation and application of natural pigments in foods. *Food Reviews International* 38(8), 1735–1790. DOI: 10.1080/87559129.2020.1837862.

Kaur, P., Singh, S., Ghoshal, G., Ramamurthy, P.C., Parihar, P. *et al.* (2022) Valorization of agri-food industry waste for the production of microbial pigments: an eco-friendly approach. In: *Advances in Agricultural and Industrial Microbiology: Volume 1: Microbial Diversity and Application in Agroindustry*. Springer Nature, Singapore, pp. 137–167. DOI: 10.1007/978-981-16-8918-5_8.

Lin, L. and Xu, J. (2022) Production of fungal pigments: Molecular processes and their applications. *Journal of Fungi* 9(1), 44. DOI: 10.3390/jof9010044.

Lin, M. and Lekhawipat, W. (2023) Key influencing factors for the success of external innovation strategies in the biotechnology industry. *Journal of Business & Industrial Marketing* 38(12), 2745–2759. DOI: 10.1108/JBIM-07-2022-0307.

Lyu, X., Lyu, Y., Yu, H., Chen, W., Ye, L. *et al.* (2022) Biotechnological advances for improving natural pigment production: A state-of-the-art review. *Bioresources and Bioprocessing* 9(1), 8. DOI: 10.1186/s40643-022-00497-4.

Malik, K., Tokkas, J. and Goyal, S. (2012) Microbial pigments: A review. *International Journal of Microbial Resource Technology* 1(4), 361–365.

Marcelli, A., Osimani, A. and Aquilanti, L. (2025) Vegetable by-products from industrial processing: From waste to functional ingredient through fermentation. *Foods* 14(15), 2704. DOI: 10.3390/foods14152704.

Mirsalami, S.M. and Mirsalami, M. (2025) Advances in genetically engineered microorganisms: Transforming food production through precision fermentation and synthetic biology. *Future Foods* 11, 100601. DOI: 10.1016/j.fufo.2025.100601.

Mitra, S. and Murthy, G.S. (2022) Bioreactor control systems in the biopharmaceutical industry: A critical perspective. *Systems Microbiology and Biomanufacturing* 2(1), 91–112. DOI: 10.1007/s43393-021-00048-6.

Mohamed, H., Shah, A.M. and Song, Y. (2021) Conversion of agro-industrial wastes into value-added products. In: *Conversion of Renewable Biomass into Bioproducts*. ACS Publications, Washington, DC, pp. 197–217.

Mulders, K.J., Lamers, P.P., Martens, D.E. and Wijffels, R.H. (2014) Phototrophic pigment production with microalgae: Biological constraints and opportunities. *Journal of Phycology* 50(2), 229–242. DOI: 10.1111/jpy.12173.

Nabeshima, E.H., Tavares, P.E. d. R., Lemos, A.L. d. S.C. and Moura, S.C.S.R. (2024) Ingredients for clean label products and food safety. *Brazilian Journal of Food Technology* 27, e2023160. DOI: 10.1590/1981-6723.016023.

Ngamwonglumlert, L., Devahastin, S. and Chiewchan, N. (2017) Natural colorants: Pigment stability and extraction yield enhancement via utilization of appropriate pretreatment and extraction methods. *Critical Reviews in Food Science and Nutrition* 57(15), 3243–3259. DOI: 10.1080/10408398.2015.1109498.

Owusu-Apenten, R. and Vieira, E. (2022) *Food safety management, GMP & HACCP*. In: *Elementary Food Science*. Springer, Cham, Switzerland, pp. 217–236. DOI: 10.1007/978-3-030-65433-7_10.

Park, H., Park, S., Yang, Y.-H. and Choi, K.-Y. (2021) Microbial synthesis of violacein pigment and its potential applications. *Critical Reviews in Biotechnology* 41(6), 879–901. DOI: 10.1080/07388551.2021.1892579.

Pathak, J., Pandey, A., Maurya, P.K., Rajneesh, R., Sinha, R.P. *et al.* (2020) Cyanobacterial secondary metabolite scytonemin: A potential photoprotective and pharmaceutical compound. *Proceedings of the National Academy of Sciences, India Section B: Biological Sciences* 90(3), 467–481. DOI: 10.1007/s40011-019-01134-5.

Patil, A.D., Kasabe, P.J. and Dandge, P.B. (2022) Pharmaceutical and nutraceutical potential of natural bioactive pigment: Astaxanthin. *Natural Products and Bioprospecting* 12(1), 25. DOI: 10.1007/s13659-022-00347-y.

Pauk, V., Barták, P. and Lemr, K. (2014) Characterization of natural organic colorants in historical and art objects by high-performance liquid chromatography. *Journal of Separation Science* 37(23), 3393–3410. DOI: 10.1002/jssc.201400650.

Paul, T., Bandyopadhyay, T.K., Mondal, A., Tiwari, O.N., Muthuraj, M. *et al.* (2022) A comprehensive review on recent trends in production, purification, and applications of prodigiosin. *Biomass Conversion and Biorefinery* 12(4), 1409–1431. DOI: 10.1007/s13399-020-00928-2.

Poorniammal, R., Prabhu, S. and Dufossé, L. (2025) Polyketides: Complex pigments from microbes with medicinal potential. In: *Microbial Colorants: Chemistry, Biosynthesis and Applications*. Wiley, Hoboken, NJ, pp. 275–285. DOI: 10.1002/9781394287888.

Pucci, C., Martinelli, C., Degl'Innocenti, A., Desii, A. and Pasquale, D. (2021) Light-activated biomedical applications of chlorophyll derivatives. *Macromolecular Bioscience* 21(9), e2100181. DOI: 10.1002/mabi.202100181.

Putnik, G., Sluga, A., ElMaraghy, H., Teti, R., Koren, Y. *et al.* (2013) Scalability in manufacturing systems design and operation: State-of-the-art and future developments roadmap. *CIRP Annals* 62(2), 751–774. DOI: 10.1016/j.cirp.2013.05.002.

Ramesh, C., Prasastha, V.R., Venkatachalam, M. and Dufossé, L. (2022) Natural substrates and culture conditions to produce pigments from potential microbes in submerged fermentation. *Fermentation* 8(9), 460. DOI: 10.3390/fermentation8090460.

Ramesh, P.V., Sangeetha Gowda, K. and More, S.S. (2025) Introduction to chemistry of microbial colorants:structures, properties, and biosynthesis. In: *Microbial Colorants: Chemistry, Biosynthesis and Applications*. Wiley, Hoboken, NJ, pp. 129–175. DOI: 10.1002/9781394287888.

Randhawa, K.S. (2024) Synthesis, properties, and environmental impact of hybrid pigments. *Scientific World Journal* 2024(1), 2773950. DOI: 10.1155/tswj/2773950.

Rao, A.S., Deka, S.P., More, S.S., Nair, A., More, V.S. *et al.* (2021) A comprehensive review on different microbial-derived pigments and their multipurpose activities 479–519. DOI: 10.1007/978-981-16-0045-6_20.

Rapson, T.D. (2004) *Bioactive 4-methoxypyrrolic natural products from two South African marine invertebrates*. Rhodes University, South Africa.

Rather, L.J., Mir, S.S. and Islam, Z. (2025) Introduction to microbial colorants. In: *Microbial Colorants: Chemistry, Biosynthesis and Applications*. Wiley, Hoboken, NJ, pp. 1–32. DOI: 10.1002/9781394287888.

Rengifo, L.R., Rosas, P., Méndez, N., Ludeña, Y., Sirvas, S. *et al.* (2022) Comparison of pigment production by filamentous fungal strains under submerged (SmF) and surface adhesion fermentation (SAF). *Journal of Fungi* 9(1), 48. DOI: 10.3390/jof9010048.

Renita, A.A., Gajaria, T.K., Sathish, S., Kumar, J.A., Lakshmi, D.S. *et al.* (2023) Progress and prospective of the industrial development and applications of eco-friendly colorants: An insight into environmental impact and sustainability issues. *Foods* 12(7), 1521. DOI: 10.3390/foods12071521.

Rodriguez-Amaya, D.B., Esquivel, P. and Meléndez-Martínez, A.J. (2023) Comprehensive update on carotenoid colorants from plants and microalgae: Challenges and advances from research laboratories to industry. *Foods* 12(22), 4080. DOI: 10.3390/foods12224080.

Rodríguez-Sifuentes, L., Marszalek, J.E., Hernández-Carbajal, G. and Chuck-Hernández, C. (2021) Importance of downstream processing of natural astaxanthin for pharmaceutical application. *Frontiers in Chemical Engineering* 2, 601483. DOI: 10.3389/fceng.2020.601483.

Sajilata, M.G., Singhal, R.S. and Kamat, M.Y. (2008) The carotenoid pigment zeaxanthin—a review. *Comprehensive Reviews in Food Science and Food Safety* 7(1), 29–49. DOI: 10.1111/j.1541-4337.2007.00028.x.

Salian, A., Dutta, S. and Mandal, S. (2021) A roadmap to UV-protective natural resources: Classification, characteristics, and applications. *Materials Chemistry Frontiers* 5(21), 7696–7723. DOI: 10.1039/D1QM00741F.

Seetharaman, B., Balu, U.R., Mudgal, G., Firdaus, S., Dash, S. *et al.* (2024) The piteous price of progress: the environmental and health costs of modern society. In: *Sustainable Development Goals Towards Environmental Toxicity and Green Chemistry: Environment and Sustainability*. Springer, Dordrecht, The Netherlands, pp. 291–313. DOI: 10.1007/978-3-031-77327-3_15.

Šelo, G., Planinić, M., Tišma, M., Tomas, S., Koceva Komlenić, D. *et al.* (2021) A comprehensive review on valorization of agro-food industrial residues by solid-state fermentation. *Foods* 10(5), 927. DOI: 10.3390/foods10050927.

Sen, T., Barrow, C.J. and Deshmukh, S.K. (2019) Microbial pigments in the food industry–challenges and the way forward. *Frontiers in Nutrition* 6, 7. DOI: 10.3389/fnut.2019.00007.

Silva, S.d.O, Mafra, A.K.C., Pelissari, F.M., Lemos, L. and Molina, G. (2025) Biotechnology in agro-industry: Valorization of agricultural wastes, by-products and sustainable practices. *Microorganisms* 13(8), 1789. DOI: 10.3390/microorganisms13081789.

Thevarajah, B., Piyatilleke, S., Nimarshana, P., Koushalya, S., Malik, A. *et al.* (2024) Exploring effective light spectral conversion techniques for enhanced production of spirulina-derived blue pigment protein, c-phycocyanin. *Bioresource Technology* 399, 130612. DOI: 10.1016/j.biortech.2024.130612.

Thilmany, D. (2012) What are niche markets? What advantages do they offer. *Assessment and Strategy Development for Agriculture* 8(13), 1–14.

Trujillo-Cayado, L.A., Sánchez-García, R.M., García-Domínguez, I., Rodríguez-Luna, A., Hurtado-Fernández, E. *et al.* (2025) Emerging trends in sustainable biological resources and bioeconomy for food production. *Applied Sciences* 15(12), 6555. DOI: 10.3390/app15126555.

Tuli, H.S., Chaudhary, P., Beniwal, V. and Sharma, A.K. (2015) Microbial pigments as natural color sources: Current trends and future perspectives. *Journal of Food Science and Technology* 52(8), 4669–4678. DOI: 10.1007/s13197-014-1601-6.

Uma, G., Jeraldine Nisha, S., Rameshbabu, D., Citarasu, T. and Immanuel, G. (2023) Algal polymers, proteins, and pigments for industrial applications 247–271.

Umrao, D., Singh, S., Singh, D.P., Maddirala, S., Sevda, S. *et al.* (2024) Treatment of solid food waste using solid-state fermentation: Its current use and future perspectives. In: *Solid Waste Management*. CRC Press, Boca Raton, FL, pp. 161–180. DOI: 10.1201/9781003229919-10.

Venil, C.K., Dufossé, L. and Renuka Devi, P. (2020) Bacterial pigments: sustainable compounds with market potential for pharma and food industry. *Frontiers in Sustainable Food Systems* 4, 100. DOI: 10.3389/fsufs.2020.00100.

Yaqoob, S., Riaz, M., Shabbir, A., Zia-Ul-Haq, M., Alwakeel, S.S. *et al.* (2021) Commercialization and marketing potential of carotenoids. In: *Carotenoids: Structure and Function in the Human Body*. Springer, Dordrecht, The Netherlands, pp. 799–826. DOI: 10.1007/978-3-030-46459-2_27.

Zhu, M.-N., Wu, G., Chen, J., Qiu, Y.-W., Zhu, C.-Q. *et al.* (2023) Ethnopharmacology, phytochemistry, and pharmacology of highland barley monascus purpureus went: A comprehensive review. *Combinatorial Chemistry & High Throughput Screening* 26(6), 1083–1092. DOI: 10.2174/1386207325666220818153054.

16 Regulatory and Safety Aspects of Using Microbial Pigments

Pavithra Sivakumar[1*], Dianambika Palaniswamy[1], Devayani Sarmah[2] and R.M. Vijaya Ramakrishnan[1]

[1]Department of Plant Pathology, Tamil Nadu Agricultural University, Coimbatore, Tamil Nadu, India; [2]Division of Plant Pathology, Indian Agricultural Research Institute, Central Rice Research Institute, Cuttack, Odisha, India

16.1 Introduction

The color of food has played a crucial role in shaping its visual appeal, often being the first sensory attribute detected and a key factor in determining consumer acceptance. Both natural and synthetic dyes are widely employed in many aspects of daily life, including the manufacture of food, textiles, paper, and agricultural techniques. Considering their harmful effects on people, animals and the environment, synthetic dyes, which replaced natural pigments in the middle of the 19th century, continue to dominate the sector (Venil *et al.*, 2013).

In recent decades, synthetic chemicals have become increasingly significant in the food sector due to the high cost and limited availability of natural ingredients. However, because of the established toxicity consequences of some synthetic substances, consumers now seek natural goods (Sowbhagya and Chitra, 2010). Plants and microbes are the primary suppliers of natural pigments. Plant pigments have limitations such as instability, limited water solubility, and seasonal availability. Microbial pigments are highly sought after for durability and easy culturing (Parekh *et al.*, 2000).

Microbial pigments are bioactive compounds produced by microorganisms, which impart color and often exhibit biofunctional properties. Common microbial pigments include carotenoids, anthraquinones, melanin, violacein, flavins, quinones, and *Monascus* pigments (Ramesh *et al.*, 2019). The World Health Organization (WHO), the European Food Standards Authority (EFSA), and the US Food and Drug Administration (FDA) have all promoted the use of these hues in food, medications, and cosmetic products at acceptable levels. Natural pigments offer eco-friendly, biodegradable colorants with antibacterial and antioxidant characteristics. They are not only employed as food additives but also in cosmetics and medicinal applications (Dufossé, 2016).

Microbial pigments are created by bacteria, fungi, and algae. Because organisms can produce bioactive substances, it is crucial to evaluate the safety of the pigment and the organism that makes it. Toxicity studies examine the consequences of long-term exposure, possible accumulation, and metabolic impacts. To make sure the pigment is pure, safe, stable and of the appropriate grade, a thorough quality assay for microbial pigments includes several testing factors. For their intended uses in the food sector, cosmetics, medicines, or

*Corresponding author: pavithrabsivakumar@gmail.com

 Biopigments of Microbial Origin: Innovations and Applications Across Industries (eds Madan *et al.*)
DOI: 10.1079/ 9781836990505.0016

textiles, these tests aid in preserving the uniformity and safety of microbial pigments.

16.2 Major Pigments Produced by Microorganisms

16.2.1 β-Carotene

Carotenoids that range in color from yellow to orange-red. Microorganisms such as *Streptomyces*, *Mycobacterium*, *Agrobacterium*, *Sulfolobus*, *Micrococcus*, and *Serratia* are used for the production of carotenoids (Kushwaha *et al.*, 2014). *Phycomyces* and *Mucor circinelloides* (wild type) are useful sources of β-carotene. The European Union Committee approved the use of β-carotene from *Blakeslea trispora* fermentation as a food coloring additive, as it is comparable to chemically created materials (Malik *et al.*, 2012). The production process adheres to GMP procedures, hygiene standards, and raw material control. The biomass and final crystalline product meet chemical and microbiological specifications, as well as the Joint FAO/WHO Expert Committee on Food Additives (JECFA) and EU specifications for food coloring (Directive 95/45/EC) (Dufossé, 2006).

The global commercial market value for carotenoids reached $520 million in 2020, and is projected to reach %780 million by 2027 (Sun *et al.*, 2023).

16.2.2 Canthaxanthin

The distinct hue and characteristics of canthaxanthin are the result of a tiny structural difference from other carotenoids like β-carotene and astaxanthin, despite their chemical similarity (Chuyen and Eun, 2017). Strains of *Bradyrhizobium* (photosynthetic) contain orange and dark pink pigments that are isolated from the stem nodules of *Aeschynomene* and *Halobacterium* spp. These microbes can generate canthaxanthin, the primary carotenoid pigment which is a strong antioxidant that prevents lipids in liposomes from oxidizing (Malik *et al.*, 2012).

16.2.3 Lycopene

Lycopene is an open chain, unsaturated carotenoid molecule that has the longest β-carotene isomer of any carotenoid. This substance, also referred to as psi-carotene, is very susceptible to heat and oxidation and is insoluble in water. Using maize fiber as a substrate, *Fusarium sporotrichioides* was used to create colorants and the antioxidant lycopene (Giovannucci *et al.*, 2002).

16.2.4 *Monascus* pigments

In Asian nations, monascus is grown on solid media to provide *anka*, a crimson colorant used as a culinary component. The word "ang-kak," or "red mold rice," was initially used in a Chinese herbal medicine text from the first century (Van Thieghem, 1884). *Monascus* pigments are a class of fungal secondary metabolites known as azaphilones that are generated by filamentous fungi belonging to the Ascomycetes group genus *Monascus*. They are red (monascorubramine and rubropunctamine), orange (rubropunctatin and monascorubrine), and yellow (ankaflavin and monascin) pigments (Vendruscolo *et al.*, 2013). *Monascus* pigments are banned in the United States and Europe owing to the presence of mycotoxins.

16.2.5 Arpink Red™

This red pigment is generated by the soil-derived *Penicillium oxalicum* strain. This pigment contains the chromophore of anthraquinone type 14. The Codex Alimentarius Commission have recommended levels of Arpink Red for a variety of food items (Kumar *et al.*, 2015). In recent years, several toxicological tests have been carried out to evaluate safety and Arpink Red was granted 2 years of permission for distribution in the EU (Dufossé *et al.*, 2005) (Table 16.1).

16.2.6 Anthraquinones

The most common pigment class that has been demonstrated to be potentially safe for human intake is the anthraquinones. The most common genera that generate anthraquinones include *Aspergillus*, *Eurotium*, *Fusarium*, *Drechslera*, *Penicillium*, *Emericella purpurea*, *Curvularia lunata*, *Mycosphaerella rubella* and *Microsporum* sp.

Table 16.1. The amount of Arpink Red recommended by the Codex Alimentarius Commission (Rotterdam meeting, March 11–15, 2002).

Products	Permissible limit
Meat products	100 mg/kg
Meat and meat product analogues	100 mg/kg
Nonalcoholic drinks	100 mg/kg
Alcoholic drinks	200 mg/kg
Milk products	150 mg/kg
Ice creams	150 mg/kg
Confectionery	300 mg/kg

16.2.7 Betalains

Betalains are made up of two primary groups: yellow-orange betaxanthines and red-purple betacyanins (Lombardelli *et al.*, 2021). They are known as E162 in the European Union and are not subject to certification in the US (Sigurdson *et al.*, 2017).

16.2.8 Riboflavin

Numerous microorganisms create the yellow, water-soluble vitamin riboflavin. Commercially viable biotechnological methods employing the ascomycete *Ashbya gossypii*, filamentous fungi like *Candida famata*, or bacteria like *Bacillus subtilis* are replacing the conventional chemical manufacturing of riboflavin (Stahmann *et al.*, 2000). Cereals, pastas, sauces, processed cheese, milk products, energy drinks, and other foods all include riboflavins as food additives (Malik *et al.*, 2012). These strains are used to extract natural yellow colorants for a range of food items such as fruit drinks, breakfast cereals, pastas, sauces, processed cheese, vitamin-enhanced milk products, and various energy beverages.

16.2.9 Violacein

Violacein is a versatile purple pigment with diverse biological properties. Several bacteria, including *Chromobacterium violaceum*, *Pseudoalteromonas*, *Collimonas*, *Janthinobacterium*, and *Microbulbifer*, are known to generate this pigment (Choi *et al.*, 2015) which is used in cosmetics, food, medicine, and textiles (Dufossé, 2016). The pigment is associated with several useful bioactivities including antibacterial, anticancer, antiviral, enzyme modulation, antiulcerogenic, and antileishmanial effects.

16.3 Good Manufacturing Practices for Quality Regulation of Microbial Pigments

Good Manufacturing Practices (GMP) guarantee that goods are consistently manufactured and regulated by quality standards. Microbial pigments, which are obtained from microorganisms such as bacteria, fungi, algae or yeasts, must adhere to GMP to guarantee that they are safe, effective, and of high quality. These pigments are often used in sectors such as food, cosmetics, medicines, and textiles, and must fulfill stringent regulatory and safety criteria to avoid contamination, assure product uniformity, and protect consumer health (Table 16.2).

16.3.1 Strain identification and validation

The microbial strain utilized for pigment synthesis must be appropriately identified (species, strain number) and verified to guarantee that the intended pigment is produced consistently. Documentation should cover strain history, genetic features, and production profiles.

16.3.2 Master seed lot

To ensure uniformity between manufacturing batches, the master seed culture should be stored frozen. Subcultures should be grown on a regular basis from this master seed culture.

16.3.3 Microbial contamination and testing

To avoid contamination by foreign microbes, microbial cultures should be handled in a sterile environment. These sterile environments can be artificially created with the help of instruments like

Table 16.2. Major pigments produced by microbial organisms.

Pigment	Sources	Applications	References
β-Carotene	*Streptomyces*, *Mycobacterium*, *Agrobacterium*, *Sulfolobus*, *Micrococcus*, *Serratia*, *Phycomyces*, *Mucor circinelloides*, *Blakeslea trispora*	Food coloring additive approved by EU, adheres to GMP and JECFA standards	(Malik *et al*., 2012; Dufossé, 2006; Kushwaha *et al*., 2014)
Canthaxanthin	*Fusarium sporotrichioides*	Antioxidant properties, prevents lipid oxidation	(Malik *et al*., 2012; Chuyen and Eun, 2017)
Lycopene	*Fusarium sporotrichioides*	Antioxidant, red pigment, used in food and pharmaceuticals	(Giovannucci *et al*., 2002)
Monascus pigments	*Monascus* spp.	Red, orange, and yellow pigments, culinary uses in Asia, banned in US/EU due to mycotoxins	(Van Thieghem, 1884; Vendruscolo *et al*., 2013)
Arpink Red	*Penicillium oxalicum*	Food coloring for meat, beverages, dairy, confectionery; subject to Codex Alimentarius limits	(Dufossé *et al*., 2005; Kumar *et al*., 2015)
Anthraquinones	*Aspergillus*, *Eurotium*, *Fusarium*, *Drechslera*, *Penicillium*, *Curvularia lunata*	Safe for human consumption, used in food and cosmetics	(Fouillaud *et al*., 2018)
Betalains	*Gluconacetobacter diazotrophicus*	Natural food color (E162 in EU), includes betaxanthines (yellow-orange) and betacyanins (red-purple)	(Sigurdson *et al*., 2017; Lombardelli *et al*., 2021)
Riboflavin	*Ashbya gossypii*, *Candida famata*, *Bacillus subtilis*	Yellow food colorant; used in cereals, sauces, cheese, milk products, and energy drinks	(Stahmann *et al*., 2000; Malik *et al*., 2012)
Violacein	*Chromobacterium violaceum*, *Pseudoalteromonas*, *Janthinobacterium*	Purple pigment used in cosmetics, food, medicine, textiles; exhibits antibacterial, anticancer, antiviral properties	(Choi *et al*., 2015; Dufossé, 2016)

Table 16.3. Microbial additives permissible in the US (Nigam and Luke, 2016).

Chemical category	Color/pigment	Pigment E-number	Microbial source
Flavin	Yellow	E101	Riboflavin (from *Bacillus subtilis, Ashbya gossypii, Candida guilliermondii*)
Carotenoid	Yellowish orange	E160a	β-Carotene (from *Blakeslea trispora*)
Carotenoid	Yellowish orange	E160a	β -Carotene (from *Dunaliella salina*)
Carotenoid	Yellowish red	E160d	Lycopene (from *Blakeslea trispora*)
Carotenoid	Yellowish red	E161j	Astaxanthin (from *Haematococcus pluvialis, Xanthophylomyes dendrorhous*)
Carotenoid	Orange, red	E161g	Canthaxanthin (from *Haematococcus lacustris, Bradyrhizobium*)

laminar flow hoods or sterile fermenters. Microbial contamination should be routinely tested to ensure that the finished product is free of hazardous bacteria. This can be analyzed using total plate count, yeast and mold count, pathogen testing, etc.

16.3.3.1 Stability testing

Pigment thermal, light, and pH stability should be studied over time to establish shelf-life and performance under various storage circumstances. Testing should be conducted to ensure that packaging materials (such as bottles and sachets) protect the pigment against deterioration and contamination throughout storage and transit.

16.3.4 Packaging and labeling

16.3.4.1 Protective packaging

To avoid deterioration, microbial pigments should be stored in air-tight, light-resistant, moisture-proof containers. Packaging materials must be nonreactive with the pigment and selected to maintain the pigment's stability.

16.3.4.2 Lot numbering

Each batch of pigment should be assigned a unique lot number to facilitate traceability in the event of quality difficulties.

16.3.4.3 Labeling

Labels should incorporate all relevant information, such as pigment identification, concentration, storage conditions, and any safety precautions. Labeling must follow appropriate rules for the target sector, for example the FDA for food, EMA for medicines, and SCCS for cosmetics.

16.3.4.4 Regulatory oversight

According to current EU regulations, 43 colorants are authorized as food additives, while about 30 additions are approved in the US; six of these 30 additives have microbiological origins. There are 19 authorized commercial food additives that are derived from microbiological sources and utilized worldwide in addition to food colorants (Table 16.3).

16.4 Quality Assay for Microbial Pigments

A quality test for microbial pigments is critical to guarantee the uniformity, purity, and safety of pigments generated by microbes, particularly for use in the food, pharmaceutical, cosmetic, and textile sectors. In a microbial pigment quality assay, purity, concentration, stability, chemical composition, and safety are frequently the most important factors (Table 16.4).

Table 16.4. Different microbial pigments, their quantity and characterization.

Microbes	Pigments	Extraction	Yield	Characterization	Reference
Lactobacillus acidophilus	Carotenoid	Fermentation	4.2 mg/100g	Spectral analysis	(Bhaskar *et al.*, 2007)
Sarcina sp.	Carotenoid	Fermentation	12.87 mg/100g	Spectral analysis	(Joshi *et al.*, 2011)
Bacillus clausii	β -Carotene	Fermentation	69.4 mg/g	Spectral and Fourier transform infrared spectroscopy	(Korumilli and Mishra, 2014)
Lactobacillus plantarum	Carotenoid	Fermentation	54.89 mg/kg dry cell	High-performance liquid chromatography (HPLC)	(Bhaskar *et al.*, 2007)
Chryseobacterium artocarpi	Flexirubin	Fermentation	540 mg/l	Chromatography, electrospray ionization mass spectrometry (ESI-MS) and attenuated total reflection	(Vidyalakshmi *et al.*, 2009)
Monascus ruber	Red and yellow pigments	Solid-state fermentation	20 mg/100g (red) and 33 mg/100g (yellow)	Column chromatography, high-performance thin layer chromatography (HPTLC), gas chromatography-mass spectrometry (GC-MS)	(Vidyalakshmi *et al.*, 2009)
Monascus purpureus	Carotenoid	Solid-state fermentation	9 units/g of dry weight	HPLC	(Kantifedaki *et al.*, 2018)
Penicillium resticulosum	Red	Fermentation	497 mg/l	Spectral analysis	(Aksu and Eren, 2005)

16.4.1 Purity analysis

To ensure the pigment is free of contaminants or degradation products, purity analysis is essential. Typical techniques for determining purity include:

- thin-layer chromatography (TLC)
- high-performance liquid chromatography (HPLC).

16.4.1.1 Thin-layer chromatography

Thin-layer chromatography is used to detect the presence of different components in a pigment sample to assess its purity. A pure pigment will leave a single spot, whereas many spots indicate contamination or the presence of additional substances (Banerjee *et al.*, 2011). A TLC test is often used to evaluate the chemical composition, purity, and presence of certain chemicals in microbial pigments. TLC is an effective and simple method for assessing the components of complicated mixtures such as biopigments and offers information on the separation, identification, and quantification of various pigment molecules.

PRINCIPLE. Thin-layer chromatography separates chemical compounds based on their affinity for a stationary phase (often a silica gel plate) and a mobile phase (a solvent or solvent combination) (Malik *et al.*, 2016). When a microbial pigment is placed on the TLC plate, the chemicals in the sample move along the plate at varying speeds based on their solubility in the solvent and their interaction with the stationary phase. Many microbial pigments such as carotenoids and anthraquinones can be seen under visible light, making them apparent even without further staining. Some pigments, such as chlorophylls, glow under UV light (Sen *et al.*, 2019). The presence of luminous substances may be detected on the TLC plate using UV light (often 254 nm or 365 nm). If the pigment is not naturally visible, the TLC plate can be sprayed with a staining reagent (iodine vapor for carotenoids) or heated to reveal the spots.

ANALYSING RESULTS. Retention factor (Rf) for each of the separated components is derived by comparing the distance traveled by the pigment to the distance traveled by the solvent front. The Rf value is unique to each molecule under certain solvent conditions and may be used to identify recognized pigments (Malik *et al.*, 2016).

Rf = Distance traveled by the compound/Distance traveled by the solvent front

SEPARATION AND PURITY. The quantity and spacing of spots on the TLC plate show how many different compounds are in the pigment mixture. Whereas several spots signify the presence of several components (contaminants or pigment degradation products), the presence of a single spot shows purity.

16.4.1.2 High-performance liquid chromatography

High-performance liquid chromatography provides a more accurate way to examine and measure each component of a pigment combination. HPLC analysis has been used to identify distinct carotenoid and phenolic components. These pigments can be dissolved in a suitable solvent and then processed through a chromatographic column (Numan *et al.*, 2018). The pigment's various components (e.g. carotenoids and chlorophylls) will elute at different periods, allowing the purity of the pigment to be identified and quantified. HPLC produces chromatograms with various peaks and retention times. Each peak, with a distinct wavelength at a given retention time, specifies a compound.

16.4.2 Determination of pigment antimicrobial properties

The antibacterial characteristics of microbial pigments are tested using many tests to ensure effectiveness. The minimum inhibitory concentration (MIC) and minimum antifungal concentration (MFC) are determined based on the lowest concentrations that influence bacteria and fungal growth rates, respectively (Salama *et al.*, 2021). The MIC is the lowest concentration required for no growth. To determine the IC_{50} for 50% inhibition of fungal growth, a regression equation was employed with mycostatin as a positive control (Díaz Dellavalle *et al.*, 2011). The red pigment isolated from the fungal medium has a minimum bactericidal concentration (MIC) of 320 μg/g against *E. coli* and *Salmonella* pathogens. This finding outperforms traditional antibacterial chemicals in terms of effectiveness. The MFC for *A. flavus* fungi was 2.75 mg/g (Salama *et al.*, 2021).

16.4.3 Determination of concentration

Measuring pigment concentration is critical for quality control, especially if the pigment is used in commercial applications. UV-Vis spectroscopy is widely used to measure microbial pigments such as carotenoids, chlorophylls, and phycocyanins. Pigments absorb light at unique wavelengths, which may be used to quantify concentration. The pigment extract is examined using a UV-Vis spectrophotometer to measure absorbance between 400 and 600 nm. Relevant wavelengths are 450 nm for carotenoids and 660 nm for chlorophyll. Red pigment has the highest absorbance at 532 nm, followed by orange at 450 nm, yellow at 468 nm, pink at 540 nm, and black at 300 nm (Malik *et al.*, 2016).

The concentration may be calculated using the Beer–Lambert law (Kazi *et al.*, 2022):

$$A = \in \cdot c \cdot l$$

where A is absorbance, ϵ is the molar absorption coefficient, c is the concentration, and l is the path length of the cell.

16.4.4 Colorimetric assay

Colorimetric tests are sometimes employed, particularly for pigments that change color when exposed to a reagent. These assays are frequently employed for large-scale pigment concentration screening in industrial settings (Salama *et al.*, 2021).

16.4.4.1 Chemical composition analysis

This test helps to validate the identification and chemical structure of the microbial pigments. There are several approaches for determining the molecular fingerprint of the pigments. MS analysis determines molecular mass and structure. It aids in determining the chemical makeup and confirming the identification of the pigment. After separation with procedures such as HPLC, the pigment is ionized and examined in a mass spectrometer to determine its molecular weight and fragments.

16.4.5 Stability and shelf-life testing

16.4.5.1 Thermal stability

The thermal stability of microbial pigments is crucial for ensuring that they retain their color, potency, and safety through time, especially during storage and processing. The pigment's ability to hold onto its color and characteristics is examined at various temperatures. This is especially crucial for pigments used in food and cosmetics. The pigment is exposed to various temperatures (e.g. 4°C, 25°C, and 40°C) for particular lengths of time, then spectrophotometry or HPLC is used to detect any changes in color or composition. *Monascus* pigments generated by *Monascus ruber* have modest stability at low pH and high temperatures (Silveira *et al.*, 2013).

16.4.5.2 Light stability

Tests can determine how light exposure affects the pigment, which is especially essential for pigments used in food and cosmetics. The pigment is exposed to different light intensities (e.g. UV or visible light) and any changes in color or chemical structure are observed over time (Kumar *et al.*, 2015).

16.4.5.3 pH Stability

At certain pH values, some pigments can deteriorate or change color. Testing pH stability is critical for pigments used in acidic or basic compositions. The pigment is exposed to various pH solutions (e.g. acidic, neutral, alkaline) and spectrophotometry is used to detect any changes in color or structure. At high temperatures, the degradation constant of red pigments rose with decreasing pH, indicating that the *Monascus* red pigment was degraded at lower pH values. Salt may also impair the durability of red pigments at lower temperatures and pH (Abdollahi *et al.*, 2021). The pH can influence pigment formation and shade. *Monascus* sp. generates colors optimally at pH 5.5–6.5, whereas *Rhodotorula* works best at pH 4.0–4.5. Lycopene creation occurs at neutral to slightly alkaline pH, while β-carotene formation occurs at acidic pH (Joshi *et al.*, 2005).

16.4.6 Microbial contaminant testing

Microbial pigments must be free of microbial contamination, especially if they are utilized in food or medicinal goods. Conventional microbiological procedures are used, including total plate count (TPC), yeast and mold count, and pathogen testing (for example, *E. coli*, *Salmonella*, or *Listeria*) (Ali *et al.*, 2023).

16.4.7 Endotoxin testing

Conventional microbiological procedures are used, including total plate count (TPC), yeast and mold count, and pathogen testing (for example, *E. coli*, *Salmonella*, or *Listeria*). The Limulus amebocyte lysate (LAL) assay is widely used to screen for endotoxins.

16.4.8 Toxicity testing

A cytotoxicity test is intended to assess the possible adverse effects of microbial pigments on live cells, to see if the pigments cause cell damage, limit cell growth or promote cell death at various doses. This is a crucial safety study, especially given the potential applications of microbial pigments in food, medicines, and cosmetics. The MTT test for cell viability and the LDH assay for cell membrane integrity and cytotoxicity are two common assays. A pure astaxanthin pigment from the marine endophytic bacterium *Pontibacter korlensis* AG6 has been demonstrated to have cytotoxic effects on the human breast cancer cell line as well as substantial antibacterial and antioxidant characteristics (Pachaiyappan *et al.*, 2021).

16.4.8.1 *Acute toxicity testing (LD_{50})*

This test determines the fatal dosage of the pigment that results in 50% death in a group of test animals (typically mice). Test animals should be given different dosages of the pigment and any symptoms of toxicity or death should be closely watched. Strict ethical rules are often followed while doing this test (Erhirhie *et al.*, 2018).

16.4.8.2 *Prevention of mycotoxin production in* Monascus *pigment*

Monascidin A was identified as citrinin. To prevent the development of this toxin, several strains of *Monascus* were examined to see if they were all toxic. Controlling the manufacture of the metabolite can help prevent the development of citrinin. *Monascus* pigments are not approved for use in the food sector, but can be nontoxic when applied carefully. *Monascus* species have long been ingested in the Far East, but this has not helped the pigment achieve approval in the EU or USA (Dufossé *et al.*, 2005).

16.4.8.3 *Brine shrimp lethality bioassay for microbial pigments*

The brine shrimp lethality bioassay (BSLA) is a simple, inexpensive, and commonly used technique for determining the toxicity of chemicals by evaluating their deadly effects on brine shrimp larvae (*Artemia salina*). This bioassay can be used as an initial screening approach for possible toxicity (Salama *et al.*, 2021). The test is based on the hypothesis that some bioactive chemicals, such as those found in microbial pigments, can be harmful to living organisms. *Artemia salina* larvae are employed because they are tiny, grow quickly, and are sensitive to a variety of chemicals.

LC_{50} VALUE. The LC_{50} value is an important toxicity indicator; a lower LC_{50} value implies that the microbial pigment is more harmful to brine shrimp at a lower dose.

- High toxicity: <100 µg/ml
- Moderate toxicity: 100–500 µg/ml
- Low toxicity: >500 µg/ml

If the brine shrimp larvae show limited mortality rate of less than 10–20% at the maximum measured dosage, the pigment can be regarded nontoxic or suitable for environmental or cosmetic application.

16.5 Regulatory Aspects of Microbial Pigments in Different Sectors

16.5.1 Food industry regulations

Foods with colorful appearances along with nutritional and health-promoting properties for increased consumer acceptance are always in demand in the food industry (Rana *et al.*, 2021). The safe improvement of food organoleptic qualities combined with the inclusion of nutraceutical qualities like aging prevention, free radical elimination, and protection against several diseases is driving the food industry's microbial carotenoids market potential to grow quickly. However, because microbial pigments are more sensitive to environmental factors, their use as colorants can also be difficult because they may fade or lose shelf-life (Sen *et al.*, 2019). Microbial pigments are frequently utilized as food additives or colorants in the food business. Strict rules are established by regulatory agencies worldwide for the safety, use, and approval of these pigments. When extracting microbial pigments, solvents should be pharmaceutical grade and suitable for food or cosmetic use if the pigments are intended for such applications. Solvents and reagents should be handled according to the Material Safety Data Sheets (MSDS).

16.5.1.1 *Key regulatory authorities*

- Food and Drug Administration (FDA) – United States
- European Food Safety Authority (EFSA) – European Union

- Joint FAO/WHO Expert Committee on Food Additives (JECFA) – global
- Food Safety and Standards Authority of India (FSSAI) – India

16.5.1.2 Approval as food additives

The use of microbial pigments in food items requires evaluation and approval. Pigments are categorized as Color Additives in the United States and require FDA approval prior to being sold. Similarly, European Regulation (EC) No 1333/2008 on food additives governs pigments in the EU (Amchova *et al.*, 2015).

16.5.1.3 GRAS status

Some microbial pigments may be considered Generally Recognized as Safe (GRAS) by the FDA. Even with GRAS classification, these pigments must undergo extensive safety testing to ensure that they pose no harm to human health at planned consumption levels (Lehto *et al.*, 2017).

16.5.1.4 Maximal permissible concentrations

Regulatory organizations also establish maximum permissible quantities of microbial pigments in food items. For example, the EFSA may establish a maximum dose for a certain biopigment used in food items, taking into account possible toxicity, allergenicity, and the amount of pigment likely to be ingested (Varsha *et al.*, 2024).

16.5.1.5 Labeling and documentation

Food items using microbial pigments must follow correct labeling standards, which may include information on the pigment's origin, microbial sources, safety data, and functional or nutritional value.

16.5.2 Cosmetics and personal care products

Microbial pigments are increasingly being employed in the cosmetic sector to color goods including skincare, haircare, and cosmetics. Regulatory bodies provide criteria to guarantee that these pigments are safe to use in cosmetics.

16.5.2.1 Key regulatory authorities

- Food and Drug Administration (FDA) – United States
- European Commission Scientific Committee on Consumer Safety (SCCS) – European Union
- Health Canada – Canada

16.5.2.2 Regulatory requirements

COSMETIC INGREDIENT SAFETY ASSESSMENT. Microbial pigments used in cosmetics must be tested by regulatory organizations to guarantee that they do not cause skin irritation, allergic responses or other negative consequences. In the United States, the FDA mandates that all cosmetic components, including microbial pigments, be safe for human use and correctly labeled.

16.5.2.3 Colorant approval

Certain cosmetic pigments, like food, may require clearance before they can be commercialized. The FDA and other regulatory organizations keep lists of permitted colorants for use in cosmetics and personal care products. These lists are regularly updated to reflect new safety data.

16.5.2.4 Product safety testing

Cosmetic manufacturers that use microbial pigments must perform dermatological testing and toxicity studies to establish the safety of their products. This involves assessing irritation, sensitization, and ocular toxicity for humans (Aruldass *et al.*, 2018).

16.5.3 Pharmaceuticals and medical products

Microbial pigments used in pharmaceutical formulations as colorants or active substances face more stringent requirements due to their possible direct influence on human health.

16.5.3.1 Key regulatory authorities

- FDA – United States
- European Medicines Agency (EMA) – European Union

- Therapeutic Goods Administration (TGA) – Australia

16.5.3.2 *Pharmaceutical-grade pigments*

Microbial pigments intended for use in pharmaceutical goods must fulfill high quality and safety criteria. These pigments are frequently subjected to GMP, which guarantee that they are made in a controlled and hygienic setting.

16.5.3.3 *Safety testing*

The safety and efficacy of microbial pigments used in pharmaceutical formulations must be evaluated rigorously through toxicological and clinical testing. This covers testing for acute and chronic toxicity, as well as mutagenicity and carcinogenicity.

16.5.3.4 *Approval as active ingredients or excipients*

Microbial pigments may be employed as either excipients (inactive substances) or active compounds in medicinal formulations. These substances require regulatory clearance depending on their intended usage and related hazards.

16.5.4 Textile and other industrial uses

Every year, over 200,000 tons of dye effluents are produced, leading to pollution and health risks (Narsing Rao *et al.*, 2017). Microbial pigments are gaining popularity in the textile industry as a safer alternative to synthetic dyes, which have environmental, health, and allergy hazards. Regulatory authorities may mandate safety testing and adherence to environmental standards. It has recently been demonstrated that the antimicrobial pigment prodigiosin, which is made from *S. marcescens* SB08, has stable and efficient dyeing capabilities for both silk and cotton textiles. The textile yarn materials drenched in pigment also showed an antibacterial function (Venil *et al.*, 2021). Biopigments produced by the fungal genera *Monascus*, *Talaromyces*, *Aspergillus*, *Fusarium*, *Trichoderma*, *Scytalidium*, *Penicillium*, *Chlorociboria*, *Cordyceps*, *Curvularia*, *Alternaria*, and *Phymatotrichum* are used to dye a variety of textiles (Santos *et al.*, 2020; Venil *et al.*, 2021).

16.5.4.1 *Key regulatory authorities*

- Environmental Protection Agency (EPA) – United States
- Registration, Evaluation, Authorization, and Restriction of Chemicals (REACH) – European Union
- ISO Standards – international

16.5.4.2 *Environmental impact*

Textile industries that use microbial pigments must follow environmental requirements for disposal and wastewater treatment. The EPA and REACH have established criteria to guarantee that biopigments used in textile goods do not harm the environment.

16.5.4.3 *Worker safety*

The possible dangers connected with the microbial pigment manufacturing process, such as dust inhalation or chemical exposure during extraction, must be addressed in accordance with local labor and health agency safety rules.

16.6 Safety Aspects of Using Microbial Pigments

Microbial pigment safety issues center on toxicological profiles, allergenicity, and the possibility of microbial contamination.

16.6.1 Toxicity and biocompatibility

16.6.1.1 *Cytotoxicity*

It is critical to evaluate the cytotoxicity of microbial pigments, particularly when utilized in medicines or cosmetics. This is often done by *in vitro* tests such as the MTT assay to evaluate cell viability or lactate dehydrogenase (LDH) leakage assays to determine cell membrane integrity (Table 16.5).

16.6.1.2 *Acute and chronic toxicity*

To determine acceptable dosage levels, microbial pigments must go through acute toxicity testing such as the LD_{50} (lethal dose for 50% of the population). Chronic toxicity, such as long-term exposure to these pigments, should

Table 16.5. Important microbial pigments and their toxicity assessment.

Fungi	Pigments	Color	Mycotoxin	Safety evaluation	Reference
Monascus purpureus	Monascorubramine, rubropunctamine	Red	Citrinin	Acute oral toxicity	(Mohankumari *et al*., 2021)
Monascus anka	Ankaflavin and monascin	Yellow	No coproduction of toxin	Acute oral toxicity	(Shi *et al*., 2017)
Fusarium graminearum	Rubrofusarin	Red	Fumonisins, zearalenone, fusaric acid, fusarins, and beauvericins	Cytotoxic in colon cells	(Cambaza, 2018)
Fusarium oxysporum	Bikaverin	Red	Fumonisins, zearalenone fusaric acid, fusarins, and beauvericins	Cytotoxic against tumour cells, apoptosis suppressor	(Santos *et al*., 2020)
Trichoderma viride	Emodin Viridol	Brown Yellow	Nonphytotoxic	Phytotoxicity assay	(Venil *et al*., 2021)
Talaromycespurpureogenus	Purpuride, monascorubrin, purpurquinone-A, ankaflavin,	Yellow and red	Nontoxic	Subacute toxicity study	(Gopal Pandit *et al*., 2020)
Rhodotorula glutinis	Carotene, torulene, and torularhodin	Red and orange	Nontoxic	Standard subchronic toxicity study	(Latha and Jeevaratanm, 2012)
Yarrowia lipolytica	Carotene	Brown	Nontoxic	Genotoxicity models and a standard subchronic toxicity study	(Groenewald *et al*., 2014)
Penicillium aculeatum	Ankaflavin	Yellow	Nontoxic, selective toxicity in cancer cells	Cytotoxicity study	(Krishnamurthy *et al*., 2020)
Penicillium purpurogenum	Purpurogenone Mitorubrin	Yellow-orange, yellow	No coproduction of toxin	Brine shrimp *Artemia salina* study	(Mapari *et al*., 2009)

also be evaluated to ensure that no negative consequences occur over time (see Table 16.5).

16.6.1.3 Mutagenicity and carcinogenicity

Some pigments may induce mutations or raise the risk of cancer. Mutagenicity tests such as the Ames test and carcinogenicity studies are frequently required by regulatory bodies to determine long-term health concerns.

16.6.2 Allergenicity and sensitization

16.6.2.1 Skin sensitization

Microbial pigments used in cosmetics must be examined for possible skin sensitivity. Some people may have allergic responses to certain colors. Patch testing and human skin testing are routinely used to assess allergy potential.

16.6.2.2 Respiratory sensitization

The inhalation of pigment dust or aerosols may cause respiratory sensitivity. This danger is especially important for industrial applications like textile manufacturing, where workers may be exposed to pigment dust. Respiratory testing is crucial for assessing this risk.

16.6.2.3 Microbial contamination

Microbial pigments are obtained from microbes, hence there is always the potential of microbial contamination during manufacturing, extraction, and storage. GMP must be followed to reduce contamination hazards. Furthermore, microbial pigments used in food and medicinal goods must fulfill strict microbiological criteria.

16.6.3 Environmental impact

Microbial pigments are frequently seen as environmentally acceptable alternatives to synthetic colors since they are biodegradable and made from renewable resources. However, their manufacturing procedures must be assessed for environmental effects. Regulatory organizations may request an examination of the life cycle of microbial pigments, including their environmental impact during manufacture, transit, and disposal.

16.7 Legal Framework Bodies Governing Microbial Pigments

The FDA, EFSA, and FSSAI have established legislative frameworks and standards to ensure the safety and quality of microbial pigments. These guidelines govern how microbial pigments can be used in food, cosmetics, and other consumer items. The following are the particular restrictions that apply in each location.

16.7.1 US Food and Drug Administration regulations

The FDA regulates microbial pigments under the Federal Food, Drug and Cosmetic Act (FFDCA). The basic requirement for food colorants is that they are either authorized color additives or GRAS. Microbial pigments that are not GRAS must go through the Color Additive Petition procedure (Lehto *et al.*, 2017). Specific safety data and toxicological research are needed. There are restrictions on the maximum allowable concentrations and labeling requirements for food colorants.

16.7.1.1 Color additive amendments

According to the Color Additive Amendments to the FFDCA, every colorant including microbial pigments used in food must be FDA approved.

- Certified Color Additives are synthetic pigments that have been certified by the FDA.
- Exemption from Certification: natural colorants, such as microbial pigments, may fall under this category if they fulfill the requisite safety criteria.

16.7.1.2 Colour additive petition process

If a novel microbial pigment is going to be utilized in food, the company must file a Color Additive Petition. This petition provides toxicological information, production procedures, and the planned use of the colorant. The FDA then determines if it is safe for human consumption (Meena *et al.*, 2022).

16.7.1.3 GRAS status

GRAS certification has been awarded to certain microbial pigments, including phycocyanin produced from *Spirulina*. GRAS status indicates that a substance is generally recognized as safe for consumption based on scientific evidence and does not require FDA approval as a food additive (Lehto *et al.*, 2017).

16.7.1.4 Regulatory approval for use

Following FDA clearance, microbial pigments are often listed in the FDA Color Additive Status List, which includes detailed restrictions relating to the foods for which they may be used and the maximum permissible concentration.

16.7.1.5 Labeling requirements

If microbial pigments are used in food items, the FDA requires producers to mention the colorants on the label. If the pigment is synthetic, it must be identified by name; for natural pigments, the word "natural color" may be used, although specifications may still be necessary depending on the circumstances.

16.7.2 European Food Safety Authority regulations

The EFSA assesses the safety of microbial pigments in the European Union (EU). Microbial pigments are regulated as food additives in the EU and their safety must be verified by the EFSA before they can be used. If a microbial pigment is found to be safe, the EFSA issues a scientific opinion on its safety, which is subsequently utilized by the European Commission to make regulatory decisions regarding its license. Microbial pigments are subject to Regulation (EC) No. 1333/2008. They must pass an EFSA risk evaluation and their Acceptable Daily Intake (ADI) is determined. If they are novel, they must apply for approval under the Novel Foods Regulation. There are specified maximum limitations on their use in certain food groups (Lehto *et al.*, 2017).

16.7.2.1 Regulation (EC) No. 1333/2008 on food additives

This law covers the use of food additives, such as colorants, in the EU. It requires that all food colorants be assessed and authorized for use before being incorporated into food items.

16.7.2.2 Risk assessment

The EFSA assesses the risks of microbial pigments using scientific evidence. If a pigment is new, i.e. it has never been used in food in the EU previously, it will be evaluated for safety under the New Foods Regulation (EU) 2015/2283 (Amchova *et al.*, 2015)

16.7.2.3 Acceptable Daily Intake

The EFSA also develops an ADI for pigments, which specifies the maximum quantity of pigment that may be eaten daily over a lifetime without causing a health concern.

16.7.2.4 Food categories

Microbial pigments that have already been approved for use in food are subject to maximum limitations defined for specific food categories.

16.7.2.5 Labeling

In the EU, food goods containing microbial pigments must be clearly labeled, with the colorant identified by its E-number or name (Dufossé, 2006). For example, a product containing phycocyanin generated from *Spirulina* would need to be labeled with the proper color code or name.

16.7.2.6 Novel foods

Microbial pigments that have not been utilized before in the EU are subject to the Novel Foods Regulation. This necessitates a comprehensive application from the producer, which includes toxicological data and evidence of safe usage in other countries. The EFSA then examines the evidence to assess safety.

16.7.3 Food Safety and Standards Authority of India regulations

Microbial pigments in India are subject to the Food Safety and Standards (Food Products Standards and Food Additives) Regulations, 2011. These rules oversee the safety of food additives, particularly colorants generated from microbes. Microbial pigments are subject to FSSAI Food Additives Regulations. To be approved, they must fulfill certain safety standards. If the pigment is new, a safety dossier must be provided. Maximum usage regulations apply and all food items containing colorants must be labeled.

16.7.3.1 Regulation 2.3.1: Food additives

This rule establishes criteria for the approval and use of food additives, which include colorants. Microbial pigments must meet the FSSAI's food additive guidelines.

16.7.3.2 Approval process for additives

New microbial pigments must undergo safety testing before being approved. If the pigment is novel (i.e. not generally used in India), the applicant must provide the FSSAI with a complete safety dossier that includes data on toxicological studies, allergenicity, and manufacturing technique safety.

16.7.3.3 Maximum permissible limits

The FSSAI sets maximum limits for the use of colorants, including microbial pigments, in various categories of food. These limits are intended to ensure consumer safety by preventing overuse or excessive exposure to the pigment.

16.7.3.4 Generally recognized As Safe

If a microbial pigment is thought to have a long history of safe usage in food, it might be categorized as GRAS. The FSSAI certifies pigments made using conventional fermentation procedures as long as they fulfill safety criteria.

16.7.3.5 Novel foods regulations

If a pigment has not previously been used in food, it may be subject to FSSAI regulation for new foods. The applicant must provide a dossier containing scientific data demonstrating its safety for human consumption. The FSSAI examines the data before issuing permission.

16.7.3.6 Labeling requirements

Like other regulatory agencies, the FSSAI requires unambiguous labeling of food goods containing microbial pigments. The food product must indicate the precise color used and whether the pigment is synthetic or natural.

16.8 Conclusion

Biopigments provide a more environmentally friendly and sustainable substitute for synthetic colors since they are generated from microorganisms such as bacteria, fungi, and algae. To prevent any health hazards, their safety and regulatory approval must be thoroughly evaluated. The regulatory and safety considerations of employing microbial pigments are critical for maintaining consumer health, environmental sustainability, and product efficacy. Before being commercialized, microbial pigments must fulfill the approval requirements established by regulatory authorities such as the FDA, EFSA, and EMA for food, cosmetics, medicines, and industrial applications. Furthermore, the safety of microbial pigments must be thoroughly assessed for toxicity, allergenicity, and contamination risk. With correct regulatory compliance and safety procedures, microbial pigments may be safely incorporated into a wide range of goods, providing an environmentally friendly alternative to typical synthetic colorants.

References

Abdollahi, F., Jahadi, M. and Ghavami, M. (2021) Thermal stability of natural pigments produced by *Monascus purpureus* in submerged fermentation. *Food Science and Nutrition* 9(9), 4855–4862. DOI: 10.1002/fsn3.2425.

Aksu, Z. and Eren, A.T. (2005) Carotenoids production by the yeast *Rhodotorula mucilaginosa*: Use of agricultural wastes as a carbon source. *Process Biochemistry* 40(9), 2985–2991. DOI: 10.1016/j.procbio.2005.01.011.

Ali, M.A., Hashish, M.H. and Fekry, M.M. (2023) Microbiological quality of some packed and unpacked bread products in Alexandria. *Egypt Journal of Food Safety* 40, e12824. DOI: 10.1186/s42506-023-00141-9.

Amchova, P., Kotolova, H. and Ruda-Kucerova, J. (2015) Health safety issues of synthetic food colorants. *Regulatory Toxicology and Pharmacology* 73(3), 914–922. DOI: 10.1016/j.yrtph.2015.09.026.

Aruldass, C.A., Dufossé, L. and Ahmad, W.A. (2018) Current perspective of yellowish-orange pigments from microorganisms – A review. *Journal of Cleaner Production* 180, 168–182. DOI: 10.1016/j.jclepro.2018.01.093.

Banerjee, D., Chatterjee, S., Banerjee, U.C., Guha, A.K. and Ray, L. (2011) Green pigment from *Bacillus cereus* m(1)(16) (MTCC 5521): Production parameters and antibacterial activity. *Applied Biochemistry and Biotechnology* 164(6), 767–779. DOI: 10.1007/s12010-011-9172-8.

Bhaskar, N., Suresh, P.V., Sakhare, P.Z. and Sachindra, N.M. (2007) Shrimp biowaste fermentation with *Pediococcus acidolactici* CFR2182: Optimization of fermentation conditions by response surface methodology and effect of optimized conditions on deproteination/demineralization and carotenoid recovery. *Enzyme and Microbial Technology* 40(5), 1427–1434. DOI: 10.1016/j.enzmictec.2006.10.019.

Cambaza, E. (2018) Comprehensive description of *Fusarium graminearum* pigments and related compounds. *Foods* 7(10), 165. DOI: 10.3390/foods7100165.

Choi, S.Y., Yoon, K.H., Lee, J.I. and Mitchell, R.J. (2015) Violacein: Properties and production of a versatile bacterial pigment. *BioMed Research International* 2015(1), 465056.

Chuyen, H.V. and Eun, J.B. (2017) Marine carotenoids: Bioactivities and potential benefits to human health. *Critical Reviews in Food Science and Nutrition* 57(12), 2600–2610. DOI: 10.1080/10408398.2015.1063477.

Dufossé, L. (2006) Microbial production of food grade pigments. *Food Technology and Biotechnology* 44, 313–321. Available at: https://hrcak.srce.hr/file/162018

Dufossé, L., Galaup, P., Yaron, A., Arad, S.M., Blanc, P. *et al.* (2005) Microorganisms and microalgae as sources of pigments for food use: A scientific oddity or an industrial reality. *Trends in Food Science & Technology* 16(9), 389–406. DOI: 10.1016/j.tifs.2005.02.006.

Dufossé, L. (2016) Current and potential natural pigments from microorganisms (bacteria, yeasts, fungi, microalgae). In: Carle, R. and Schweiggert, R.M. (eds) *Handbook on Natural Pigments in Food and Beverages*. Woodhead Publishing, Sawston, UK, pp. 337–354.

Díaz Dellavalle, P., Cabrera, A., Alem, D., Larrañaga, P., Ferreira, F. *et al.* (2011) Antifungal activity of medicinal plant extracts against phytopathogenic fungus *Alternaria* spp. *Chilean Journal of Agricultural Research* 71(2), 231–239. DOI: 10.4067/S0718-58392011000200008.

Erhirhie, E.O., Ihekwereme, C.P. and Ilodigwe, E.E. (2018) Advances in acute toxicity testing: Strengths, weaknesses and regulatory acceptance. *Interdisciplinary Toxicology* 11(1), 5–12. DOI: 10.2478/intox-2018-0001.

Fouillaud, M., Caro, Y., Venkatachalam, M., Grondin, I. and Dufossé, L. (2018) Anthraquinones. In: *Phenolic Compounds in Food*. CRC Press, Boca Raton, FL, pp. 131–172.

Giovannucci, E., Rimm, E.B., Liu, Y., Stampfer, M.J. and Willett, W.C. (2002) A prospective study of tomato products, lycopene, and prostate cancer risk. *CancerSpectrum Knowledge Environment* 94(5), 391–398. DOI: 10.1093/jnci/94.5.391.

Gopal Pandit, S., Honganoor Puttananjaiah, M., Serva Peddha, M. and Appasaheb Dhale, M. (2020) Corrigendum to “Safety efficacy and chemical profiling of water-soluble *Talaromyces purpureogenus* CFRM02 pigment”. *Food Chemistry* 317, 126403. DOI: 10.1016/j.foodchem.2020.126403.

Groenewald, M., Boekhout, T., Neuvéglise, C., Gaillardin, C., van Dijck, P.W.M. *et al.* (2014) *Yarrowia lipolytica*: Safety assessment of an oleaginous yeast with a great industrial potential. *Critical Reviews in Microbiology* 40(3), 187–206. DOI: 10.3109/1040841X.2013.770386.

Joshi, V.K., Attri, D. and Rana, M.S. (2005) Optimization of apple pomace based medium and fermentation conditions for pigment production by *Sarcina* sp. *Indian Journal of Natural Products and Resources* 2, 421–427.

Joshi, V.K., Attri, D. and Rana, N.S. (2011) Optimization of apple pomace based medium and fermentation conditions for pigment production by Sarcina sp. *Indian Journal of Natural Products and Resources*. Dec 1;2(4):421-7.

Kantifedaki, A., Kachrimanidou, V., Mallouchos, A., Papanikolaou, S. and Koutinas, A.A. (2018) Orange processing waste valorisation for the production of bio-based pigments using the fungal strains *Monascus purpureus* and *Penicillium purpurogenum*. *Journal of Cleaner Production* 185, 882–890. DOI: 10.1016/j.jclepro.2018.03.032.

Kazi, Z., Hungund, B.S., Yaradoddi, J.S., Banapurmath, N.R., Yusuf, A.A. *et al*. (2022) Production, characterization, and antimicrobial activity of pigment from *Streptomyces* species. *Journal of Nanomaterials* 2022(1), 3962301. DOI: 10.1155/2022/3962301.

Korumilli, T. and Mishra, S. (2014) Carotenoid production by *Bacillus clausii* using rice powder as the sole substrate: Pigment analyses and optimization of key production parameters. *Journal of Biochemical Technology* 5(4), 788–794. DOI: 10.51847/PRp8ic1.

Krishnamurthy, S., Narasimha Murthy, K. and Thirumale, S. (2020) Characterization of ankaflavin from *Penicillium aculeatum* and its cytotoxic properties. *Natural Product Research* 34(11), 1630–1635. DOI: 10.1080/14786419.2018.1522633.

Kumar, A., Vishwakarma, H.S., Singh, J., Dwivedi, S. and Kumar, M. (2015) Microbial pigments: Production and their applications in various industries. *International Journal of Pharmaceutical, Chemical and Biological Sciences* 5, 203–212. DOI: http://www.ijpcbs.com.

Kushwaha, K., Saini, A., Saraswati, P., Agarwal, M.K. and Saxena, J. (2014) Colorful world of microbes: Carotenoids and their applications. *Advances in Biology* 837891. DOI: 10.1155/2014/837891.

Latha, B.V. and Jeevaratanm, K. (2012) Thirteen-week oral toxicity study of carotenoid pigment from *Rhodotorula glutinis* DFR-PDY in rats. *Indian Journal of Experimental Biology* 50(9), 645–651. DOI: https://pubmed.ncbi.nlm.nih.gov/23140023.

Lehto, S., Buchweitz, M., Klimm, A., Strassburger, R., Bechtold, C. *et al*. (2017) Comparison of food colour regulations in the EU and the US: A review of current provisions. *Food Additives & Contaminants* 34(3), 335–355. DOI: 110.1080/19440049.2016.127443.

Lombardelli, C., Benucci, I., Mazzocchi, C. and Esti, M. (2021) Betalain extracts from beetroot as food colorants: Effect of temperature and UV-light on storability. *Plant Foods for Human Nutrition* 76(3), 347–353. DOI: 10.1007/s11130-021-00915-6.

Malik, K., Tokkas, J. and Goyal, S. (2012) Microbial pigments: A review. *International Journal of Microbial Resource Technology* 1, 361–365.

Malik, K., Tokas, J. and Anand, R.C. (2016) Characterization and cytotoxicity assay of pigment producing microbes. *International Journal of Microbiology and Applied Sciences* 5(6), 370–376. Available at: https://www.ijcmas.com/5-6-2016/Kamla%20Malik,%20et%20al.pdf

Mapari, S.A.S., Meyer, A.S., Thrane, U. and Frisvad, J.C. (2009) Identification of potentially safe promising fungal cell factories for the production of polyketide natural food colorants using chemotaxonomic rationale. *Microbial Cell Factories* 8, 24. DOI: 10.1186/1475-2859-8-24.

Meena, N.K., Meena, V.S., Verma, M. and Mishra, S. (2022) Plant extracts as coloring agents. In: *Plant Extracts: Applications in the Food Industry*. Elsevier, Amsterdam, The Netherlands, pp. 187–207. DOI: 10.1016/B978-0-12-822475-5.00012-0.

Mohankumari, H.P., Naidu, K.A., Narasimhamurthy, K. and Vijayalakshmi, G. (2021) Bioactive pigments of *Monascus purpureus* attributed to antioxidant, HMG-CoA reductase inhibition and anti-atherogenic functions. *Frontiers in Sustainable Food Systems* 5, 590427. DOI: 10.3389/fsufs.2021.590427.

Narsing Rao, M.P., Xiao, M. and Li, W.-J. (2017) Fungal and bacterial pigments: Secondary metabolites with wide applications. *Frontiers in Microbiology* 8, 1113. DOI: 10.3389/fmicb.2017.01113.

Nigam, P.S. and Luke, J.S. (2016) Food additives: Production of microbial pigments and their antioxidant properties. *Current Opinion in Food Science* 7, 93–100. DOI: 10.1016/j.cofs.2016.02.004.

Numan, M., Bashir, S., Mumtaz, R., Tayyab, S., Ullah, I. *et al*. (2018) Chemical profile and *in-vitro* pharmacological activities of yellow pigment extracted from *Arthrobacter gandavensis*. *Process Biochemistry* 75, 74–82. DOI: 10.1016/j.procbio.2018.08.033.

Pachaiyappan, A., Sadhasivam, G., Kumar, M. and Muthuvel, A. (2021) Biomedical potential of astaxanthin from novel endophytic pigment producing bacteria *Pontibacter korlensis AG6*. *Waste and Biomass Valorization* 12(4), 2119–2129. DOI: 10.1007/s12649-020-01169-0.

Parekh, S., Vinci, V.A. and Strobel, R.J. (2000) Improvement of microbial strains and fermentation processes. *Applied Microbiology and Biotechnology* 54(3), 287–301. DOI: 10.1007/s002530000403.

Ramesh, C., Vinithkumar, N.V., Kirubagaran, R., Venil, C.K. and Dufossé, L. (2019) Multifaceted applications of microbial pigments: Current knowledge, challenges and future directions for public health implications. *Microorganisms* 7(7), 186. DOI: 10.3390/microorganisms7070186.

Rana, B., Bhattacharyya, M., Patni, B., Arya, M. and Joshi, G.K. (2021) The realm of microbial pigments in the food color market. *Frontiers in Sustainable Food Systems* 5, 603892. DOI: 10.3389/fsufs.2021.603892.

Salama, H.A., Badr, A.N., Elkhadragy, M.F., Hussein, A.M.S., Shaban, I.A.-S. *et al.* (2021) New antifungal microbial pigment applied to improve safety and quality of processed meat-products. *Microorganisms* 9(5), 989. DOI: 10.3390/microorganisms9050989.

Santos, M.C. dos, Mendonça, M. de L. and Bicas, J.L. (2020) Modeling bikaverin production by *Fusarium oxysporum* CCT7620 in shake flask cultures. *Bioresources and Bioprocessing* 7(1), 13. DOI: 10.1186/s40643-020-0301-5.

Sen, T., Barrow, C.J. and Deshmukh, S.K. (2019) Microbial pigments in the food industry-challenges and the way forward. *Frontiers in Nutrition* 6, 7. DOI: 10.3389/fnut.2019.00007.

Shi, K., Tang, R., Huang, T., Wang, L. and Wu, Z. (2017) Pigment fingerprint profile during extractive fermentation with *Monascus anka* gim 3.592. *BMC Biotechnology* 17, 46. DOI: 10.1186/s12896-017-0367-0.

Sigurdson, G.T., Tang, P. and Giusti, M.M. (2017) Natural colorants: Food colorants from natural sources. *Annual Review of Food Science and Technology* 8(1), 261–280. DOI: 10.1146/annurev-food-030216-025923.

Silveira, S.T., Daroit, D.J., Sant'Anna, V. and Brandelli, A. (2013) Stability modeling of red pigments produced by *Monascus purpureus* in submerged cultivations with sugarcane bagasse. *Food and Bioprocess Technology* 6(4), 1007–1014. DOI: 10.1007/s11947-011-0710-8.

Sowbhagya, H.B. and Chitra, V.N. (2010) Enzyme-assisted extraction of flavorings and colorants from plant materials. *Critical Reviews in Food Science and Nutrition* 50(2), 146–161. DOI: 10.1080/10408390802248775.

Stahmann, K.P., Revuelta, J.L. and Seulberger, H. (2000) Three biotechnical processes using *Ashbya gossypii*, *Candida famata*, or *Bacillus subtilis* compete with chemical riboflavin production. *Applied Microbiology and Biotechnology* 53(5), 509–516. DOI: 10.1007/s002530051649.

Sun, H., Wang, Y., He, Y., Liu, B., Mou, H. *et al.* (2023) Microalgae-derived pigments for the food industry. *Marine Drugs* 21(2), 82.

Van Thieghem, P. (1884) *Monascus*, genre nouveau de l'ordre des ascomycètes. *Bulletin de La Société de Botanique de France* 31, 226–229. DOI: 10.1080/00378941.1884.10828230.

Varsha, V.S., Boreda, T., Kamana, A.R., Pailla, S.R., Kambhampati, Y. *et al.* (2024) Food colours toxicology and food safety. In: *Microbial Pigments*. CRC Press, Boca Raton, FL, pp. 216–235. DOI: 10.1201/9781003353980-11.

Vendruscolo, F., Luise Müller, B., Esteves Moritz, D., de Oliveira, D., Schmidell, W. *et al.* (2013) Thermal stability of natural pigments produced by *Monascus ruber* in submerged fermentation. *Biocatalysis and Agricultural Biotechnology* 2(3), 278–284. DOI: 10.1016/j.bcab.2013.03.008.

Venil, C.K., Zakaria, Z.A. and Ahmad, W.A. (2013) Bacterial pigments and their applications. *Process Biochemistry* 48(7), 1065–1079. DOI: 10.1016/j.procbio.2013.06.006.

Venil, C.K., Dufossé, L., Velmurugan, P., Malathi, M. and Lakshmanaperumalsamy, P. (2021) Extraction and application of pigment from *Serratia marcescens* SB08, an insect enteric gut bacterium, for textile dyeing. *Textiles* 1(1), 21–36. DOI: 10.3390/textiles1010003.

Vidyalakshmi, R., Paranthaman, R., Murugesh, S. and Singaravadivel, K. (2009) Microbial bioconversion of rice broken to food grade pigments. *Global Journal of Biotechnology and Biochemistry* 4, 84–87. Available at: http://www.idosi.org/gjbb/gjbb4(2)09/3.pdf

17 Case Studies and Success Stories of Microbial Pigment Applications

Rahul Kumar[1], Ayush Madan[2,3] and Debasis Mitra[1*]
[1]Department of Microbiology, Graphic Era (Deemed to be University), Dehradun, Uttarakhand, India; [2]Department of Biotechnology, School of Research & Technology (SORT), People's University, Bhanpur, Bhopal, Madhya Pradesh, India; [3]Institute of Tropical Aquaculture and Fisheries (AKUATROP), Universiti Malaysia Terengganu, Kuala Nerus, Terengganu, Malaysia

17.1 Introduction

Microbial pigments are natural colorants produced by microorganisms such as bacteria, fungi, and algae (Dufossé, 2018). These pigments are secondary metabolites that serve other purposes, including protection from environmental stresses, attraction of symbiotic organisms, and defense against predators (Behera *et al.*, 2021a). Over recent decades, microbial pigments have gained considerable attention as sustainable alternatives to synthetic dyes and colorants in various industrial applications, including food, textiles, cosmetics, and pharmaceuticals (Aman Mohammadi *et al.*, 2022).

Microbial pigments are organic compounds synthesized by microorganisms that can absorb light at various wavelengths and confer colors (Orlandi *et al.*, 2022). Generally, they are water soluble, and thus have numerous uses, and have antioxidant, antimicrobial, or antiinflammatory properties (Budiman *et al.*, 2023). Microbial pigments can be classified into several types based on chemical structure, type of microorganism, and function. The major categories include carotenoids, anthraquinones, phycobilins, flavonoids, and bacterial pigments (Tang *et al.*, 2025).

Carotenoids are generally produced by bacteria, fungi, and algae. They protect cells from oxidative stress via their antioxidant properties, which are also used in photosynthesis. Anthraquinones are generally aromatic pigments, produced by fungi including *Monascus* spp. and *Aspergillus* spp. that are stable in nature and exhibit bioactivity related to antifungal and antibacterial properties (Lin and Xu, 2020). Phycobilins are water-soluble pigments present mainly in cyanobacteria and red algae and are also the building blocks of the photosynthetic apparatus in these organisms (Assunção *et al.*, 2022). Phycocyanin is a blue phycobilin which has received much attention because of its bright blue color and antioxidant activity. Applications for phycocyanin include use in the food and cosmetic industries as a blue colorant, combating oxidative stress and reducing inflammation. Flavonoids are a group of polyphenolic compounds produced by bacteria, fungi, and plants, exhibit a range of colors, including yellow, red, and purple, and are known for their antioxidant

*Corresponding author: debasismitra3@gmail.com

 Biopigments of Microbial Origin: Innovations and Applications Across Industries (eds Madan *et al.*)
DOI: 10.1079/ 9781836990505.0017

and antimicrobial properties (Álvarez-Martínez *et al.*, 2020). Melanin is a complex pigment produced by various microorganisms, such as fungi and bacteria, that is typically brown or black in color and protects the microorganism from UV radiation, oxidative stress, and chemical damage. Its applications include drug delivery systems, antimicrobial and antioxidative properties, and skin-protective benefits (Mohd Zaid *et al.*, 2023).

Bacterial pigments can produce a wide range of brilliant colors like red, orange, yellow, and green produced by *Pseudomonas*, *Serratia*, and *Chromobacterium* spp. (Sajjad *et al.*, 2020). Additional potential pharmaceutical properties include anticancer and antifungal activity. These hold great promise for various industrial and commercial applications, providing enormous benefits that are increasingly valued in the context of sustainability and eco-friendly practices.

Environmental sustainability, health benefits, biotechnological applications, market demand for natural products, advances in production technology (Nethravathy *et al.*, 2019), and diversity and versatility all make microbial pigments a worthwhile alternative to synthetic dyes and colorants.

17.1.1 Microbial pigments in industry

Microbial pigments offer several advantages over synthetic colorants, such as biodegradability, eco-friendliness, and potential health benefits. These pigments are gaining increasing usage in various industries such as food, textiles, cosmetics, agriculture, and pharmaceuticals (Table 17.1) (Venil *et al.*, 2020). The food industry is one of the largest consumers of colorants, where microbial pigments are increasingly seen as natural and safer alternatives to synthetic dyes. Natural pigments produced from microorganisms include a spectrum of colors from yellow and orange through to red and blue. The microbial pigments most used in food are the carotenoids, which include β-carotene, astaxanthin, and lutein.

These pigments provide significant health benefits due to their antioxidant properties (Sharma *et al.*, 2021). For instance, *Haematococcus pluvialis* produces astaxanthin, a red-orange pigment highly valued for its potent antioxidant effect (Ruiz-Domínguez *et al.*, 2019). In the food industry, astaxanthin has many applications, such as beverages, dietary supplements, and even aquaculture feed to help enhance the coloration of fish. Phycocyanin is another microbial pigment widely used in the food industry (Kumar *et al.*, 2024). This is a blue pigment that is produced by cyanobacteria, especially *Spirulina*. Phycocyanin has been used as a natural food colorant for various products including smoothies, ice creams, candies, and pharmaceuticals (Bürck *et al.*, 2024). It is prized not only for its brilliant blue color but also for its antioxidant, antiinflammatory, and neuroprotective properties.

Microbial pigments have been developed for application in textile dyeing and consequently provide a potential solution for the environmental problems associated with synthetic dyes. Prodigiosin, a red pigment, has been evaluated for its efficiency in imparting color to cotton and wool fibers, while violacein, a violet pigment, had been applied for sustainable dyeing of fabrics (Guryanov and Naumenko, 2024). These pigments provide unique and attractive colors while reducing the environmental footprint associated with conventional synthetic dyes. Besides the environmental advantage, microbial pigments for textiles can also provide extra value in the form of antimicrobial activity.

Microbial pigments are gaining significant importance in response to the global demand for natural and cruelty-free cosmetics (Nhani *et al.*, 2024). They are mostly sourced from renewable sources and generally nontoxic, so are ideal for sensitive skin formulations. Carotenoids and anthraquinones are two important classes of pigments employed in cosmetics. Carotenoids, including β-carotene and astaxanthin, are used in skin care due to their antioxidant properties; these remove oxygen from the skin, thus protecting the skin from the harmful effects of UV radiation and reducing signs of aging. Astaxanthin, for example, is used because it protects the skin from oxidative stress-induced damage and is therefore a popular ingredient in antiaging creams and sunscreen formulations (Alugoju *et al.*, 2023). Flavonoids are used for their skin-lightening and antiinflammatory properties. Melanin is a microbial pigment that

Table 17.1. Types of microbial pigments and their sources, functions and applications (Author's own table)

Pigments of microbial origin	Sources	Name of pigment	Functions	Applications
Serratamolide	*Serratia marcescens*	Polyketide	Bioengineering of compounds for drug development	Biotechnology, bioengineering
Fusidic acid	*Fusidium* spp.	Steroid-derived pigment	Antibacterial, antiinflammatory, antiviral	Skincare, pharmaceutical applications
Cyanidin	*Monascus* spp., *Aspergillus* spp.	Flavonoid	Antioxidant, antiinflammatory	Nutraceuticals, food colorants, textiles
Anthraquinones	*Aspergillus* spp., *Penicillium* spp.	Anthraquinone	Anticancer, biopesticide, antimicrobial	Pharmaceuticals, bioremediation, textiles
Carminic acid	*Cochineal* insect (though microbial sources are being explored)	Quinone	Antioxidant, natural dye	Cosmetics, food colorants, textiles
Rhodopsin	*Rhodobacter sphaeroides, Halobacterium*	Retinal-based pigment	Photoelectric conversion, light sensitivity	Biosensors, bioelectronics
Monascus pigments	*Monascus purpureus*	Anthraquinone	Antioxidant, cholesterol-lowering antimicrobial	Biopesticides, food colorants, textiles
Melanin	*Aspergillus, Bipolaris, Exophiala*	Polyphenolic pigment	Antiaging, UV protection, antioxidant	Bioremediation, medical treatments, cosmetics (UV protection)
Prodigiosin	*Serratia marcescens*	Red pigment	Antiinflammatory, Antimicrobial, anticancer	Bioremediation, textile dyeing, antimicrobial applications
Violacein	*Chromobacterium violaceum*	Indole-derived pigment	Antioxidant, antibacterial, anticancer, antifungal	Antimicrobial applications, food colorants, textiles
Phycocyanin	*Spirulina platensis, Arthrospira*	Phycobilin	Cognitive health, antioxidant, antiinflammatory	Nutraceuticals, cosmetics, food colorants, beverages
β-Carotene	*Blakeslea trispora, Xanthophyllomyces dendrorhous*	Carotenoid	Immune booster, vitamin A precursor, antioxidant	Dietary supplements, food colorants, nutraceuticals
Astaxanthin	*Haematococcus pluvialis, Chlorella, Phaffia rhodozyma*	Carotenoid	Enhances fish color, antioxidant, neuroprotective, antiinflammatory	Aquaculture, cosmetics, dietary supplements, food colorants

has been used in skincare products because of its ability to protect against UV radiation and pigmentation disorders (Tran-Ly *et al.*, 2020).

Microbial pigments are also highly useful in agriculture, particularly for biopesticides and plant growth promotion. Many of the microbially synthesized pigments display antimicrobial, antifungal, and herbicidal properties and are thus useful in sustainable agriculture practices. For instance, anthraquinones are potential natural biopesticides of fungi like *Monascus* spp. and *Aspergillus* spp. Carotenoids from various bacterial species like *Bacillus* have been proven to improve the growth of plants and resist the adverse conditions brought about by environmental stresses, thereby enhancing crop health (Ramesh *et al.*, 2022).

Most microbial pigments have shown extreme bioactivity that makes them invaluable in pharmaceutical and biotechnological applications. Astaxanthin, violacein, and prodigiosin are microbial pigments that have promising medicinal uses, such as antimicrobial, anticancer, and antiinflammatory effects (Agarwal *et al.*, 2023). Astaxanthin, obtained from *Haematococcus pluvialis* and other algae, hisof particular interest as an antioxidant and antiinflammatory agent for health-related studies (Mapelli-Brahm *et al.*, 2023). Violacein is a purple pigment that is secreted by the bacterium *Chromobacterium violaceum*, which is currently being developed for the treatment of drug-resistant bacterial infections and specific types of cancer, which are exciting new avenues for potential pharmaceutical applications in the future (Ahmed *et al.*, 2021).

17.2 Microbial Pigments Classification

Bacterial pigments are the products of secondary metabolism by bacterial species in protecting themselves from UV radiation, antimicrobial activity, and contact with other organisms (Behera *et al.*, 2021a). They include carotenoids, quinones, indole derivatives, and melanoid, extensively used in pharmaceutical industries, food industry, and textiles (Fig. 17.1). Fungal pigments, produced by *Monascus*, *Aspergillus*, *Penicillium*, and *Cladosporium*, are known for their complex chemical structures, including anthraquinones, carotenoids, flavonoids, and melanin (Lin and Xu, 2020). These pigments are used in food, cosmetics, and pharmaceuticals for their antimicrobial and anticancer activities. Algal pigments, produced by marine and freshwater species, are crucial for photosynthesis and UV radiation protection (Patel *et al.*, 2022). They are rich in carotenoids, chlorophylls, and phycobilins and have antioxidant, antiinflammatory, and anticancer properties. Algal pigments, particularly phycocyanin and astaxanthin, are used as natural colorants in the food industry, antiaging treatments in the cosmetic industry, and in nutraceuticals due to their potential health-promoting effects (Patel *et al.*, 2022).

17.3 Case Study 1: Microbial Pigments in the Food and Beverage Industry

The food industry has been affected by consumer preferences for aesthetically pleasing, healthy, and safe products. Artificial colorants have long been used to enhance the appearance of foods (Malabadi *et al.*, 2022). In relation to growing health concerns and demand for natural and organic products, consumers increasingly seek out food with natural colorants. Consumers are increasingly aware of the health risks associated synthetic chemicals and additives contained in foods, including allergies, hyperactivity in children, and even cancers (Kraemer *et al.*, 2022). With high-profile increases in health-consciousness, consumers are looking for foods that are free of synthetic chemicals and have natural ingredients.

Regulators and governments around the world are also looking at synthetic food colorants and in some cases have banned or restricted them due to safety issues (Mota *et al.*, 2023). Banned or warning-colored artificial colorants include several that have been prohibited or alerted with warning notices in the European Union (Evans and Schmidt, 1990). Due to their dangers to human health, the US FDA has controlled the use of certain synthetic colors (Dey and Nagababu, 2022).

With advances in extraction and production technologies like sophisticated fermentation techniques and biotechnological innovations, natural colorants can be manufactured efficiently and effectively (Wang *et al.*, 2023), thus

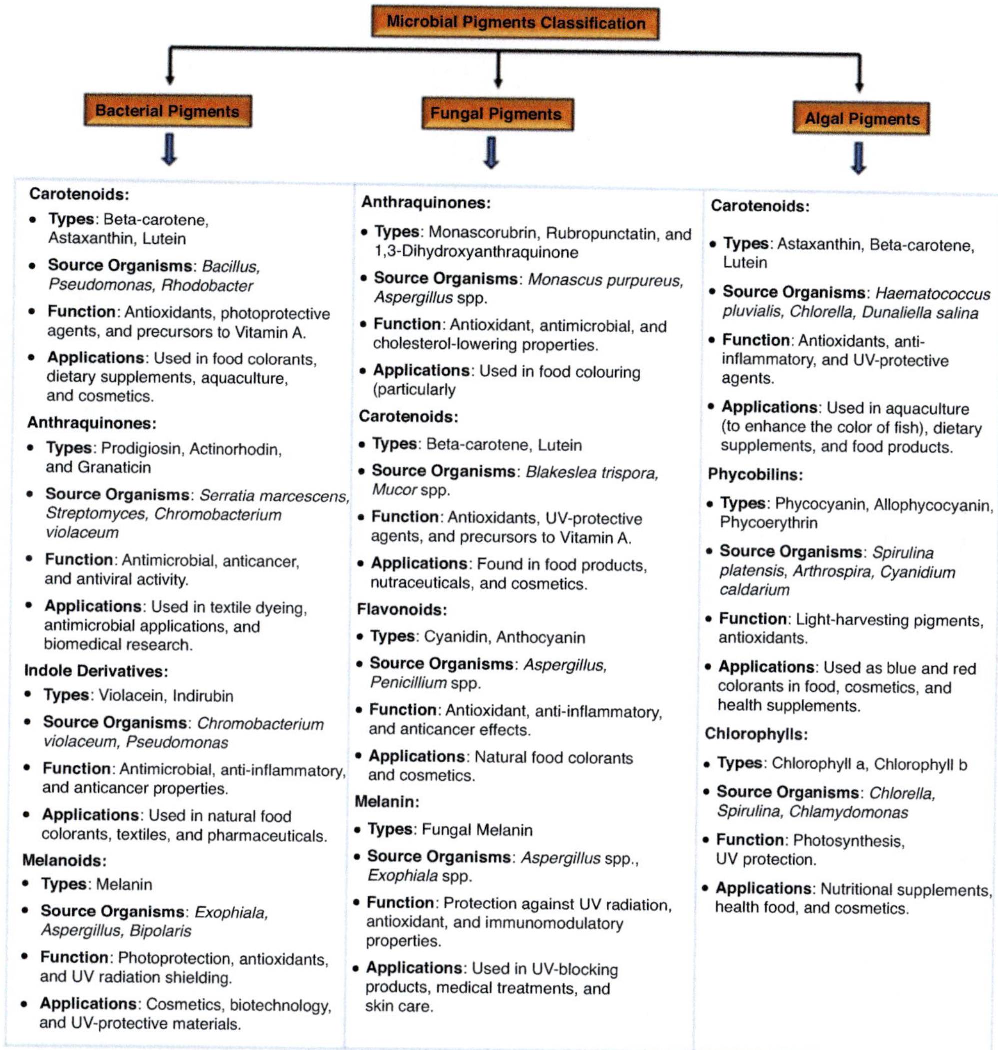

Fig. 17.1. Microbial pigment classification.

bringing these colorants more within reach of the food industry in an economical manner.

17.3.1 Benefits of natural colorants

Natural colorants are increasingly being used for their health and safety benefits, clean labeling, sustainability, and consumer trust (Saraiva *et al.*, 2020). They are considered safer than synthetic dyes and are often rich in antioxidants and bioactive compounds, which may offer health benefits such as antiinflammatory, anticancer, and antimicrobial properties (Olas *et al.*, 2021). They also meet the growing demand for "clean-label" foods, using simple, recognizable, and natural ingredients. Natural colorants, especially those derived from microbial or plant sources, are considered more sustainable than synthetic dyes (Fried *et al.*, 2022), as they can be produced through fermentation processes, which are more environmentally friendly and require fewer resources. Consumers are more likely to trust brands using natural ingredients, as they reflect a company's commitment to quality and sustainability (Manguerra-Mahusay and Liu, 2024).

17.3.2 Examples of natural colorants in the food industry

The use of natural colorants in the food industry to replace synthetic dyes is on the rise. Beet juice, a source of betalains, is used for its red color and antioxidant properties in juices, ice creams, candies, and sauces (Nabi *et al.*, 2023). Carrot juice, obtained from β-carotene, is an orange colorant in beverages, snacks, and dairy products. The blue pigment derived from the algae *Spirulina* is used in beverages, dairy products, and confectionery (Hassanzadeh *et al.*, 2023). Turmeric is a yellow colorant used in dairy and savory food products and beverages. Carmine is a red pigment derived from the cochineal insect. It is used in food and beverages but has ethical concerns and consumer sensitivity. Annatto is a yellow to orange pigment, used in dairy products, cheese, snacks, and meats (Renita *et al.*, 2023).

17.3.3 Challenges of using natural colorants

Natural colorants, while offering numerous benefits, face several challenges in the food industry. They are often more expensive to produce than synthetic dyes due to the cost of extraction or cultivation processes, which can be a barrier for manufacturers in price-sensitive markets (Teli *et al.*, 2020). Natural pigments are less stable and may degrade when exposed to heat, light, or air, affecting color consistency and shelf-life. They also have a limited color range, as natural pigments are known to produce bright, intense hues. The availability of natural colorants is also subject to fluctuations because of climate change, crop failures, or sourcing limitations, creating uncertainties in the supply chain for food manufacturers (Niinimäki *et al.*, 2020).

17.3.4 Future of natural colorants in the food industry

The demand for natural colorants is expected to increase due to consumer preferences and regulatory pressures. Future research will focus on enhancing the stability, cost-effectiveness, and diversity of colors for natural pigments (Novais *et al.*, 2022). Biotechnological advancements, including microbial fermentation and plant breeding, can be innovative sources. The growth of natural colorants will be driven by sustainable, health-conscious, and transparent practices in the food industry (Galanakis, 2024). Manufacturers embracing these trends will contribute to a healthier, more sustainable food system.

17.4 Case Study 2: Pigments in the Textile Industry

Synthetic dyes are used in textiles, food, and cosmetics. They provide bright colors but also pose environmental and health concerns (Dave *et al.*, 2022). This case study examines the environmental impact, toxicity, and the need for sustainable alternatives. It uses examples to illustrate the problem and potential solutions, emphasizing the need for sustainable alternatives in the industry (Charter and Tischner, 2017).

17.4.1 Environmental impact of synthetic dyes

Synthetic dyes are used in industries such as textiles, printing, leather, cosmetics, and food. The textile industry is one of the largest consumers of these dyes, with millions of tons produced and used annually (Islam *et al.*, 2022). However, their disposal and manufacturing methods can have severe environmental consequences. Water pollution is a significant issue, as dyeing processes often involve large quantities of water contaminated with residual dyes, salts, and chemicals (Kishor *et al.*, 2021). These dyes are nonbiodegradable and they can exist in the environment for a long period, thus contaminating aquatic systems. In India, about 20% of water pollution is caused by textile dyeing, and one of the prominent dyes in this regard is azo dyes (Selvaraj *et al.*, 2021). Soil contamination occurs when wastewater discharged without proper care or used for irrigation finds its way into the soil, thereby changing the soil's composition and, consequently, its health

(Singh, 2021). Some dyes contain heavy metal toxicities, which can contribute to poisoning of the food chain. There is also significant energy consumption in making synthetic dyes because they involve high temperatures and chemical processes on a large scale.

17.4.2 Toxicity and health risks of synthetic dyes

Synthetic dyes have adverse health implications to humans. They are highly risky to workers and consumers in terms of human health. Their long-term toxic effects on human health are a major cause of concern (Lippmann, 2000). Workers, especially in developing countries, are directly exposed to synthetic dyes in the production of textiles and leather. Azo dyes, for instance, are reported to be skin irritants, allergenic, and carcinogenic (Keshava *et al.*, 2023). Certain azo dyes and their degradation products have been identified by the World Health Organization as causing bladder cancer. Other hazardous chemicals such as formaldehyde, phthalates, and chlorinated solvents may be absorbed into the bodies of workers through the skin or via inhalation and can lead to serious diseases including respiratory disorders, kidney and liver damage, and neurological conditions (Patra and Pariti, 2022). Consumers are indirectly exposed to synthetic dyes through products like dyed textiles, food, and cosmetics, with some dyes associated with hyperactivity and allergic reactions, particularly in children. The European Food Safety Authority (EFSA) and the US FDA have also raised concerns over certain food colorants, recommending their withdrawal due to the potential to cause cancer and other health issues (Amchova *et al.*, 2024).

17.4.3 Examples of synthetic dye challenges

Bangladesh, one of the world's biggest garment-producing nations, faces very severe environmental pollution due to the heavy consumption of synthetic dyes within the textile industries (Uddin *et al.*, 2023). Some of the most dangerous synthetic dyes are poorly treated before being discharged into water bodies. Lead, mercury, and chromium all have dangerous water samples recorded around the area near Dhaka's Buriganga River (Hossain *et al.*, 2024).

Environmental groups have helped to raise awareness concerning the adverse effects of dye effluent on communities around factories. In some manufacturing units, pilot projects have brought in environmentally friendly dyeing and effluent treatment technologies (Li *et al.*, 2019). The use of red cochineal extract, known as carmine, in food, cosmetics, and textiles has become a cause of public uproar because of its animal origin and the health risk that it may pose. People have voiced concerns over the ethics of the source of synthetic food dyes, and severe allergic reactions in some individuals. In 2009, a major US supermarket chain was accused of using synthetic dyes instead of carmine, highlighting the challenges faced by the food industry in balancing consumer demand for natural or safe dyes with the use of cheaper synthetic alternatives (Ghosh *et al.*, 2024).

To mitigate environmental and health issues brought about by synthetic dyes, food, textile, and cosmetics companies are looking towards the utilization of natural colorants from plants, algae, fungi, and bacteria (Saravanan *et al.*, 2024). Safe, biodegradable alternatives currently gaining popularity include violacein and carotenoids from bacteria and phycocyanin from algae. Green chemistry innovations, including enzyme-based dyeing techniques and closed-loop wastewater treatment systems, are reducing the impact of industries by lowering energy consumption, minimizing water usage, and recycling dye wastewater (Behera *et al.*, 2021b).

17.5 Case Study 3: Microbial Pigments in Cosmetics and Personal Care

In the past few years, the cosmetics industry has shifted toward natural and cruelty-free products. This move is attributed to the increasing concern over the health hazards from synthetic chemicals used in products that have been proved to be involved in endocrine disruption, irritation of the skin, and so on (Metcalfe *et al.*, 2022). The most banned chemicals are parabens, phthalates, and sulfate. Health-conscious consumers seek beauty and personal care products that contribute to their overall wellbeing. Natural ingredients such as essential oils, plant extracts, botanical oils, and minerals are considered safer compared to synthetic compounds (Khursheed *et al.*, 2022). Aloe vera, lavender oil, tea tree oil, coconut oil, and green tea extract have emerged as important ingredients in most skincare, haircare, and make-up products as they are thought to improve the condition of skin health, possess antiaging properties, and be a good source of antioxidants (Aqil *et al.*, 2020).

Natural cosmetics are poised to witness substantial growth over the coming years at an estimated annual growth rate of around 5–7%, led by consumer interest in clean and nontoxic products (Alnuqaydan, 2024). The trend of natural cosmetics demands transparency and brands are embracing clean labeling by indicating on packaging that no harmful ingredients are used, appealing to consumers who want products that share their values. Sephora's Clean at Sephora initiative has gained consumer trust, showing how much consumers are demanding nontoxic, clean-label products.

17.5.1 Cruelty-free movement

Cruelty-free cosmetics, products not tested on animals in the development process, are in high demand in the beauty industry (Magano *et al.*, 2022), due to growing concerns regarding animal welfare as well as ethical consumerism. Consumption patterns among contemporary consumers have changed because of social media and other related platforms including Instagram, YouTube, and TikTok, which help them determine the ethical dimensions of their purchase decisions. Influencers and activists in the beauty space are also advocating for products that match ethical values and urging consumers to choose products that do not contribute to animal testing. The Humane Society International estimated that over 40% of global consumers now care about animal testing, particularly in the context of the beauty industry. To cater to consumer demand, many cosmetic brands are seeking to obtain certifications from well-known groups like Leaping Bunny, PETA's Beauty Without Bunnies, and Cruelty-Free International. Some of the most popular cosmetic brands that show off their cruelty-free certification labels include Urban Decay, Too Faced, and Tarte Cosmetics (Gottfried, 2017).

Additionally, many customers are looking for vegan cosmetics which do not contain animal byproducts such as beeswax, lanolin, or keratin. The demand for vegan beauty products has risen by 175% in the last five years, with industry leaders such as Milk Makeup, Fenty Beauty, and Lush formulating cruelty-free vegan products. This conjunction gives consumers the opportunity to act in accordance with both ethical and environmental desires, further increasing the desirability scale of natural, cruelty-free products.

17.5.2 Environmental sustainability and eco-friendly packaging

The increase in natural and cruelty-free beauty movements has raised concerns about environmental sustainability (Varma and Ray, 2023). Consumers have become more conscious of brands that contribute less to environmental degradation, particularly from plastic wastes. Many brands focus on environmentally sustainable sourcing of ingredients, such as certified organic oils and Fair Trade ingredients, that reduce deforestation and environmental degradation (Bom *et al.*, 2019). Brands such as Burt's Bees and The Body Shop have partnered with Fair Trade Certified suppliers and set the industry standard for sustainability. Packaging is another area of challenge for the cosmetics industry, as many products come in plastic containers that are not easily recyclable (Cinelli *et al.*, 2019). Brands are moving towards eco-friendly solutions, including biodegradable materials, refillable containers, and recycled plastic. Lush Cosmetics has taken the "naked" concept, which is products without packaging, and Aveda and Kiehl's have implemented post-consumer recycled plastic.

Social media has played a significant role in driving the rise of natural, cruelty-free cosmetics. Beauty influencers, vloggers, and activists have used platforms like Instagram, YouTube, and

TikTok to promote cruelty-free brands and educate their audiences about the environmental and ethical implications of their purchases. Online communities and forums dedicated to sustainable beauty, such as Reddit's Green Beauty community (Fox *et al.*, 2021), provide sources of information and reviews of natural, cruelty-free, and vegan products, and empower consumers to make decisions based on shared values, fostering a culture of consumption that spreads responsibility.

17.6 Case Study 4: Microbial Pigments in Agriculture and Plant Protection

The production of microbial pigments by bacteria, fungi, and algae has become an area of interest in agriculture, as they have the potential to be used as natural plant growth promoters and biopesticides (Gonçalves, 2021). These pigments, which include carotenoids, anthraquinones, flavonoids, and violacein, have biological properties that promote plant health, increase yields, and offer eco-friendly alternatives to conventional chemical fertilizers and pesticides (Vlaiculescu and Varrone, 2022).

In this case study, the use of microbial pigments for agricultural applications is addressed, with special interest in their potential to stimulate plant growth and act as biopesticides (Thakur *et al.*, 2020). These pigments exhibit antioxidant, antifungal, antibacterial, and phytohormonal activities that benefit plant health.

17.6.1 Mechanisms of plant growth promotion

Microbial pigments play a significant role in plant growth by producing plant growth regulators (PGRs) (Du *et al.*, 2020). These stimulate root growth and better uptake of nutrients, and enhance resistance to stress. PGRs like auxins, cytokinin, and gibberellins produced by microorganisms such as *Pseudomonas putida* improve root growth and hence plant development (Timofeeva *et al.*, 2024). Solubilization of phosphorus, a vital nutrient, also enhances nutrient availability. They produce organic acids and enzymes that break down soil-bound nutrients, making them more accessible to plant roots. Carotenoids and anthocyanins, through their antioxidant activity, help plants cope with oxidative stress caused by environmental factors such as drought, high salinity, and UV radiation (Hussain *et al.*, 2019). These pigments protect plants from cellular damage and promote growth even under adverse conditions. Some microbial pigments, such as violacein, have antimicrobial activity, combating pathogenic microorganisms and thereby reducing the application of chemical pesticides and making plants healthier (Ahmed *et al.*, 2021).

17.6.2 Examples of plant growth promotion using microbial pigments

Bacillus subtilis, one of the most well-known bacteria that produce carotenoids, has been established to improve growth in tomato by its antioxidant action, ensuring a stronger root system and superior growth (Kumar *et al.*, 2024). Additionally, the pigment from the microbes increases tolerance to environmental stresses such as drought and high temperature. *Pseudomonas putida*, a frequent bacterium in the environment with the ability to generate violacein, synthesizes the phytohormone IAA which has strong effects on *Arabidopsis thaliana* and wheat plants including root elongation, nutrient acquisition, and enhancement of plant fitness (Chaudhary *et al.*, 2022). Violacein may additionally offer protection against diseases caused by hostile pathogens. Fungal pigments, including ferrichrome, have improved iron availability with better growth outcomes, especially where the soil iron content is already low (Lewis *et al.*, 2019).

17.6.3 Microbial pigments as biopesticides

Microbial pigments are an eco-friendly substitute for chemical pesticides and are also plant growth promoters. These pigments can be used for the protection of plants against many pests and pathogens due to their antimicrobial and antifungal properties (Narsing Rao *et al.*, 2017). Violacein, produced by *Chromobacterium violaceum*, is an antimicrobial pigment that inhibits disease caused by soil-borne pathogens such as *Pythium*, *Fusarium*, etc. (Saikia *et al.*, 2022). Carotenoids, produced by *Rhodobacter* and *Rhodospirillum* bacteria, have antifungal activity and protect crops against fungal infections such as powdery mildew and gray mold. Anthraquinones, produced by fungi such as *Aspergillus* and *Penicillium*, have antifungal as well as insecticidal activity, controlling aphids, whiteflies, and caterpillars in crops such as tomatoes and cotton (Chaudhary *et al.*, 2024). These pigments provide an alternative to chemical pesticides that is sustainable and friendly to the environment.

17.6.4 Examples of biopesticide applications

Researchers at the University of São Paulo tested violacein as a biocontrol agent against *Fusarium oxysporum* in banana and tomato crops, and noted decreased disease incidence (Doan *et al.*, 2020). Carotenoids from carotenoid-producing bacteria like *Rhodobacter sphaeroides* reduced botrytis blight in strawberry plants, using their natural pigments as growth promoters and protectants against *Botrytis cinerea* (Daub *et al.*, 2013). Anthraquinones from *Penicillium* species reduced the whitefly population on cotton plants, demonstrating the potential of these fungal pigments as eco-friendly alternatives to chemical insecticides (Chaudhary *et al.*, 2024).

17.6.5 Advantages of microbial pigments as biopesticides

Microbial pigments are nontoxic, ecofriendly, and sustainable alternatives to chemical pesticides for pest and disease management (Pathak *et al.*, 2022). They have specific activity against pathogens and pests and reduce the damage caused to nontarget organisms, thus maintaining ecological balance and biodiversity in agricultural ecosystems (Khan *et al.*, 2023). Microbial pigments are cost-effective and scalable because they can be produced through fermentation processes, making them suitable for large-scale agricultural applications and easy integration into existing practices (Nguyen *et al.*, 2022). This makes them a sustainable and eco-friendly solution for pest and disease management.

17.6.6 Challenges and future prospects

Microbial pigments have the potential to be plant growth promoters and biopesticides. However, common challenges they face are standardization and consistency in quality (Aamir *et al.*, 2020). Factors such as environmental conditions, concentration, and application method affect how effective these pigments can be. To ensure consistent quality and effectiveness, microbial pigments need regulatory frameworks (Sandybayeva *et al.*, 2022). Despite their benefits, they face competition from cheaper chemical pesticides and fertilizers. Widespread adoption demands greater awareness of biopesticides, investment in research and development, and large-scale commercial methods of production (Wyckhuys *et al.*, 2018). Biopesticides require a lengthy, intensive approval process, including safety and efficacy testing, thereby delaying their introduction into the marketplace and limiting availability in some geographic areas.

17.7 Case Study 5: Microbial Pigments in the Pharmaceutical and Biomedical Sectors

Secondary metabolites, found in various organisms, plants, and animals, have shown promising potential in medicine and pharmacology (Velu *et al.*, 2018). These compounds, which participate in photosynthesis, UV protection, and communication, have significant bioactive properties that have been exploited in the development of natural medicines. These pigments, including carotenoids, anthocyanins, betalains, and flavonoids, have been thoroughly studied for their therapeutic applications (Madadi *et al.*, 2020). They can be broadly categorized based on chemical structure and color: carotenoids with their orange-yellow pigments, anthocyanins with their red-blue colors, and betalains that contribute red-yellow pigments. Every class of pigment demonstrates different biological activities that find use in medical applications. This section reviews this group's potential use in medicine and pharmacology and considers their therapeutic application, mode of action, and clinical utility (Guardado Yordi *et al.*, 2024).

17.7.1 Pigments with medicinal properties

Carotenoids are fat-soluble pigments responsible for the yellow, orange, and red colors of fruits and vegetables (Langi *et al.*, 2018). These pigments include β-carotene, lutein, zeaxanthin, lycopene, and astaxanthin, which are known for their antioxidant properties, protecting cells from oxidative stress and free radical damage (Black *et al.*, 2020). Carotenoids have therapeutic properties, including antioxidant activity, preventing diseases associated with aging, such as neurodegenerative disorders, cardiovascular diseases, and cancer (Cho *et al.*, 2018). They also possess anticancer properties, in the form of scavenging free radicals, gene expression modulation, and preventing cell division of cancer cells. The carotenoids lutein and zeaxanthin have proved useful for the eye by protecting the retina from harmful ultraviolet blue light and oxidative stress and indirectly preventing AMD and cataract formation (Rodella *et al.*, 2023). Carotenoids such as astaxanthin, for example, have antiinflammatory properties, meaning they can affect inflammatory pathways, reduce cytokines, and eventually treat diseases like arthritis and inflammatory bowel disease (Bas, 2024).

Anthocyanins can be found in fruits, vegetables, and flowers, giving these plants red, blue, or purple colors. They have been shown to possess strong antioxidant capacities, neutralize free radicals, and reduce inflammation, which can then protect against diseases such as cardiovascular diseases, diabetes, and neurodegenerative disorders (Teleanu *et al.*, 2019). Intake of anthocyanin-rich foods can reduce the risk of atherosclerosis and improve heart function. Furthermore, anthocyanins have neuroprotective effects, enhancing cognitive function and memory, possibly preventing neurodegenerative diseases such as Alzheimer's and Parkinson's by reducing oxidative stress in the brain and promoting neuronal health.

Betalains are water-soluble pigments in plants including beetroot, Swiss chard, and amaranth (Phukan *et al.*, 2023). They include betacyanin (red violet) and betaxanthins (yellow orange). Betalains contain antioxidants and antiinflammatory agents, helping reduce oxidative stress and inflammation to prevent the risk of heart diseases, diabetes, and cancer. The antioxidants reduce damage to the liver caused by toxins and oxidative stress, promoting efficient detoxification and ensuring healthy functioning of the liver (Unsal *et al.*, 2020). The betalains induce apoptosis in cancerous cells, inhibiting the growth of the tumor. There has been some research suggesting the ability of beetroot-derived betalains to combat cancer such as colon cancer.

Flavonoids are polyphenolic compounds that are present in fruits, vegetables, and beverages such as green tea, grapes, apples, and onions. They possess antioxidant, antiinflammatory, and antimicrobial properties (Garg *et al.*, 2019). They protect the cardiovascular system by improving blood vessel function, reducing blood pressure, and lowering the risk of stroke and heart disease. Flavonoids also have anticancer effects through antioxidant activity, modulation of cell signaling pathways, and inhibition of angiogenesis. The epigallocatechin gallate in green tea has been found to suppress the growth of breast and prostate cancer cells (Sharifi-Rad *et al.*, 2020). Flavonoids are also neuroprotective, protecting the brain from oxidative stress and neurodegeneration, potentially improving cognitive function, reducing the risk of Alzheimer's disease, and supporting overall brain health (Lee *et al.*, 2020).

17.7.2 Clinical applications and use in modern medicine

Pigments are used in various medical treatments, including dietary supplements, topical treatments, and pharmaceutical formulations (Al Fahad *et al.*, 2024). Carotenoids are used for eye health, while anthocyanin-rich extracts improve vascular health and reduce diabetes and heart disease risk. Blueberry extract is known for its cognitive-enhancing properties. Beetroot-based supplements improve exercise performance by improving blood flow and oxygen delivery to muscles. Beneficial flavonoid-rich products include green tea and grape seed extract. They contribute antioxidant, antiinflammatory, and cardiovascular benefits so, overall, pigments can form the backbone for several medical treatments (Khan *et al.*, 2021).

17.7.3 Sustainability and future potential of microbial pigments

Microbial pigments have many environmental, economic, and health benefits compared to traditional synthetic dyes, particularly in textiles, food, cosmetics, and pharmaceuticals (Sen *et al.*, 2019). The sustainability of microbial pigments can be analyzed from several perspectives: their production, environmental impact, economic feasibility, and societal implications.

Microbial pigments have significant sustainable advantages including biodegradability, reduced toxicity, decreased water requirements, valorization of waste products, reduced carbon footprints, and energy inputs (Cassani *et al.*, 2022). Traditional synthetic dyes synthesis relies on the use of harmful chemicals and metals from which toxic effluents arise. Typically, microbial pigments are made via fermentation-based production processes and use less water in their overall manufacture. Agri-byproducts, industrial wastes, and cheap carbon sources in the form of sugars can be used to valorize waste (Solehin *et al.*, 2022). The production of microbial pigment has a substantially lower carbon footprint than synthetic traditional pigments, most especially so when the electricity used comes from renewable energy sources. The carbons used in this microbial pigment formulation, such as glucose and glycerol, are obtained through biomas, which can thus close the carbon loop in sustainable ways. Other innovative approaches incorporate carbon capture technologies with microbial pigment production, which make the entire process more environmentally friendly (Nandhini *et al.*, 2023). In addition, microbial pigments can be used as biopesticides and plant growth promoters, thereby contributing to sustainability in agriculture.

The use of microbial pigments reduces dependency on synthetic fertilizers and pesticides, thus reducing the environmental footprint of agriculture and promoting more sustainable farming practices. For instance, violacein exhibits strong antimicrobial activity, making it a promising tool for controlling soil-borne pathogens and plant diseases (Awadelkareem *et al.*, 2022).

17.8 Future Potential of Microbial Pigments

The future of microbial pigments is looking bright, as they have great potential across various industries. Recent breakthroughs in synthetic biology and metabolic engineering open vistas to improving production of microbial pigments by manipulating pathways in microorganisms. Genetic alterations in bacteria, fungi, or algae can increase efficiency through modifications that increase carotenoids and anthraquinone production. Synthetic biology tools enable the design of new pigments with improved antioxidant or antimicrobial activities, thus opening new avenues for medicinal, cosmetic, and textile applications. Optimization of fermentation can reduce the cost of microbial pigment production by improving culture conditions, nutrient utilization, and fermentation time.

Microbial pigments have the potential to revolutionize various industries, from food to cosmetics and pharmaceuticals, contributing to sustainability by replacing harmful synthetic chemicals with eco-friendly alternatives. In the food industry, microbial pigments can replace synthetic colorants like tartrazine and sunset yellow, which have raised health concerns due to their potential toxicity. Cosmetics and personal care are increasingly shifting towards cruelty-free, natural, and sustainable products, with pigments like anthocyanins, carotenoids, and flavonoids found in formulations for skincare, lipsticks, nail polishes, and hair dyes. In the pharmaceutical industry, various microbial pigments have been identified for use in therapeutic treatments, including anticancer, antimicrobial, and antiinflammatory agents. The use of violacein, carotenoids, and anthocyanins offers great promise in terms of drug production, opening pathways to more ecologically friendly, safer pharmaceutical products. The use of microbial pigments can also present an opportunity to the textile industries as a future source of greener alternatives for synthetic dyes, which can be produced in substantial quantities, thus giving a more ecological, scalable solution.

However, several challenges need to be addressed to make microbial pigment production truly sustainable and commercially viable. The current production costs are higher than those of synthetic dyes due to high fermentation costs. In the future, improvements in fermentation technology, optimization of microbial strains, and use of cheap carbon sources may lower production costs. Scalability requires further development in bioprocess engineering and fermentation techniques, and regulatory approval is essential for consumer safety and industry acceptance.

17.9 Challenges and Limitations

Microbial pigments could change the course of several industries, but adoption is not going to be an easy feat. Challenges in terms of cost-effectiveness, stability, regulatory issues, and skepticism from consumers all play a part. Production challenges in microbial pigments include costs related to fermentation processes, optimization of yield, raw material, economies of scale, and formulating. The fermentation costs are the primary cost in the production of microbial pigments because it is sustainable, though more expensive than conventional synthetic dye production. Yield and productivity are essential to reduce costs in the production of microbial pigments. Genetic modifications or optimization of growth conditions must be applied to make the process economically viable. High-priced carbon sources and nutrients such as raw material costs can add to the final price of pigment and depend on market demand. Low-cost feedstocks for production, especially those derived from agricultural byproducts or waste products from other industries, are gaining greater interest.

For industrial-scale production, economies of scale are fundamental. Large-scale facilities with optimized processes, efficient bioreactors, and integrated downstream processing technologies can dramatically reduce production costs. Advances in bioprocess engineering, metabolic engineering, and automation promise to make microbial pigment production more cost-effective. Researchers are also looking into ways of using biorefinery approaches to coproduce pigments along with other valuable compounds such as biofuels or bioplastics for further optimization of the economics of production.

Another major challenge is the stability and consistency of microbial pigments in different applications, especially in the food, cosmetic, and textile industries. Although microbial pigments are more sustainable and eco-friendly than synthetic alternatives, their properties can vary greatly depending on temperature, pH, light exposure, and storage conditions. Microbial pigments are very sensitive to environmental conditions, which can cause color degradation and loss of functional properties. Shelf-life and product stability are critical concerns in the food and cosmetics industries, as many microbial pigments degrade over time or under certain conditions, making them unsuitable for long-term use unless stabilizing agents or specific formulations are introduced.

Other regulatory obstacles exist in several industries. Food industrial regulations will influence the use of microbial pigments as food colors. Cosmetic regulations are extensive to ensure safety in terms of skin irritation, allergic properties, and long-term effects. Textiles regulations face difficulty in standardizing and quality checking during manufacturing processes. The need to overcome consumer skepticism about novel products in different markets affects the success of microbial pigments. Factors include awareness about the safety and benefits of microbial pigments, perceived novelty and safety concerns, transparency and education, and product labeling and certification. The sustainability of microbial pigments and their eco-friendly properties can make them mainstream alternatives in food, cosmetics, and textiles by offering certifications that highlight the sustainability of microbial pigments.

References

Aamir, M., Rai, K.K., Zehra, A., Dubey, M.K., Kumar, S. *et al.* (2020) Microbial bioformulation-based plant biostimulants: A plausible approach toward next generation of sustainable agriculture. In: *Microbial Endophytes*. Woodhead Publishing, Sawston, UK, pp. 195–225.

Agarwal, H., Bajpai, S., Mishra, A., Kohli, I., Varma, A. *et al.* (2023) Bacterial pigments and their multifaceted roles in contemporary biotechnology and pharmacological applications. *Microorganisms* 11(3), 614. DOI: 10.3390/microorganisms11030614.

Ahmed, A., Ahmad, A., Li, R., Al-Ansi, W., Fatima, M. *et al.* (2021) Recent advances in synthetic, industrial and biological applications of violacein and its heterologous production. *Journal of Microbiology and Biotechnology* 31(11), 1465–1480. DOI: 10.4014/jmb.2107.07045.

Al Fahad, M.A., Ahamed, R., Ahmed, T., Jahan, N., Mia, R. *et al.* (2024) Pharmaceutical applications of natural dyes and pigments. In: *Renewable Dyes and Pigments*. Elsevier, New York, pp. 165–175.

Alnuqaydan, A.M. (2024) The dark side of beauty: An in-depth analysis of the health hazards and toxicological impact of synthetic cosmetics and personal care products. *Frontiers in Public Health* 12, 1439027. DOI: 10.3389/fpubh.2024.1439027.

Alugoju, P., Krishna Swamy, V.K.D., Anthikapalli, N.V.A. and Tencomnao, T. (2023) Health benefits of astaxanthin against age-related diseases of multiple organs: A comprehensive review. *Critical Reviews in Food Science and Nutrition* 63(31), 10709–10774. DOI: 10.1080/10408398.2022.2084600.

Álvarez-Martínez, F.J., Barrajón-Catalán, E., Encinar, J.A., Rodríguez-Díaz, J.C. and Micol, V. (2020) Antimicrobial capacity of plant polyphenols against gram-positive bacteria: A comprehensive review. *Current Medicinal Chemistry* 27(15), 2576–2606. DOI: 10.2174/0929867325666181008115650.

Aman Mohammadi, M., Ahangari, H., Mousazadeh, S., Hosseini, S.M. and Dufossé, L. (2022) Microbial pigments as an alternative to synthetic dyes and food additives: A brief review of recent studies. *Bioprocess and Biosystems Engineering* 45(1), 1–12. DOI: 10.1007/s00449-021-02621-8.

Amchova, P., Siska, F. and Ruda-Kucerova, J. (2024) Food safety and health concerns of synthetic food colors: An update. *Toxics* 12(7), 466. DOI: 10.3390/toxics12070466.

Aqil, M., Chaudhuri, A. and Qadir, A. (2020) Herbal cosmeceuticals: New opportunities in cosmetology. *Trends in Phytochemical Research* 4(3), 117–142.

Assunção, J., Amaro, H.M., Malcata, F.X. and Guedes, A.C. (2022) Cyanobacterial pigments: Photosynthetic function and biotechnological purposes. In: *The Pharmacological Potential of Cyanobacteria*. Academic Press, Washington, DC, pp. 201–256.

Awadelkareem, A.M., Al-Shammari, E., Elkhalifa, A.O., Adnan, M., Siddiqui, A.J. *et al.* (2022) Biosynthesized silver nanoparticles from *Eruca sativa* miller leaf extract exhibits antibacterial, antioxidant, anti-quorum-sensing, antibiofilm, and anti-metastatic activities. *Antibiotics* 11(7), 853. DOI: 10.3390/antibiotics11070853.

Bas, T.G. (2024) Bioactivity and bioavailability of carotenoids applied in human health: Technological advances and innovation. *International Journal of Molecular Sciences* 25(14), 7603. DOI: 10.3390/ijms25147603.

Behera, H.T., Mojumdar, A., Nivedita, S. and Ray, L. (2021a) Microbial pigments: Secondary metabolites with multifaceted roles. In: *Microbial Polymers: Applications and Ecological Perspectives*. Springer, Singapore, pp. 631–654.

Behera, M., Nayak, J., Banerjee, S., Chakrabortty, S. and Tripathy, S.K. (2021b) A review on the treatment of textile industry waste effluents towards the development of efficient mitigation strategy: An integrated system design approach. *Journal of Environmental Chemical Engineering* 9(4), 105277. DOI: 10.1016/j.jece.2021.105277.

Black, H.S., Boehm, F., Edge, R. and Truscott, T.G. (2020) The benefits and risks of certain dietary carotenoids that exhibit both anti- and pro-oxidative mechanisms-a comprehensive review. *Antioxidants* 9(3), 264. DOI: 10.3390/antiox9030264.

Bom, S., Jorge, J., Ribeiro, H.M. and Marto, J.O.A.N.A. (2019) A step forward on sustainability in the cosmetics industry: A review. *Journal of Cleaner Production* 225, 270–290. DOI: 10.1016/j.jclepro.2019.03.255.

Budiman, A., Rusdin, A. and Aulifa, D.L. (2023) Current techniques of water solubility improvement for antioxidant compounds and their correlation with its activity: Molecular pharmaceutics. *Antioxidants* 12(2), 378. DOI: 10.3390/antiox12020378.

Bürck, M., Fratelli, C., Campos Assumpção de Amarante, M. and Braga, A.R.C. (2024) Unveiling the potential of spirulina biomass–a glimpse into the future circular economy using green and blue ingredients. *Biomass* 4(3), 704–719. DOI: 10.3390/biomass4030039.

Cassani, L., Marcovich, N.E. and Gomez-Zavaglia, A. (2022) Valorization of fruit and vegetables agro-wastes for the sustainable production of carotenoid-based colorants with enhanced bioavailability. *Food Research International* 152, 110924. DOI: 10.1016/j.foodres.2021.110924.

Charter, M. and Tischner, U. (eds) (2017) *Sustainable Solutions: Developing Products and Services for the Future*. Routledge, Abingdon, UK. DOI: 10.4324/9781351282482.

Chaudhary, P., Agri, U., Chaudhary, A., Kumar, A. and Kumar, G. (2022) Endophytes and their potential in biotic stress management and crop production. *Frontiers in Microbiology* 13, 933017. DOI: 10.3389/fmicb.2022.933017.

Chaudhary, R., Nawaz, A., Khattak, Z., Butt, M.A., Fouillaud, M. *et al.* (2024) Microbial bio-control agents: A comprehensive analysis on sustainable pest management in agriculture. *Journal of Agriculture and Food Research* 18, 101421. DOI: 10.1016/j.jafr.2024.101421.

Cho, K.S., Shin, M., Kim, S. and Lee, S.B. (2018) Recent advances in studies on the therapeutic potential of dietary carotenoids in neurodegenerative diseases. *Oxidative Medicine and Cellular Longevity* 2018(1), 4120458. DOI: 10.1155/2018/4120458.

Cinelli, P., Coltelli, M.B., Signori, F., Morganti, P. and Lazzeri, A. (2019) Cosmetic packaging to save the environment: Future perspectives. *Cosmetics* 6(2), 26. DOI: 10.3390/cosmetics6020026.

Daub, M.E., Herrero, S. and Chung, K.R. (2013) Reactive oxygen species in plant pathogenesis: The role of perylenequinone photosensitizers. *Antioxidants & Redox Signaling* 19(9), 970–989. DOI: 10.1089/ars.2012.5080.

Dave, S., Das, J., Varshney, B. and Sharma, V.P. (2022) Dyes and pigments: Interventions and how safe and sustainable are colors of life. In: *Trends and Contemporary Technologies for Photocatalytic Degradation of Dyes*. Springer International Publishing, Cham, Switzerland, pp. 1–20.

Dey, S. and Nagababu, B.H. (2022) Applications of food color and bio-preservatives in the food and its effect on the human health. *Food Chemistry Advances* 1, 100019. DOI: 10.1016/j.focha.2022.100019.

Doan, H.K., Maharaj, N.N., Kelly, K.N., Miyao, E.M., Davis, R.M. *et al.* (2020) Antimycotal activity of *Collimonas* isolates and synergy-based biological control of fusarium wilt of tomato. *Phytobiomes Journal* 4(1), 64–74. DOI: 10.1094/PBIOMES-05-19-0027-R.

Du, H., Ren, J., Li, Z., Zhang, H., Wang, K. *et al.* (2020) Plant growth regulators affect biomass, protein, carotenoid, and lipid production in *Botryococcus braunii*. *Aquaculture International* 28(3), 1319–1340. DOI: 10.1007/s10499-020-00528-x.

Dufossé, L. (2018) Microbial pigments from bacteria, yeasts, fungi, and microalgae for the food and feed industries. In: *Natural and Artificial Flavoring Agents and Food Dyes*. Academic Press, Washington, DC, pp. 113–132. DOI: 10.1016/B978-0-12-811518-3.00004-1.

Evans, D.L. and Schmidt, J.O. (eds) (1990) *Insect Defenses: Adaptive Mechanisms and Strategies of Prey and Predators*. SUNY Press, New York.

Fox, N., Graham, L.J., Eigenbrod, F., Bullock, J.M. and Parks, K.E. (2021) Reddit: A novel data source for cultural ecosystem service studies. *Ecosystem Services* 50, 101331. DOI: 10.1016/j.ecoser.2021.101331.

Fried, R., Oprea, I., Fleck, K. and Rudroff, F. (2022) Biogenic colourants in the textile industry – a promising and sustainable alternative to synthetic dyes. *Green Chemistry* 24(1), 13–35. DOI: 10.1039/D1GC02968A.

Galanakis, C.M. (2024) The future of food. *Foods* 13(4), 506. DOI: 10.3390/foods13040506.

Garg, S.K., Shukla, A. and Choudhury, S. (2019) Polyphenols and flavonoids. In: *Nutraceuticals in Veterinary Medicine*. Springer, Cham, Switzerland, pp. 187–204.

Ghosh, S., Sarkar, T., Chakraborty, R., Shariati, M.A. and Simal-Gandara, J. (2024) Nature's palette: An emerging frontier for coloring dairy products. *Critical Reviews in Food Science and Nutrition* 64(6), 1508–1552. DOI: 10.1080/10408398.2022.2117785.

Gonçalves, A.L. (2021) The use of microalgae and cyanobacteria in the improvement of agricultural practices: A review on their biofertilising, biostimulating and biopesticide roles. *Applied Sciences* 11(2), 871. DOI: 10.3390/app11020871.

Gottfried, M. (2017) *Vegan Love: Dating and Partnering for the Cruelty-Free Gal, with Fashion, Makeup & Wedding Tips*. Simon and Schuster, New York.

Guardado Yordi, E., Pérez Martínez, A., Radice, M., Scalvenzi, L., Abreu-Naranjo, R. *et al.* (2024) Seaweeds as source of bioactive pigments with neuroprotective and/or anti-neurodegenerative activities: Astaxanthin and fucoxanthin. *Marine Drugs* 22(7), 327. DOI: 10.3390/md22070327.

Guryanov, I. and Naumenko, E. (2024) Bacterial pigment prodigiosin as multifaceted compound for medical and industrial application. *Applied Microbiology* 4(4), 1702–1728. DOI: 10.3390/applmicrobiol4040115.

Hassanzadeh, H., Lotfi, P. and Ghanbarzadeh, B. (2023) An overview of the applications of spirulina platensis algae in improving the nutritional and functional properties of dairy products and its use in the recycling of dairy industry waste. *Journal of Food Science and Technology* 20(141), 175–199.

Hossain, M.M., Jahan, I., Dar, M.A., Dhanavade, M.J., Mamtaz, A.F.B. *et al.* (2024) A review of potentially toxic elements in sediment, water, and aquatic species from the river ecosystems. *Toxics* 13(1), 26. DOI: 10.3390/toxics13010026.

Hussain, S., Rao, M.J., Anjum, M.A., Ejaz, S., Zakir, I. *et al.* (2019) Oxidative stress and antioxidant defense in plants under drought conditions. In: Hasanuzzaman, M., Hakeem, K.R., Nahar, K. and Alharby, H.F. (eds) *Plant Abiotic Stress Tolerance*. Springer Nature, Cham, Switzerland, p. 207. DOI: 10.1007/978-3-030-06118-0_9.

Islam, Md., Islam, T., Islam, T. and Repon, Md. R. (2022) Synthetic dyes for textile colouration: Process, factors and environmental impact. *Textile & Leather Review* 5, 327–373. DOI: 10.31881/TLR.2022.27.

Keshava, C., Nicolai, S., Vulimiri, S.V., Cruz, F.A., Ghoreishi, N. *et al.* (2023) Application of systematic evidence mapping to identify available data on the potential human health hazards of selected market-relevant azo dyes. *Environment International* 176, 107952. DOI: 10.1016/j.envint.2023.107952.

Khan, B.A., Nadeem, M.A., Nawaz, H., Amin, M.M., Abbasi, G.H. *et al.* (2023) Pesticides: Impacts on agriculture productivity, environment, and management strategies. In: *Emerging Contaminants and Plants: Interactions, Adaptations and Remediation Technologies*. Springer International Publishing, Cham, Switzerland, pp. 109–134.

Khan, J., Deb, P.K., Priya, S., Medina, K.D., Devi, R. *et al.* (2021) Dietary flavonoids: Cardioprotective potential with antioxidant effects and their pharmacokinetic, toxicological and therapeutic concerns. *Molecules* 26(13), 4021. DOI: 10.3390/molecules26134021.

Khursheed, A., Rather, M.A., Jain, V., Wani, A.R., Rasool, S. *et al.* (2022) Plant based natural products as potential ecofriendly and safer biopesticides: A comprehensive overview of their advantages over conventional pesticides, limitations and regulatory aspects. *Microbial Pathogenesis* 173(Pt A), 105854. DOI: 10.1016/j.micpath.2022.105854.

Kishor, R., Purchase, D., Saratale, G.D., Saratale, R.G., Ferreira, L.F.R. *et al.* (2021) Ecotoxicological and health concerns of persistent coloring pollutants of textile industry wastewater and treatment approaches for environmental safety. *Journal of Environmental Chemical Engineering* 9(2), 105012. DOI: 10.1016/j.jece.2020.105012.

Kraemer, M.V.D.S., Fernandes, A.C., Chaddad, M.C.C., Uggioni, P.L., Rodrigues, V.M. *et al.* (2022) Food additives in childhood: A review on consumption and health consequences. *Revista de Saude Publica* 56, 32. DOI: 10.11606/s1518-8787.2022056004060.

Kumar, A., Naroju, S.P., Kumari, N., Arsey, S., Kumar, D. *et al.* (2024) The role of drought response genes and plant growth promoting bacteria on plant growth promotion under sustainable agriculture: A review. *Microbiological Research* 286, 127827. DOI: 10.1016/j.micres.2024.127827.

Langi, P., Kiokias, S., Varzakas, T. and Proestos, C. (2018) Carotenoids: From plants to food and feed industries. *Microbial Carotenoids: Methods and Protocols* 57–71. DOI: 10.1007/978-1-4939-8742-9_3.

Lee, K.H., Cha, M. and Lee, B.H. (2020) Neuroprotective effect of antioxidants in the brain. *International Journal of Molecular Sciences* 21(19), 7152. DOI: 10.3390/ijms21197152.

Lewis, R.W., Islam, A., Opdahl, L., Davenport, J.R. and Sullivan, T.S. (2019) Comparative genomics, siderophore production, and iron scavenging potential of root zone soil bacteria isolated from "Concord" grape vineyards. *Microbial Ecology* 78(3), 699–713. DOI: 10.1007/s00248-019-01324-8.

Li, W., Mu, B. and Yang, Y. (2019) Feasibility of industrial-scale treatment of dye wastewater via bio-adsorption technology. *Bioresource Technology* 277, 157–170. DOI: 10.1016/j.biortech.2019.01.002.

Lin, L. and Xu, J. (2020) Fungal pigments and their roles associated with human health. *Journal of Fungi* 6(4), 280. DOI: 10.3390/jof6040280.

Lippmann, M. (ed.) (2000) Environmental toxicants: Human exposures and their health effects. *Occupational and Environmental Medicine* 57, 502D.

Madadi, E., Mazloum-Ravasan, S., Yu, J.S., Ha, J.W., Hamishehkar, H. *et al.* (2020) Therapeutic application of betalains: A review. *Plants* 9(9), 1219. DOI: 10.3390/plants9091219.

Magano, J., Au-Yong-Oliveira, M., Ferreira, B. and Leite, Â. (2022) A cross-sectional study on ethical buyer behavior towards cruelty-free cosmetics: What consequences for female leadership practices? *Sustainability* 14(13), 7786. DOI: 10.3390/su14137786.

Malabadi, R.B., Kolkar, K.P. and Chalannavar, R.K. (2022) Plant natural pigment colorants-health benefits: Toxicity of synthetic or artificial food colorants. *International Journal of Innovative Research and Scientific Studies* 4, 3418–3429.

Manguerra-Mahusay, S.C. and Liu, G. (2024) Consumer perceptions on food production companies' brand strategy and customer experience alignment. *Social Science and Humanities Journal* 8(7), 4371–4399. DOI: 10.18535/sshj.v8i07.1209.

Mapelli-Brahm, P., Gómez-Villegas, P., Gonda, M.L., León-Vaz, A., León, R. *et al.* (2023) Microalgae, seaweeds and aquatic bacteria, archaea, and yeasts: Sources of carotenoids with potential antioxidant and anti-inflammatory health-promoting actions in the sustainability era. *Marine Drugs* 21(6), 340. DOI: 10.3390/md21060340.

Metcalfe, C.D., Bayen, S., Desrosiers, M., Muñoz, G., Sauvé, S. *et al.* (2022) An introduction to the sources, fate, occurrence and effects of endocrine disrupting chemicals released into the environment. *Environmental Research* 207, 112658. DOI: 10.1016/j.envres.2021.112658.

Mohd Zaid, N.A., Sekar, M., Bonam, S.R., Gan, S.H., Lum, P.T. *et al.* (2023) Promising natural products in new drug design, development, and therapy for skin disorders: An overview of scientific evidence and understanding their mechanism of action. *Drug Design, Development and Therapy* 16, 23–66. DOI: 10.2147/DDDT.S326332.

Mota, I.G.C., Neves, R.A.M.D., Nascimento, S.S.D.C., Maciel, B.L.L., Morais, A.H.D.A. *et al.* (2023) Artificial dyes: Health risks and the need for revision of international regulations. *Food Reviews International* 39(3), 1578–1593. DOI: 10.1080/87559129.2021.1934694.

Nabi, B.G., Mukhtar, K., Ahmed, W., Manzoor, M.F., Ranjha, M.M.A.N. *et al.* (2023) Natural pigments: Anthocyanins, carotenoids, chlorophylls, and betalains as colorants in food products. *Food Bioscience* 52, 102403. DOI: 10.1016/j.fbio.2023.102403.

Nandhini, R., Sivaprakash, B., Rajamohan, N. and Vo, D.V.N. (2023) Carbon-free hydrogen and bioenergy production through integrated carbon capture and storage technology for achieving sustainable and circular economy – a review. *Fuel* 342, 126984. DOI: 10.1016/j.fuel.2022.126984.

Narsing Rao, M.P., Xiao, M. and Li, W.J. (2017) Fungal and bacterial pigments: Secondary metabolites with wide applications. *Frontiers in Microbiology* 8, 1113. DOI: 10.3389/fmicb.2017.01113.

Nethravathy, M.U., Mehar, J.G., Mudliar, S.N. and Shekh, A.Y. (2019) Recent advances in microalgal bioactives for food, feed, and healthcare products: Commercial potential, market space, and sustainability. *Comprehensive Reviews in Food Science and Food Safety* 18(6), 1882–1897. DOI: 10.1111/1541-4337.12500.

Nguyen, T.H., Wang, S.L. and Nguyen, V.B. (2022) Recent advances in eco-friendly and scaling-up bioproduction of prodigiosin and its potential applications in agriculture. *Agronomy* 12(12), 3099. DOI: 10.3390/agronomy12123099.

Nhani, G.B.B., Di Filippo, L.D., de Paula, G.A., Mantovanelli, V.R., da Fonseca, P.P. *et al.* (2024) High-tech sustainable beauty: Exploring nanotechnology for the development of cosmetics using plant and animal by-products. *Cosmetics* 11(4), 112. DOI: 10.3390/cosmetics11040112.

Niinimäki, K., Peters, G., Dahlbo, H., Perry, P., Rissanen, T. *et al.* (2020) The environmental price of fast fashion. *Nature Reviews Earth & Environment* 1(4), 189–200. DOI: 10.1038/s43017-020-0039-9.

Novais, C., Molina, A.K., Abreu, R.M.V., Santo-Buelga, C., Ferreira, I.C.F.R. *et al.* (2022) Natural food colorants and preservatives: A review, a demand, and a challenge. *Journal of Agricultural and Food Chemistry* 70(9), 2789–2805. DOI: 10.1021/acs.jafc.1c07533.

Olas, B., Białecki, J., Urbańska, K. and Bryś, M. (2021) The effects of natural and synthetic blue dyes on human health: A review of current knowledge and therapeutic perspectives. *Advances in Nutrition* 12(6), 2301–2311. DOI: 10.1093/advances/nmab081.

Orlandi, V.T., Martegani, E., Giaroni, C., Baj, A. and Bolognese, F. (2022) Bacterial pigments: A colorful palette reservoir for biotechnological applications. *Biotechnology and Applied Biochemistry* 69(3), 981–1001. DOI: 10.1002/bab.2170.

Patel, A.K., Albarico, F.P.J.B., Perumal, P.K., Vadrale, A.P., Nian, C.T. *et al.* (2022) Algae as an emerging source of bioactive pigments. *Bioresource Technology* 351, 126910. DOI: 10.1016/j.biortech.2022.126910.

Pathak, V.M., Verma, V.K., Rawat, B.S., Kaur, B., Babu, N. *et al.* (2022) Current status of pesticide effects on environment, human health and its eco-friendly management as bioremediation: A comprehensive review. *Frontiers in Microbiology* 13, 962619. DOI: 10.3389/fmicb.2022.962619.

Patra, A.K. and Pariti, S.R.K. (2022) Restricted substances for textiles. *Textile Progress* 54(1), 1–101. DOI: 10.1080/00405167.2022.2101302.

Phukan, B.C., Roy, R., Paul, R., Mazumder, M.K., Nath, J. *et al.* (2023) Traversing through the cell signaling pathways of neuroprotection by betanin: Therapeutic relevance to alzheimer's disease and parkinson's disease. *Metabolic Brain Disease* 38(3), 805–817. DOI: 10.1007/s11011-023-01177-8.

Ramesh, C., Prasastha, V.R., Venkatachalam, M. and Dufossé, L. (2022) Natural substrates and culture conditions to produce pigments from potential microbes in submerged fermentation. *Fermentation* 8(9), 460. DOI: 10.3390/fermentation8090460.

Renita, A.A., Gajaria, T.K., Sathish, S., Kumar, J.A., Lakshmi, D.S. *et al.* (2023) Progress and prospective of the industrial development and applications of ECO-friendly colorants: An insight into environmental impact and sustainability issues. *Foods* 12(7), 1521.

Rodella, U., Honisch, C., Gatto, C., Ruzza, P. and D'Amato Tóthová, J. (2023) Antioxidant nutraceutical strategies in the prevention of oxidative stress related eye diseases. *Nutrients* 15(10), 2283. DOI: 10.3390/nu15102283.

Ruiz-Domínguez, M.C., Espinosa, C., Paredes, A., Palma, J., Jaime, C. *et al.* (2019) Determining the potential of *Haematococcus pluvialis* oleoresin as a rich source of antioxidants. *Molecules* 24(22), 4073. DOI: 10.3390/molecules24224073.

Saikia, B., Gogoi, S., Savani, A.K. and Bhattacharyya, A. (2022) Metabolites and peptides of endophytic origin in plant growth promotion and defense reactions in solanaceous crop tomato. In: *New and Future Developments in Microbial Biotechnology and Bioengineering*. Elsevier, New York, pp. 89–110.

Sajjad, W., Din, G., Rafiq, M., Iqbal, A., Khan, S. *et al.* (2020) Pigment production by cold-adapted bacteria and fungi: Colorful tale of cryosphere with wide range applications. *Extremophiles* 24(4), 447–473. DOI: 10.1007/s00792-020-01180-2.

Sandybayeva, S.K., Kossalbayev, B.D., Zayadan, B.K., Sadvakasova, A.K., Bolatkhan, K. *et al.* (2022) Prospects of cyanobacterial pigment production: Biotechnological potential and optimization strategies. *Biochemical Engineering Journal* 187, 108640. DOI: 10.1016/j.bej.2022.108640.

Saraiva, A., Carrascosa, C., Raheem, D., Ramos, F. and Raposo, A. (2020) Natural sweeteners: The relevance of food naturalness for consumers, food security aspects, sustainability and health impacts. *International Journal of Environmental Research and Public Health* 17(17), 6285. DOI: 10.3390/ijerph17176285.

Saravanan, T.S., Gopikrishnan, V., Manikkam, R., Kaari, M., Annamalai, K.K. *et al.* (2024) Translational value of natural pigments with reference to healthcare and cosmetic applications. In: *Translational Research in Biomedical Sciences: Recent Progress and Future Prospects*. Springer Nature Singapore, Singapore, pp. 235–250.

Selvaraj, V., Swarna Karthika, T., Mansiya, C. and Alagar, M. (2021) An over review on recently developed techniques, mechanisms and intermediate involved in the advanced azo dye degradation for industrial applications. *Journal of Molecular Structure* 1224, 129195. DOI: 10.1016/j.molstruc.2020.129195.

Sen, T., Barrow, C.J. and Deshmukh, S.K. (2019) Microbial pigments in the food industry-challenges and the way forward. *Frontiers in Nutrition* 6, 7. DOI: 10.3389/fnut.2019.00007.

Sharifi-Rad, M., Pezzani, R., Redaelli, M., Zorzan, M., Imran, M. *et al.* (2020) Preclinical pharmacological activities of epigallocatechin-3-gallate in signaling pathways: An update on cancer. *Molecules* 25(3), 467. DOI: 10.3390/molecules25030467.

Sharma, S., Katoch, V., Kumar, S. and Chatterjee, S. (2021) Functional relationship of vegetable colors and bioactive compounds: Implications in human health. *Journal of Nutritional Biochemistry* 92, 108615. DOI: 10.1016/j.jnutbio.2021.108615.

Singh, A. (2021) A review of wastewater irrigation: Environmental implications. *Resources, Conservation and Recycling* 168, 105454. DOI: 10.1016/j.resconrec.2021.105454.

Solehin, S.N., Kamarudin, K.R., Badrulhisham, N.S. and Rehan, A.M. (2022) The potential use of papaya and banana peels as substrate to enhance the pigment production of Gram-positive bacterial strain isolated from Holothuria (*Lessonothuria*) pardalis. *Malaysian Journal of Microbiology* 18(2). DOI: 10.21161/mjm.211253.

Tang, Q., Li, Z., Chen, N., Luo, X. and Zhao, Q. (2025) Natural pigments derived from plants and microorganisms: Classification, biosynthesis, and applications. *Plant Biotechnology Journal* 23(2), 592–614. DOI: 10.1111/pbi.14522.

Teleanu, R.I., Chircov, C., Grumezescu, A.M., Volceanov, A. and Teleanu, D.M. (2019) Antioxidant therapies for neuroprotection – a review. *Journal of Clinical Medicine* 8(10), 1659. DOI: 10.3390/jcm8101659.

Teli, M.D., Pandit, P. and Singha, K. (2020) Economical and sustainable price sensitive fashion and apparels marketplace. In: *Recycling from Waste in Fashion and Textiles: A Sustainable and Circular Economic Approach*. Scrivener Publishing, Beverly, MA, pp. 277–297.

Thakur, N., Kaur, S., Tomar, P., Thakur, S. and Yadav, A.N. (2020) Microbial biopesticides: Current status and advancement for sustainable agriculture and environment. In: *New and Future Developments in Microbial Biotechnology and Bioengineering*. Elsevier, New York, pp. 243–282.

Timofeeva, A.M., Galyamova, M.R. and Sedykh, S.E. (2024) How do plant growth-promoting bacteria use plant hormones to regulate stress reactions? *Plants* 13(17), 2371. DOI: 10.3390/plants13172371.

Tran-Ly, A.N., Reyes, C., Schwarze, F.W.M.R. and Ribera, J. (2020) Microbial production of melanin and its various applications. *World Journal of Microbiology & Biotechnology* 36(11), 170. DOI: 10.1007/s11274-020-02941-z.

Uddin, M.A., Begum, M.S., Ashraf, M., Azad, A.K., Adhikary, A.C. *et al.* (2023) Water and chemical consumption in the textile processing industry of Bangladesh. *PLOS Sustainability and Transformation* 2(7), e0000072. DOI: 10.1371/journal.pstr.0000072.

Unsal, V., Dalkıran, T., Çiçek, M. and Kölükçü, E. (2020) The role of natural antioxidants against reactive oxygen species produced by cadmium toxicity: A review. *Advanced Pharmaceutical Bulletin* 10(2), 184–202. DOI: 10.34172/apb.2020.023.

Varma, A. and Ray, S. (2023) Revolutionizing the Indian market through eco-friendly sustainable products: The rise of vegan beauty inspired by nature. *International Journal of Research in Marketing Management and Sales* 5(2), 18–26. DOI: 10.33545/26633329.2023.v5.i2a.128.

Velu, G., Palanichamy, V. and Rajan, A.P. (2018) Phytochemical and pharmacological importance of plant secondary metabolites in modern medicine. In: *Bioorganic Phase in Natural Food: An Overview*. Springer, Cham, Switzerland, pp. 135–156.

Venil, C.K., Dufossé, L. and Renuka Devi, P. (2020) Bacterial pigments: Sustainable compounds with market potential for pharma and food industry. *Frontiers in Sustainable Food Systems* 4, 100. DOI: 10.3389/fsufs.2020.00100.

Vlaiculescu, A. and Varrone, C. (2022) Sustainable and eco-friendly alternatives to reduce the use of pesticides. In: *Pesticides in the Natural Environment*. Elsevier, New York, pp. 329–364.

Wang, Y.-H., Zhang, R.-R., Yin, Y., Tan, G.-F., Wang, G.-L. *et al.* (2023) Advances in engineering the production of the natural red pigment lycopene: A systematic review from a biotechnology perspective. *Journal of Advanced Research* 46, 31–47. DOI: 10.1016/j.jare.2022.06.010.

Wyckhuys, K.A.G., Bentley, J.W., Lie, R., Nghiem, L.T.P. and Fredrix, M. (2018) Maximizing farm-level uptake and diffusion of biological control innovations in today's digital era. *BioControl* 63(1), 133–148. DOI: 10.1007/s10526-017-9820-1.

18 Challenges and Future Prospects in the Field of Microbial Pigments

Avani Joshi[1], Dishank Purandare[1], Adityaraj Bhope[1], Shradha Chaudhari[2], Arpana Patil[3] and Satish Polshettiwar[1*]

[1]School of Health Sciences and Technology, Dr Vishwanath Karad MIT World Peace University, Pune, Maharashtra, India; [2]Department of Biosciences and Technology, School of Sciences and Environmental Studies, Faculty of Sciences and Health Sciences, Dr Vishwanath Karad MIT World Peace University, Pune, Maharashtra, India; [3]STES's Smt. Kashibai Navale College of Pharmacy, Pune, Maharashtra, India

18.1 Introduction

Microbial pigments are naturally occurring metabolites synthesized by microorganisms, including bacteria, fungi, actinomycetes, and algae (Anshi *et al.*, 2024). They have garnered significant attention due to their diverse applications in industrial, pharmaceutical, and environmental sectors. With growing environmental and health concerns associated with synthetic dyes (Lellis *et al.*, 2019), microbial pigments are emerging as an eco-friendly and sustainable alternative (Rajendran *et al.*, 2023). Historically, natural dyes derived from plants, minerals, and insects were widely used in ancient civilizations for textiles, art, and rituals (Alegbe and Uthman, 2024). The Industrial Revolution of the 19th century marked a shift to synthetic dyes, which became the primary source of colorants due to their low cost, vibrant colors, and scalability. However, the production and disposal of synthetic dyes have led to widespread ecological damage, including water pollution, nonbiodegradable waste, and toxic effects on human health and aquatic ecosystems (Tkaczyk *et al.*, 2020; Al-Tohamy *et al.*, 2022). Consequently, the demand for safe, natural, and sustainable alternatives has paved the way for microbial pigments to rise in prominence.

Microbial pigments offer unique advantages over both synthetic dyes and traditional natural colorants (Mummaleti *et al.*, 2024). They are biodegradable, nontoxic, and renewable, addressing the significant drawbacks of synthetic counterparts. Additionally, microbial pigments often exhibit bioactive properties (Ramesh *et al.*, 2019) such as antimicrobial, anticancer, and antioxidant effects, which make them valuable in fields beyond coloration, including pharmaceuticals, nutraceuticals, and cosmetics (Devi *et al.*, 2024). Unlike plant- or animal-derived pigments, which face challenges such as seasonal variability, limited yield, and agricultural dependency, microbial pigments can be produced through fermentation processes in controlled environments (Numan *et al.*,

*Corresponding author: satish.polshettiwar@mitwpu.edu.in

DOI: 10.1079/ 9781836990505.0018

2018). This allows for consistent quality and scalability while utilizing inexpensive substrates, including agro-industrial waste, which further enhances their sustainability and economic feasibility. The spectrum of microbial pigments is diverse, ranging from carotenoids (yellow to orange-red) and melanin (brown to black) to prodigiosin (red) and violacein (violet) (Orlandi *et al.*, 2022).

These pigments not only cater to the aesthetic needs of various industries but also provide functional benefits. For instance, carotenoids derived from *Rhodotorula* and *Blakeslea* species are potent antioxidants with significant applications in cosmetics and healthcare (Tang *et al.*, 2019; Ochoa-Viñals *et al.*, 2024) while prodigiosin produced by *Serratia marcescens* has demonstrated promising anticancer activity (Anwar *et al.*, 2022; Nguyen *et al.*, 2022). Similarly, melanin pigments, known for their UV-absorbing and photoprotective properties, are increasingly being explored for use in sunscreens and other personal care products (Brenner and Hearing, 2008; Solano, 2020; Zamudio Díaz *et al.*, 2024). The bioactivities of these pigments expand their market potential, making them highly sought after in sectors that prioritize both functionality and sustainability.

The market for natural pigments, particularly microbial pigments, is experiencing significant growth due to increased consumer awareness of the environmental impact of synthetic dyes and the preference for safer, eco-friendly products (Nonglait and Gokhale, 2024). Industries such as food, cosmetics, textiles, and pharmaceuticals are actively incorporating microbial pigments to meet the demand for natural alternatives. For example, in the food industry, pigments like astaxanthin and canthaxanthin, shown in Fig. 18.1, are used as colorants and dietary supplements (Barreto *et al.*, 2023), while the textile industry explores microbial pigments as sustainable dyes to reduce chemical pollution during dyeing processes. The cosmetics industry has also embraced microbial pigments for their vibrant colors and additional benefits such as antiaging and UV protection properties (Barreto *et al.*, 2023).

However, microbial pigments face several challenges that need to be addressed to realize their full commercial and industrial potential. One of the primary hurdles is the cost of production which, although decreasing with advancements in fermentation technology, remains higher than that of synthetic dyes. Fermentation processes require optimization to enhance yields and reduce costs, particularly for large-scale production (Schmidt, 2005). Additionally, the extraction, purification, and stabilization of microbial pigments can be technically challenging, especially for applications requiring stringent quality standards, such as food and pharmaceuticals. Regulatory hurdles also play a significant role, as microbial pigments must meet safety and efficacy standards set by organizations like the FDA and EFSA.

Advancements in biotechnology, particularly in metabolic engineering and synthetic biology, are paving the way for overcoming these challenges. Genetic modifications of

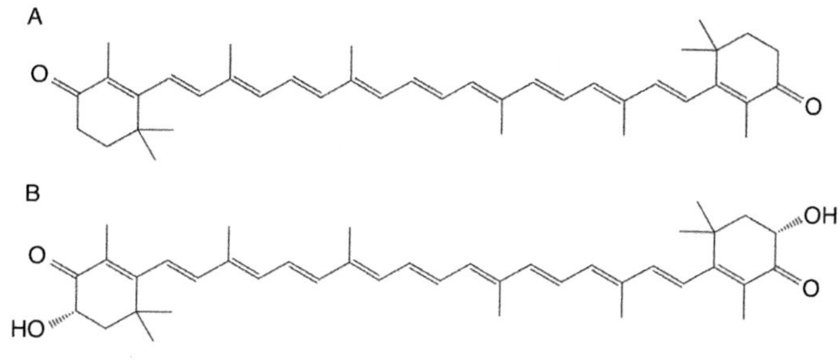

Fig. 18.1. (A) Structure of canthaxanthin. (B) Structure of astaxanthin.

pigment-producing microorganisms can enhance production yields, improve stability, and tailor pigment properties for specific industrial applications. For instance, CRISPR-Cas9 technology has been employed to engineer pathways in microorganisms like *Escherichia coli* and *Saccharomyces cerevisiae* to produce high-value pigments such as carotenoids and flavins (Zhao *et al.*, 2020). Furthermore, the integration of artificial intelligence and machine learning into bioprocess optimization enables precise control over fermentation parameters, resulting in cost-effective production and reduced environmental impact (Saha *et al.*, 2024; Zhou *et al.*, 2024).

The future prospects of microbial pigments are promising, with ongoing research focusing on expanding their applications, improving production processes, and addressing market challenges. Innovations in renewable feedstocks, such as agricultural and food waste, as fermentation substrates are expected to reduce costs further and align with circular economy principles. Collaborative efforts between academia, industry, and regulatory bodies are essential to streamline the commercialization of microbial pigments and ensure their safety and acceptance in global markets. Additionally, integrating microbial pigments into a broader framework of sustainable practices, such as green chemistry and eco-friendly manufacturing, will be crucial in establishing them as a viable alternative to synthetic dyes. Thus, microbial pigments represent a transformative opportunity to address the dual challenges of environmental sustainability and industrial functionality. Their unique properties, including vibrant colors, bioactivity, and renewable production, position them as a cornerstone of green biotechnology. While challenges such as cost, scalability, and regulatory compliance persist, advances in biotechnological innovation and increasing consumer demand for sustainable products are driving the rapid growth and adoption of microbial pigments across industries.

As we delve deeper into the challenges and future prospects of this field, it becomes evident that microbial pigments hold immense potential to revolutionize the way we approach colorants, contributing to a more sustainable and eco-conscious world.

18.2 Applications of Microbial Pigments

Microbial pigments have garnered increasing attention across diverse industries due to their multifaceted properties, eco-friendly production, and functional benefits that extend beyond coloration (Ramesh *et al.*, 2019). Their applications span sectors such as food, cosmetics, pharmaceuticals, and textiles, each leveraging the unique characteristics of these natural colorants, as shown in Fig. 18.2.

In the food industry, microbial pigments serve as natural colorants that align with clean-label trends and the rising consumer preference for safe, natural additives. Pigments like carotenoids, such as astaxanthin and β-carotene, derived from microorganisms like *Haematococcus pluvialis* and *Rhodotorula glutinis* (Ambati *et al.*, 2014; Mussagy *et al.*, 2021; Nair *et al.*, 2023; Zhou *et al.*, 2024), are widely used for their vibrant hues and antioxidant properties. These pigments not only enhance the visual appeal of food products but also confer health benefits, such as boosting immune function and reducing oxidative stress. Similarly, microbial-derived pigments like curcumin analogues (Yixuan *et al.*, 2021) and phycocyanin from cyanobacteria (Roy *et al.*, 2024) are employed in beverages, confectionery, and dairy products, offering stability under various processing conditions while adhering to stringent safety regulations. Recent innovations in fermentation technology have also enabled the cost-effective production of microbial pigments, paving the way for their wider adoption in the food sector.

In the cosmetics industry, microbial pigments are prized for their dual functionality, offering vibrant colors alongside therapeutic properties (Kiki, 2023). Carotenoids, violacein, and melanin are among the most extensively studied pigments in this domain. Carotenoids like lutein and zeaxanthin, obtained from microalgae and yeast, are incorporated into antiaging formulations for their potent antioxidant properties, which help neutralize free radicals and prevent premature skin aging (Kumar *et al.*, 2024). Melanin, a natural UV absorber produced by fungi and actinomycetes, is increasingly utilized in sunscreens and photoprotective cosmetics, offering broad-spectrum

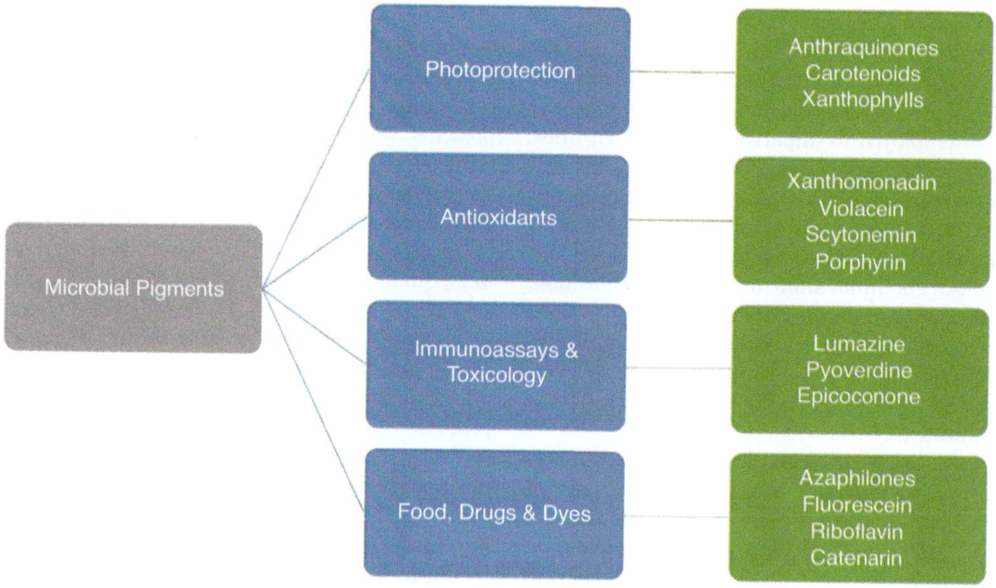

Fig. 18.2. Broad applications of microbial pigments. Source: Anshi *et al.* (2024).

protection against harmful ultraviolet radiation. Violacein, derived from *Chromobacterium violaceum*, has gained attention for its antimicrobial and antiinflammatory properties, making it a valuable ingredient in skincare products targeting acne and other dermatological conditions (Durán *et al.*, 2012). Additionally, the demand for sustainable and ethical products in the beauty industry has driven interest in microbial pigments as replacements for synthetic dyes and animal-derived colorants, providing an eco-friendly and cruelty-free alternative without compromising on quality or performance.

Pharmaceutical applications of microbial pigments highlight their bioactivity and therapeutic potential. Many pigments possess inherent antimicrobial, anticancer, antiinflammatory, and immunomodulatory properties, making them valuable candidates for drug development. For instance, prodigiosin, a red pigment produced by *Serratia marcescens*, exhibits strong anticancer activity against various cancer cell lines, including those derived from leukemia, lung, and colon cancers (Lu *et al.*, 2024). This pigment also shows immunosuppressive activity, with potential applications in organ transplantation and autoimmune disease management. Similarly, violacein demonstrates broad-spectrum antimicrobial effects against pathogenic bacteria and fungi, as well as antileishmanial activity, making it a promising candidate for infectious disease therapies. Carotenoids, beyond their antioxidant role, have shown potential in reducing the risk of chronic diseases such as cardiovascular conditions and certain cancers. Recent research has also explored microbial pigments like flexirubin from *Flavobacterium* species for their role in combating drug-resistant tuberculosis, providing new avenues for therapeutic intervention (Kim *et al.*, 2019). The integration of microbial pigments into the pharmaceutical pipeline is further supported by advancements in metabolic engineering, which enable the production of these bioactive compounds in scalable and cost-efficient systems.

The textile industry, one of the largest contributors to synthetic dye pollution, is also exploring the potential of microbial pigments as a sustainable alternative for eco-friendly coloring. Synthetic dyes have been associated with water pollution, toxic effluent discharge, and long-term environmental degradation. In contrast, microbial pigments offer a biodegradable, nontoxic, and renewable solution. For example, pigments like indigoidine, a blue pigment produced by certain *Streptomyces* species, have been successfully used to dye fabrics without the need for harsh chemical processes (Starr *et al.*, 1966). Similarly, prodigiosin and melanin have been employed to impart red and brown hues, respectively, to textiles, offering a vibrant and durable finish (Lu *et al.*, 2024). These pigments not only reduce the environmental footprint of the textile industry but also provide opportunities for innovation in sustainable fashion. Moreover, the use of microbial pigments aligns with growing consumer demand for ethical and eco-conscious products, creating new market opportunities for brands prioritizing sustainability.

Recent advancements in biotechnology have expanded the potential applications of microbial pigments in emerging fields such as bioplastics and electronic devices. For instance, melanin's electrical conductivity has sparked interest in its use for bioelectronics and sensors, while its biocompatibility makes it suitable for biomedical implants and drug delivery systems. The versatility of microbial pigments is further enhanced by their ability to be produced using low-cost and renewable substrates, including agricultural waste, which aligns with the principles of a circular economy. This not only reduces production costs but also minimizes environmental impact, making microbial pigments a key component of sustainable industrial practices. Thus, microbial pigments are redefining the landscape of natural colorants, offering a sustainable, multifunctional, and eco-friendly alternative to synthetic dyes. As research and technology continue to advance, the scope of microbial pigments is likely to expand further, unlocking new possibilities in industrial applications and contributing to a more sustainable future.

18.3 Challenges in Production

The production of microbial pigments, despite their immense potential as eco-friendly and sustainable alternatives to synthetic dyes, faces many challenges that hinder their large-scale commercialization and widespread adoption across industries. At the forefront of these challenges is the difficulty in scaling up production processes. Microbial pigments are typically produced efficiently in controlled laboratory environments using optimized conditions. However, replicating these results at an industrial scale presents significant hurdles, often leading to reduced yields, increased costs, and operational inefficiencies. Scaling up requires the transition from small-scale laboratory fermenters to large bioreactors, which introduces variables such as inadequate mixing, oxygen transfer limitations, and nutrient availability. These issues can result in uneven microbial growth, lower pigment yields, and increased production times, thereby affecting the overall economic viability of the process.

Variability in pigment yield is another major challenge, as production is highly sensitive to several factors. The type of microorganism used, the composition of the growth medium, fermentation parameters such as pH, temperature, and aeration, and environmental conditions all influence the quantity and quality of pigment produced. Even slight deviations in these factors can result in inconsistent pigment production, making it difficult to achieve the uniformity required for industrial applications. For instance, while specific strains of *Monascus* are well known for producing red pigments, their yield can vary significantly depending on the carbon and nitrogen sources provided in the medium (Gong *et al.*, 2023). This variability poses challenges for industries such as food and pharmaceuticals, where strict consistency in color and quality is paramount.

Cost inefficiencies further complicate the large-scale production of microbial pigments. The fermentation process, while eco-friendly, often involves high operational costs associated with raw materials, energy inputs, and downstream processing. Raw materials such as glucose, peptone, or other substrates required for microbial growth can be expensive, particularly when high-purity inputs are used to ensure optimal pigment production (Kieliszek and Misiewicz, 2014). Moreover, downstream processing, including extraction, purification, and stabilization of pigments, is labor-intensive and resource-demanding. Traditional methods such as solvent extraction or filtration are not only costly but can also result in pigment losses, thereby reducing overall efficiency. Additionally, the high energy demand for maintaining fermentation parameters, such as temperature and aeration, further drives up costs, making microbial pigments less competitive than synthetic dyes, which are produced at a fraction of the cost using chemical synthesis (Zhou *et al.*, 2018).

Stability issues of microbial pigments under varying environmental and processing conditions pose yet another significant challenge. Many microbial pigments are sensitive to factors such as light, heat, pH, and oxygen, which can lead to color fading, degradation, or loss of functionality over time. For example, carotenoids, widely used for their vibrant colors and antioxidant properties, are prone to oxidative degradation and photo-bleaching when exposed to light or air. Similarly, pigments like violacein and prodigiosin, while offering exceptional bioactivity, may lose their structural integrity under extreme pH conditions or elevated temperatures, limiting their application in industries such as food processing or cosmetics, where stability is critical (Darshan and Manonmani, 2015). These stability concerns necessitate the development of advanced stabilization techniques, such as encapsulation or chemical modification, to ensure the pigments maintain their desired properties throughout their shelf-life and usage.

Maintaining consistent pigment quality across production batches is a pressing concern, particularly for industries with stringent quality and safety regulations. In the pharmaceutical and food industries, for instance, microbial pigments must meet precise standards regarding purity, safety, and bioactivity to gain regulatory approval. However, contamination risks during fermentation, such as the introduction of undesirable microorganisms or toxins, can compromise the quality of the pigments. Additionally, the production process may result in undesired byproducts, which require thorough purification to meet industry standards. Ensuring reproducibility and quality control on an industrial scale is a complex task that often requires significant investment in monitoring and analytical technologies. One of the fundamental challenges lies in the selection of suitable microorganisms for pigment production. While many microorganisms naturally produce pigments, not all strains are suitable for industrial applications. Some may produce pigments at low yields, while others may require specific and costly substrates for optimal growth. Additionally, certain pigment-producing strains may be pathogenic or generate harmful byproducts, necessitating the identification or engineering of safe, high-yielding strains. Genetic engineering has emerged as a promising tool to address these limitations, allowing researchers to modify microbial strains for enhanced pigment production, improved yield consistency, and reduced byproduct formation. However, the use of GMOs in production raises regulatory and public acceptance challenges, particularly in the food and pharmaceutical sectors (Bawa and Anilakumar, 2013).

Environmental concerns related to microbial pigment production also need to be addressed. While microbial pigments are often considered sustainable, their production processes can still have environmental impacts if not carefully managed. The use of energy-intensive fermentation processes, the disposal of spent fermentation broth, and the use of organic solvents in pigment extraction can contribute to environmental pollution. Therefore, developing more sustainable production methods, such as the use of renewable feedstocks, energy-efficient bioreactors, and green extraction techniques, is essential to minimize the ecological footprint of microbial pigment production. Efforts to overcome these challenges are under way, driven by advancements in biotechnology and bioprocess engineering. The optimization of fermentation processes, for instance, has shown promise in enhancing pigment yields and reducing production costs. Techniques such as fed-batch fermentation, where nutrients are added incrementally to maintain optimal growth conditions, have been successfully employed to improve the efficiency of microbial pigment production. Additionally, the use of low-cost and renewable substrates, such as agricultural and food waste, as carbon and nitrogen sources, offers a sustainable and economically viable alternative to traditional inputs.

Research into metabolic engineering is also yielding promising results, with genetically modified strains demonstrating significantly enhanced pigment production. For example, the introduction of targeted mutations or the overexpression of specific biosynthetic pathways can increase pigment yields several fold, making production more cost-effective and scalable.

Advances in downstream processing are also contributing to the resolution of cost and quality challenges. The development of novel extraction and purification methods, such as supercritical fluid extraction or membrane-based separation, offers efficient and environmentally friendly alternatives to conventional techniques. These methods not only improve the recovery of pigments but also reduce solvent use and waste generation, aligning with the principles of green chemistry. Encapsulation technologies, such as microencapsulation or nanoencapsulation, are being explored to enhance the stability and shelf-life of microbial pigments. These techniques protect pigments from environmental stressors, such as light and heat, while preserving their functional properties. To ensure consistent quality, industries are increasingly adopting advanced analytical and quality control methods. Techniques such as HPLC, mass spectrometry, and spectrophotometry are employed to monitor pigment purity and concentration, enabling the identification of contaminants and byproducts. The integration of real-time monitoring systems in fermentation processes further ensures the reproducibility and reliability of pigment production.

Regulatory and public acceptance challenges related to microbial pigments, particularly those produced using genetic engineering, require proactive strategies to build trust and compliance. Transparent communication regarding the safety, environmental benefits, and ethical considerations of microbial pigments is essential to gain consumer acceptance and regulatory approval. Collaborative efforts between academia, industry, and regulatory bodies are crucial to establish standardized guidelines for the production and application of microbial pigments, ensuring their safe and effective use across industries.

Thus, while the production of microbial pigments faces significant challenges, including scalability, yield variability, cost inefficiencies, stability concerns, and quality control, ongoing advancements in biotechnology and bioprocess engineering are paving the way for their widespread adoption. By addressing these challenges through innovative solutions, microbial pigments can fulfill their potential as sustainable and eco-friendly alternatives to synthetic dyes, contributing to a greener and more sustainable future.

18.4 Regulatory and Safety Considerations

The commercialization and widespread adoption of microbial pigments require strict adherence to regulatory frameworks and safety standards, given their diverse applications across industries. These pigments, while offering sustainable and bioactive alternatives to synthetic dyes, must undergo comprehensive safety and toxicity evaluations to ensure their suitability for industrial and consumer use. Regulatory and safety considerations encompass an intricate blend of scientific scrutiny, legal compliance, and public trust, all of which are crucial for the successful integration of microbial pigments into various market segments. This section delves into the regulatory frameworks that govern microbial pigments, the safety assessments necessary for their approval, and the importance of fostering consumer trust to support their commercial viability.

18.4.1 Regulatory frameworks governing microbial pigments

The approval and usage of microbial pigments are governed by stringent regulatory frameworks that vary across industries and regions. In the food sector, where microbial pigments are often used as natural colorants, regulatory bodies such as the US FDA, the EFSA, and the Codex Alimentarius Commission set the standards for their evaluation and approval. These agencies require detailed dossiers on the pigment's composition, source, production process, and safety profile before granting approval. For example, microbial pigments intended for food use must demonstrate safety through rigorous toxicological studies, including assessments of acute toxicity, genotoxicity, carcinogenicity, and allergenicity (Rana *et al.*, 2021). The FDA classifies food colorants under categories such as "certified" and "exempt from certification," with microbial pigments typically falling into the latter group due to their natural origin. However, this classification does not exempt them from the requirement to meet safety benchmarks.

In the pharmaceutical industry, microbial pigments with therapeutic properties are subject to regulatory scrutiny similar to that of new drug entities. Agencies such as the FDA, EFSA, and the EMA mandate preclinical and clinical trials to evaluate the efficacy, safety, and pharmacokinetics of these pigments. For example, a pigment like prodigiosin, which exhibits anticancer properties, must undergo a thorough investigation to assess its therapeutic index, bioavailability, and potential side-effects. Its structure is shown in Fig. 18.3. The regulatory approval process for pharmaceuticals is typically more complex and time-intensive, reflecting the critical need to ensure patient safety and treatment efficacy.

In cosmetics, microbial pigments are regulated by bodies such as the FDA and the EU's Regulation on Cosmetic Products (EC No. 1223/2009) (Dhull *et al.*, 2015). These regulations emphasize safety assessments of the pigments, focusing on their stability, dermal absorption, and potential to cause irritation or sensitization. Additionally, cosmetic formulations containing microbial pigments must comply with labeling requirements, including clear declarations of the pigment's origin, concentration, and intended use. Regulatory bodies also require manufacturers to provide data on the environmental impact of the pigments, especially if they are produced using GMOs. In the textile industry, microbial pigments are subjected to regulations addressing environmental safety and sustainability. Certifications such as GOTS and OEKO-TEX assess the environmental impact of the pigments used in dyeing processes, ensuring they meet criteria for biodegradability, nontoxicity, and minimal water pollution. Regulatory frameworks in this domain often overlap with environmental regulations, reflecting the industry's focus on reducing its ecological footprint.

18.4.2 Safety and toxicity evaluations

The safety evaluation of microbial pigments involves a multidisciplinary approach encompassing toxicology, microbiology, and analytical chemistry. Toxicity assessments are integral to the regulatory approval process, as they determine the pigment's potential to cause harm to humans or the environment. Acute toxicity studies assess the short-term effects of exposure, while chronic toxicity studies evaluate long-term safety under repeated use. For pigments intended for ingestion, such as food colorants, these studies must address the potential for bioaccumulation and metabolic breakdown products that may pose health risks.

Genotoxicity studies are conducted to evaluate whether the pigment or its metabolites can damage genetic material, a critical concern for pigments with therapeutic or cosmetic applications. Carcinogenicity studies, typically carried out over extended periods, assess whether the pigment has the potential to induce cancer. For example, pigments like violacein and prodigiosin, which demonstrate bioactivity against cancer cells, must also be evaluated to ensure they do not promote carcinogenesis in healthy tissues. Allergenicity and immunogenicity are additional aspects of safety evaluations, particularly for

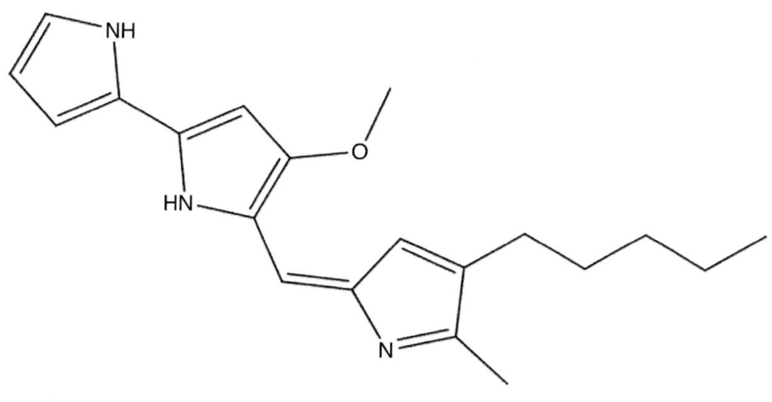

Fig. 18.3. Structure of prodigiosin.

pigments used in cosmetics and pharmaceuticals (Bialas *et al.*, 2023). These assessments focus on the potential for microbial pigments to elicit allergic reactions or immune responses. Dermal absorption studies are essential for pigments intended for topical applications, as they determine the extent to which the pigment penetrates the skin and enters the systemic circulation.

Another critical component of safety evaluations is the assessment of residuals and contaminants in the pigment preparation. Fermentation processes used to produce microbial pigments can sometimes result in residual solvents, endotoxins, or unintentional byproducts, which must be identified and quantified. Advanced analytical techniques such as HPLC, GC-MS, and inductively coupled plasma mass spectrometry are employed to ensure that pigments meet purity standards and are free from harmful contaminants.

18.4.3 Regulatory challenges with genetically modified organisms

The use of genetically modified microorganisms to enhance pigment yield or tailor pigment properties introduces additional regulatory challenges. While GMOs can significantly improve the efficiency and scalability of microbial pigment production, their use is often met with heightened scrutiny due to potential risks to human health and the environment (Buiatti *et al.*, 2013). Regulatory agencies require detailed documentation of the genetic modifications, including information on the introduced genes, their origin, and the potential for horizontal gene transfer. Furthermore, the production process must demonstrate containment measures to prevent the release of GMOs into the environment.

18.4.4 Building consumer trust and ensuring compliance

Beyond meeting regulatory requirements, fostering consumer trust is crucial for the successful market integration of microbial pigments. Consumers are increasingly aware of and concerned about the safety and environmental impact of the products they use, prompting companies to adopt transparent practices and clear labeling. Educating consumers about the benefits of microbial pigments, such as their natural origin, biodegradability, and reduced environmental footprint, can help build confidence in their use. Certification labels, such as "organic," "non-GMO," or "eco-friendly," can further reassure consumers about the product's safety and sustainability.

Compliance with regulatory and safety standards is not merely a legal obligation but also a competitive advantage in the marketplace. Companies that invest in rigorous safety testing, sustainable production methods, and transparent communication are more likely to succeed in gaining regulatory approval and consumer acceptance. Collaborative efforts between industry stakeholders, regulatory bodies, and academic researchers are essential to establish standardized testing protocols and guidelines for microbial pigments. This not only streamlines the approval process but also ensures a level playing field for manufacturers.

Regulatory and safety considerations play a pivotal role in the development, approval, and commercialization of microbial pigments. From food and pharmaceuticals to cosmetics and textiles, these pigments must meet stringent standards to ensure their safety, efficacy, and environmental sustainability. Comprehensive toxicity evaluations, robust regulatory frameworks, and proactive consumer engagement are essential components of this process. As microbial pigments continue to gain traction as sustainable alternatives to synthetic dyes, addressing regulatory challenges and fostering public trust will be critical to unlocking their full potential. By aligning innovation with safety and compliance, microbial pigments can contribute to a more sustainable and health-conscious future.

18.5 Advances in Technology

The field of microbial pigment production has seen significant technological advancements aimed at overcoming the challenges of low yields, high production costs, and stability issues. These innovations span metabolic engineering, recombinant DNA technology, bioprocess optimization, and novel extraction methods, which collectively enhance pigment yields and reduce the economic burden associated with large-scale production. By integrating these cutting-edge approaches, researchers aim to achieve a balance between sustainability and industrial feasibility.

18.5.1 Metabolic engineering for enhanced pigment production

Metabolic engineering involves manipulating microbial metabolic pathways to increase the production of desired secondary metabolites, including pigments. By identifying key biosynthetic genes and regulatory networks, researchers can reprogram microbes to channel more resources towards pigment synthesis. For example, the production of carotenoids, a class of microbial pigments with significant antioxidant properties, has been enhanced by overexpressing genes in the mevalonate pathway and introducing heterologous genes from other microorganisms. Recent studies have demonstrated the success of redirecting precursor fluxes by blocking competitive pathways. For instance, in *Monascus purpureus*, a well-known producer of red pigments, targeted gene knockouts have reduced the synthesis of unwanted byproducts, leading to higher pigment yields. Moreover, advancements in CRISPR-Cas9 technology allow precise editing of microbial genomes, enabling the fine-tuning of biosynthetic pathways to optimize pigment production while maintaining microbial viability (Cao *et al.*, 2023).

18.5.2 Recombinant DNA technology and synthetic biology

Recombinant DNA technology has emerged as a cornerstone for producing microbial pigments at an industrial scale. Through this approach, pigment biosynthetic genes from natural producers are cloned and expressed in more robust microbial hosts such as *Escherichia coli* or *Saccharomyces cerevisiae*. These hosts are preferred for their rapid growth, ease of genetic manipulation, and well-characterized metabolic pathways. Synthetic biology takes this concept further by designing and assembling artificial biosynthetic pathways tailored for specific pigments. For instance, violacein, a pigment with antimicrobial and anticancer properties, has been successfully produced in nonnative hosts by assembling its biosynthetic pathway using modular genetic circuits. This approach allows for the scalable and consistent production of high-value pigments, addressing the challenges posed by natural variability in pigment-producing strains (Khan *et al.*, 2016).

18.5.3 Bioprocess optimization

Optimizing bioprocesses is crucial to ensure the economic viability of microbial pigment production. This includes refining growth media, adjusting environmental conditions, and designing efficient bioreactors (Mandal and Majumdar, 2023).

18.5.3.1 Nutritional optimization

The choice of carbon and nitrogen sources has a profound impact on pigment yields. Low-cost substrates such as agricultural residues and industrial byproducts have been explored as alternatives to reduce production costs. For example, using molasses, a byproduct of sugar production, as a substrate for *Monascus* spp. has resulted in significant cost reductions without compromising pigment yield.

18.5.3.2 Environmental factors

Factors such as pH, temperature, and aeration play critical roles in microbial metabolism. Adaptive laboratory evolution has been employed to develop strains that can tolerate suboptimal conditions while maintaining high pigment productivity.

18.5.3.3 Bioreactor design

Advances in bioreactor technology, including the use of fed-batch and continuous fermentation systems, have improved process scalability. Novel designs such as bubble columns and airlift bioreactors enhance oxygen transfer and mixing efficiency, which are particularly important for aerobic pigment-producing microbes.

18.5.4 Innovative fermentation techniques

Fermentation remains the cornerstone of microbial pigment production, and recent innovations have focused on improving its efficiency. SmF and SSF are the two primary methods used, each with distinct advantages (Augustin *et al.*, 2024).

18.5.4.1 Submerged fermentation (SmF)

This technique involves growing microbes in liquid media and is widely used for large-scale production due to its ease of monitoring and automation. Recent advancements in SmF include the use of microbubble technologies to enhance oxygen transfer and the development of dynamic feeding strategies to maintain optimal nutrient levels throughout the fermentation process.

18.5.4.2 Solid-state fermentation (SSF)

Solid-state fermentation involves cultivating microbes on solid substrates with minimal free water. It is particularly suited for pigment production from filamentous fungi such as *Monascus* and *Penicillium*. The use of agro-industrial residues as substrates not only reduces production costs but also contributes to waste valorization. Advances in SSF include the use of bioreactor designs that allow better temperature and humidity control, addressing the traditional challenges of scaling up this process.

18.5.5 Extraction and purification techniques

Efficient extraction and purification of microbial pigments are essential to ensure product quality and reduce downstream processing costs. Traditional solvent extraction methods often involve the use of organic solvents, which can pose environmental and safety concerns. Recent advancements have focused on greener and more sustainable alternatives.

18.5.5.1 Aqueous two-phase systems (ATPS)

This method employs water-soluble polymers and salts to create two immiscible aqueous phases. Pigments preferentially partition into one of the phases, enabling their separation with minimal solvent use. ATPS has been successfully applied to extract pigments such as astaxanthin and carotenoids with high purity and recovery rates.

18.5.5.2 Ultrasound-assisted extraction (UAE)

Ultrasound-assisted extraction uses ultrasonic waves to disrupt microbial cells, enhancing the release of intracellular pigments. This method is rapid, energy efficient, and compatible with various solvents, making it an attractive option for large-scale applications.

18.5.5.3 Supercritical fluid extraction (SFE)

Supercritical fluid extraction uses supercritical carbon dioxide as a solvent to extract pigments. This technique is particularly effective for nonpolar pigments such as carotenoids and offers the advantages of being nontoxic and environmentally friendly.

18.5.6 Emerging trends in pigment stabilization

The stability of microbial pigments is a critical factor for their industrial application, particularly in food and cosmetics. Microencapsulation techniques, such as spray drying and freeze drying, have been employed to enhance pigment stability against light, heat, and oxidation. Encapsulation also improves the solubility and bioavailability of pigments, broadening their application potential. Nanotechnology offers another promising avenue for pigment stabilization. By encapsulating pigments in nanocarriers such as liposomes or polymeric nanoparticles, their stability and bioactivity can be significantly enhanced. This approach has been particularly effective for pigments like curcumin and anthocyanins, which are prone to degradation under environmental stress (Wijesekara and Xu, 2024).

18.5.7 Role of artificial intelligence and big data

The integration of AI and big data analytics is revolutionizing the field of microbial pigment production. Machine learning algorithms are being used to predict optimal culture conditions, design efficient metabolic pathways, and identify potential pigment-producing microbes from genomic data. By leveraging these tools, researchers can significantly accelerate the discovery and optimization of microbial pigments.

18.6 Novel Microbial Strains and Discovery

The exploration of novel microbial strains represents a pivotal area in the advancement of microbial pigment production, driven by the need to diversify sources, enhance yields, and uncover unique functionalities.

Extremophiles, microorganisms adapted to extreme conditions such as high salinity, temperature, or pH, have emerged as promising candidates for pigment production due to their ability to synthesize robust and stable pigments under harsh conditions. For example, halophilic archaea and bacteria are known to produce carotenoids like bacterioruberin, which exhibit potent antioxidant and photoprotective properties, making them ideal for applications in cosmetics and food preservation. Similarly, thermophilic species, such as *Thermus* spp., are capable of producing thermally stable pigments, expanding their utility in industrial processes requiring high-temperature conditions. Marine-derived microbes, encompassing a diverse array of bacteria, fungi, and microalgae, also present an untapped reservoir of bioactive pigments. Marine microbes like *Vibrio* spp. and *Pseudoalteromonas* spp. produce pigments with antimicrobial and anticancer properties (Eze *et al.*, 2023), while microalgae such as *Dunaliella salina* and *Haematococcus pluvialis* are renowned for their production of commercially valuable pigments like β-carotene and astaxanthin, respectively (Reshma *et al.*, 2021).

Advances in genomics and high-throughput screening technologies have accelerated the discovery of these novel strains, enabling researchers to identify and characterize pigment-producing microbes from diverse ecosystems. Additionally, the development of genetically modified strains has revolutionized pigment production by enabling the incorporation of biosynthetic pathways from high-yielding natural producers into fast-growing and easily cultivable hosts such as *Escherichia coli* and *Saccharomyces cerevisiae*. Synthetic biology approaches, including modular pathway engineering and genome editing using CRISPR-Cas9, allow the optimization of metabolic flux towards pigment biosynthesis while minimizing the production of unwanted by-products. These engineered strains not only enhance pigment yields but also enable the tailoring of pigment properties to meet specific industrial requirements, such as improved solubility, stability, and bioactivity. Furthermore, interdisciplinary efforts integrating bioinformatics, proteomics, and metabolomics are facilitating the identification of novel pigments and their biosynthetic genes, paving the way for the discovery of unique compounds with potential applications in pharmaceuticals, textiles, and environmental remediation.

By coupling innovative bioprospecting strategies with advanced genetic engineering, the field is poised to unlock the full potential of microbial pigments, addressing global demands for sustainable, functional, and eco-friendly colorants.

18.7 Future Prospects and Innovations

The future of microbial pigment research is marked by promising innovations and emerging trends aimed at addressing global demands for sustainable and multifunctional products. The integration of green chemistry practices is expected to play a pivotal role in advancing pigment production processes, focusing on eco-friendly cultivation, minimal waste generation, and the use of renewable substrates such as agricultural residues. Green extraction techniques, including aqueous-based systems and supercritical fluid extraction, are gaining traction as environmentally benign alternatives to traditional methods that often rely on organic solvents. Beyond sustainability, the development of multifunctional pigments with bioactive properties is a burgeoning area of research. By harnessing the natural antimicrobial, antioxidant, and antiinflammatory activities of microbial pigments, these compounds are being tailored for use in diverse fields, including pharmaceuticals, cosmetics, and food preservation. For instance, pigments like violacein and prodigiosin are being investigated not only for their vibrant colors but also for their therapeutic potential in combating bacterial infections and cancer.

A particularly exciting avenue lies in the creation of personalized or industry-specific pigments. Advances in synthetic biology and metabolic engineering are enabling researchers to customize pigments for specific applications. For example, pigments can be engineered to exhibit enhanced thermal stability for use in high-temperature industrial processes or tailored solubility to improve their incorporation into food and beverage products. Similarly, personalized pigments for cosmetic and pharmaceutical applications could be designed to align with individual skin types or therapeutic needs, further expanding their market potential.

Emerging trends also include the exploration of new markets, such as the incorporation of microbial pigments in bioplastics, smart textiles, and bioinks for 3D printing (Burns *et al.*, 2025). These novel applications could transform pigments from simple colorants into functional materials with added value, such as antimicrobial coatings for medical devices or UV-protective films for environmental applications.

The convergence of microbial pigment research with advancements in omics technologies is expected to accelerate the discovery of novel pigments and their bioactive properties. Genomic and transcriptomic studies are facilitating the identification of biosynthetic gene clusters in previously unexplored microorganisms, such as extremophiles and marine-derived species. Coupled with metabolomics, these tools enable a deeper understanding of pigment biosynthesis pathways and their regulation, providing a foundation for the rational design of high-yield production systems. Furthermore, artificial intelligence and machine learning are being increasingly adopted to optimize fermentation parameters, predict pigment stability, and identify potential new markets. Collaborations across disciplines, including biotechnology, materials science, and environmental engineering, will be crucial in driving innovation and fostering the widespread adoption of microbial pigments. By addressing current challenges such as scalability, cost-effectiveness, and regulatory hurdles, these collaborative efforts have the potential to revolutionize industries and create a new paradigm for sustainable colorant production.

The future holds immense potential for microbial pigments to evolve into multifunctional, industry-specific solutions that not only fulfill aesthetic purposes but also contribute to health, sustainability, and environmental conservation.

18.8 Conclusion

Microbial pigments stand at the forefront of sustainable innovation, offering a versatile and eco-friendly alternative to synthetic dyes and traditional natural colorants. These pigments not only address the environmental concerns posed by synthetic dye production but also exhibit remarkable bioactive properties, including antimicrobial, antioxidant, and anticancer effects, expanding their potential beyond mere coloration. Through advancements in metabolic engineering, synthetic biology, and innovative fermentation techniques, significant strides have been made in improving production efficiency, scalability, and stability, overcoming many traditional challenges associated with microbial pigment commercialization. The discovery of novel pigment-producing strains, particularly extremophiles and marine-derived microbes, coupled with genome editing technologies, has further enriched the diversity and functionality of microbial pigments.

These developments have opened new avenues in various industries, such as food, cosmetics, textiles, and pharmaceuticals, where microbial pigments are increasingly valued for their safety, biodegradability, and multifunctionality. Moreover, the integration of green chemistry practices and the use of renewable substrates are reinforcing their role in promoting a circular economy, aligning with global sustainability goals.

Looking ahead, the future of microbial pigments is intertwined with cutting-edge innovations such as artificial intelligence for process optimization, personalized pigment development, and their incorporation into emerging markets like smart textiles and bioplastics. Collaborative efforts among academia, industry, and regulatory bodies will be crucial to address remaining challenges, including cost reduction, regulatory compliance, and consumer acceptance.

Ultimately, microbial pigments represent a paradigm shift in the colorant industry, transforming it into a more sustainable, health-conscious, and environmentally responsible sector. Their unique combination of vibrant hues, bioactivity, and ecological benefits positions them as a cornerstone of green biotechnology, poised to redefine the future of colorants across diverse applications.

Acknowledgments

The authors would like to express their gratitude to MIT WPU and STES's Smt. Kashibai Navale College of Pharmacy, for providing the necessary facilities for completing this book chapter.

References

Alegbe, E.O. and Uthman, T.O. (2024) A review of history, properties, classification, applications and challenges of natural and synthetic dyes. *Heliyon* 10(13), e33646. DOI: 10.1016/j.heliyon.2024.e33646.

Al-Tohamy, R., Ali, S.S., Li, F., Okasha, K.M., Mahmoud, Y.A.-G. *et al.* (2022) A critical review on the treatment of dye-containing wastewater: Ecotoxicological and health concerns of textile dyes and possible remediation approaches for environmental safety. *Ecotoxicology and Environmental Safety* 231, 113160. DOI: 10.1016/j.ecoenv.2021.113160.

Ambati, R., Phang, S.-M., Ravi, S. and Aswathanarayana, R. (2014) Astaxanthin: Sources, extraction, stability, biological activities and its commercial applications—a review. *Marine Drugs* 12(1), 128–152. DOI: 10.3390/md12010128.

Anshi, K.S., Goswami, L. and Sharma, V. (2024) Unveiling the intricacies of microbial pigments as sustainable alternatives to synthetic colorants: Recent trends and advancements. *Micro* 4(4), 621–640. DOI: 10.3390/micro4040038.

Anwar, M.M., Albanese, C., Hamdy, N.M. and Sultan, A.S. (2022) Rise of the natural red pigment 'prodigiosin' as an immunomodulator in cancer. *Cancer Cell International* 22(1), 419–439. DOI: 10.1186/s12935-022-02815-4.

Augustin, M.A., Hartley, C.J., Maloney, G. and Tyndall, S. (2024) Innovation in precision fermentation for food ingredients. *Critical Reviews in Food Science and Nutrition* 64(18), 6218–6238. DOI: 10.1080/10408398.2023.2166014.

Barreto, J.V. de O., Casanova, L.M., Junior, A.N., Reis-Mansur, M.C.P.P. and Vermelho, A.B. (2023) Microbial pigments: Major groups and industrial applications. *Microorganisms* 11(12), 2920. DOI: 10.3390/microorganisms11122920.

Bawa, A.S. and Anilakumar, K.R. (2013) Genetically modified foods: Safety, risks and public concerns-a review. *Journal of Food Science and Technology* 50(6), 1035–1046. DOI: 10.1007/s13197-012-0899-1.

Bialas, I., Zelent-Kraciuk, S. and Jurowski, K. (2023) The skin sensitisation of cosmetic ingredients: Review of actual regulatory status. *Toxics* 11(4), 392. DOI: 10.3390/toxics11040392.

Brenner, M. and Hearing, V.J. (2008) The protective role of melanin against UV damage in human skin. *Photochemistry and Photobiology* 84(3), 539–549. DOI: 10.1111/j.1751-1097.2007.00226.x.

Buiatti, M., Christou, P. and Pastore, G. (2013) The application of gmos in agriculture and in food production for a better nutrition: Two different scientific points of view. *Genes & Nutrition* 8(3), 255–270. DOI: 10.1007/s12263-012-0316-4.

Burns, N., Rajesh, A., Manjula-Basavanna, A. and Duraj-Thatte, A. (2025) 3D extrusion bioprinting of microbial inks for biomedical applications. *Advanced Drug Delivery Reviews* 217, 115505. DOI: 10.1016/j.addr.2024.115505.

Cao, K., Cui, Y., Sun, F., Zhang, H., Fan, J. *et al.* (2023) Metabolic engineering and synthetic biology strategies for producing high-value natural pigments in microalgae. *Biotechnology Advances* 68, 108236. DOI: 10.1016/j.biotechadv.2023.108236.

Darshan, N. and Manonmani, H.K. (2015) Prodigiosin and its potential applications. *Journal of Food Science and Technology* 52(9), 5393–5407. DOI: 10.1007/s13197-015-1740-4.

Devi, M., Ramakrishnan, E., Deka, S. and Parasar, D.P. (2024) Bacteria as a source of biopigments and their potential applications. *Journal of Microbiological Methods* 219, 106907. DOI: 10.1016/j.mimet.2024.106907.

Dhull, K., Tripathy, S. and Dureja, H. (2015) Cosmetics: Regulatory scenario in USA, EU and India. *Journal of Pharmaceutical Technology, Research and Management* 3, 127–139. DOI: 10.15415/jptrm.2015.32010.

Durán, M., Ponezi, A.N., Faljoni-Alario, A., Teixeira, M.F.S., Justo, G.Z. *et al.* (2012) Potential applications of violacein: A microbial pigment. *Medicinal Chemistry Research* 21(7), 1524–1532. DOI: 10.1007/s00044-011-9654-9.

Eze, O.C., Berebon, D.P., Emencheta, S.C., Evurani, S.A., Okorie, C.N. *et al.* (2023) Therapeutic potential of marine probiotics: A survey on the anticancer and antibacterial effects of *Pseudoalteromonas* spp. *Pharmaceuticals* 16(8), 1091. DOI: 10.3390/ph16081091.

Gong, P., Shi, R., Liu, Y., Luo, Q., Wang, C. *et al.* (2023) Recent advances in monascus pigments produced by *Monascus purpureus*: Biosynthesis, fermentation, function, and application. *LWT* 185, 115162. DOI: 10.1016/j.lwt.2023.115162.

Khan, S., Ullah, M.W., Siddique, R., Nabi, G., Manan, S. *et al.* (2016) Role of recombinant DNA technology to improve life. *International Journal of Genomics* 1–14. DOI: 10.1155/2016/2405954.

Kieliszek, M. and Misiewicz, A. (2014) Microbial transglutaminase and its application in the food industry. A review. *Folia Microbiologica* 59(3), 241–250. DOI: 10.1007/s12223-013-0287-x.

Kiki, M.J. (2023) Biopigments of microbial origin and their application in the cosmetic industry. *Cosmetics* 10(2), 47. DOI: 10.3390/cosmetics10020047.

Kim, I., Kim, J., Chhetri, G. and Seo, T. (2019) *Flavobacterium humi* sp. nov., a flexirubin-type pigment producing bacterium, isolated from soil. *Journal of Microbiology* 57(12), 1079–1085. DOI: 10.1007/s12275-019-9350-x.

Kumar, P., Banik, S.P., Ohia, S.E., Moriyama, H., Chakraborty, S. *et al.* (2024) Current insights on the photoprotective mechanism of the macular carotenoids, lutein and zeaxanthin: Safety, efficacy and bio-delivery. *Journal of the American Nutrition Association* 43(6), 505–518. DOI: 10.1080/27697061.2024.2319090.

Lellis, B., Fávaro-Polonio, C.Z., Pamphile, J.A. and Polonio, J.C. (2019) Effects of textile dyes on health and the environment and bioremediation potential of living organisms. *Biotechnology Research and Innovation* 3(2), 275–290. DOI: 10.1016/j.biori.2019.09.001.

Lu, Y., Liu, D., Jiang, R., Li, Z. and Gao, X. (2024) Prodigiosin: Unveiling the crimson wonder – a comprehensive journey from diverse bioactivity to synthesis and yield enhancement. *Frontiers in Microbiology* 15, 1412776. DOI: 10.3389/fmicb.2024.1412776.

Mandal, D. and Majumdar, S. (2023) Bacteria as biofactory of pigments: Evolution beyond therapeutics and biotechnological advancements. *Journal of Bioscience and Bioengineering* 135(5), 349–358. DOI: 10.1016/j.jbiosc.2023.01.008.

Mummaleti, G., Udo, T., Mohan, A. and Kong, F. (2024) Synthesis, characterization and application of microbial pigments in foods as natural colors. *Critical Reviews in Food Science and Nutrition* 1–30. DOI: 10.1080/10408398.2024.2417802.

Mussagy, C.U., Guimarães, A.A.C., Rocha, L.V.F., Winterburn, J., Santos-Ebinuma, V. de C. *et al.* (2021) Improvement of carotenoids production from *Rhodotorula glutinis* CCT-2186. *Biochemical Engineering Journal* 165, 107827. DOI: 10.1016/j.bej.2020.107827.

Nair, A., Ahirwar, A., Singh, S., Lodhi, R., Lodhi, A. *et al.* (2023) Astaxanthin as a king of ketocarotenoids: Structure, synthesis, accumulation, bioavailability and antioxidant properties. *Marine Drugs* 21(3), 176. DOI: 10.3390/md21030176.

Nguyen, S.L.T., Nguyen, T.C., Do, T.T., Vu, T.L., Nguyen, T.T. *et al.* (2022) Study on the anticancer activity of prodigiosin from variants of *Serratia marcescens* QBN VTCC 910026. *BioMed Research International* 2022(1), 4053074. DOI: 10.1155/2022/4053074.

Nonglait, D.L. and Gokhale, J.S. (2024) Review insights on the demand for natural pigments and their recovery by emerging microwave-assisted extraction (MAE). *Food and Bioprocess Technology* 17(7), 1681–1705. DOI: 10.1007/s11947-023-03192-0.

Numan, M., Bashir, S., Mumtaz, R., Tayyab, S., Rehman, N.U. *et al.* (2018) Therapeutic applications of bacterial pigments: A review of current status and future opportunities. *3 Biotech* 8(4), 207–222. DOI: 10.1007/s13205-018-1227-x.

Ochoa-Viñals, N., Alonso-Estrada, D., Pacios-Michelena, S., García-Cruz, A., Ramos-González, R. *et al.* (2024) Current advances in carotenoid production by *Rhodotorula* sp. *Fermentation* 10(4), 190. DOI: 10.3390/fermentation10040190.

Orlandi, V.T., Martegani, E., Giaroni, C., Baj, A. and Bolognese, F. (2022) Bacterial pigments: A colorful palette reservoir for biotechnological applications. *Biotechnology and Applied Biochemistry* 69(3), 981–1001. DOI: 10.1002/bab.2170.

Rajendran, P., Somasundaram, P. and Dufossé, L. (2023) Microbial pigments: Eco-friendly extraction techniques and some industrial applications. *Journal of Molecular Structure* 1290, 135958. DOI: 10.1016/j.molstruc.2023.135958.

Ramesh, C., Vinithkumar, N., Kirubagaran, R., Venil, C. and Dufossé, L. (2019) Multifaceted applications of microbial pigments: Current knowledge, challenges and future directions for public health implications. *Microorganisms* 7(7), 186. DOI: 10.3390/microorganisms7070186.

Rana, B., Bhattacharyya, M., Patni, B., Arya, M. and Joshi, G.K. (2021) The realm of microbial pigments in the food color market. *Frontiers in Sustainable Food Systems* 5, 603892. DOI: 10.3389/fsufs.2021.603892.

Reshma, R., Chitra Devi, K., Dinesh Kumar, S., Santhanam, P., Perumal, P. *et al.* (2021) Enhancement of pigments production in the green microalga *Dunaliella salina* (PSBDU05) under optimized culture condition. *Bioresource Technology Reports* 14, 100672. DOI: 10.1016/j.biteb.2021.100672.

Roy, D., Verma, S., Das, S., Chakdar, H. and Pabbi, S. (2024) Production and applications of cyanobacterial phycocyanin: Trends and prospects. *Indian Journal of Microbiology*. DOI: 10.1007/s12088-024-01332-4.

Saha, R., Chauhan, A. and Rastogi Verma, S. (2024) Machine learning: An advancement in biochemical engineering. *Biotechnology Letters* 46(4), 497–519. DOI: 10.1007/s10529-024-03499-8.

Schmidt, F.R. (2005) Optimization and scale up of industrial fermentation processes. *Applied Microbiology and Biotechnology* 68(4), 425–435. DOI: 10.1007/s00253-005-0003-0.

Solano, F. (2020) Photoprotection and skin pigmentation: Melanin-related molecules and some other new agents obtained from natural sources. *Molecules* 25(7), 1537–1555. DOI: 10.3390/molecules25071537.

Starr, M.P., Cosens, G. and Knackmuss, H.-J. (1966) Formation of the blue pigment *indigoidine* by phytopathogenic *erwinia*. *Applied Microbiology* 14(6), 870–872. DOI: 10.1128/am.14.6.870-872.1966.

Tang, W., Wang, Y., Zhang, J., Cai, Y. and He, Z. (2019) Biosynthetic pathway of carotenoids in *Rhodotorula* and strategies for enhanced their production. *Journal of Microbiology and Biotechnology* 29(4), 507–517. DOI: 10.4014/jmb.1801.01022.

Tkaczyk, A., Mitrowska, K. and Posyniak, A. (2020) Synthetic organic dyes as contaminants of the aquatic environment and their implications for ecosystems: A review. *Science of the Total Environment* 717, 137222. DOI: 10.1016/j.scitotenv.2020.137222.

Wijesekara, T. and Xu, B. (2024) A critical review on the stability of natural food pigments and stabilization techniques. *Food Research International* 179, 114011. DOI: 10.1016/j.foodres.2024.114011.

Yixuan, L., Qaria, M.A., Sivasamy, S., Jianzhong, S. and Daochen, Z. (2021) Curcumin production and bioavailability: A comprehensive review of curcumin extraction, synthesis, biotransformation and delivery systems. *Industrial Crops and Products* 172, 114050. DOI: 10.1016/j.indcrop.2021.114050.

Zamudio Díaz, D.F., Busch, L., Kröger, M., Klein, A.L., Lohan, S.B. *et al.* (2024) Significance of melanin distribution in the epidermis for the protective effect against UV light. *Scientific Reports* 14(1), 3488. DOI: 10.1038/s41598-024-53941-0.

Zhao, J., Fang, H. and Zhang, D. (2020) Expanding application of CRISPR-Cas9 system in microorganisms. *Synthetic and Systems Biotechnology* 5(4), 269–276. DOI: 10.1016/j.synbio.2020.08.001.

Zhou, D., Fei, Z., Liu, G., Jiang, Y., Jiang, W. *et al.* (2024) The bioproduction of astaxanthin: A comprehensive review on the microbial synthesis and downstream extraction. *Biotechnology Advances* 74, 108392. DOI: 10.1016/j.biotechadv.2024.108392.

Zhou, Y., Han, L.-R., He, H.-W., Sang, B., Yu, D.-L. *et al.* (2018) Effects of agitation, aeration and temperature on production of a novel glycoprotein GP-1 by *Streptomyces kanasenisi* ZX01 and scale-up based on volumetric oxygen transfer coefficient. *Molecules* 23(1), 125. DOI: 10.3390/molecules23010125.

19 Sustainability and Environmental Impact of Biopigment Production

Priyanka[1]* and Ayush Madan[2]

[1]Graduate Institute of Biomedical Sciences, China Medical University Hospital, Taichung City, Taiwan; [2]Department of Biotechnology, School of Research and Technology, People's University, Bhopal, Madhya Pradesh, India

19.1 Introduction

19.1.1 Overview of biopigments: definitions and importance

Biopigments, also referred as natural pigments, are bioactive compounds produced by a diverse range of living organisms, including plants, microalgae, fungi, and various bacteria. These pigments are responsible for the vibrant colors observed in nature, and they play pivotal roles beyond mere aesthetic qualities. Biopigments possess unique ecological, industrial, and health-related benefits that make them increasingly relevant in multiple fields. Unlike synthetic dyes, which may have adverse effects on health and the environment, biopigments are characterized by their biocompatibility, low toxicity, and natural biodegradability. This makes them particularly suitable for sustainable applications across a variety of industries, including food, cosmetics, textiles, and pharmaceuticals. Significant research has been conducted to explore the potential of biopigments as eco-friendly alternatives to synthetic dyes. For example, studies conducted by Suwannarach *et al.* (2019) and Bhatnagar *et al.* (2022) delve into the use of fungal pigments, showcasing their ability to create sustainable dyeing processes while maintaining color quality and reducing environmental hazards.

The escalating interest in biopigments can be primarily attributed to their multifunctionality. In addition to providing coloration, biopigments, such as carotenoids, anthocyanins, and chlorophylls, have been identified as possessing antioxidant properties, acting as antiinflammatory agents and offering protection from harmful ultraviolet (UV) radiation.

Similar to how biopigments derived from microalgae and fungi enhance plant stress tolerance through their antioxidant and UV-protective properties, Abdelkhalik *et al.* (2024) demonstrated that combining Moringa leaf extract with beneficial microorganisms significantly improves sweet potato resilience when grown in cadmium-contaminated saline soils. Furthermore, observed variations in photosynthesis and isoprenoid-mediated photoprotection in different Douglas fir provenances under varying soil water availability illustrate the complex adaptive mechanisms plants employ to cope with environmental stresses (Junker *et al.*, 2017). These insights into photoprotection complement the antioxidant functions of biopigments, reinforcing their vital role in plant resilience and health. Research highlights

*Corresponding author: priyankas19991202@gmail.com

 Biopigments of Microbial Origin: Innovations and Applications Across Industries (eds Madan *et al.*)
DOI: 10.1079/ 9781836990505.0019

microalgae as promising sources of high-value pigments that not only mitigate environmental damage but also promote human health and wellness. Notably, advancements in the production of fucoxanthin, a valuable carotenoid pigment, have been achieved using the haptophyte marine microalga *Pavlova* sp., demonstrating effective biotechnological methods for pigment extraction and utilization (Kanamoto *et al.*, 2021). Furthermore, the extraction of these pigments from sustainable and renewable feedstocks aligns seamlessly with the principles of a circular bioeconomy, thus contributing to resource conservation and environmental stewardship.

The ongoing exploration of biopigments opens exciting new opportunities for biotechnological innovations. Microalgae species like *Chlorella* and *Chlamydomonas reinhardtii* have been recognized for their ability to produce significant amounts of carotenoids and chlorophyll under controlled conditions. Recent research on the cultivation of *Chlorella protothecoides* under various growth modes and its application in oil/water emulsions further highlights the versatility of microalgae in sustainable biotechnological processes (Caporgno *et al.*, 2019). In addition, fungal sources like *Nigrospora aurantiaca* are demonstrating promising applications in the textile sector. These advancements underscore the escalating significance of biopigments as sustainable alternatives, addressing the growing global emphasis on reducing dependency on synthetic chemicals and mitigating environmental degradation.

19.1.2 Historical context: transition from synthetic to biopigments

The evolution of pigment usage can be traced back centuries, with ancient civilizations depending predominantly on natural resources such as plants, minerals, and insects to obtain colors for artistic, textile, and food applications. However, the introduction of synthetic dyes in the 19th century marked a transformative period, characterized by the provision of vibrant, durable, and cost-effective alternatives. While synthetic pigments gained widespread acceptance for their economic advantages, they also generated significant environmental and health challenges, contributing to pollution, depletion of natural resources, and exposure to toxic substances. The environmental repercussions of synthetic pigment production underscores the urgent need for more sustainable alternatives.

The transition from synthetic to biopigments marks a critical paradigm shift, largely driven by ecological concerns and changing consumer preferences towards sustainability. Advances in bioprocessing technologies, which enable the mass cultivation of microalgae and fungi, have facilitated the large-scale production of biopigments. Bioprocessing approaches for macroalgae like *Palmaria palmata*, including metabolite fractionation and ensiling techniques for long-term storage, have expanded the possibilities for using marine biomass in sustainable bio-industries (Gallagher *et al.*, 2021). Additionally, various studies illustrate innovative approaches to incorporating waste valorization into pigment extraction processes, thereby promoting a zero waste philosophy that aligns with circular bioeconomy principles. Significant historical milestones reveal the growing acceptance of biopigments across different industries. For instance, the textile sector is increasingly opting for fungal and plant-based pigments as a response to stringent regulations on chemical discharge and the rising consumer demand for environmentally friendly products. In parallel, the food industry has begun embracing microalgae-derived pigments, such as astaxanthin and phycocyanin, recognizing their dual functionality as natural colorants and valuable nutritional supplements.

Although the initial challenges around scalability and cost-effectiveness persist, ongoing research and optimization of bioprocessing methods are progressively addressing these barriers. Furthermore, the transition towards biopigments is being supported by global policies advocating for sustainability and environmental responsibility, as emphasized in several comprehensive review articles.

19.1.3 Scope of sustainability in biopigment production

The production of biopigments is emerging as a cornerstone for promoting sustainability across various industries. Unlike their synthetic counterparts, biopigments are sourced from renewable biological materials and are often produced through low-energy, eco-friendly processes. This practice aligns closely with the objectives of the BioCircularGreen (BCG) economy, which prioritizes waste minimization, resource efficiency, and reduced ecological impact.

One of the most promising aspects of biopigment production is its capacity to integrate seamlessly with waste valorization systems. For example, as discussed by Saba *et al.* (2023), by utilizing food processing byproducts and agricultural residues as feedstocks for the production of microbial and algal pigments, biopigments can emerge as high-value products while simultaneously reducing waste. Moreover, cultivating microalgae on nonarable land and using wastewater not only conserve precious freshwater resources but also contribute to decreasing carbon emissions. Sustainability in biopigment production thus represents a multifaceted opportunity encompassing environmental conservation, resource optimization, and economic viability. This crucial intersection addresses the dire need for sustainable practices in industry while fostering innovations that can yield both environmental and health benefits.

Through these ongoing developments, biopigments are poised to revolutionize various sectors by offering a sustainable, eco-friendly alternative to synthetic dyes and pigments, aligning with a vision of a healthier planet and a more responsible approach to consumption and production.

19.2 Biopigment Sources and Production Processes

19.2.1 Microbial sources: bacteria, fungi, and algae

Microbial sources, encompassing bacteria, fungi, and algae, play a crucial role in the biopigment production landscape due to their remarkable metabolic diversity and ability to generate high-value pigments in controlled cultivation settings (Table 19.1; Fig. 19.1). Among these groups, microalgae are particularly noteworthy for their capacity to produce a wide range of pigments, including carotenoids, chlorophylls, and phycobiliproteins. Recent research highlights microalgae's inherent versatility, not only in pigment accumulation but also in contributing to more sustainable biobased production systems.

19.2.1.1 Bacteria as pigment producers

Bacteria have gained recognition for their ability to synthesize a diverse array of pigments, including carotenoids, melanin, and prodigiosin. Specific bacterial species, notably from the genera *Pseudomonas* and *Serratia*, have been genetically engineered to optimize pigment production, yielding bright red and yellow pigments with significant applications in food coloration and cosmetics. In addition to their aesthetic properties, these bacterial pigments are noted for their antimicrobial and antioxidant characteristics, which enhance their utility in various pharmaceutical applications. Ongoing research into metabolic pathways and gene expression is further illuminating ways to increase yield and efficiency in microbial pigment production.

19.2.1.2 Fungal pigments: natural and synthetic applications

Fungi, especially endophytic species, have been the subject of extensive research for their pigment-producing capabilities. For example, Suwannarach *et al.* (2019) investigated the red pigment sourced from *Nigrospora aurantiaca*, demonstrating its potential use as a natural dye for textiles. Fungal pigments, such as anthraquinones and flavins, provide eco-friendly alternatives to synthetic dyes, appealing especially to industries focused

Table 19.1. Microbial pigment types and applications.

Microorganism source	Pigment type	Applications
Bacteria	Carotenoids	Food colorants, cosmetics
Fungi	Anthraquinones	Textiles, eco-friendly dyes
Microalgae	Chlorophylls	Nutraceuticals, biofuels

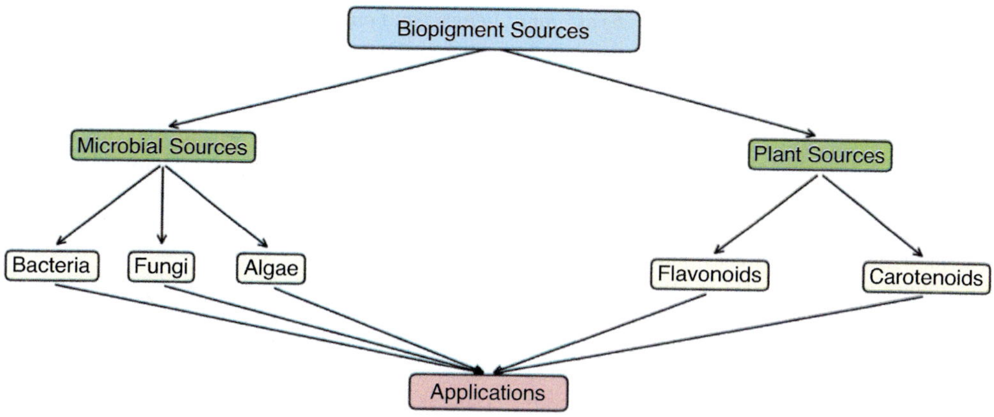

Fig. 19.1. Sources and pathways of microbial and plant-derived pigment production, with their applications.

Table 19.2. Potential of microalgae in industrial practices.

Industry	Application	Impact
Energy	Biodiesel production	Reduced reliance on fossil fuels
Chemicals	Bio-based chemicals	Safer, biodegradable options
Textiles	Natural pigments	Reduced water pollution
Aquaculture	Fish feed	Sustainable food production
Food packaging	Biodegradable materials	Reduced plastic waste

on sustainability. Advances in fermentation technology are enabling large-scale production of these pigments, thus improving cost-effectiveness and ecological viability. The exploration of fungal metabolic pathways is paving the way for the development of customized pigments suitable for specific industrial applications.

19.2.1.3 Algal pigments: a focus on microalgae

Microalgae are unicellular organisms recognized for their remarkable ability to generate significant quantities of high-value pigments (Table 19.2). Notably, species belonging to the Chlorophyta, such as *Chlamydomonas reinhardtii*, are recognized as prolific producers of carotenoids and other bioactive compounds. Researchers are actively exploring mass cultivation techniques and stress-induced biosynthesis strategies to maximize pigment output. According to studies by Sayanova *et al.* (2017), environmental stressors such as light intensity and nutrient deprivation can significantly enhance both lipid and pigment biosynthesis in diatoms and other microalgal species.

Additionally, microalgae are an integral component of the BCG economy. Cheirsilp *et al.* (2023) examined zero waste strategies in microalgal biorefineries, which facilitate the simultaneous production of pigments alongside valuable coproducts such as biofuels and biofertilizers. This approach not only enhances economic viability but also addresses sustainability challenges by tapping into a circular economy model.

19.2.2 Enhancing pigment production in microorganisms

Innovative techniques, including adjustments to light systems and photoperiod management, have shown promise in improving pigment synthesis in microalgae. Tailoring light conditions can significantly boost the production of specific pigments, optimizing both their yield and quality.

Emerging research has shed light on the complex interactions between algae and bacteria, which play crucial roles in nutrient cycling, biofilm formation, and the production of bioactive compounds. These interactions significantly influence algal growth, pigment production, and environmental adaptability, presenting opportunities for harnessing algal–bacterial consortia in biotechnological applications (Nowruzi *et al.*, 2022). Understanding these relationships enhances the potential for optimizing biopigment yields and developing sustainable bioprocesses. Additionally, coupled cultivation systems that integrate dark fermentation with microalgal growth have been identified as effective means to enhance productivity while reducing overall energy consumption.

19.2.3 Plant-derived pigments: flavonoids and carotenoids

Plant-derived pigments, with a particular focus on flavonoids and carotenoids, have a long history of utilization across various industries, including food, cosmetics, and pharmaceuticals (Table 19.3). These pigments not only contribute vibrant colors but also offer valuable health benefits such as antioxidant and antiinflammatory properties.

19.2.3.1 Flavonoids: diversity and applications

Flavonoids encompass a diverse group of plant pigments responsible for a spectrum of colors, including yellow, red, and blue hues. These compounds are predominantly sourced from fruits, vegetables, and flowers, with anthocyanins being the most well known, as they contribute to the rich colours found in red lettuce and various berries. Recent reviews delve into the intricate genetic and environmental conditions influencing anthocyanin biosynthesis. Optimizing these factors in controlled plant factory environments has been demonstrated to significantly enhance flavonoid yield while preserving the nutritional integrity of the products. Along with that, plant waste generated from food processing operations, such as vegetable peels, has emerged as a valuable resource for the extraction of bioactive flavonoids. Sharma *et al.* (2021) championed the valorization of such waste as a means to extract high-value pigments, aligning with the principles of a circular bioeconomy. This process not only reduces waste but also creates economic opportunities through the generation of premium products.

19.2.3.2 Carotenoids: functional and commercial importance

Carotenoids, including well-known varieties such as β-carotene, lutein, and zeaxanthin, are essential pigments synthesized by an array of plants. These compounds are most commonly found in foods such as carrots, sweet potatoes, and an assortment of leafy greens.

Beyond their role in food coloration, carotenoids are critically important for human health, contributing to vision support and bolstering immune function. Recent studies, including those by Rengel *et al.* (2022), have illustrated the potential for simultaneous production of carotenoids alongside other beneficial biochemicals from plant biomass, showcasing the feasibility of integrated bioprocessing systems.

The ongoing research into optimized cultivation and extraction methods holds promise for maximizing both yield and economic viability within the realm of plant-derived pigments. Such advancements will likely play a pivotal role in enhancing the sustainability and efficiency of the entire biopigment production system.

Table 19.3. Applications of biopigments.

Industry	Example pigments	Application	Benefits
Food	β-Carotene	Natural food coloring	Nontoxic, antioxidant
Pharmaceuticals	Melanin	Antimicrobial agents	Functional and bioactive
Textiles	Anthraquinones	Natural dyes	Sustainable, eco-friendly
Cosmetics	Prodiginines	Skin formulations	Antioxidant, aesthetic appeal

19.2.4 Cultivation systems: open ponds, photobioreactors, and fermenters

The rising global demand for natural pigments is primarily driven by their extensive applications across diverse sectors, including food, pharmaceuticals, cosmetics, and textiles. Microalgae and fungi have surfaced as promising and versatile sources of biopigments, enabling sustainable production through a variety of cultivation systems (Table 19.4). These systems include open ponds, photobioreactors and fermenters, each come with their unique sets of benefits and challenges.

19.2.4.1 Open ponds

Open ponds are a traditional and cost-effective method for cultivating microalgae, particularly advantageous for large-scale pigment production. Usually established outdoors, these systems rely on natural sunlight and ambient CO_2 to facilitate photosynthesis. According to recent research, open ponds are predominantly utilized for the mass production of pigments such as carotenoids and chlorophylls due to their simplicity and low operational costs. However, several challenges impede their efficiency, including contamination from other microorganisms, significant water loss due to evaporation, and lower productivity levels that can arise under suboptimal environmental conditions. The reliance on weather patterns and sunlight can also lead to inconsistent yields, making it difficult for producers to meet fluctuating market demands.

19.2.4.2 Photobioreactors

Photobioreactors (PBRs) are innovative, closed systems that offer enhanced control over the cultivation environment, which is crucial for optimizing microalgae growth and pigment production. According to Parveen *et al.* (2023), PBRs provide the ability to fine-tune factors such as light exposure, nutrient delivery, and CO_2 supply to maximize yield and minimize the risk of contamination. These systems often incorporate advanced technologies, including sophisticated lighting set-ups that simulate ideal sunlight conditions and adjustable photoperiods to promote efficient growth cycles.

By maintaining a controlled environment, PBRs not only improve pigment yields but also enable the cultivation of sensitive algal strains that might otherwise be vulnerable in open systems. Nevertheless, the initial set-up and maintenance costs of photobioreactors are substantially higher than open ponds, posing a financial barrier for some producers.

19.2.4.3 Fermenters

Fermentation systems are gaining traction for producing pigments derived from fungal sources and heterotrophic algae. A study by Suwannarach *et al.* (2019) highlighted the potential of using fermentation techniques to cultivate organisms such as *Nigrospora aurantiaca*, renowned for its production of high-value red pigments in a controlled setting. Fermenters can utilize various substrates, including agro-industrial waste, which not only improves cost-effectiveness but also aligns with sustainability goals by repurposing waste materials. Furthermore, these systems allow for greater control over growth conditions, such as temperature, pH, and nutrient availability, which are critical for optimizing pigment synthesis in fungi and algae alike.

Table 19.4. Comparison of cultivation systems.

Parameter	Open ponds	Photobioreactors (PBRs)	Fermenters
Set-up cost	Low	High	Medium
Yield	Moderate	High	High
Contamination risk	High	Low	Low
Scalability	High	Moderate	Moderate
Key applications	Large-scale algal pigment	Controlled microalgae growth	Fungal pigment production

19.2.5 Advances in genetic engineering for pigment optimization

The integration of genetic engineering into biopigment production has marked a significant breakthrough, enhancing yield, stability, and the functional properties of pigments. Recent advancements demonstrate how the strategic manipulation of genetic material in both microalgae and fungi can optimize their metabolic pathways, facilitating targeted pigment synthesis (Table 19.5).

19.2.5.1 Engineering microalgae

Microalgae species, such as *Chlamydomonas reinhardtii*, have undergone extensive genetic modifications aimed at enhancing the overexpression of genes associated with carotenoid biosynthesis. Research conducted by Rengel *et al.* (2022) indicates that such genetic interventions not only boost pigment production but simultaneously facilitate the generation of other valuable metabolites, thereby decreasing dependence on fossil-based chemicals for industrial processes. The application of cutting-edge techniques like CRISPR-Cas9 and synthetic biology frameworks has streamlined the gene editing process, empowering researchers and producers to customize microalgal strains for specific pigment outputs with greater efficiency.

19.2.5.2 Engineering fungi

Fungal systems represent an exciting frontier for pigment production, thanks to their inherent ability to synthesize a wide array of bioactive compounds. In their exploration, Suwannarach *et al.* (2019) investigated how optimizing gene clusters responsible for secondary metabolite production can significantly increase the yield of desirable pigments, such as anthraquinones and azaphilones. The utilization of omics-based technologies, including transcriptomics and metabolomics, has been instrumental in identifying critical regulatory genes that govern these metabolic pathways, thus opening up new avenues for enhancing pigment production in fungi.

19.2.5.3. Challenges and future directions

Despite the immense potential that genetic engineering offers, several critical challenges remain. Ethical and regulatory issues surrounding genetically modified organisms (GMOs) pose significant hurdles that must be addressed, particularly concerning public perception and ecological impacts. Furthermore, scaling up genetically engineered strains for commercial applications is another area requiring innovative advancements in bioprocessing technologies.

To fully realize the benefits of genetically engineered pigments, ongoing research and development efforts must focus on refining these technologies while ensuring that they are economically viable and sustainable in the long term.

19.3 Sustainability in Feedstock Utilization

The production of pigments from sustainable resources is increasingly being recognized as a vital component of modern biotechnology. This shift is driven by the growing demand for eco-friendly processes within the bioeconomy, which seeks to harmonize economic growth with environmental sustainability. By integrating agro-industrial waste as substrates, adopting circular bioeconomy strategies, and making

Table 19.5. Genetic engineering techniques in biopigment production.

Organism type	Targeted genes	Outcome	Techniques used
Microalgae	Carotenoid biosynthesis	Higher carotenoid yield	CRISPR-Cas9, synthetic biology
Fungi	Secondary metabolite genes	Enhanced pigment production	Omics-based technologies

resource-efficient substrate selections, we can pave the way for sustainable practices in pigment production that also enhance cost-effectiveness. This section delves deeper into these concepts, drawing on insights from recent studies focused on biorefineries, microbial bioprospecting, and waste valorization (Table 19.6).

19.3.1 Agro-industrial waste as substrates for pigment production

Agro-industrial waste emerges as a promising option for providing renewable, abundant, and cost-effective substrates for the production of microbial and algal pigments. These types of waste encompass byproducts generated during food processing, crop residues, and materials rich in lignocellulose, all of which are abundant in carbohydrates, proteins, and essential micronutrients. Such characteristics make them suitable feedstock for microbial fermentation and cultivation processes. Importantly, the utilization of these agro-industrial wastes not only diminishes environmental pollution but also facilitates resource recovery by transforming previously discarded materials into valuable products.

Recent studies have showcased the potential of various microorganisms, including microalgae and fungi, to effectively utilize agro-industrial residues as growth substrates. For example, Udayan *et al.* (2022) clearly demonstrated the viability of cultivating microalgae using effluents derived from food waste as a nutrient source, successfully achieving high pigment yields in the process. Similarly, Suwannarach *et al.* (2019) optimized the extraction of red pigments from the fungus *Nigrospora aurantiaca* by using agricultural byproducts, thus realizing significant reductions in production costs while enhancing the overall sustainability of the process.

The valorization of lignocellulosic biomass is another pivotal aspect of this movement. As highlighted by Lu *et al.* (2022), this process often involves breaking down complex lignocellulosic structures into fermentable sugars through enzymatic hydrolysis, which microorganisms can then use for pigment biosynthesis. This method aligns closely with the principles of a circular economy, actively transforming waste streams into high-value outputs while minimizing the environmental footprint (Fig. 19.2).

19.3.2 Circular bioeconomy integration: zero waste approaches

The circular bioeconomy is predicated on the efficient use of resources by emphasizing the recovery and valorization of waste within production systems, effectively creating a closed loop process. Zero waste approaches in pigment production prioritize the co-utilization of all the available biomass components. This strategy ensures minimal resource wastage while maximizing economic returns. Microalgae-based biorefineries serve as exemplary models of integrating circular bioeconomy practices. According to Cheirsilp *et al.* (2023), the development of biorefinery systems that leverage microalgal biomass to produce multiple products, such as pigments, biofuels, and biofertilizers, is essential for sustainable production. For instance, employing advanced pretreatment techniques like ultrasonically assisted deep eutectic solvent extraction has enabled researchers to successfully produce biodiesel and pigments,

Table 19.6. Advantages and challenges of feedstock sources.

Feedstock source	Advantages	Challenges
Agro-industrial waste	Cost-effective, abundant, reduces pollution	Variability in composition, need for preprocessing
Lignocellulosic biomass	Rich in fermentable sugars, aligns with circular economy principles	Complex structure requires enzymatic hydrolysis
Wastewater	Low cost, nutrient rich, minimizes environmental impact	Potential contamination, requires pretreatment

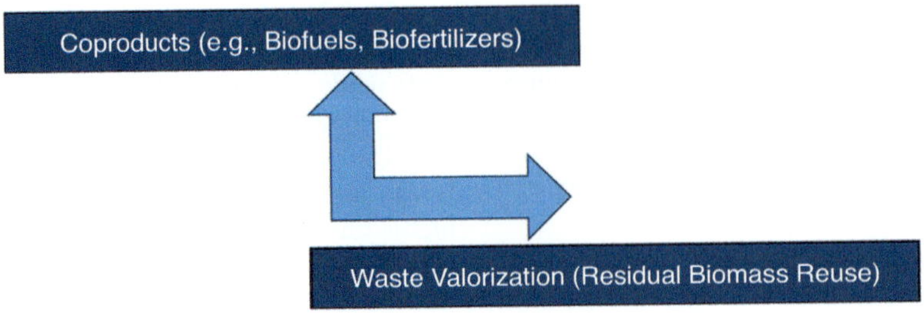

Fig. 19.2. Circular bioeconomy model.

while simultaneously recovering residual biomass for biofertilizer production.

In addition, innovative dark fermentation systems that are paired with microalgae cultivation have been explored to attain zero waste objectives. Lacroux *et al.* (2023) elaborated on how these integrated systems can effectively utilize organic waste substrates for biogas production, with the effluent generated serving as a nutrient-rich medium for subsequent microalgae cultivation. The microalgal biomass harvested from these systems can then undergo further processing to extract pigments and other valuable metabolites, embodying the principles of resource efficiency and sustainability.

19.3.3 Cost and resource efficiency in substrate selection

The process of selecting substrates is crucial in ensuring that pigment production processes are both economically feasible and sustainable. The costs associated with substrates can significantly impact the overall economic viability of production, making it imperative to utilize low-cost, renewable, and locally available feedstock. Microalgae and fungi have exhibited impressive adaptability to a wide variety of substrates, including wastewater, agricultural residues, and byproducts from food processing. Ahmad *et al.* (2022) conducted research that indicated microalgae can effectively utilize wastewater as a growth medium, leading to reduced production costs and minimized environmental impacts. Equally, Sharma *et al.* (2021) highlighted the considerable potential for food waste valorization in microbial pigment production, achieving high yield rates while simultaneously addressing waste management challenges.

In terms of resource efficiency, careful substrate selection is paramount. Rengel *et al.* (2022) investigated the simultaneous production of carotenoids and chemical building blocks from *Chlamydomonas reinhardtii*, underscoring the importance of optimizing both growth conditions and substrate utilization to maximize overall productivity. The utilization of lignocellulosic biomass as a substrate, as discussed by Lu *et al.* (2022) further emphasizes the need for effective pretreatment and fermentation technologies aimed at enhancing sugar availability and reducing processing costs. Moreover, alongside the economic considerations, the environmental sustainability of substrate selection plays a critical role. Leveraging agro-industrial waste not only diverts potentially harmful waste away from landfills but also effectively mitigates greenhouse gas emissions associated with waste decomposition. As pointed out by Cheirsilp *et al.* (2023), the integration of sustainable and circular practices in the development of feedstock utilization strategies is essential to fostering a more resilient bioeconomy that addresses both economic and environmental challenges.

19.4 Environmental Impact Assessment of Biopigment Production

The environmental sustainability of biopigment production, particularly when comparing microalgae-derived pigments to synthetic dyes, necessitates a thorough evaluation of various factors. These include resource consumption, waste generation, and overall environmental emissions. With the growing need for sustainable production methods, microalgae are gaining attention as a promising alternative due to their intrinsic ability to utilize waste streams and their significantly lower carbon footprint relative to conventional synthetic pigment manufacturing processes.

19.4.1 Life cycle assessment of biopigments versus synthetic dyes

A life cycle assessment (LCA) serves as a comprehensive framework for evaluating the environmental impacts associated with every stage of a product's life, from raw material extraction through production, use, and disposal. In the context of biopigments sourced from microalgae compared to synthetic dyes, the LCA uncovers numerous sustainability advantages associated with biopigments.

Biopigments such as carotenoids and chlorophyll are derived from microalgae through a natural process of photosynthesis, where carbon dioxide serves as the primary input. This form of cultivation allows for production with minimal external inputs, thus requiring significantly less water and energy. In contrast, traditional synthetic dye production is heavily reliant on petroleum-based raw materials, which necessitates substantial water and energy resources and results in toxic chemical byproducts. Moreover, microalgae cultivation presents an opportunity for integration into a biocircular green economy. This innovative approach enables the cultivation of microalgae on nonarable land, utilizing wastewater and CO_2 emissions, thereby alleviating pressure on freshwater resources and fostering a circular approach to waste management. The zero waste principle inherent in microalgal biomass production ensures that all byproducts, such as proteins, lipids, and carbohydrates, can be valorized and further processed for use in various sectors, including biofuels, biofertilizers, and nutraceuticals. These practices collectively contribute to a significantly reduced environmental footprint compared to synthetic dye manufacturing, which is characterized by the generation of hazardous waste.

Recent studies, including those conducted by Cheirsilp *et al.* (2023), reveal that microalgae production systems can be optimized as biorefineries. In such systems, different fractions of algal biomass are utilized in a cascading format that enhances resource efficiency and minimizes waste generation. Thus, microalgae emerge as an exemplary and sustainable alternative to synthetic dyes.

19.4.2 Carbon footprint and energy consumption

The carbon footprint and energy consumption linked to biopigment production from microalgae are markedly lower than those resulting from the production of synthetic dyes (Table 19.7). This reduction in carbon footprint is attributable to the inherent capability of algae to sequester carbon dioxide during the photosynthetic process. The optimization of this process can be further enhanced by integrating carbon capture technologies, allowing for dual benefits: the production of biopigments while simultaneously mitigating greenhouse gas emissions. In optimal conditions, the production process can reach carbon neutral status or even become carbon negative in certain scenarios.

From an energy consumption perspective, algal cultivation systems require significantly less energy compared to traditional synthetic dye production. The latter often entails high temperature processes, chemical synthesis, and the generation of toxic byproducts, all of which demand substantial energy input. Conversely, biopigment production predominantly relies on sunlight or efficient artificial lighting systems that dramatically cut energy consumption. Recent advancements in the optimization of light systems for microalgae cultivation, highlighted by Parveen *et al.* (2023), emphasize the potential for further reductions in energy use. Microalgae thrive on photosynthetically active radiation (PAR), which is abundant and less energy intensive compared to the chemical processes involved in synthetic dye production. In addition, microalgal cultivation can benefit

Table 19.7. Carbon footprint and energy consumption.

Metric	Biopigments from microalgae	Synthetic dyes
Carbon footprint (kg CO_2/kg)	0.5–1.0	5–10
Energy consumption (MJ/kg)	50–100	300–500
Waste generated (kg/kg)	Minimal	High (toxic byproducts)

from integration into closed loop systems. As demonstrated in recent studies, waste heat from industrial processes or power generation can be repurposed to warm algal cultivation ponds, thus enhancing energy efficiency and sustainability in biopigment production.

On the other hand, the production of synthetic dyes is associated with a significantly higher carbon footprint. The reliance on petrochemical feedstocks, the complexity of chemical synthesis processes, and the high energy demands contribute to this elevated environmental impact. Additionally, the large quantities of synthetic dyes required in industries such as textiles lead to severe environmental consequences when disposed of improperly, resulting in extensive water contamination and lasting detrimental effects on aquatic ecosystems and public health.

19.4.2.1 Sustainable processing and circular economy models

A key aspect of the sustainability of microalgae as a source of biopigments is their alignment with circular economy principles. The biomass generated from microalgae is rich in bioactive compounds and can be repurposed for various applications beyond pigment production, including biofuels, bioplastics, and animal feed. Moreover, *S. betaceum* fruit waste has been identified as a valuable source of bioactive compounds with potential uses in both food and nonfood industries (Isla *et al.*, 2022). This multifaceted value-added approach not only minimizes waste generation but also optimizes resource utilization, creating a comprehensive and sustainable system. Microalgae can thrive in waste environments, such as industrial effluents, municipal wastewater, and CO_2-rich emissions, allowing for the effective recycling of resources and promoting environmental remediation.

By contributing to a more sustainable industrial ecology, microalgal production systems exemplify the potential of integrating biological processes into the broader framework of sustainable manufacturing and resource management, ultimately paving the way for greener alternatives in the pigment industry.

19.4.3 Water usage and effluent management

Microalgae have emerged as a highly promising resource for biopigment production due to their remarkable water efficiency compared to traditional agricultural crops and synthetic pigment manufacturing processes. One of the key advantages of utilizing microalgae is that their cultivation often employs innovative closed loop systems or semicontinuous production methods. These methods significantly minimize water wastage by recycling water within the growth system, making them substantially more efficient than typical farming practices that require large volumes of fresh water.

Certain microalgal species, notably *Chlorella* and *Spirulina*, possess the unique ability to flourish in less than ideal water conditions, such as brackish water or untreated wastewater. This characteristic not only supports large-scale biopigment production but also provides an effective means of wastewater treatment. By using wastewater as a growth medium, these microalgae can help purify polluted water sources, thus offering a dual benefit: reducing the demand for fresh water while simultaneously treating a byproduct that would typically be discarded.

Effective effluent management is essential in the context of microalgal biopigment production. The introduction of wastewater into microalgal cultures can potentially lead to the accumulation of harmful pollutants and excess nutrients in the resulting effluent. Therefore, it is crucial to implement robust management systems designed to treat these effluents before their release into natural water bodies. If not adequately treated, such effluents may contribute to environmental degradation, including nutrient pollution that can lead to harmful algal blooms. Recent research has focused on developing zero waste production approaches that integrate biorefining systems. In these systems, residual biomass and waste can be further processed to create additional valuable products, including biodiesel, biofertilizers, or high-value metabolites. This circular economy model enhances resource efficiency and mitigates the environmental impact associated with waste disposal. Furthermore, the integration of biorefinery technologies into biopigment production allows for the utilization of nonarable land and saline water sources. This strategy minimizes the competition for freshwater resources that is particularly critical in regions facing water scarcity and reduces the overall environmental footprint of production. Consequently, the implementation of integrated systems and closed loop water management practices represents a compelling pathway towards sustainable and environmentally responsible biopigment production.

19.4.4 Environmental benefits of biopigment biodegradability

The biodegradability of biopigments derived from microalgae and fungi presents a significant environmental advantage over synthetic dyes, which are notorious for their persistence in the environment. Synthetic dyes often contribute to severe soil and water pollution, with many substances proving toxic to aquatic life and requiring intensive, costly treatment methods to remove from effluent streams. In contrast, biopigments demonstrate a far more favorable environmental profile, as they typically undergo faster and more complete biodegradation in natural conditions. For instance, naturally occurring pigments like carotenoids and phycocyanin, commonly produced by various microalgae, have shown a strong capacity to break down in soil and aquatic environments, thus reducing the long-term environmental impacts associated with their use. The biodegradation process is often facilitated by indigenous microorganisms that metabolize these pigment compounds into harmless substances, thus significantly lowering the need for expensive clean-up operations.

Along with their inherent biodegradability, the incorporation of biopigments into eco-friendly products such as biodegradable packaging, natural dyes for textiles, and organic cosmetics further contributes to the reduction of environmental pollution. As these natural pigments lead to the creation of products that are compostable or recyclable, they align with the principles of the circular economy which aims to close the material use loop. This characteristic also enhances their market appeal in sectors like food packaging, textiles, and cosmetics, where consumers increasingly prioritize sustainability.

Innovative edible packaging systems derived from food processing byproducts offer sustainable solutions for food preservation and safety (Hamed *et al.*, 2022). Probiotic and synbiotic edible films and coatings have been shown to effectively functionalize and bio-preserve processed food products, enhancing their safety and nutritional properties (Hellebois *et al.*, 2020). Recent advances in sustainable materials have led to the development of innovative bio-based films combining bacterial nanocellulose with poly(hydroxyoctanoate), activated by prodigiosin pigment. These double-layer films exhibit excellent transparency alongside strong UV-blocking capabilities, presenting eco-friendly alternatives for packaging and protective applications (Malagurski *et al.*, 2024). Such materials align with the increasing demand for biodegradable, nontoxic, and multifunctional biopigment applications, further expanding the role of biopigments beyond coloration to functional material science.

Beyond the processing and disposal stage, the environmental advantages of biopigment biodegradability extend to reducing toxicity levels associated with various industrial processes. The harmful effects of petrochemical-based dyes on ecosystems and human health can be significantly mitigated by replacing these synthetic pigments with biodegradable alternatives. As industries adopt biopigments, they play a crucial role in decreasing the prevalence of persistent pollutants, endorsing cleaner production methodologies, and promoting a more sustainable product life cycle.

In summary, the production of biopigments from microalgae and fungi offers an array of environmental benefits, particularly concerning water usage, wastewater management, and biodegradability. By embracing principles such as the circular economy and sustainable biorefining practices, the environmental footprint of biopigment production can be substantially reduced. This strategic approach positions biopigments as an essential component in the transition toward a more sustainable, biobased economy, offering both ecological advantages and commercial viability.

Factors in sustainable processing include the following.

- *Purpose*: the resource flow in sustainable biopigment production.
- *Inputs*: CO_2, wastewater, sunlight.
- *Processes*: cultivation → extraction → biorefining → valorization.
- *Outputs*: pigments, biofuels, biofertilizers.

19.5 Economic and Industrial Feasibility of Biopigment Production from Microalgae

19.5.1 Cost-effectiveness of scaling biopigment production

The scaling of biopigment production from microalgae is a promising avenue for sustainable industrial applications, particularly in sectors like food, cosmetics, and textiles. However, the economic feasibility of large-scale biopigment production relies heavily on reducing production costs and optimizing resource utilization (Table 19.8). Research including Udayan *et al.* (2022) and Cheirsilp *et al.* (2023) provides insights into the potential and challenges of large-scale microalgae cultivation. These studies emphasize the necessity of integrating efficient biorefinery practices that utilize microalgae biomass for multiple products, including pigments.

A significant portion of the production cost is tied to cultivation and harvesting methods (Fig. 19.3). Traditional methods, such as open ponds and photobioreactors, require substantial energy input and maintenance. Recent advancements in optimizing light systems and photoperiods, as discussed by Parveen *et al.* (2023), suggest that bioreactor design and operational conditions play a crucial role in enhancing yield while lowering costs. Additionally, improving biomass yield per unit area can significantly reduce land and resource usage, thereby contributing to cost-effectiveness.

Another critical factor is the reduction of downstream processing costs. The extraction of biopigments, particularly carotenoids and chlorophylls, is labor intensive and requires solvents or mechanical methods that increase operational costs. However, methods such as enzyme-assisted extraction or novel green solvents could potentially reduce these expenses, as highlighted in Rengel *et al.* (2022). Furthermore, coupling the pigment production with other valuable coproducts like biofuels and biofertilizers, as seen in zero waste biorefining, can improve overall profitability by maximizing the utilization of raw materials and byproducts.

Table 19.8. Economic comparison between biopigment production and synthetic dye production.

Aspect	Biopigments (microalgae)	Synthetic dyes
Raw material cost	Low (e.g. wastewater)	High (petroleum based)
Energy requirements	Low to moderate	High (chemical synthesis)
Waste generation	Minimal (zero waste)	Significant (toxic byproducts)
Revenue streams	Multiple (e.g. coproducts)	Single
Carbon footprint	Low (carbon negative potential)	High
Regulatory complexity	Moderate	High

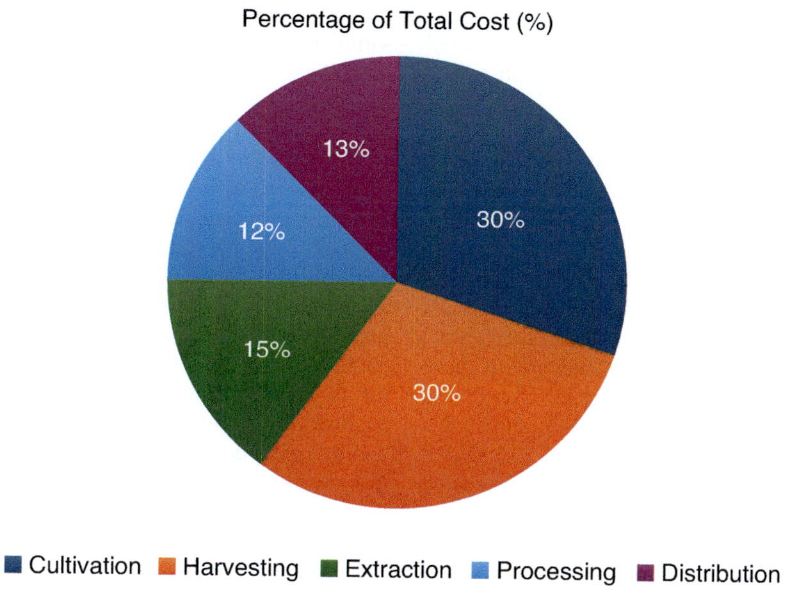

Fig. 19.3. Cost distribution of biopigment production.

19.5.2 Coproducts and byproducts in integrated biorefineries

Microalgae-based biorefineries, which integrate the production of various high-value coproducts along with pigments, offer an opportunity to improve the economic feasibility of biopigment production. These coproducts include biofuels, biofertilizers, lipids, proteins, and polyunsaturated fatty acids. A zero waste approach is vital for reducing environmental impact and maximizing resource efficiency. By converting the residual biomass into valuable coproducts, the biorefinery can offset production costs, making the biopigment business more economically viable.

The biorefinery model proposed in Rengel *et al.* (2022) demonstrates how multiple outputs can be derived from a single microalgal strain. For instance, while carotenoids are extracted for use as pigments, other valuable compounds such as starch and lipids can be harvested for use in biodiesel production. This integrated system not only ensures the sustainability of the production process but also enhances the economic feasibility by generating diverse revenue streams. Furthermore, food waste valorization, as discussed in Sharma *et al.* (2021), offers an additional avenue for integrating microalgae cultivation with other waste to value processes. By incorporating food waste as a feedstock for microalgae, the cost of biomass production can be significantly lowered, as food waste is often a low-cost and abundant resource. This approach aligns with circular economy principles, ensuring the sustainability of both the algae-based production system and the food industry.

19.5.3 Regulatory and policy landscape

The regulatory and policy landscape surrounding the commercialization of biopigments derived from microalgae is multifaceted and plays a pivotal role in either facilitating or obstructing progress in this innovative field. Government regulations and policies significantly influence the viability of scaling up production by affecting critical elements such as funding opportunities, market accessibility, and adherence to environmental standards.

Microalgae cultivation and their subsequent biorefining processes are subject to stringent regulatory frameworks aimed at ensuring safety and compatibility with environmental standards. Specifically, when it comes to applications in food and cosmetic products, the use of biopigments must align with established food safety regulations and environmental health guidelines. Regulatory bodies, such as the European Food Safety Authority (EFSA) and the US Food and Drug Administration (FDA), have delineated comprehensive guidelines governing the safe application of microalgal pigments in food products.

Recent research, including findings from Anum *et al.* (2024), emphasizes how regulatory frameworks can dictate the manufacturing processes employed, particularly focusing on those methods that minimize negative environmental impacts. In an era of shifting policies toward a circular bioeconomy, as emphasized in research such as Cheirsilp *et al.* (2023), there has been a marked increase in governmental initiatives aimed at bolstering sustainable bio-based industries.

This proactive political approach is vital for fostering investment from the private sector into algae-based biorefineries, thereby creating a more robust economic incentive for stakeholders. However, significant challenges persist, notably in the realm of harmonizing regulatory standards across diverse geographical locations. The commercialization of microalgae-based products, particularly biopigments, often encounters regulatory obstacles associated with product certification and the safety assessments necessary for nonfood applications, including uses in textiles or cosmetics. The study by Suwannarach *et al.* (2019) illustrates that the approval processes for natural pigments across varying industries mandate comprehensive safety and quality evaluations, which can substantially delay the timeline for market entry. Furthermore, the regulatory environment concerning the sustainability of algae production is growing increasingly stringent. For example, the overapplication of fertilizers or pesticides in algae cultivation poses inherent risks to local ecosystems and biodiversity.

Nevertheless, recent advancements in sustainable cultivation methodologies, such as leveraging waste streams as nutrient sources for algae growth, discussed in Leong and Chang (2022), have started to address and alleviate these ecological concerns. Such innovations underscore the importance of a well-structured regulatory framework that incentivizes and rewards sustainable practices, which in turn can stimulate the development and expansion of the microalgae biopigment sector.

In summary, the economic viability and industrial feasibility of producing biopigments from microalgae are fundamentally intertwined with advancements in cost-effective cultivation and extraction technologies, the strategic integration of byproducts and coproducts, and a favorable regulatory milieu that champions sustainability. The research findings from the aforementioned studies illustrate that while challenges are evident, the prospect of microalgae as a cornerstone of a biocircular green economy is significant, bolstered by ongoing technological developments and evolving policy frameworks that advocate for environmentally conscious growth.

19.6 Challenges in Sustainable Biopigment Production

The production of biopigments, particularly those sourced from microalgae and fungi, presents a compelling alternative to synthetic dyes, characterized by their versatility across a multitude of industries. Environmental contaminants can disrupt biopigment-producing organisms; Bouchnak *et al.* (2024) demonstrated that polyethylene microplastics alter the toxicity of pentachlorophenol to the microalga *Isochrysis galbana*, showing how pollutant interactions may impact pigment yield and quality. Despite their considerable potential, several significant challenges impede the large-scale sustainable production of these natural pigments. This section elaborates on three principal challenges: optimizing yield and process efficiency, ensuring stability while meeting application-specific requirements, and navigating market competition with established synthetic dyes (Fig. 19.4).

19.6.1 Yield optimization and process efficiency

A foremost challenge in sustainable biopigment production lies in effectively optimizing pigment yields. The primary biological sources of biopigments are microalgae, fungi, and plants that exhibit considerable variability in their growth

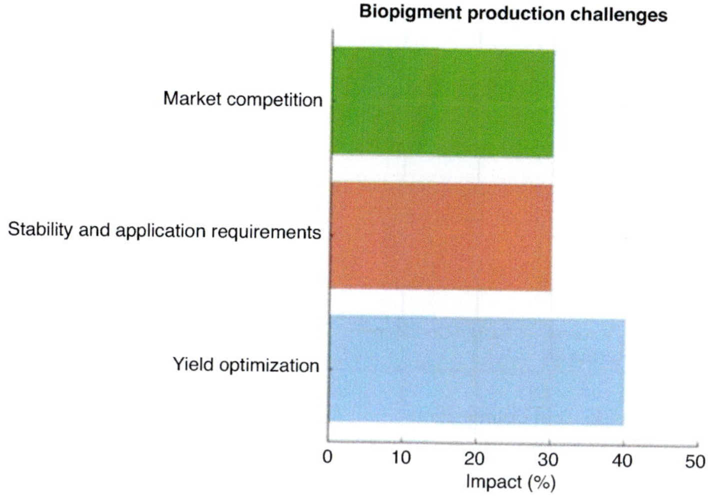

Fig. 19.4. The proportion of challenges faced in yield optimization, stability, and market competition.

and pigment synthesis, which are often influenced by a multitude of factors. These include nutrient availability, varying light conditions, temperature fluctuations, and distinct cultivation methodologies. For instance, the microalga *Chlamydomonas reinhardtii* is recognized for its ability to produce valuable pigments such as carotenoids and chlorophyll. However, the yield of these pigments can dramatically fluctuate based on the specific cultivation conditions employed, as evidenced by recent studies that highlight such variability (Figs. 19.5 and 19.6).

In pursuit of enhanced pigment production, researchers are exploring various cultivation techniques. These include photobioreactors that maximize light exposure and nutrient uptake, large-scale open pond systems that leverage natural sunlight, and innovative bioreactor designs that aim to create optimal growth conditions. Nonetheless, scaling these systems presents its own set of challenges, as maintaining optimal growth conditions over extensive areas or large volumes often leads to inefficiencies. Ultrasound-assisted extraction of betalain from prickly pear has been optimized to improve *in vitro* digestive stability, enabling the development of functional food products such as gummies with enhanced bioactive properties (Mehta *et al.*, 2024). This approach represents a promising direction in maximizing the utility of natural pigments in health-promoting applications. Furthermore, the processing steps necessary for extracting biopigments from biomass, particularly in microalgae, frequently include energy-intensive procedures. These processes can significantly compromise the overall environmental benefits associated with utilizing natural dyes. For instance, extracting carotenoids and other bioactive compounds from microalgae typically requires methods such as solvent extraction or mechanical disruption of cellular structures. These techniques can be neither cost-effective nor environmentally friendly, thereby diminishing the sustainability aspect of biopigment production.

Recent advances in plant physiology have shown that the combined application of physcion and chitosan-oligosaccharide (COS) significantly improves crop yield by enhancing photosynthetic efficiency and resilience. Specifically, in wheat (*Triticum aestivum* L.), this synergistic treatment boosts the plant's ability to withstand environmental stresses while increasing productivity (Li *et al.*, 2023). These findings support the growing interest in using natural bioactive compounds to improve crop performance sustainably. Therefore, it is imperative to optimize the entire biopigment production process, encompassing every stage from cultivation to extraction, to achieve reduced production costs while enhancing sustainability. Advancements in metabolic engineering,

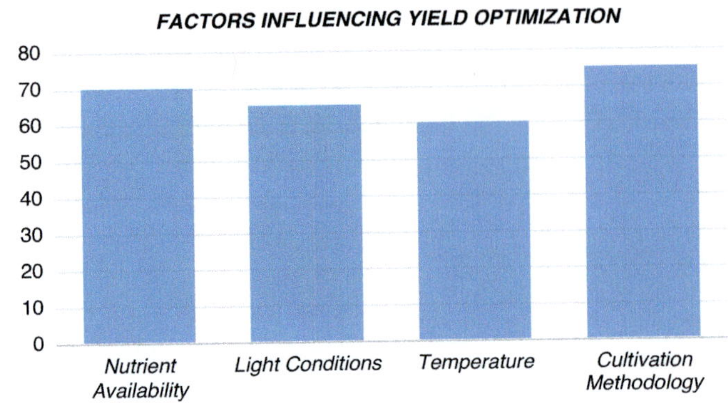

Fig. 19.5. Factors influencing yield optimization.

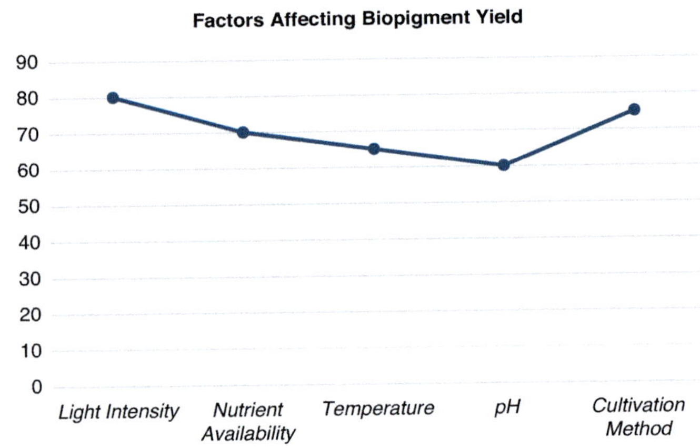

Fig. 19.6. Key factors impacting biopigment yield.

synthetic biology, and bioprocess optimization hold promise for improving both the yield and efficiency of biopigment production. By focusing on innovative strategies and techniques, the sector can work towards overcoming these critical challenges and unlocking the full potential of biopigments.

19.6.2 Stability, color diversity, and application-specific requirements

A significant hurdle in the production of biopigments involves ensuring their stability after synthesis. Biopigments often encounter variability in color stability when subjected to various environmental factors. Key factors such as UV radiation, temperature fluctuations, and shifts in pH levels can significantly affect the stability and performance of these pigments. This aspect is particularly crucial in industries like food coloring and textile dyeing, where the aesthetic quality and longevity of the product are paramount. If biopigments degrade under these conditions, the resultant fading or discoloration can severely impact consumer satisfaction and product viability.

Beyond plants and microorganisms, pigmentation plays a significant role in functional agro-biodiversity, notably illustrated by the genetic variability in coat color genes among goat populations. Such genetic diversity not only affects aesthetic traits but also has implications for adaptation and breed conservation (Nicoloso *et al.*, 2012).

Despite being widely studied, the application of biopigments in food and cosmetics is often limited due to their susceptibility to color fading over time. This instability not only jeopardizes the visual appeal of products but also diminishes their nutritional value and functional efficacy. Additionally, the spectrum of colors achievable through biopigments often falls short compared to that of synthetic dyes, which boast a more extensive and varied palette. This limitation necessitates the development of improved biopigment production systems that can not only ensure better color stability but also broaden the range of available hues tailored to specific applications.

In the textile industry, there is a growing trend toward utilizing natural pigments derived from diverse biological sources, such as fungi, algae, and plants, as environmentally friendly alternatives to traditional synthetic dyes. While this movement aligns with increasing consumer demand for sustainable products, significant challenges remain. The durability of these natural dyes is a crucial concern; their resistance to factors such as light exposure, washing, and abrasion is often inferior compared to synthetic counterparts. The incorporation of modified lignin into bio-bitumen coating sheets has been shown to significantly enhance their aging properties, thus improving durability and environmental performance. Ren *et al.* (2020) demonstrated that such modifications lead to better resistance against degradation factors, supporting the development of more sustainable and longer-lasting bio-based materials. This approach complements the growing emphasis on utilizing natural biopolymers to enhance the functionality of eco-friendly industrial products.

Ensuring that biopigments can meet the stringent requirements of different sectors, including food-grade purity, long-lasting color, and exceptional stability, is essential for widespread adoption. Continued research is needed to better understand the performance of these pigments under varying environmental conditions, thereby optimizing their practical applications across different industries.

19.6.3 Addressing market competition with synthetic dyes

One of the most formidable challenges facing the sustainable biopigment industry is the overwhelming market competition posed by synthetic dyes (Fig. 19.7). Synthetic dyes currently dominate the global market due to their low production costs, ease of manufacturing, and extensive array of available colors. Typically produced from petrochemical feedstocks, the manufacturing processes for these dyes often involve harmful chemicals and pollutants, raising significant concerns about environmental sustainability. Despite their known drawbacks, synthetic dyes remain the first choice in many industrial applications because of their cost-effectiveness and reliable performance.

For biopigments to gain a foothold in this competitive landscape, they must be produced at commercial scale while also striving to maintain or reduce overall production costs. Achieving commercial viability represents a considerable challenge, as the production processes associated with natural pigments tend to be more resource intensive and less predictable than those used for synthetic dyes. The biopigment market is still in its nascent stages, and the absence of established infrastructure and processing technologies adds another layer of complexity to their commercial success. While there is a growing interest among consumers for natural and sustainable products, many may hesitate to pay a premium for biopigments unless their advantages are demonstrably communicated in terms of sustainability, safety, and overall product quality. This presents a challenge for industries looking to pivot from synthetic dyes to biopigments, as it necessitates a multifaceted approach that combines technological innovations, reductions in production costs, and educational initiatives that highlight the long-term environmental and health benefits of switching to natural pigments.

The potential for biopigments derived from natural sources is immense when considering the shift toward sustainable and eco-friendly colorants. However, substantial challenges surrounding yield optimization, pigment stability, and competition with synthetic dyes must be decisively addressed for biopigments to emerge as a mainstream alternative in various industries (Fig. 19.8). Advances in biotechnology, such as the genetic modification of microorganisms and enhancements in bioreactor systems, can significantly improve both the efficiency and cost-effectiveness of biopigment production, as explained by Banerjee *et al.* (2020) who showed that genome-scale metabolic rewiring can dramatically increase titers, rates, and yields of the nonnative product indigoidine at industrial scale.

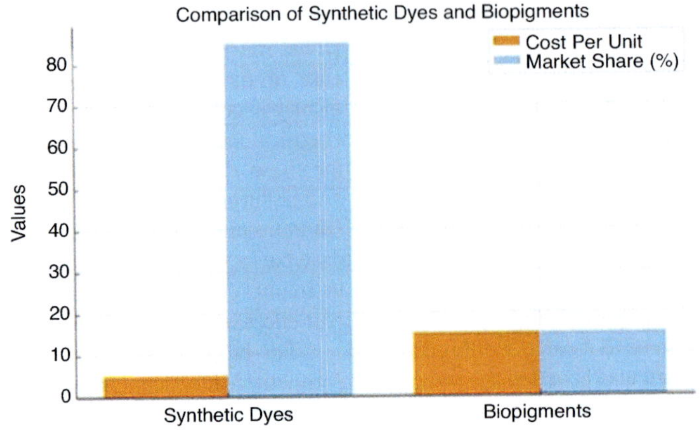

Fig. 19.7. Comparison of synthetic dyes and biopigments in cost per unit and market share.

Furthermore, undertaking further research into the stability and diversity of biopigments can unlock new opportunities for their application in an even broader range of products. By overcoming these multifaceted challenges, biopigments can make a meaningful contribution to the biocircular green economy, positioning themselves as key players in the pursuit of a more sustainable and environmentally conscious future.

19.7 Applications of Biopigments in the Green Industry

Biopigments have emerged as a promising alternative in multiple sectors of the green industry. These pigments are not only sustainable and biodegradable, but they also offer numerous functional benefits that contribute positively to environmental health, human wellness, and economic development. This section delves into the diverse applications of biopigments, particularly in the food industry, cosmetics, and textiles, highlighting their significant role in promoting sustainability and tackling pressing environmental issues.

19.7.1 Food industry: natural colorants for enhanced nutrition

In the food industry, biopigments have garnered substantial attention as natural colorants that can effectively replace synthetic dyes, which are often linked to various health risks and environmental concerns. Notably, microalgae stand out as a particularly rich source of bioactive pigments such as carotenoids, chlorophylls, and phycobiliproteins. These compounds not only enhance the visual appeal of food products through vibrant colors but also contribute to improved flavor and nutritional value. The rising consumer demand for clean-label products has catalyzed the transition toward these natural alternatives. For instance, biopigments such as astaxanthin and lutein, which are extracted from specific strains of microalgae, are increasingly used in the food industry due to their remarkable health benefits. Astaxanthin is known for its strong antioxidant properties, which can help mitigate oxidative stress, while lutein is celebrated for its role in promoting eye health and reducing the risk of age-related macular degeneration.

Advances in green chemistry have fostered the development of nature-inspired catalytic systems such as porphyrin N-pincer Pd(II) complexes, which enable efficient oxidative self-coupling reactions of potassium aryltrifluoroborates in water without the need for a base. This base-free protocol operates under open-air conditions, exemplifying sustainable approaches to chemical synthesis with reduced environmental impact (Prasad *et al.*, 2021). Such innovations align with the broader goals of environmentally friendly bioprocessing and chemical production. Moreover, microalgae-derived pigments serve as natural preservatives, enhancing the shelf-life of food products while minimizing the need for synthetic preservatives. For example, carotenoids

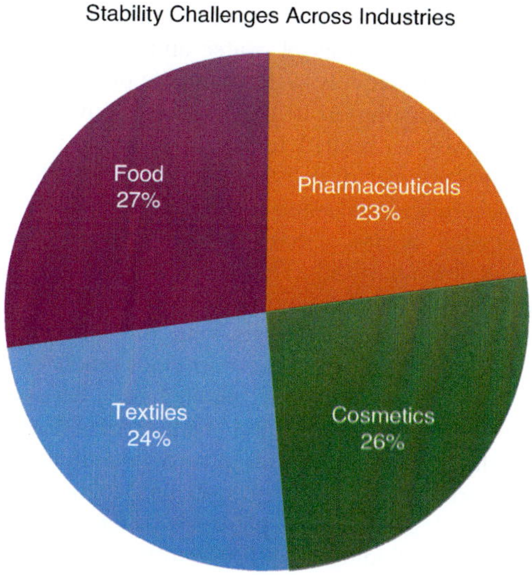

Fig. 19.8. Stability challenges across various industries.

like β-carotene not only impart vibrant colors to various products, including dairy items, beverages, and snack foods, but also provide essential nutrients that boost immune function and skin health.

The shift towards utilizing biopigments is intricately linked to the principles of the BCG. By adopting sustainable production methods that prioritize renewable resources and resource efficiency, the food industry can significantly reduce its environmental footprint. This transition not only minimizes reliance on synthetic chemicals but also fosters a more sustainable model of production that emphasizes resource conservation. According to recent studies, microalgae play a crucial role in advancing a zero waste approach in food production. Beyond being a source of pigments, they produce a wealth of other high-value compounds such as lipids and proteins. These byproducts can be integrated into biorefinery systems, creating a comprehensive waste reduction strategy that maximizes the utility of every component harvested from microalgae.

In summary, the integration of biopigments into the food industry not only enhances product appeal and nutritional quality but also aligns with a broader movement toward sustainability, health, and waste minimization, key tenets for future food systems.

19.7.2 Cosmetics: biocompatible and hypoallergenic applications

The cosmetics industry has traditionally relied on synthetic dyes for coloring products, but recent trends indicate a transformative shift towards biocompatible, natural alternatives. Biopigments are emerging as eco-friendly substitutes that not only reduce the risk of skin irritation and allergic reactions but also have a lower impact on the environment. Natural pigments such as those obtained from microalgae exemplify this trend. Phycocyanin, a striking blue pigment found in spirulina, is increasingly featured in skincare formulations due to its remarkable antioxidant and antiinflammatory properties. These properties work to defend the skin from damage associated with oxidative stress, which can be exacerbated by factors such as UV radiation, pollution, and natural aging. The incorporation of biopigments like carotenoids, known for their ability to enhance skin hydration and minimize the appearance of fine lines and wrinkles, is becoming commonplace in a variety of products, from antiaging creams to sunscreens.

In addition to their ecological and industrial significance, biopigments and plant extracts have demonstrated valuable bioactive properties under environmental stress. For instance, *in vitro* cultures of *Thymus lotocephalus* exhibited adaptive responses to drought stress, with green extracts showing promising antioxidant activity and potential for use in cosmetic formulations (Mansinhos *et al.*, 2022). This highlights the multifaceted applications of plant-derived compounds, extending their relevance beyond traditional uses into innovative sectors such as cosmetics.

As the "Green Beauty" movement gains traction, consumer demand for safe, sustainable, and chemical-free products continues to rise. Biopigments align perfectly with these consumer values, being derived from natural and renewable resources while offering superior skin benefits without the harmful side-effects commonly associated with synthetic additives. Scientific research has further highlighted the potential of optimizing the conditions under which microalgae are cultivated by using different light systems and photoperiods to enhance the yield and quality of the extracted pigments.

This emphasis on cultivation and extraction not only improves health benefits but also promotes the sustainable and ethical development of ingredients within the cosmetics sector.

19.7.3 Textiles: alternatives to hazardous synthetic dyes

The textile industry is heavily reliant on synthetic dyes, which are notorious for their toxicity and nonbiodegradability, contributing extensively to environmental degradation. In stark contrast, biopigments are emerge as a promising alternative, epitomizing sustainability through their biodegradable nature and sourcing from renewable materials. Key contributors to this initiative include fungi and algae, with ongoing research dedicated to optimizing their production processes for large-scale application.

Fungal pigments, like those derived from *Nigrospora aurantiaca*, exhibit vibrant red hues that are highly suitable for textile dyeing. These pigments are produced under meticulously controlled conditions, ensuring high yield, color stability, and commercial viability as a substitute for synthetic dyes. Additionally, microalgal pigments sourced from species such as *Chlorella* and *Spirulina* are being used to color fabrics in a manner that is both eco-friendly and efficacious. These natural pigments not only produce rich, desirable colours but are also nontoxic, lacking the heavy metals and harmful substances prevalent in many synthetic dyes but they can also benefit from advanced nanoscale approaches, such as the confinement of photosystem I within metal-organic frameworks (Bennett *et al.*, 2019), for enhancing photosynthetic processes that could improve biopigment extraction efficiency.

Adopting biopigments in the textile industry is in alignment with the principles of a circular economy. This shift reduces reliance on petrochemical-based dyes and helps minimize the overall waste generated in the dyeing process. Microalgae, for instance, have a remarkable ability to thrive in wastewater conditions, turning a potential environmental hazard into an opportunity for benefit. This dual function allows for the cleaning of polluted water while simultaneously harvesting valuable pigments, thus significantly lessening the environmental footprint of textile production compared to traditional methods that often consume substantial amounts of water and chemicals. The biodegradability of biopigments presents a sustainable solution to the long-term environmental issues related to textile waste. Unlike many synthetic dyes, which can take decades to decompose and often leach into waterways, harming aquatic ecosystems, biopigments break down naturally. This trait contributes to a textile industry that is not only environmentally responsible but also aligns with the growing consumer awareness of eco-conscious practices.

The integration of biopigments across the food, cosmetics, and textile industries highlights a profound shift towards sustainable, eco-friendly, and health-conscious production methodologies. These pigments offer many benefits, including enhanced nutritional profiles, biocompatibility with human skin, and a significant reduction in overall environmental impact. As the global demand for natural and biodegradable products escalates, the green industry is poised to incorporate biopigments as key components in its sustainability initiatives.

Ongoing research and innovation in biopigments, particularly through advancements in biorefinery systems and zero waste approaches, will be pivotal in driving progress within these sectors. By tapping into the natural capabilities of these resources, the green industry can effectively mitigate its ecological footprint while providing consumers with not just safer, but also more sustainable alternatives to synthetic chemicals. This holistic approach to the production and consumption underscores the transformative potential of biopigments, emphasizing their role in fostering a more sustainable and circular economy.

19.8 Future Directions and Innovations in Microalgal-Based Production Systems

19.8.1 Advances in bioreactor technologies

Bioreactors have become a pivotal component in the large-scale cultivation of microalgae, enabling efficient and sustainable biomass production. As the demand for microalgae-derived products, such as biofuels, pigments, and pharmaceuticals, grows, advances in bioreactor technologies are essential to maximize productivity, optimize resource usage, and reduce operational costs.

Recent developments in bioreactor design focus on enhancing light utilization, improving gas exchange, and integrating sensors for real time monitoring. One emerging trend is the development of photobioreactors (PBRs) with better light management systems. Traditional PBRs often suffer from issues related to light attenuation and photoinhibition in dense cultures. To address this, innovations like the use of optical fibers, internal light-emitting diodes (LEDs), and adjustable external reflectors are being explored to optimize light distribution. Studies have also shown that the incorporation of novel materials such as transparent and flexible substrates can improve light penetration and increase microalgae growth rates in PBRs. Additionally, new types of bioreactors, such as bubble column and airlift reactors, are being integrated with automated control systems to maintain optimal conditions for microalgal growth (Table 19.9). These systems ensure the precise regulation of parameters such as pH, temperature, and nutrient concentrations.

The use of artificial intelligence (AI) for process optimization has also gained attention, as it allows for continuous monitoring and adjustments based on sensor data, thereby enhancing productivity while minimizing human intervention.

19.8.2 Emerging trends in synthetic biology for pigment production

Synthetic biology is playing an increasingly important role in enhancing the bioperspective of microalgae, particularly in the production of high-value pigments. Pigments such as carotenoids, chlorophylls, and phycocyanin have applications in food, cosmetics, and pharmaceuticals. However, the commercial production of these pigments is often limited by the natural yield and efficiency of pigment biosynthesis pathways in algae. To overcome this challenge, synthetic biology approaches are being developed to engineer microalgae for higher pigment yields.

One promising avenue involves the modification of metabolic pathways to increase the flow of carbon towards pigment production. By introducing or enhancing specific genes involved in pigment biosynthesis, researchers have successfully improved pigment accumulation in microalgae including *Chlamydomonas reinhardtii* and *Scenedesmus* species. Genetic engineering techniques, such as CRISPR-Cas9 and synthetic promoters enable fine-tuning of pigment production pathways to enhance output without compromising growth rates. For example, the overexpression of genes involved in carotenoid biosynthesis has been shown to increase carotenoid production in several microalgal strains, such as *Dunaliella*. Additionally, the use of synthetic biology to integrate heterologous biosynthesis pathways is another innovative approach (Fig. 19.9). By introducing genes from other organisms, such as fungi or plants, into microalgae, it is possible to produce novel pigments with improved properties. This approach not only enhances the production of natural pigments but also paves the way for the synthesis of entirely new bio-based colorants for use in a variety of industries.

The field of synthetic biology also has potential for optimizing pigment production under specific environmental conditions. Advances in light management systems, including the integration of LED technology and automated light intensity regulation, can be used to optimize the growth environment for microalgae, ensuring that pigment production occurs under ideal light conditions. This targeted approach can help maximize pigment yields while reducing the need for expensive and resource-intensive cultivation systems.

19.8.3 Role of machine learning and AI in process optimization

The integration of machine learning (ML) and AI in microalgal biomass production is a game-changer for the optimization of bioreactor processes and the management of cultivation systems. As microalgal cultivation becomes more sophisticated and large scale, the need for automated systems that can predict and optimize various parameters such as nutrient concentrations, temperature, and light exposure is becoming more critical.

Table 19.9. Comparison of bioreactor types.

Bioreactor type	Key features	Advantages	Limitations
Flat panel photobioreactor (PBR)	High surface area, simple design	Good for light penetration	Limited scalability
Bubble column	Gas sparging for mixing	Low energy consumption	Moderate productivity
Airlift reactor	Circulation via air bubbles	Improved mixing and aeration	Higher maintenance costs
LED-integrated PBR	Internal/external light sources	High pigment yield, precise control	Expensive set-up

Machine learning algorithms are being utilized to model the growth dynamics of microalgae, predict their metabolic responses to changing environmental conditions, and identify optimal operating conditions. These models can incorporate a vast amount of data, including environmental factors (light, temperature, CO_2 levels) and cellular parameters (growth rate, pigment production). AI can process this data to predict how changes in one factor will affect the overall production process, helping operators make data-driven decisions for better outcomes. Furthermore, AI-driven control systems are being integrated with bioreactors to enable continuous real-time monitoring and adjustments (Table 19.10). For example, smart sensors coupled with machine learning models can predict when to adjust light intensity or nutrient levels to optimize pigment production or biomass yield. This type of automation can significantly reduce energy consumption and costs while maintaining a steady and high-quality output of microalgal products.

Another area where AI is making an impact is in the development of predictive maintenance systems for bioreactors. By analyzing sensor data, AI can predict when equipment may need maintenance, preventing costly breakdowns and improving overall system efficiency. This integration of AI not only improves the operational efficiency of microalgae production systems but also contributes to their sustainability by reducing waste and resource consumption.

The future of microalgal biomass production and pigment synthesis is poised for significant advancements, driven by innovations in bioreactor technologies, synthetic biology, and AI-based process optimization. By improving bioreactor designs, optimizing pigment biosynthesis through genetic engineering, and integrating machine learning for real-time monitoring, researchers are creating more efficient and sustainable systems for microalgal cultivation. These innovations are not only enhancing the economic viability of microalgal-based products but also supporting the transition toward a circular bioeconomy, where waste is minimized and resources are optimized. The continued development of these technologies holds immense potential for revolutionizing industries ranging from food and pharmaceuticals to textiles and renewable energy.

19.9 Conclusion

19.9.1 Summary of sustainability and environmental impacts

The integration of microalgae in the BCG economy marks a significant advancement in sustainable practices, emphasizing zero waste approaches for microalgal biomass production and biorefineries. Microalgae contribute to sustainability by producing biofuels, biodegradable chemicals, and essential metabolites while requiring fewer resources compared to conventional crops (Table 19.11). For instance, microalgae thrive with minimal water, which makes them a highly attractive option in water-scarce environments. Furthermore, these organisms offer significant environmental benefits, including the reduction of greenhouse gases and waste recycling. They can capture carbon dioxide, contributing to climate change mitigation efforts, and they provide a sustainable feedstock for

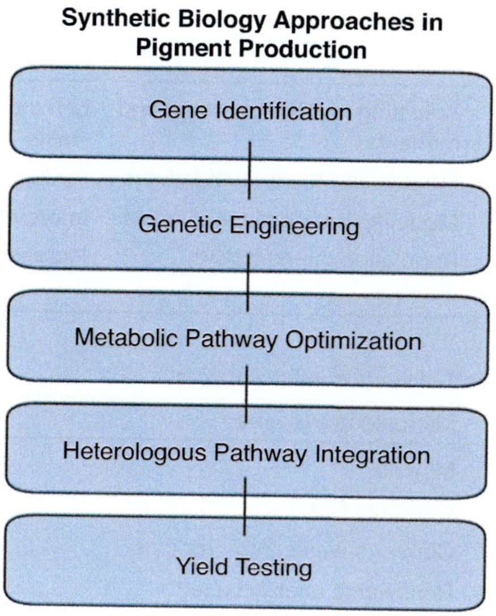

Fig. 19.9. Flowchart demonstrating synthetic biology approaches in pigment production.

bio-based products like biodiesel, biofertilizers, and valuable chemicals.

By utilizing food waste and industrial byproducts, microalgal cultivation complements circular economy models that reduce resource depletion and waste generation. Additionally, microalgae cultivation systems, when coupled with other waste-processing technologies, enhance waste valorization by converting organic materials into high-value products. This innovative approach aligns with environmental sustainability goals, positioning microalgae as key players in reducing the ecological footprint of industrial activities and fostering a cleaner environment.

19.9.2 Potential for transforming industrial practices

Microalgae possess the transformative potential to revolutionize various industrial sectors, from agriculture to bioenergy production. Their versatility as a sustainable source of biofuels, biochemicals, and nutraceuticals positions them as a critical alternative to fossil-derived resources. Microalgae can replace fossil fuels in several applications, such as biodiesel production, without the environmental consequences associated with traditional oil extraction processes. Essential oils such as jasmine and lavender, as well as mustard fixed oil, have been shown to possess significant acaricidal properties against two-spotted spider mites and to positively influence the growth and yield of eggplants (Farouk *et al.*, 2021). This shift could significantly reduce carbon emissions and foster a transition toward a greener energy landscape. Moreover, the bio-based chemicals produced by microalgae are nontoxic and biodegradable, and can be used to replace hazardous synthetic chemicals in industries including food packaging, pharmaceuticals, and cosmetics. For example, natural pigments derived from fungi and microalgae can replace synthetic dyes in the textile industry, reducing pollution caused by the toxic discharge of chemical dyes into water bodies. The enantioselective ecotoxicity of pharmaceuticals, such as venlafaxine, poses significant risks to aquatic organisms. Ribeiro *et al.* (2022) investigated the differential toxic effects of venlafaxine enantiomers on species like *Daphnia* and zebrafish, revealing varied sensitivity and ecological impacts. The widespread adoption of microalgal-based biorefining systems could thus streamline industrial processes, reduce waste, and help achieve more circular, resource-efficient production models across various sectors.

In addition to bioenergy and chemical production, microalgae could also reshape the aquaculture industry. As a sustainable alternative to

Table 19.10. Applications of AI in microalgal cultivation.

AI application	Functionality	Benefit
Process optimization	Adjusting pH, temperature, and nutrients	Enhanced productivity, reduced waste
Predictive maintenance	Forecasting equipment failures	Reduced downtime, cost savings
Growth dynamics prediction	Modeling metabolic responses	Improved decision making
Light management automation	Regulating intensity and duration	Increased pigment yield

Table 19.11. Comparative analysis of sustainability impacts.

Factor	Microalgae systems	Conventional systems
Water usage	Minimal	High
Carbon sequestration	Captures CO_2	Emits CO_2
Waste recycling	Converts waste into products	Limited recycling
Energy efficiency	Renewable energy based	Fossil fuel dependent
Environmental impact	Low	High (pollution, emissions)

fishmeal, microalgae are increasingly recognized for their potential to support the growing demand for seafood while reducing the strain on marine ecosystems. The use of microalgae in fish feed aligns with sustainable food security goals by ensuring that the aquaculture industry grows in an environmentally responsible manner.

19.9.3 Strategic steps for wider adoption

To facilitate the wider adoption of microalgal technologies, several strategic steps need to be taken. First, scaling up the production of microalgae to meet industrial demands requires addressing the challenges of large-scale cultivation and harvesting. Innovations in biorefining processes, such as the integration of ultrasonic-assisted methods and efficient algal harvesting techniques, are critical to improving the economic viability of microalgae-based systems. Investments in research and development, particularly in optimizing growth conditions and enhancing biomass yields, are essential to driving these advancements.

Second, building an infrastructure that supports the integration of microalgae into existing industries is crucial. This includes establishing efficient supply chains for raw materials, processing facilities, and distribution networks. Governments and private enterprises must collaborate to create policies and incentives that promote the adoption of biocircular practices, providing financial support and regulatory frameworks that encourage the development of microalgal-based solutions.

Third, public awareness and education around the environmental and economic benefits of microalgal systems are necessary to accelerate their acceptance. Engaging stakeholders from academia, industry, and the general public in dialogue can help overcome resistance to adopting new technologies and demonstrate the long-term advantages of sustainable practices. It is important to highlight the environmental and economic returns on investment associated with microalgae, such as reduced waste and emissions, as well as the potential to create new, high-value jobs in the bioeconomy sector.

Lastly, collaborative efforts between industries, research institutions, and policymakers are vital in advancing the commercialization of microalgae-based products. Developing a strong ecosystem that supports innovation and addresses technical, financial, and regulatory hurdles will pave the way for the widespread adoption of microalgal technologies. This collaboration could foster a thriving bio-based economy, unlocking the full potential of microalgae in creating a sustainable and circular future (Fig. 19.10).

In conclusion, microalgae stand at the forefront of sustainable industrial transformation, offering a versatile and eco-friendly solution to many of the challenges faced by current production systems. Through strategic investment, technological innovation, and widespread adoption, microalgae can drive the shift toward a circular, green economy, reducing waste and environmental impact while promoting sustainable industrial practices.

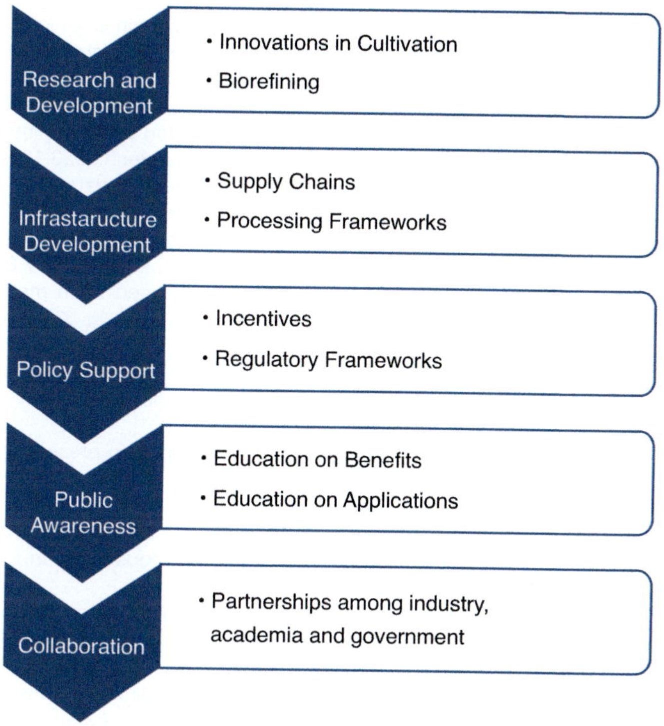

Fig. 19.10. Strategic steps for implementation.

References

Abdelkhalik, A., Abdou, N.M., Gyushi, M.A.H., Shaaban, A., Abd El-Mageed, S.A. *et al.* (2024) Enhancing sweet potato (*Ipomoea batatas*) resilience grown in cadmium-contaminated saline soil: A synergistic approach using moringa leaf extract and effective microorganisms application. *Environmental Science and Pollution Research International* 31(22), 32464–32479. DOI: 10.1007/s11356-024-33295-w.

Ahmad, A., Hassan, S. and Banat, F. (2022) An overview of microalgae biomass as a sustainable aquaculture feed ingredient: Food security and circular economy. *Bioengineered* 13(4), 9521–9547. DOI: 10.1080/21655979.2022.2061148.

Anum, H., Li, K., Tabusam, J., Saleh, S.A.A., Cheng, R.-F. *et al.* (2024) Regulation of anthocyanin synthesis in red lettuce in plant factory conditions: A review. *Food Chemistry* 458, 140111. DOI: 10.1016/j.foodchem.2024.140111.

Banerjee, D., Eng, T., Lau, A.K., Sasaki, Y., Wang, B. *et al.* (2020) Genome-scale metabolic rewiring improves titers rates and yields of the non-native product indigoidine at scale. *Nature Communications* 11(1), 5385. DOI: 10.1038/s41467-020-19171-4.

Bennett, T.H., Vaughn, M.D., Davari, S.A., Park, K., Mukherjee, D. *et al.* (2019) Jolly green MOF: Confinement and photoactivation of photosystem I in a metal-organic framework. *Nanoscale Advances* 1(1), 94–104. DOI: 10.1039/c8na00093j.

Bhatnagar, S., Kobori, T., Ganesh, D. and Aoyagi, H. (2022) Fungal pigment-assisted silver nanoparticle synthesis and their antimicrobial and cytotoxic potential. *Methods in Molecular Biology* 2469, 65–78. DOI: 10.1007/978-1-0716-2185-1_6.

Bouchnak, R., El Ayari, T., Rabeh, I., Salhi, O., Aloui, F. *et al.* (2024) Polyethylene microplastic modulates the toxicity of pentachlorophenol to the microalgae *Isochrysis galbana*, clone t-ISO. *Chemosphere* 367, 143588. DOI: 10.1016/j.chemosphere.2024.143588.

Caporgno, M.P., Haberkorn, I., Böcker, L. and Mathys, A. (2019) Cultivation of *Chlorella protothecoides* under different growth modes and its utilisation in oil/water emulsions. *Bioresource Technology* 288, 121476. DOI: 10.1016/j.biortech.2019.121476.

Cheirsilp, B., Maneechote, W., Srinuanpan, S. and Angelidaki, I. (2023) Microalgae as tools for bio-circular-green economy: Zero-waste approaches for sustainable production and biorefineries of microalgal biomass. *Bioresource Technology* 387, 129620. DOI: 10.1016/j.biortech.2023.129620.

Farouk, S., Almutairi, A.B., Alharbi, Y.O. and Al-Bassam, W.I. (2021) Acaricidal efficacy of jasmine and lavender essential oil or mustard fixed oil against two-spotted spider mite and their impact on growth and yield of eggplants. *Biology* 10(5), 4doi:10. DOI: 10.3390/biology10050410.

Gallagher, J.A., Adams, J.M.M., Turner, L.B., Kirby, M.E., Toop, T.A. *et al.* (2021) Bio-processing of macroalgae *Palmaria palmata*: Metabolite fractionation from pressed fresh material and ensiling considerations for long-term storage. *Journal of Applied Phycology* 33(1), 533–544. DOI: 10.1007/s10811-020-02295-x.

Hamed, I., Jakobsen, A.N. and Lerfall, J. (2022) Sustainable edible packaging systems based on active compounds from food processing byproducts: A review. *Comprehensive Reviews in Food Science and Food Safety* 21(1), 198–226. DOI: 10.1111/1541-4337.12870.

Hellebois, T., Tsevdou, M. and Soukoulis, C. (2020) Functionalizing and bio-preserving processed food products via probiotic and synbiotic edible films and coatings. *Advances in Food and Nutrition Research* 94, 161–221. DOI: 10.1016/bs.afnr.2020.06.004.

Isla, M.I., Orqueda, M.E., Moreno, M.A., Torres, S. and Zampini, I.C. (2022) *Solanum betaceum* fruits waste: A valuable source of bioactive compounds to be used in foods and non-foods applications. *Foods* 11(21), 3363. DOI: 10.3390/foods11213363.

Junker, L.V., Kleiber, A., Jansen, K., Wildhagen, H., Hess, M. *et al.* (2017) Variation in short-term and long-term responses of photosynthesis and isoprenoid-mediated photoprotection to soil water availability in four douglas-fir provenances. *Scientific Reports* 7, 40145. DOI: 10.1038/srep40145.

Kanamoto, A., Kato, Y., Yoshida, E., Hasunuma, T. and Kondo, A. (2021) Development of a method for fucoxanthin production using the haptophyte marine microalga pavlova sp. opms 30543. *Marine Biotechnology* 23(2), 331–341. DOI: 10.1007/s10126-021-10028-5.

Lacroux, J., Llamas, M., Dauptain, K., Avila, R., Steyer, J.-P. *et al.* (2023) Dark fermentation and microalgae cultivation coupled systems: Outlook and challenges. *Science of the Total Environment* 865, 161136. DOI: 10.1016/j.scitotenv.2022.161136.

Leong, Y.K. and Chang, J.S. (2022) Integrated role of algae in the closed-loop circular economy of anaerobic digestion. *Bioresource Technology* 360, 127618. DOI: 10.1016/j.biortech.2022.127618.

Li, J., Tian, Z., Li, J., Askari, K., Han, A. *et al.* (2023) Physcion and chitosan-oligosaccharide (COS) synergistically improve the yield by enhancing photosynthetic efficiency and resilience in wheat (*Triticum aestivum* L.). *Plant Physiology and Biochemistry* 203, 107993. DOI: 10.1016/j.plaphy.2023.107993.

Lu, H., Yadav, V., Bilal, M. and Iqbal, H.M.N. (2022) Bioprospecting microbial hosts to valorize lignocellulose biomass – environmental perspectives and value-added bioproducts. *Chemosphere* 288(Pt 2), 132574. DOI: 10.1016/j.chemosphere.2021.132574.

Malagurski, I., Lazic, J., Ilic-Tomic, T., Salevic, A., Guzik, M. *et al.* (2024) Double layer bacterial nanocellulose-poly(hydroxyoctanoate) film activated by prodigiosin as sustainable, transparent, UV-blocking material. *International Journal of Biological Macromolecules* 279(Pt 1), 135087. DOI: 10.1016/j.ijbiomac.2024.135087.

Mansinhos, I., Gonçalves, S., Rodríguez-Solana, R., Duarte, H., Ordóñez-Díaz, J.L. *et al.* (2022) Response of thymus lotocephalus *in vitro* cultures to drought stress and role of green extracts in cosmetics. *Antioxidants* 11(8), 1475. DOI: 10.3390/antiox11081475.

Mehta, D., Kuksal, K., Yadav, K., Kumar Yadav, S., Zhang, Y. *et al.* (2024) Ultrasound-assisted extraction and encapsulation of betalain from prickly pear: Process optimization, in-vitro digestive stability, and development of functional gummies. *Ultrasonics Sonochemistry* 108, 106975. DOI: 10.1016/j.ultsonch.2024.106975.

Nicoloso, L., Negrini, R., Ajmone-Marsan, P. and Crepaldi, P. (2012) On the way to functional agro biodiversity: Coat colour gene variability in goats. *Animal* 6(1), 41–49. DOI: 10.1017/S175173111100139X.

Nowruzi, B., Shishir, M.A., Porzani, S.J. and Ferdous, U.T. (2022) Exploring the interactions between algae and bacteria. *Mini Reviews in Medicinal Chemistry* 22(20), 2596–2607. DOI: 10.2174/1389557522666220504141047.

Parveen, A., Bhatnagar, P., Gautam, P., Bisht, B., Nanda, M. *et al.* (2023) Enhancing the bio-prospective of microalgae by different light systems and photoperiods. *Photochemical & Photobiological Sciences* 22(11), 2687–2698. DOI: 10.1007/s43630-023-00471-9.

Prasad, S.S., Naidu, B.R., Hanafiah, M.M., Lakshmidevi, J., Marella, R.K. *et al.* (2021) Porphyrin *n*-pincer pd(II)-complexes in water: A base-free and nature-inspired protocol for the oxidative self-coupling of potassium aryltrifluoroborates in open-air. *Molecules* 26(17), 5390. DOI: 10.3390/molecules26175390.

Ren, Y., Cao, H., Xu, H., Xiong, X., Krastev, R. *et al.* (2020) Improved aging properties of bio-bitumen coating sheets by using modified lignin. *Journal of Environmental Management* 274, 111178. DOI: 10.1016/j.jenvman.2020.111178.

Rengel, R., Giraldez, I., Díaz, M.J., García, T., Vigara, J. *et al.* (2022) Simultaneous production of carotenoids and chemical building blocks precursors from chlorophyta microalgae. *Bioresource Technology* 351, 127035. DOI: 10.1016/j.biortech.2022.127035.

Ribeiro, O., Félix, L., Ribeiro, C., Castro, B., Tiritan, M.E. *et al.* (2022) Enantioselective ecotoxicity of venlafaxine in aquatic organisms: Daphnia and Zebrafish. *Environmental Toxicology and Chemistry* 41(8), 1851–1864. DOI: 10.1002/etc.5338.

Saba, B., Bharathidasan, A.K., Ezeji, T.C. and Cornish, K. (2023) Characterization and potential valorization of industrial food processing wastes. *Science of the Total Environment* 868, 161550. DOI: 10.1016/j.scitotenv.2023.161550.

Sayanova, O., Mimouni, V., Ulmann, L., Morant-Manceau, A., Pasquet, V. *et al.* (2017) Modulation of lipid biosynthesis by stress in diatoms. *Philosophical Transactions of the Royal Society of London. Series B, Biological Sciences* 372(1728), 20160407. DOI: 10.1098/rstb.2016.0407.

Sharma, P., Gaur, V.K., Sirohi, R., Varjani, S., Hyoun Kim, S. *et al.* (2021) Sustainable processing of food waste for production of bio-based products for circular bioeconomy. *Bioresource Technology* 325, 124684. DOI: 10.1016/j.biortech.2021.124684.

Suwannarach, N., Kumla, J., Nishizaki, Y., Sugimoto, N., Meerak, J. *et al.* (2019) Optimization and characterization of red pigment production from an endophytic fungus, nigrospora aurantiaca CMU-ZY2045, and its potential source of natural dye for use in textile dyeing. *Applied Microbiology and Biotechnology* 103(17), 6973–6987. DOI: 10.1007/s00253-019-09926-5.

Udayan, A., Sirohi, R., Sreekumar, N., Sang, B.I. and Sim, S.J. (2022) Mass cultivation and harvesting of microalgal biomass: Current trends and future perspectives. *Bioresource Technology* 344(Pt B), 126406. DOI: 10.1016/j.biortech.2021.126406.

20 Emerging Technologies and Innovations in Biopigment Research

Gundala Prasada Babu[1], Arifullah Mohammed[2], Ee Lee Gaik[3], Mukul Jain[4] and Chinthala Paramageetham[1*]

[1]Department of Microbiology, Sri Venkateswara University, Tirupati, Andhra Pradesh, India; [2]Department of Biotechnology, Koneru Lakshmaiah Education Foundation (KLEF), Vaddeswaram Campus, Guntur, Andhra Pradesh, India; [3]Institute of Tropical Biodiversity and Sustainable Development, Universiti Malaysia Terengganu, Kuala Nerus, Malaysia; [4]Department of Life Sciences, Parul Institute of Applied Sciences, Parul University, Vadodara, Gujarat, India

20.1 Introduction

Biopigments are naturally occurring compounds produced by various organisms, including plants, algae, fungi, and bacteria, which impart color due to their ability to selectively absorb and reflect light. These pigments play crucial roles in biological systems, such as photosynthesis, protection against oxidative stress, and communication within ecosystems (Kumar *et al.*, 2020). Broadly categorized, biopigments include carotenoids, such as β-carotene and lutein, which are responsible for yellow to orange hues; chlorophylls, the green pigments essential for photosynthesis; and phycobiliproteins, water-soluble pigments found in cyanobacteria and red algae. Other important classes include flavonoids and anthocyanins, which contribute to the diverse coloration of fruits, flowers, and leaves, and melanins, dark pigments with protective functions (Prabhu and Umesh Hebbar, 2021).

The significance of biopigments extends beyond their natural roles, as they have gained immense attention for their applications in food, cosmetics, pharmaceuticals, and environmental sectors. Unlike synthetic dyes, biopigments are biodegradable, nontoxic, and derived from renewable resources, aligning with the global emphasis on sustainability. For instance, carotenoids such as astaxanthin exhibit potent antioxidant properties, making them valuable in health and nutrition (Sathasivam and Ki, 2018). Moreover, microbial sources of biopigments, such as those derived from *Monascus* and *Spirulina*, offer scalable and eco-friendly alternatives to conventional pigment production (Ali *et al.*, 2022).

Biopigments have emerged as versatile and sustainable alternatives to synthetic dyes, finding applications across diverse industries due to their natural origin, safety profile, and functional properties. In the food industry, biopigments such as carotenoids (e.g. β-carotene,

*Corresponding author: paramageetham@yahoo.com

DOI: 10.1079/ 9781836990505.0020

lycopene) and anthocyanins are extensively used as natural colorants to enhance the visual appeal of products while offering health benefits due to their antioxidant properties (Martins *et al.*, 2019). Phycocyanin, derived from *Spirulina*, is particularly popular as a blue pigment in beverages and confectioneries. In pharmaceuticals, biopigments like astaxanthin and curcumin are utilized for their antiinflammatory, antioxidant, and anticancer activities, contributing to the development of nutraceuticals and therapeutic formulations (Ramesh *et al.*, 2020).The textile industry has also embraced biopigments for eco-friendly dyeing processes. Pigments from microbial sources, such as *Monascus* and *Chromobacterium*, are gaining attention for their ability to produce vibrant and biodegradable dyes with reduced environmental impact compared to synthetic counterparts (Venil *et al.*, 2020). Similarly, in the cosmetics sector, biopigments like lutein and flavonoids are incorporated into skincare and make-up products for their aesthetic and functional benefits, including UV protection and antiaging properties. These applications underscore the growing demand for sustainable, nontoxic, and multifunctional biopigments in modern industries (Ali *et al.*, 2022).

20.1.1 Importance of sustainable and eco-friendly alternatives to synthetic pigments

The shift towards sustainable and eco-friendly alternatives to synthetic pigments has become imperative due to the environmental and health concerns associated with synthetic dye production and use. Synthetic pigments, often derived from petrochemicals, are notorious for their toxic byproducts, high energy consumption during production, and contribution to water pollution through industrial effluents (Chandanshive *et al.*, 2020). These issues have heightened the demand for biopigments, which are biodegradable, nontoxic, and sourced from renewable resources.

Biopigments offer significant advantages as they are produced using sustainable methods, such as fermentation or extraction from plants and microbes, which have minimal environmental impact (Venil *et al.*, 2020). Moreover, their potential for dual functionality, combining coloring properties with bioactive effects such as antioxidant and antimicrobial activities, makes them highly desirable in industries such as food, cosmetics, and textiles (Sathasivam and Ki, 2018). Transitioning to biopigments not only aligns with global sustainability goals but also helps industries adhere to stricter environmental regulations while catering to consumer preferences for natural and green products (Martins *et al.*, 2019).

20.2 Current State of Biopigment Research

20.2.1 Sources of biopigments

Biopigments are derived from diverse natural sources, including microorganisms, plants, and algae, each offering unique advantages for sustainable pigment production (Table 20.1). Microbial sources, particularly bacteria, fungi, and cyanobacteria, are highly valued for their ability to produce a wide range of pigments, such as carotenoids, melanin, and prodigiosin, under controlled fermentation conditions. Microbial pigments are advantageous due to their scalability, rapid growth rates, and ability to thrive on inexpensive substrates, making them a promising alternative for industrial applications (Venil *et al.*, 2020). Plant-based sources, such as fruits, flowers, and leaves, are rich in anthocyanins, flavonoids, and betalains, which are extensively used in the food, cosmetic, and textile industries. For example, beetroot-derived betalains are widely recognized for their vibrant color and antioxidant properties (Martins *et al.*, 2019).

Algae, including microalgae and macroalgae, are another crucial source of biopigments. Pigments such as chlorophyll, carotenoids, and phycobiliproteins are abundantly produced by species like *Spirulina* and *Chlorella*, which can be cultivated in diverse environments, including wastewater and saline conditions. Algal pigments are not only sustainable but also multifunctional, offering applications in nutraceuticals, pharmaceuticals, and biofuels (Sathasivam and Ki,

Table 20.1. Types of biopigments and their sources.

Biopigment	Source	Chemical structure	Primary applications	Benefits
Carotenoids	Microbial, algae, plants	Hydrocarbons (e.g. β-carotene)	Food colorants, pharmaceuticals	Antioxidant properties, vitamin A precursor
Phycobiliproteins	Algae (e.g. *Spirulina*)	Protein-bound pigments (e.g. phycocyanin)	Food, cosmetics, biomedical research	High stability, protein based
Anthocyanins	Fruits, vegetables, plants	Flavonoid compounds	Food, cosmetics, natural dyes	Antiinflammatory, antioxidant
Chlorophyll	Algae, plants	Porphyrin complex	Food colorants, pharmaceuticals	Photosynthesis, antioxidant

2018). These diverse natural sources highlight the potential of biopigments as eco-friendly and renewable resources for various industries.

20.2.2 Notable biopigments

Biopigments such as carotenoids, phycobiliproteins, flavonoids, and anthocyanins are among the most widely studied and commercially utilized natural pigments due to their vibrant colors and bioactive properties. Carotenoids, including β-carotene, lycopene, and astaxanthin, are lipid-soluble pigments found in plants, algae, and some microorganisms. These pigments are known for their yellow, orange, and red hues and are widely used as food colorants, nutritional supplements, and antioxidants (Sathasivam and Ki, 2018). Phycobiliproteins, such as phycocyanin and phycoerythrin, are water-soluble pigments produced by cyanobacteria and red algae. These fluorescent pigments are extensively utilized in food, cosmetics, and analytical applications such as fluorescence-based assays (Kuddus, 2021).

Flavonoids, including quercetin, luteolin, and kaempferol, are polyphenolic compounds predominantly found in fruits, vegetables, and flowers. Known for their yellow coloration and strong antioxidant activity, flavonoids are widely used in nutraceuticals and cosmetics, and as natural colorants in food (Martins *et al.*, 2019). Anthocyanins, responsible for red, purple, and blue hues in many fruits and flowers, are another important class of water-soluble pigments. They are extensively used in the food and beverage industry as natural colorants and are valued for their health-promoting properties, including antiinflammatory and cardioprotective effects (Khoo *et al.*, 2017). The multifunctionality and eco-friendly nature of these pigments underline their importance in sustainable industrial applications.

20.2.3 Challenges in large-scale production

Large-scale production of biopigments faces several challenges that limit their widespread adoption across industries. One significant hurdle is achieving high yield under scalable conditions, as many natural sources produce limited quantities of pigments. Optimizing growth conditions, substrates, and production processes is crucial but can be resource intensive (Venil *et al.*, 2020). Another challenge is stability, as biopigments are often sensitive to environmental factors such as light, temperature, and pH. For instance, carotenoids are prone to oxidative degradation, while anthocyanins may lose their color under alkaline conditions, making their industrial application complex (Khoo *et al.*, 2017).

Cost-effectiveness is another critical issue, as the production, extraction, and purification of biopigments often involve expensive and labor-intensive processes. For microbial pigments, fermentation costs, nutrient requirements, and downstream processing contribute significantly to production expenses (Kuddus, 2021). Similarly, plant- and algae-based pigments require extensive cultivation and harvesting efforts, which can further escalate costs. Addressing these challenges through advancements in metabolic engineering, bioprocess optimization, and green extraction technologies is essential for making biopigment production more economically viable and sustainable (Martins *et al.*, 2019).

20.3 Emerging Technologies in Biopigment Research

20.3.1 Bioprocess optimization

Bioprocess optimization is critical for enhancing the efficiency and cost-effectiveness of biopigment production. This involves improving various stages of the production process, including microbial strain selection, culture media composition, and environmental conditions such as pH, temperature, and aeration. For example, metabolic engineering of microbial strains has been employed to enhance pigment yields by redirecting metabolic fluxes or introducing novel biosynthetic pathways (Venil *et al.*, 2020). Similarly, the use of low-cost and renewable substrates, such as agro-industrial byproducts, has been explored to reduce production costs and improve sustainability (Kuddus, 2021).

Optimization also extends to downstream processes, where efficient extraction and purification methods are vital to maintain pigment quality and stability. Techniques such as solvent extraction, supercritical fluid extraction, and membrane filtration are increasingly being refined for higher efficiency and reduced environmental impact (Martins *et al.*, 2019). Advances in bioprocess automation, including the use of bioreactors with real-time monitoring and control systems, further enhance scalability and reproducibility.

By integrating these strategies, bioprocess optimization not only increases productivity but also aligns with the principles of green and sustainable manufacturing (Chandanshive *et al.*, 2020) (Fig. 20.1).

20.3.1.1 Submerged, solid-state, and fed-batch fermentation techniques

Fermentation is a critical process in biopigment production, with different techniques offering distinct advantages based on the target pigment and production requirements. Submerged fermentation (SmF) is the most widely used technique, where microorganisms grow in a liquid medium with controlled parameters such as pH, temperature, and aeration. This method is particularly suitable for large-scale production of water-soluble pigments like phycocyanin and anthocyanins due to its scalability and ease of nutrient distribution (Venil *et al.*, 2020). Solid-state fermentation (SSF), on the other hand, involves microbial growth on solid substrates with low moisture content. SSF is ideal for producing pigments such as prodigiosin and monascin, offering advantages such as lower energy requirements, reduced water usage, and utilization of agro-industrial residues as substrates (Kuddus, 2021).

Fed-batch fermentation combines the benefits of both SmF and SSF, where nutrients are added incrementally during the fermentation process. This technique helps in maintaining optimal growth conditions and preventing substrate inhibition, thereby enhancing pigment yields. Fed-batch systems are particularly advantageous for producing carotenoids and other secondary metabolites that require precise control over metabolic pathways (Martins *et al.*, 2019).

By selecting the appropriate fermentation technique and optimizing process parameters, industries can achieve higher efficiency, sustainability, and economic feasibility in biopigment production.

20.3.1.2 Metabolic engineering and synthetic biology for enhanced production

Metabolic engineering and synthetic biology have emerged as transformative approaches for enhancing biopigment production by enabling precise manipulation of biosynthetic pathways.

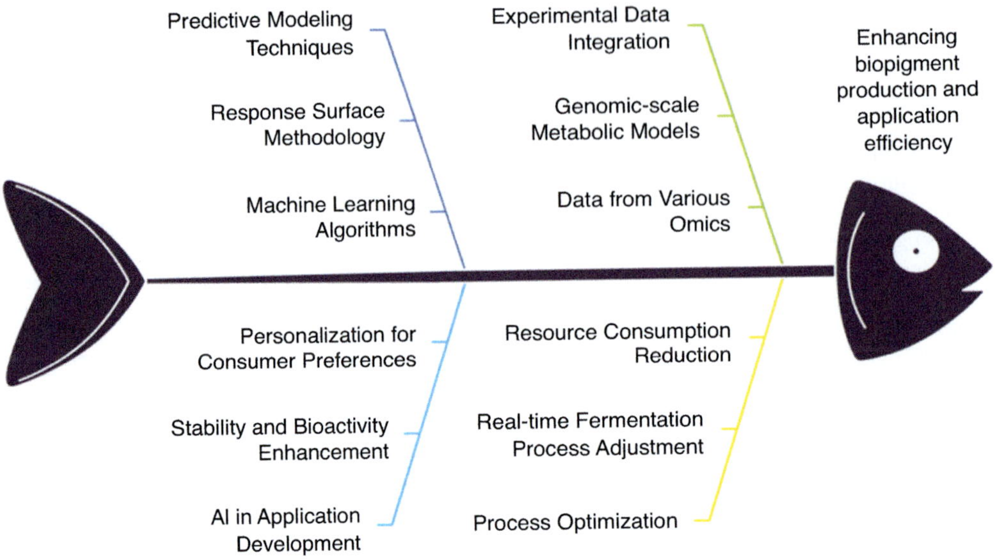

Fig. 20.1. Optimizing biopigment production and application.

Through metabolic engineering, microbial strains can be optimized to overexpress genes involved in pigment biosynthesis while downregulating competing pathways, thereby increasing yield. For example, carotenoid production in *Escherichia coli* and *Saccharomyces cerevisiae* has been significantly improved by engineering key enzymes in the mevalonate pathway (Venil *et al.*, 2020). Similarly, advances in synthetic biology allow the design of novel biosynthetic routes by introducing heterologous genes from diverse organisms into robust microbial hosts, such as *Corynebacterium glutamicum* and *Bacillus subtilis* (Martins *et al.*, 2019).

In addition to enhancing production, synthetic biology tools like CRISPR-Cas9 enable precise genome editing, facilitating the development of high-yield strains in a shorter time. Systems biology approaches, including omics technologies and computational modeling, further complement these strategies by providing insights into cellular metabolism and guiding the optimization of production pathways (Chandanshive *et al.*, 2020). These advancements not only improve the scalability and cost-effectiveness of biopigment production but also open avenues for creating pigments with tailored properties for specific industrial applications.

20.3.2 Nanotechnology

20.3.2.1 Nanoencapsulation for stability and delivery

Nanoencapsulation is a promising strategy for improving the stability, bioavailability, and controlled delivery of biopigments. By encapsulating biopigments in nanocarriers, such as liposomes, solid lipid nanoparticles (SLNs), or polymeric nanoparticles, the degradation of sensitive pigments, including carotenoids and anthocyanins, can be significantly reduced under environmental stressors such as light, oxygen, and heat (Kuddus, 2021). These nanocarriers act as protective barriers, ensuring that the pigments remain stable over longer periods, thus enhancing their shelf-life and applicability in food, cosmetics, and pharmaceutical formulations. Additionally, nanoencapsulation allows for the targeted and controlled release of biopigments, optimizing their bioavailability and therapeutic efficacy when used in nutraceuticals or drug delivery systems (Martins *et al.*, 2019).

The use of nanoencapsulation also offers benefits in improving the solubility of hydrophobic biopigments, which are typically challenging to formulate in aqueous systems. Nanoparticles improve the dispersibility of these pigments, facilitating their integration into water-based products without compromising their color or functionality. Moreover, the controlled release of encapsulated biopigments ensures a sustained effect, making them suitable for various applications, such as functional foods or antioxidant therapies (Venil *et al.*, 2020).

Through these advanced formulations, nanoencapsulation not only enhances the performance of biopigments but also contributes to more sustainable and eco-friendly production methods.

20.3.2.2 Applications of nanoparticles in biopigment extraction and utilization

Nanoparticles have gained attention for their potential in improving the extraction and utilization of biopigments, offering innovative and sustainable solutions. In the extraction process, nanomaterials such as magnetic nanoparticles, carbon nanotubes, and mesoporous silica nanoparticles are employed to selectively adsorb biopigments from natural sources, enhancing the efficiency and selectivity of pigment recovery (Martins *et al.*, 2019). These nanoparticles can be functionalized with specific ligands to selectively capture pigments, such as carotenoids or anthocyanins, from complex matrices, reducing the need for harsh solvents and energy-intensive methods. Magnetic nanoparticles, in particular, enable easy recovery of pigments using external magnetic fields, facilitating a greener and more efficient extraction process (Chandanshive *et al.*, 2020).

Once extracted, nanoparticles are also utilized to enhance the stability and delivery of biopigments. For example, biopigments encapsulated in nanoparticles are protected from degradation caused by environmental factors (Kuddus, 2021). These encapsulated pigments can be incorporated into a variety of formulations, such as food, cosmetics, or pharmaceuticals, where they maintain their color, nutritional value, and antioxidant properties over time. Moreover, nanoparticle-based delivery systems provide controlled and targeted release, improving the bioavailability and efficacy of the biopigments when used in functional foods or therapeutic applications (Venil *et al.*, 2020).

Through these applications, nanoparticles significantly enhance the sustainability, cost-effectiveness, and versatility of biopigments in various industries.

20.3.3 Omics technologies

20.3.3.1 Genomics, transcriptomics, and proteomics for strain improvement

The integration of genomics, transcriptomics, and proteomics has revolutionized strain improvement strategies for enhanced biopigment production. Genomics provides comprehensive insights into the genetic make-up of microorganisms, allowing researchers to identify and manipulate genes involved in biopigment biosynthesis. By sequencing the genome of pigment-producing strains, key enzymes and regulatory pathways can be targeted for optimization through gene editing or overexpression (Martins *et al.*, 2019). For instance, the identification of carotenoid biosynthetic genes in *Escherichia coli* and *Saccharomyces cerevisiae* has facilitated metabolic engineering approaches that significantly increase pigment yield (Venil *et al.*, 2020).

Transcriptomics, which analyzes gene expression patterns, complements genomics by providing insights into the regulation of pigment biosynthetic pathways under different growth conditions. By profiling mRNA expression, researchers can identify genes that are upregulated or downregulated during pigment production, enabling the design of more efficient fermentation strategies (Kuddus, 2021). Similarly, proteomics allows for the study of proteins and their modifications, helping to identify key enzymes in the pigment biosynthesis pathway. It also provides information on protein folding, posttranslational modifications, and enzyme activity, which are critical for optimizing pigment production (Chandanshive *et al.*, 2020).

Combining these omics technologies provides a holistic understanding of cellular metabolism and facilitates more precise strain improvement, resulting in higher yields and more sustainable production processes.

20.3.3.2 *Systems biology approaches to pathway elucidation*

Systems biology approaches play a crucial role in elucidating metabolic pathways involved in biopigment biosynthesis, offering a comprehensive, integrative framework to understand complex cellular processes. These approaches combine experimental techniques, such as genomics, transcriptomics, and proteomics, with computational modeling to predict and simulate metabolic networks (Martins *et al.*, 2019). By constructing detailed metabolic models, researchers can identify key enzymes, regulatory factors, and potential bottlenecks in biopigment production pathways, facilitating the design of more efficient and optimized production processes (Chandanshive *et al.*, 2020). For example, metabolic flux analysis (MFA) and constraint-based optimization models are used to quantify the flow of metabolites through biosynthetic pathways, allowing for the identification of rate-limiting steps in pigment production (Venil *et al.*, 2020).

Additionally, systems biology incorporates the use of genetic and metabolic networks, providing insights into the dynamic regulation of gene expression and enzyme activity under various growth conditions. This allows for the identification of potential genetic modifications or regulatory interventions that could lead to enhanced biopigment production. Furthermore, the integration of omics data with computational tools such as genome-scale metabolic models (GEMs) enables the prediction of optimal pathways for maximizing pigment yields, guiding strain engineering strategies (Kuddus, 2021).

By leveraging these systems biology approaches, researchers can not only elucidate the underlying pathways of biopigment biosynthesis but also apply these insights to improve the sustainability, cost-effectiveness, and scalability of production processes.

20.3.4 Machine learning and AI

20.3.4.1 *Predictive modeling for production optimization*

Predictive modeling is an essential tool for optimizing the production of biopigments, enabling researchers to forecast and enhance various aspects of the biosynthesis process. By using mathematical models and computational algorithms, predictive modeling integrates data from genomics, transcriptomics, proteomics, and metabolic flux analysis to predict the optimal conditions for maximum pigment yield (Chandanshive *et al.*, 2020). These models can simulate the effects of different variables, such as nutrient concentrations, temperature, pH, and oxygen levels, on biopigment production.

One commonly used approach is response surface methodology (RSM), which helps identify the ideal conditions for production by analyzing the interactions between multiple factors (Martins *et al.*, 2019). Additionally, machine learning algorithms are increasingly being applied to model complex biological systems, improving the prediction of metabolic shifts and potential bottlenecks in pigment production pathways (Venil *et al.*, 2020) (Fig. 20.2).

By combining experimental data with computational tools, predictive modeling allows for the optimization of fermentation processes in real time, reducing trial-and-error experiments and speeding up strain development. GEMs also play a vital role in this process by simulating entire microbial metabolic networks and identifying key genes or enzymes that need to be modified to enhance pigment yields (Kuddus, 2021).

These predictive tools not only improve the efficiency and scalability of biopigment production but also facilitate more sustainable practices by reducing resource consumption and waste generation.

20.3.4.2 *AI in pigment application development*

Artificial intelligence has emerged as a transformative tool in the development of biopigment applications, enhancing both the innovation and efficiency of product formulations. AI algorithms, particularly machine learning and deep learning, are employed to analyze large datasets, which can include the chemical structure of pigments, their interactions with different substrates, and their stability under various conditions (Martins *et al.*, 2019). These models help identify the most suitable biopigments for specific applications in food, cosmetics, and pharmaceuticals, by predicting their behavior

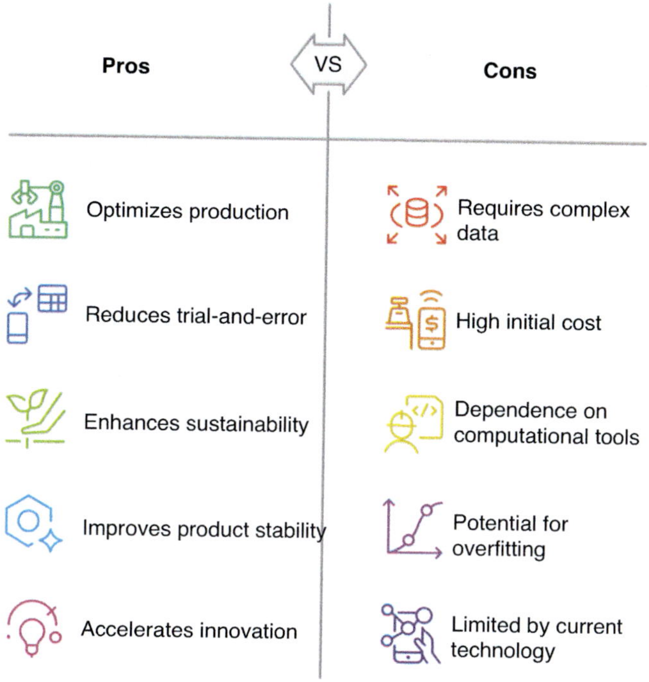

Fig. 20.2. Predictive modeling and AI in biopigment production.

and performance without the need for exhaustive experimental trials. AI-driven optimization techniques can also assist in improving the stability, color intensity, and bioactivity of pigments in different formulations, addressing challenges such as degradation due to light or temperature exposure (Venil *et al.*, 2020).

Furthermore, AI can be applied to predictive modeling in biopigment production, enabling the identification of optimal conditions for microbial growth and pigment synthesis, thus reducing resource consumption and improving the scalability of production processes (Chandanshive *et al.*, 2020). In addition to streamlining production, AI technologies can enhance product personalization, particularly in the cosmetics industry, where AI models can help formulate pigments tailored to specific consumer preferences, such as skin tone compatibility or long-lasting effects (Kuddus, 2021). By integrating AI into biopigment application development, researchers can accelerate the design of sustainable and eco-friendly products, ultimately advancing the commercial viability and consumer acceptance of natural pigments (see Fig. 20.2).

20.4 Innovative Extraction and Purification Techniques

20.4.1 Green extraction methods: supercritical fluid extraction, ultrasound-assisted extraction, and ionic liquids

Green extraction methods have become essential in biopigment recovery, offering more sustainable, efficient, and eco-friendly alternatives to traditional extraction techniques that often rely on harmful solvents. Supercritical fluid extraction (SFE), particularly with supercritical CO_2, has gained significant attention due to its ability to selectively extract pigments without the use of toxic organic solvents. SFE operates under high pressure and temperature, utilizing CO_2 in its supercritical state, which allows it to penetrate plant and microbial cells, extracting pigments such as carotenoids and anthocyanins with high efficiency and minimal environmental impact (Martins *et al.*, 2019). This method also benefits from non-toxicity, recyclability of CO_2, and low energy requirements compared to traditional solvent-based methods (Fig. 20.3).

Another innovative green method is ultrasound-assisted extraction (UAE), which utilizes high-frequency sound waves to create microscopic bubbles in the solvent, causing cavitation that enhances the breakdown of cell walls and improves the release of biopigments. UAE offers several advantages, including reduced extraction times, lower solvent usage, and the ability to extract a wider range of pigments with higher yields (Venil *et al.*, 2020). The technique is particularly effective for extracting pigments from complex matrices such as plant tissues and microbial biomass.

Ionic liquids (ILs), which are salts that remain in liquid form at or near room temperature, have also emerged as promising green solvents for pigment extraction. ILs exhibit excellent solubilizing properties, allowing for efficient extraction of biopigments from a variety of sources, including algae and fungi, with minimal toxicity and the ability to recycle solvents (Kuddus, 2021). Furthermore, they offer the advantage of tamability, where the composition of the ionic liquid can be adjusted to optimize pigment extraction from specific raw materials.

These green methods not only provide high extraction efficiency but also align with sustainable practices by minimizing environmental impact and reducing the need for hazardous chemicals (Table 20.2).

20.4.2 Chromatographic and nonchromatographic purification techniques

Purification of biopigments is a crucial step in ensuring their high quality and functionality for various applications, and both chromatographic and nonchromatographic techniques are commonly employed for this purpose. Chromatographic techniques, such as high-performance liquid chromatography (HPLC) and thin-layer chromatography (TLC), are widely used for the separation, identification, and purification of biopigments due to their high specificity, precision, and ability to resolve complex mixtures (Chandanshive *et al.*, 2020). HPLC is particularly effective for purifying pigments such as carotenoids, anthocyanins, and chlorophylls, as it allows for fine separation based on the pigments' chemical properties such as polarity and molecular size. Additionally, TLC is a simpler and more cost-effective method that is often used for preliminary purification and identification, particularly in large-scale screenings (Martins *et al.*, 2019). These chromatographic techniques ensure high purity of biopigments and are often followed by additional analysis, such as spectrophotometry or mass spectrometry, to confirm the identity and concentration of the isolated compounds.

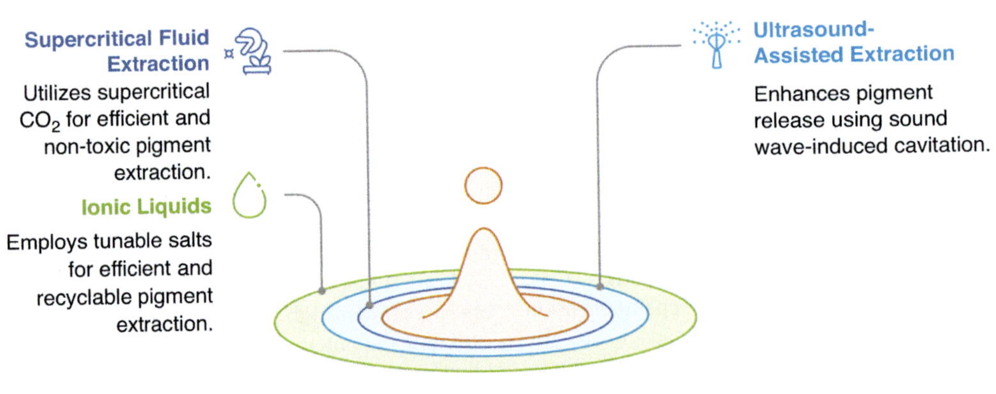

Fig. 20.3. Green extraction methods.

Table 20.2. Comparison of extraction techniques for biopigments.

Extraction method	Biopigments extracted	Advantages	Disadvantages
Solvent extraction	Carotenoids, flavonoids, anthocyanins	Simple, widely used	Solvent residues, low yield
Supercritical fluid extraction	Carotenoids, chlorophyll, phycobiliproteins	Efficient, environmentally friendly	High set-up costs, specialized equipment
Ultrasound-assisted extraction	Carotenoids, flavonoids	High yield, fast processing	High energy consumption, costly equipment
Enzyme-assisted extraction	Carotenoids, phycobiliproteins	Eco-friendly, mild conditions	Longer processing times

On the other hand, nonchromatographic purification techniques, such as precipitation, ultrafiltration, and adsorption, are often preferred for large-scale applications due to their simplicity and scalability. Precipitation using solvents like acetone or ethanol is a straightforward method that allows for the separation of biopigments based on solubility differences. Ultrafiltration, a type of membrane filtration, is widely used to concentrate and purify biopigments by separating them from larger molecules, such as proteins and polysaccharides, based on size exclusion (Venil *et al.*, 2020). Adsorption, using materials such as activated carbon or resins, is another effective nonchromatographic technique that selectively binds pigments, allowing for easy separation and purification. These nonchromatographic methods are less resource intensive and often more suited for large-scale pigment production processes, although they may require additional purification steps to achieve the desired level of purity.

20.4.3 Advances in downstream processing

Downstream processing (DSP) plays a critical role in the production of biopigments, focusing on the recovery, purification, and formulation of pigments post fermentation or extraction. Over the years, several advances in DSP have been made to increase efficiency, reduce costs, and improve the quality of biopigments for industrial applications. One significant development is the integration of membrane filtration technologies, such as microfiltration, ultrafiltration, and nanofiltration, which offer high efficiency in separating pigments from biomass and other impurities. These technologies are particularly advantageous because they are scalable, can operate continuously, and avoid the use of chemical solvents, making them eco-friendly alternatives to traditional methods (Venil *et al.*, 2020). Moreover, electrochemical separation techniques, which apply an electric field to separate biopigments from their culture medium, are gaining attention due to their simplicity, low energy requirements, and minimal environmental impact (Chandanshive *et al.*, 2020).

Recent innovations in centrifugation and liquid-liquid extraction have also enhanced the recovery of biopigments from complex microbial and algal biomass. High-speed centrifugation has been optimized to improve the separation of pigment-rich cell fractions, while liquid-liquid extraction techniques are being refined to improve selectivity, resulting in purer extracts with higher yields (Kuddus, 2021). In addition, nanotechnology-based methods, including the use of nanomaterials and magnetic nanoparticles, have shown promise in enhancing the purification of biopigments by offering higher binding affinities and faster processing times (Martins *et al.*, 2019). These advances have not only increased the recovery and purity of biopigments but also contributed to the scalability and sustainability of production processes, facilitating their commercialization in sectors like food, cosmetics, and textiles.

20.5 Case Studies and Success Stories

20.5.1 Examples of commercialized biopigments

Several biopigments have been successfully commercialized and are now used in a wide range of applications, including food, cosmetics, textiles, and pharmaceuticals. β-Carotene, one of the most well-known carotenoids, is widely used as a natural food colorant and dietary supplement due to its antioxidant properties and role in vitamin A synthesis. It is commercially extracted from algae, such as *Dunaliella salina*, and is used in products ranging from beverages and dairy items to health supplements (Venil *et al.*, 2020). Another prominent example is astaxanthin, a red-orange pigment found in marine organisms including *Haematococcus pluvialis* and Salmonidae species. It is extensively used in aquaculture to color fish, and has gained popularity as a dietary supplement due to its potent antioxidant properties (Martins *et al.*, 2019).

In the textile industry, indigo, a blue pigment originally derived from the plant *Indigofera tinctoria*, has also been produced using microbial fermentation processes, providing a sustainable and eco-friendly alternative to synthetic dyes (Kuddus, 2021). This biopigment has been commercialized for use in denim production, reducing the environmental impact associated with conventional chemical-based dyeing methods. Additionally, phycobiliproteins, such as c-phycocyanin from cyanobacteria, have found applications in cosmetics and pharmaceuticals. These pigments are valued not only for their bright blue color but also for their bioactive properties, including antioxidant and antiinflammatory activities (Chandanshive *et al.*, 2020).

These examples highlight the growing commercialization of biopigments, demonstrating their versatility and sustainability in various industries.

20.5.2 Innovative start-ups and industries leveraging biopigment technology

The growing demand for sustainable and eco-friendly alternatives to synthetic pigments has led to the emergence of innovative start-ups and industries that leverage biopigment technology. One notable example is Algatech, an Israeli start-up that specializes in the production of natural astaxanthin from *Haematococcus pluvialis*. Algatech has positioned itself as a leader in the biotechnology space, utilizing proprietary fermentation processes to produce high-quality astaxanthin for use in the dietary supplement and cosmetics industries (Venil *et al.*, 2020). Another innovative company, Mikroalgae, focuses on producing carotenoids and other biopigments from microalgae. Their products are used in a variety of sectors, including food, pharmaceuticals, and cosmetics, highlighting the versatility of biopigments in different markets (Martins *et al.*, 2019).

In the textile industry, companies such as DyeCoo are at the forefront of promoting sustainable dyeing technologies. DyeCoo uses supercritical CO_2 as a dyeing agent, and is exploring the use of biopigments as part of its eco-friendly dyeing processes. This approach significantly reduces water usage and chemical waste compared to traditional dyeing methods (Kuddus, 2021). Moreover, start-ups like Aether Biotech are focusing on the bioproduction of natural dyes for the textile industry by cultivating microorganisms that produce pigments such as indigo and anthocyanins, providing a sustainable alternative to petrochemical-based synthetic dyes (Chandanshive *et al.*, 2020).

These start-ups and industries are not only making strides in the commercial application of biopigments but are also pushing the boundaries of sustainability and innovation in sectors ranging from food and cosmetics to textiles and pharmaceuticals.

20.5.3 Academic and industry collaborations driving innovations

Collaborations between academia and industry are essential for driving innovation in biopigment technology, as they foster the exchange of knowledge, resources, and expertise that lead to new discoveries and applications. Many universities and research institutes are partnering with companies to accelerate the development of biopigment production processes, enhance pigment yields, and optimize extraction and purification methods. For example, the collaboration between the University of Tokyo and industry leaders in biotechnology has led to the successful commercialization of phycobiliproteins, which are used in the food and cosmetics industries (Martins *et al.*, 2019). Such collaborations facilitate the translation of basic research into practical solutions and improve the scalability of biopigment production.

Industry players like DuPont™ have also partnered with academic institutions to explore innovative methods for enhancing pigment stability and exploring alternative sources of biopigments. These partnerships have led to breakthroughs in the use of biopigments in textile and food applications, such as the development of natural, biodegradable colorants that align with sustainable consumer trends (Venil *et al.*, 2020). Another example is BASF, a global leader in chemicals, which has collaborated with universities to develop biotechnological processes for pigment production from algae, significantly reducing the environmental footprint of traditional pigment manufacturing processes (Kuddus, 2021).

These collaborations between academia and industry not only promote the commercialization of biopigments but also stimulate the growth of green technologies and innovations in a wide range of industries.

20.6 Future Perspectives

20.6.1 Potential of CRISPR and gene-editing technologies in biopigment research

The advent of CRISPR-Cas9 and other gene-editing technologies has opened new frontiers in biopigment research, offering unprecedented precision in manipulating the genetic pathways responsible for pigment biosynthesis. These technologies enable researchers to optimize microbial strains, algae, and plants for enhanced pigment production, improving yield, stability, and cost-effectiveness. For example, CRISPR has been applied to modify genes involved in carotenoid biosynthesis in microorganisms such as *Escherichia coli* and *Saccharomyces cerevisiae*, leading to the development of more efficient microbial production systems for carotenoids like β-carotene and astaxanthin (Zhou *et al.*, 2020). By targeting key enzymes in the biosynthetic pathways, gene editing can accelerate the production of high-value pigments, allowing for the fine-tuning of pigment profiles to meet specific industrial needs.

Synthetic biology, when combined with CRISPR technologies, offers an even more powerful tool for creating engineered strains that can produce novel or rare pigments. For instance, researchers have successfully utilized CRISPR-based techniques to introduce genes from different organisms into microbial hosts to produce pigments such as phycocyanin and anthocyanins (Kuddus, 2021). These innovations not only improve the efficiency of pigment production but also provide new avenues for the sustainable production of biopigments, reducing reliance on traditional agricultural sources or synthetic chemical methods.

As these technologies continue to evolve, CRISPR and gene-editing tools are expected to play a pivotal role in driving the next generation of biopigment production systems, with the potential to revolutionize industries such as food, cosmetics, and textiles by providing cleaner, more efficient, and eco-friendly alternatives to synthetic dyes and pigments (Fig. 20.4).

20.6.2 Integration of renewable energy sources in biopigment production

The integration of renewable energy sources in biopigment production is gaining increasing attention as industries strive to reduce their carbon footprint and improve the sustainability of their processes. The production of biopigments, particularly through microbial fermentation and algal cultivation, is energy intensive, and the use of traditional fossil fuels has significant environmental implications. To address this, many research initiatives are exploring the use of renewable energy sources, such as solar and wind energy, to power biopigment production systems. For instance, algae-based biopigment production can benefit from solar energy, as algae require light for growth, and harnessing solar power can significantly lower energy costs while promoting a sustainable process (Venil *et al.*, 2020). This approach

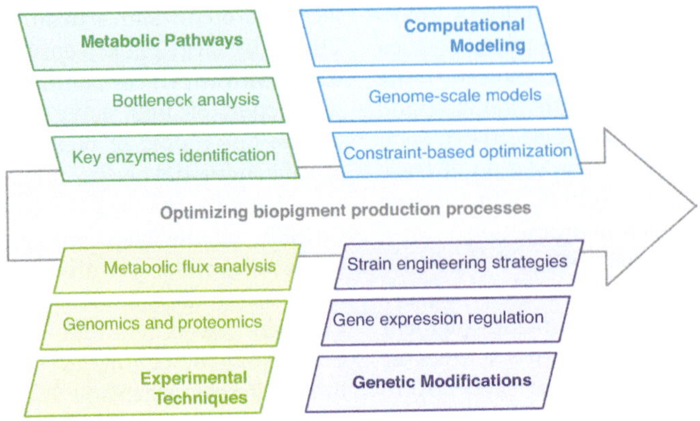

Fig. 20.4. Enhancing bioproduction through systems biology.

not only reduces energy consumption from nonrenewable sources but also aligns with the principles of green biotechnology.

Biogas produced from organic waste is another renewable energy source that can be used to power fermentation processes for biopigment production. Studies have shown that the integration of biogas from agricultural or industrial waste into fermentation systems can replace conventional energy sources, making the entire biopigment production process more sustainable (Martins *et al.*, 2019). Additionally, the use of hydropower and geothermal energy has been proposed for large-scale biopigment production, particularly in areas where these renewable resources are abundant.

The combination of renewable energy with biopigment production not only minimizes the environmental impact but also enhances the economic viability of biopigment industries, making them more competitive with synthetic pigment manufacturing. As renewable energy technology continues to improve, its integration into biopigment production is expected to become a key component of sustainable biotechnological processes.

20.6.3 Ethical considerations and regulatory frameworks

As biopigment production technologies evolve, ethical considerations and regulatory frameworks become increasingly important to ensure that the development and commercialization of biopigments are aligned with sustainability goals, public health, and safety standards. One key ethical issue is the environmental impact of biopigment production, especially regarding the cultivation of microorganisms or algae, which may involve the use of genetically modified organisms (GMOs). While GMOs can enhance biopigment production, their use raises concerns related to biodiversity and ecosystem balance (Kuddus, 2021). Regulatory bodies, such as the Food and Drug Administration (FDA) and the European Food Safety Authority (EFSA), play crucial roles in evaluating the safety of biopigments derived from GMOs before they are approved for commercial use in food and cosmetics (Martins *et al.*, 2019).

In addition to food safety, ethical concerns also surround labor practices in biopigment production, especially in large-scale, resource-intensive farming and fermentation facilities. Ensuring fair wages and working conditions is critical in promoting socially responsible biopigment production (Venil *et al.*, 2020).

Regulatory frameworks for biopigments are still developing, particularly for the use of renewable or bio-based pigments in various applications. Governments and international organizations must establish comprehensive regulations that govern the approval, labeling, and monitoring of biopigments, ensuring transparency and consumer protection. The European Union's Regulation (EC) No 1333/2008, which provides guidelines for food additives, colors, and preservatives, includes provisions for the use of natural pigments, setting clear criteria for their safety and purity (Kuddus, 2021). Furthermore, regulatory policies must address the life cycle impacts of biopigment production, promoting the use of renewable resources and ensuring that production processes are sustainable and free from harmful chemicals. Thus, a balanced approach that integrates ethical considerations with robust regulatory frameworks is essential for the sustainable development of the biopigment industry.

20.7 Key Advances

Recent advances in biopigment research have transformed the landscape of pigment production, with a growing emphasis on sustainability, efficiency, and eco-friendly alternatives to synthetic dyes. Key advances include the development of novel biopigments derived from microbial, algal, and plant-based sources, which are gaining traction due to their natural origin, safety, and versatility. Innovations in genetic engineering and synthetic biology, such as the use of CRISPR and gene-editing technologies, have significantly improved the yields and stability of biopigments, allowing for the optimization of microbial and algal strains for high-efficiency production (Zhou *et al.*, 2020). These advances have made biopigments a viable alternative in industries ranging from food and cosmetics to textiles and pharmaceuticals.

Furthermore, renewable energy integration in biopigment production processes, such as the use of solar, wind, and biogas energy, is helping reduce the carbon footprint of biopigment manufacturing while promoting sustainable practices (Venil *et al.*, 2020). Advances in extraction and purification techniques, including green methods like supercritical fluid extraction and ultrasound-assisted extraction, are enhancing the efficiency and scalability of biopigment production (Martins *et al.*, 2019). Moreover, innovative start-ups and industry–academic collaborations are driving forward the commercialization of biopigments, with companies leveraging biotechnology to introduce eco-friendly colorants into the market (Kuddus, 2021).

Despite these promising developments, challenges such as cost-effectiveness, scalability, and regulatory frameworks still need to be addressed to fully realize the potential of biopigments on a commercial scale. Nonetheless, the ongoing research in these areas is poised to revolutionize the biopigment industry, contributing to a more sustainable and greener future for a variety of sectors (Table 20.3).

20.8 Future of Biopigment Research

The future of biopigment research holds immense promise, driven by growing consumer demand for natural, sustainable, and eco-friendly alternatives to synthetic pigments. As industries continue to prioritize sustainability and reduce their environmental impact, biopigments are poised to play a key role in transforming sectors such as food, cosmetics, pharmaceuticals, and textiles. The future vision for biopigment research focuses on several key areas, including enhanced production efficiency, sustainability, and novel applications.

One of the most significant advancements will be in metabolic engineering and synthetic biology to create high-yielding, robust microbial and algal strains that can produce a wider variety of pigments in a cost-effective manner (Kuddus, 2021). The integration of CRISPR-based gene editing and genomic tools will continue to optimize biosynthetic pathways and introduce novel colorants that were previously unavailable. Furthermore, biopigment production systems will increasingly rely on renewable energy sources, such as solar, wind, and biogas, making the entire process more sustainable and reducing the environmental footprint of pigment manufacturing (Venil *et al.*, 2020).

Another crucial area of growth will be the development of bioprocess optimization, including green extraction and purification methods, which will further enhance the scalability and efficiency of pigment production (Martins *et al.*, 2019). The convergence of nanotechnology, particularly in nanoencapsulation and nanoparticle delivery systems, will provide new ways to improve the stability, solubility, and bioavailability of biopigments, expanding their potential

Table 20.3. Key advances in biopigment production technologies.

Technology	Biopigments produced	Benefits	Applications
Genetic engineering (CRISPR)	Carotenoids, phycobiliproteins	Increased yield, genetic diversity	Food industry, cosmetics, pharmaceuticals
Metabolic engineering	Carotenoids, anthocyanins, flavonoids	Enhanced biosynthesis, reduced costs	Pharmaceuticals, cosmetics
Renewable energy integration	All types of biopigments	Reduced carbon footprint, sustainability	Industrial production, algae farming
Nanotechnology (nanoencapsulation)	Carotenoids, chlorophyll	Increased stability, controlled release	Food, pharmaceuticals, cosmetics

applications in various industries (Zhou *et al.*, 2020).

Looking forward, interdisciplinary collaborations between academia, industry, and government will be vital for addressing regulatory challenges, ensuring the safe use of biopigments in consumer products, and accelerating the commercialization of biopigments (Kuddus, 2021). With increasing interest from start-ups and large corporations alike, the future of biopigments will involve the scaling up of production, greater market adoption, and continuous innovation in pigment applications, contributing to a more sustainable and eco-conscious global economy.

References

Ali, H., Shabir, G. and Qureshi, T.M. (2022) Advances in microbial pigment production and its potential applications. *Critical Reviews in Biotechnology* 42(3), 1234–1256. DOI: 10.1080/07388551.2022.2086404.

Chandanshive, V.V., Kadam, S.R. and Khandare, R.V. (2020) Synthetic dye decolorization by microbial communities: Challenges and sustainable solutions. *Environmental Technology Reviews* 9(1), 113–125. DOI: 10.1080/09593330.2020.1736905.

Khoo, H.E., Azlan, A., Tang, S.T. and Lim, S.M. (2017) Anthocyanidins and anthocyanins: Colored pigments as food, pharmaceutical ingredients, and the potential health benefits. *Food & Nutrition Research* 61(1), 1361779. DOI: 10.1080/16546628.2017.1361779.

Kuddus, M. (2021) Phycobiliproteins from marine cyanobacteria and red algae: Applications and challenges. *Frontiers in Marine Science* 8, 690. DOI: 10.3389/fmars.2021.690.

Kumar, S., Gupta, R. and Kumari, R. (2020) Microbial pigments: Production and their applications in various industries. *Biotechnology Reports* 28, e00574. DOI: 10.1016/j.btre.2020.e00574.

Martins, N., Roriz, C.L., Morales, P., Barros, L. and Ferreira, I.C. (2019) Coloring attributes of natural food pigments: A review of structure–activity relationships. *Trends in Food Science & Technology* 84, 1–10. DOI: 10.1016/j.tifs.2018.12.004.

Prabhu, S. and Umesh Hebbar, H. (2021) Biopigments: Properties and applications in food, pharmaceuticals, and cosmetics. *Comprehensive Reviews in Food Science and Food Safety* 20(2), 1342–1367. DOI: 10.1111/1541-4337.12739.

Ramesh, C., Vinithkumar, N.V. and Kirubagaran, R. (2020) Marine bioprospecting of microbial pigments and their applications. *Critical Reviews in Microbiology* 46(6), 765–787. DOI: 10.1080/1040841X.2020.1827590.

Sathasivam, R. and Ki, J.S. (2018) A review of the biological activities and biosynthesis of pigments in microalgae. *Molecules* 23(11), 2685. DOI: 10.3390/molecules23112685.

Venil, C.K., Zakaria, Z.A. and Ahmad, W.A. (2020) Biopigments as sustainable alternatives in textile coloration. *Frontiers in Sustainable Materials* 3, 12. DOI: 10.3389/fsust.2020.00012.

Zhou, X., Wang, C., Zhang, F., Li, H. and Liu, Y. (2020) CRISPR-based genome editing in microbial production of carotenoids. *Biotechnology Advances* 38, 10746. DOI: 10.1016/j.biotechadv.2020.107460.

Index

Academia and industry collaborations, 374–375
Acceptable daily intake, 239
Acidithiobacillus ferrooxidans, 84
Acidophiles, 84
Actinobacterium, 269
Actinomycetes, 77
Actinomycetes-derived
 melanins, 81
 pigments, 80–81
Actinorhodin, 78, 163
Acute and chronic toxicity, 293, 295
Acute toxicity testing, 291
ADMET (Absorption, Distribution, Metabolism, Excretion, Toxicity) properties, 146
Adsorption/desorption technique, 210
Age-related macular degeneration, 178
Agricultural applications, 248–249, 256–259
 biofertilizers and growth enhancers, 252–253
 carotenoids and anthocyanins in boosting plant health, 253–254
 pest and disease management, 254
 seed treatment and yield improvement, 254–255
Agriculture and plant protection, 309–310
Agrobacterium aurantiacum, 23
Agro-industrial waste, 343
Alcoholic drinks, 168–169
Algae, 137, 365–366
Algal, 339
 biopigments, 81–82, 138, 271
 and cyanobacterial pigments, 230–232
Alkaliphiles, 84
Alkaloid derivatives, 38–41
Allomelanins, 44
Allophycocyanin, 78
α-Carotene, 4–5
Alternaria, 138
Amino acids and proteins, 668
Ammonium, 68
Ammonium sulfate precipitation, 126, 128
And photoprotection, 251–252
Ankaflavin, 195
Anthocyanidin synthase (ANS), 6
Anthocyanin pigments, 252
Anthocyanins, 6–7, 158, 170, 173, 174, 177, 187–188, 246, 253–254, 311
Anthraquinones, 42–43, 79, 237, 284
Antiaging compounds and skincare products, 191, 193–194
Anticancer properties, 176–178
Antifouling and material protection, 237
Antiinflammatory, 176–178
Antimicrobial properties, 289
Antioxidant, 176–178
Antioxidant and antimicrobial properties, 234, 236–237
 acceptable daily intake, 239
 applications of, 234–235, 237
 bioactive in healthcare, food, and cosmetics, 229
 dual, 237–238
 ecological balance, 228
 industrial and clinical implications, 238
 industrial and economic relevance, 229
 mechanisms of, 233–234, 236
 nanotechnology, 241
 novel applications, 240–241
 potential, 236–237
 regulatory approval process, 239–240
 representative pigments, 234
 sustainability and eco-friendly, 229
 synergistic effects, 238
 technological advancements in production, 240
 toxicological assessment, 239

Antioxidant-rich functional foods, 173
Aqueous two-phase systems (ATPS), 330
Arabidopsis thaliana, 309
Archaea, 137
Armillaria cepistipes, 57
Arpink Red, 284
Art and aesthetic expression, 61
Artemia salina, 291
Artificial colorants, 186–187
Artificial intelligence (AI), 330–331, 357–360, 370–371
Aspergillus glaucus, 196
Aspergillus niger, 97, 253
Astaxanthin, 10, 12, 30, 82, 83, 96, 190, 193, 194–195, 233, 234, 237, 240, 255, 304
Azaphilones, 41, 79
Azo dyes, 76

Bacillus subtilis, 198, 253, 309
Bacillus thuringiensis, 198
Bacteria, 69, 137, 338
Bacteria-derived pigments, 138
Bacterial biopigments, 78–79
Bacterial pigments, 23, 136, 162–163, 230, 269
Bacterioruberin, 236–238
β-Cryptoxanthin, 193
Beer-Lambert law, 289
β-Carotene, 4–5, 30, 36, 157, 171, 173, 175, 178, 190, 284
Betalain, 174
Betalains, 5–6, 39, 159, 172, 173, 285, 311
Beta vulgaris, 159
Betaxanthins, 159
Beverage industry, 168–169
Big data, 330–331
Bilin pigments, 38
Bilirubin, 9
Binding efficiency, 209
BioCircularGreen (BCG) economy, 338
Biodegradability, 217, 347–348
Bioelectronics and smart materials, 87
Biofertilizers, 249
    and growth enhancers, 252–253
Biofilm formation, 236
Biogas, 376
Bioindicators, 248
Biological pigments, 22
Biomordanting, 212
Biopesticides, 248, 309–310
Biopigments, 1, 14–15
    of animal origin, 7, 11
        bilirubin, 9
        carotenoproteins, 9
        haemoglobin, 9–10
        melanin, 7
    carotenoids, 340
    challenges, 15–16, 350–354
    classification of, 156–160
    cultivation systems, 341
    definitions and importance, 336–337
    as eco-friendly alternatives, 247–248
    environmental
        benefits, 261–262
        impacts, 358–359
        sustainability, 344–348
    flavonoids, 340
    future directions and innovations, 356–358
    genetic engineering, 342
    in green industry, 354–356
    history, 337
    improvements, 16–17
    industrial practices, 359–360
    microalgae, production from, 348–350
    microbial sources, 338–339
    of microorganism origin, 10
        astaxanthin, 10, 12
        lycopene, 12
        monascin, 12–13
        violacein, 13–14
    pigment production in microorganisms, 339–340
    of plant origin, 1–2, 8
        anthocyanins, 6–7
        betalains, 5–6
        carotenoids, 4–5
        chlorophyll, 2–4
    production challenges, 255
    regulatory and safety concerns, 255, 260
    research and development, 260–261, 377–378
    sources of, 365–366
    stability and degradation, 260
    sustainability in, 337–338, 342–344, 358–359
    sustainable agricultural practices, 261
    wider adoption, 360–361
Bioprocess, 240, 329, 367–368
Bioreactor technologies, 356–357
Biorefinery model, 349
Bioremediation, 86–87, 248
Biosynthesis genes and enzymes, 97
Biosynthesis of microbial pigments, 49–50
    carotenoid biosynthesis, 51–52
    melanin, 55–57
    phenazines, 53–55
    polyketide biosynthesis, 52
    porphyrin biosynthesis, 52–53
    regulatory mechanisms, 57–58
Biosynthetic pathways, 70, 231, 233
    and genetic regulation, 96–97
Biotechnology/biotechnological, 220, 324, 353
    applications, 86
    innovations, 91
    strategy, 98–99
    techniques, 216
Bradyrhizobium, 23

Brine shrimp lethality bioassay (BSLA), 291
Butylated hydroxytoluene (BHT), 141

Caenorhabditis elegans, 40
Cakes, pastries, and puddings, 170–171
Camouflage and attraction, 61
Candies and sweets, 170
Canthaxanthin, 284
Carbon footprint, 345–346
Carbon source, 67–68
Carcinogenicity, 295
Carotenes, 51
Carotenoids, 4–5, 23, 36, 78, 79, 81–83, 97, 141, 148, 157–158, 173, 190, 206, 231, 234, 246, 250, 251, 253–254, 301, 311, 323, 340, 354–355
  and anthocyanins, 68
  biosynthesis, 51–52, 70
Carotenoids block nuclear factor-XB (NF-XB) activation, 176
Carotenoproteins, 9
C50 carotenoids, 234
Cell membrane integrity, 236
Chaetoviridins, 41
Chalcone synthase (CHS), 6
Cheese, 171–172
Chemical
  composition analysis, 290
  inertness, 64
  modifications for stability enhancement, 144
  stability, 63
  structures, 231, 271
Chitosan-oligosaccharide (COS), 351
Chlamydomonas reinhardtii, 342, 344, 351
Chlorella protothecoides, 337
Chlorophyllin, 3
Chlorophylls, 2–4, 38, 68, 81–82, 96, 156–157, 246, 251, 255
Chlorophyll synthesis genes, 70
Chlorophytes, 138
Chlorosis, 68–69
Chocolates and sugar-coated confectionery, 170
Chromatographic and nonchromatographic purification techniques, 372–373
Chromatographic purification techniques, 140–141
Chromatographic separation, 142
Chromatography, 121, 127, 141–142
  high-performance liquid chromatography, 122–124
  ion exchange chromatography, 124
  preparative thin-layer chromatography, 121–122
  size exclusion chromatography, 124–126
Chromobacterium, 40
Chromobacterium violaceum, 96, 100, 138, 147–148, 189, 206, 230, 250, 269
Chryseobacterium, 208
Circular bioeconomy, 343–344
Cladosporium, 138
Clean-label marketing, 275
Codex Alimentarius Commission, 326
Cognitive health and brain function, 173
Color additive amendments, 295
Colorant approval, 292
Color fastness, 219
Colorimetric assay, 289–290
Color of food, 283
Color shades, 64–65
Colour additive petition process, 295–296
Colour production, 61
Colwellia psychrerythraea, 84
Combinatorial biosynthesis, 220
Commercialized biopigments, 374
Confectionery and desserts, 169–171
Consumer perception and market readiness, 277
Consumer preferences and market trends, 169
Cosmetic products, 234
Cosmetics, 355–356
  industry, 240
  and personal care, 275, 307–309
  and personal care products, 292
Cost-effectiveness, 366–367
  of scaling, 348–349
Cost-efficiency and yield inconsistency, 275–276
Cost inefficiencies, 324
Cost structure and profitability indicators, 273–274
CRISPR-Cas9, 98, 210, 219, 322, 368, 375
CRISPR-Cas systems, 240
Crop protection, 249
Cruelty-free movement, 308
Cryptococcus neoformans, 198
Crystallization, 128
Crystallography, 141–142
Cultivation systems, 341
Cyanobacteria, 138, 231, 271
Cyanobium sp., 194
Cytotoxicity, 293

Dairy products, 171–172
Degradation, 141
3-deoxy-d-arabinoheptuloson ate-7-phosphate (DAHP) synthase, 54
Dihydrokaempferol (DHK), 6
Dihydroquercetin (DHQ), 6
Dimethylallyl diphosphate (DMAPP), 4, 9
Direct dyeing, 211–212
Dispersion, 64
Diversity of microbial pigments, 35
  alkaloid derivatives, 38–41
  flavins, 36–37
  isoprenoid derivatives, 35–36
  melanin, 43–44
  phenol-containing pigments, 44
  polyketides, 41–43
  tetrapyrrole derivatives, 38
DOPA-melanin/eumelanin, 56

Downstream processing (DSP), 373
Downstream processing and purification, 99–100
DPPH (2,2-diphenyl-1-picrylhydrazyl) assay, 176, 233
Dual antioxidant-antimicrobial properties, 237–238
Dual bioactivity, 238
Dunaliella salina, 160, 193, 246
DuPont™, 375
Dyes and pigments, 61–62

Eco-friendly
- alternatives, 247–248
- dyes, 248
- packaging, 308–309

Ecological balance, 228
Economic analysis, 273–274
Electrospray ionization (ESI)-MS, 142
Emodin, 43
Encapsulation and protective delivery systems, 144
Encapsulation methods, 160
Endogenous antioxidant systems, 233–234
Endotoxin testing, 290
Energy consumption, 345–346
Energy efficiency, 217–218
Engineering fungi, 342
Engineering microalgae, 342
Enhance pigment yield, 210–211
Environmental
- benefits, 261–262
- and economic considerations, 169
- stress resistance, 144
- sustainability, 308–309, 344–348
- and textile use, 149

Environmental applications, 248–250
- pollutant degradation, 250
- soil health and enrichment, 251
- UV protection and photoprotection, 251–252
- wastewater treatment, 250

Environment sensing genes, 70
Enzymatic extraction, 164–165
Enzymatic pretreatment, 212
Enzyme-assisted extraction, 141
Enzyme-assisted extraction (EAE), 118
Eumelanin, 43–44
European Food Safety Authority (EFSA), 239, 292, 296–297
Eurotium cristatum, 196
Extraction and purification techniques, 330
Extremophile pigments, 84–85
Extremophiles, 148
Eye health and cardiovascular function, 178–179

Fed-batch fermentation, 367
Federal Food, Drug and Cosmetic Act (FFDCA), 295
Fermentation, 68, 166–168, 211, 330, 341
Flavin adenine dinucleotide (FAD), 37, 208
Flavin mononucleotide (FMN), 37, 208
Flavins, 36–37, 78, 79, 83, 208
Flavobacterium, 23
Flavonoids, 301, 311, 340
Food and beverage industry, 274, 304–306
Food and cosmetics industry, 148–149
Food and Drug Administration (FDA), 239, 269, 295–296
Food industry, 81, 291–292
Food preservation and packaging, 237
Food Safety and Standards Authority of India (FSSAI), 297
Formulation stability, 144
Fourier transform infrared spectroscopy (FTIR), 142
Functional foods, 172–173
Functional group profiling, 144
Fungal biopigments, 79–80
Fungal pigments, 22–23, 138, 161–162, 230, 269, 271
Fungal systems, 342
Fungi, 69, 137, 148, 338–339
Fusarium sporotrichioides, 284
Fusarium spp., 43

Gas chromatography (GC), 142
GEMs, 370
Generally recognized as safe (GRAS), 76, 92, 239, 292, 295–296
Genes and enzymes, 97
Genetically modified organisms (GMOs), 197, 327, 328, 376
Genetic engineering, 71, 72, 98, 167, 219–220, 260–261, 342
Genetic factors, 70–71
Genetic regulation, 58
Genomics, 369
Genotoxicity, 239, 327
Geobacillus, 84
Geranylgeranyl diphosphate (GGPP), 12
Good Manufacturing Practices (GMP), 239, 273, 285–287
Gram-negative bacteria, 236
Grand View Research, 76
Green and sustainable extraction technologies, 141
Green Beauty movement, 355
Green bioprocessing, 141
Green chemistry, 354
Green extraction methods, 371–372
Green technology, 185
Growth phase, 69–70

Haematococcus pluvialis, 82, 160, 253, 302
Haemoglobin, 9–10
Halobacterium, 23, 137
Halobacterium salinarum, 84
Haloferax alexandrinus, 84
Halophiles, 84
Hazard analysis critical control point (HACCP), 273

Healthcare, food, and cosmetics, 229
Heat exposure, 174–175
Heavy metals, 70
Heme, 38
Heterologous expression, 219, 240
High chemical resistance, 63
High-performance liquid chromatography (HPLC), 99, 122–124, 141, 142, 289, 372
Horizontal gene transfer, 70–71
HPLC-DAD, 122
HPLC-FLD, 122
HPLC-MS/MS, 124
Humane Society International, 308
Hydroxymethyl bilane (HMB), 53

Ice cream and frozen desserts, 170–171
Ideal pigment, 62–64
Immune support and antiinflammatory effects, 173
Indigo, 206, 374
Industrial and clinical implications, 238
Industrial and economic relevance, 229
Inorganic carbon, 67–68
Inorganic pigments, 62
Integrated pest management (IPM), 254
International Council for Harmonisation (ICH) guidelines, 239
Ion exchange chromatography (IEC), 124
Ionic liquid-assisted extraction (ILAE), 119
Ionic liquids (ILs), 372
Isopentenyl diphosphate (IPP), 4, 9
Isoprenoid derivatives, 35–36
Isoprenoid pathway, 96

Janthinobacterium lividum, 206
Joint FAO/WHO Expert Committee on Food Additives (JECFA), 239
Juices and smoothies, 168

Laccase-derived pigments, 79–80
Large-scale production, 102–103, 366–367
LC_{50} value, 291
L-cysteine (L-Cys), 216
L-3,4-dihydroxyphenylalanine (L-DOPA), 5
Lichens, 230
Life cycle assessment (LCA), 345
Light, 69, 99
    absorption, 61
    exposure, 174–176
    stability, 290
Lightfastness, 63
Light-responsive genes, 70
Linseed oil, 214
Lipid peroxidation, 233
Liquid-liquid partitioning (LLP), 128
Listeria monocytogenes, 238
Livestock feed additives, 249
Lot numbering, 287
Low-cost and waste substrates, 273
Lush Cosmetics, 308
Lutein, 81, 157
Lycopene, 12, 157–158, 190, 193, 284

Machine learning (ML), 331, 357–358, 370–371
Macromolecular synthesis, 236
Maize steep liquor, 209
Market trends and commercial success, 274
Mass spectrometry, 142
Master seed culture, 285
Material Safety Data Sheets (MSDS), 291
Matrix metalloproteinases (MMPs), 177
Maximum covering power, 63
Meat and processed foods, 172
Melanin-coated microbial cells, 198
Melanins, 7, 43–44, 55–57, 78, 79, 81, 83, 148, 149, 198, 208, 210, 246, 250, 251, 302, 322–324
Membrane affinity, 238
Metabolic engineering, 219, 329, 367–368
Metabolic engineering, 260–261
Metal chelation, 233
Metal ion interaction, 238
4-methoxy-2,2'-bipyrrole-5-carbo xaldehyde (MBC), 40
Methylglyoxal (MG), 12
Microalgae, 230–231, 246, 271, 339, 346, 354, 359
Microalgae-based biorefineries, 349
Microalgal technologies, 360–361
Microbial contamination and testing, 285, 287, 290, 295
Microbial-derived carotenoids, 191
Microbial dyes, 100
Microbial hues, 252
Microbial pigments, 21–22, 90–91, 109, 136, 197–198, 228, 247, 268–269, 320–322
    advantages of, 91–92
    in agriculture and plant protection, 309–310
    ancient and medieval uses, 23–24
    applications of, 322–324
    bacterial pigments, 23
    biosynthetic pathways and genetic regulation, 96–97
    challenges and future prospects, 102–103, 196–197, 324–326
    challenges and limitations, 275–277, 313
    chemical structure, 138
    classification of, 92–95, 187–188, 231, 233, 269–271, 304
    commercial applications, 26, 28, 274–275
    cosmeceutical applications, 189–196
    in cosmetics and personal care, 307–309
    economic analysis, 273–274
    extraction and purification techniques, 139–140
    food and beverage industry, 304–306

Microbial pigments (*Continued*)
- fungal pigments, 22–23
- future of, 277, 312–313, 331–332
- genetic engineering, 98
- Good Manufacturing Practices, 285–287
- guidelines, 102
- and history, 185–186
- industrial applications, 26, 100
- in industry, 302–304
- legal framework bodies, 295–297
- in medicine, 28–29
- as natural colorants, 30–31
- novel microbial strains and discovery, 331
- pathogenicity, 197–199
- in the pharmaceutical and biomedical sectors, 310–312
- pharmaceutical applications, 187–189
- production, 29–30, 69, 98, 271–273
- quality assay for, 287–291
- quality control, 273
- regulatory and safety considerations, 326–328
- regulatory aspects, 291–293
- safety aspects of, 293–295
- safety, toxicity, and regulatory status, 100–102
- scientific discovery and classification, 24–27
- sources and types of, 230–231, 269–271
- strain improvement, 31
- in technology, 328–331
- in textile industry, 306–307
- textile industry applications, 28
- yeast pigments, 23

Micrococcus, 30
Microcystis, 252
Microorganisms, biopigments from, 30, 77–78, 96, 137
- actinomycetes, 80–81
- advantages, 85–86
- algae, 81–82
- bacteria, 78–79
- extremophiles, 84–85
- fungi, 79–80
- future, 86–87
- yeast, 83–84

Microwave-assisted extraction (MAE), 116–118
Minimum antifungal concentration (MFC), 289
Minimum inhibitory concentration (MIC), 289
Mixed-origin pigments, 231
Molecular modeling and docking, 144, 146
Monascidin A, 291
Monascin, 12–13
Monascus, 30, 77, 96, 167, 230, 253, 275, 291, 324
Monascus fermentation, 273
Monascus pigments (MPs), 80, 161–162, 236, 239, 240, 271, 284
Monascus purpureus, 31, 161, 329
Monascus ruber, 189
Mordanting, 212
Mordant type and concentration, 214
Mutagenicity, 239, 295

Nanoencapsulation, 368–369
Nanoparticles, 369
Nanotechnology, 216, 241, 330, 368–369
Naphthoquinones, 43
Natural colorants, 86, 156, 305–306, 354–355
Natural cosmetics, 308
Natural pigments, 21, 91–92, 186–187, 283
Natural quinones, 42
Nernst Distribution Law, 110, 115
Next-generation sequencing (NGS) technologies, 98
Nigrospora aurantiaca, 337
Nitrate, 68
Nitrogen deficiencies, 68–69
Nitrogen limitation, 69
Nitrogen source, 68–69
NMR spectroscopy, 141–143
Nocardia, 81
Nonchromatographic purification, 126
- ammonium sulfate precipitation, 126, 128
- crystallization, 128
- liquid-liquid partitioning, 128

Nonribosomal peptide synthetase (NRPS) genes, 70
Nontoxic and safe, 63
No-observed-adverse-effect level (NOAEL), 239
Notable biopigments, 366
Novel Foods Regulation, 296–297
Novel microbial strains, 331
Nutraceutical pigments, 275

Omics technologies, 146–147, 369–370
Omics tools, 98
Open ponds, 341
Optimal particle size, 63
Organic carbon, 67
Organic pigments, 62
Oxidative stress, 70
Oxygen availability, 69

Packaging and labeling, 287
Palmaria palmata, 337
Penicillium, 77, 162
Penicillium bilaii, 196
Pest and disease management, 254
pH, 69, 143–146, 174–176, 213, 290
Phaffia rhodozyma, 96, 253
Pharmaceuticals
- applications, 148
- and biomedical sectors, 310–312
- -grade pigments, 293
- and medical products, 292–293
- and nutraceuticals, 275

and therapeutic applications, 237
Phenazine-1-carboxylic acid (PCA), 97
Phenazine derivatives, 236, 240
Phenazines, 40, 53–55, 96–97, 141, 239
Phenol-containing pigments, 44
Phenolic pigments, 44, 234
Pheomelanin, 44
Photobioreactors (PBRs), 341, 357
Photoprotection and antioxidant capacity, 189–191
Photoprotective efficacy, 192–193
Photosynthesis, 67–68
Phycobilins, 252–253
Phycobiliproteins, 82, 138, 159–160, 195, 230–231, 234, 250
Phycocyanin, 78, 159, 160, 177, 195, 233, 238, 239, 253, 301
Physcion, 351
Pigment extraction techniques, 110–114, 209–210
enzyme-assisted extraction, 118
ionic liquid-assisted extraction, 119
microwave-assisted extraction, 116–118
pulsed electric field, 120–121
solvent liquid extraction, 110, 115
supercritical fluid extraction, 118–119
ultrasonic extraction, 115–116
Pigment-producing bacteria, 206
Pigments, 1, 60
from algae, 160–161
classification of, 62
and color shades, 64–65
concentration, 289
and dyes, 61–62
ecological and evolutionary implications, 71
environmental conditions, 69
functions of, 61
genetic factors, 70–71
growth phase, 69–70
ideal, 62–64
in industrial areas, 64, 66
mechanism, 65–67
modification and genetic engineering, 86
nature's color architects, 60–61
nutrient availability
carbon source, 67–68
nitrogen source, 68–69
organic *vs.* inorganic, 62
oxygen availability, 69
properties of, 64
stress condition, 70
substances, 195–196
uses and prospects, 71–72
Plant-based pigments, 91
Plant-derived pigments, 340
Plant growth promotion, 309
Plant growth regulators (PGRs), 309
Pollutant degradation, 250
Polycyclic aromatic pigments, 271
Polydopamine (PDA), 220
Polyketide biosynthesis, 52
Polyketide-derived pigments, 231
Polyketides, 41–43, 79
Polyketide synthases (PKSs), 52, 70
Polyunsaturated fatty acids, 233
Porphobilinogen (PBG), 53
Porphyrin biosynthesis, 52–53
Postharvest preservation, 249
Potassium aluminum sulfate (alum), 212
Predictive modeling, 370
Predictive toxicology, 146
Preparative thin-layer chromatography (PTLC), 121–122
Prodigiosin, 41, 78, 147, 188, 208, 236–238, 250, 304, 324
Product safety testing, 292
Proteomics, 369
Pseudomonas, 30, 96, 230
Pseudomonas aeruginosa, 96–98, 137, 198, 209
Pseudomonas putida, 309
Pseudomonas syringae, 253
Psychrophiles, 84
Pulcherrimin, 83
Pulsed electric field (PEF), 120–121
Purity analysis, 288–289
Pyocyanin, 78
Pyomelanin, 219, 250

Quality assay, 287–291
Quality control, 273
Quinone pigments, 234
Quinones, 41–43, 231
Quorum sensing, 58, 70

Radical scavenging activity, 233
Ralstonia solanacearum, 253
Reactive oxygen species (ROS), 40, 176
Recombinant DNA technology, 329
Redox cycling, 238
Red yeast rice (RYR), 189
Regulatory and policy landscape, 349–350
Regulatory and safety concerns, 255, 260
Regulatory approval process, 239–240
Regulatory challenges, 240
Regulatory genes, 70
Regulatory oversight, 287
Renewable energy, 375–377
Respiratory sensitization, 295
Response surface methodology (RSM), 370
Reusability of carbon dioxide, 212–213
Rhizoctonia solani, 253
Rhodobacter sphaeroides, 253
Rhodococcus erythropolis, 84
Rhodopsins, 78, 178
Rhodotorula, 26, 30, 78, 96, 97
Rhodotorula glutinis, 23, 253
Riboflavin, 37, 285
ROS-mediated microbial killing, 236

Saccharomyces cerevisiae, 98
Safety, scalability, and regulatory gaps, 241
Safety, toxicity, and regulatory aspects, 100–102, 238–240
Salt concentration, 70
Sarcina, 30
Scanning and transmission electron microscopy (SEM/TEM), 137, 143
Scytalidium cuboideum, 214
Secondary metabolites, 70, 310
Seed coating, 249
Seed treatment and yield improvement, 254–255
Serratia marcescens, 41, 70, 96, 138, 147, 162, 209, 230, 250, 253, 269, 321, 323
Shared reactive interfaces, 238
Shikimate-derived pigments, 231
Size exclusion chromatography (SEC), 124–126
Skin depigmentation, 194
Skin sensitization, 295
Social media, 308–309
Soil health and enrichment, 251
Soil health indicators, 249
Solar energy, 218
Solid-state fermentation (SSF), 272, 330, 367
Solvent-based extraction methods, 138, 140
Solvent extraction, 163–164
Solvent liquid extraction (SLE), 110, 115
Spectroscopy, 141–142
Spirulina, 167, 302
Spirulina-derived phycocyanin, 274
Spirulina platensis, 84, 160
Stability and degradation, 260, 276
Start-ups and industries, 374
Strain identification and validation, 285
Streptomyces, 70, 81, 138, 148, 162–163, 324
Streptomyces bellus, 195
Streptomyces coelicolor, 81, 163
Streptomyces davawensis, 37
Streptomyces globisporus, 81
Stress condition, 70, 72
Stress responses, 58
Structure-activity relationship (SAR), 146
Submerged fermentation (SmF), 272, 330, 367
Substrate selection, 344
Supercritical CO_2, 166
Supercritical fluid extraction (SFE), 118–119, 141, 166, 210, 330, 371–372
Surface modification and crosslinking, 216
Sustainability and eco-friendly, 229
Sustainability and future potential, 312
Sustainable agricultural practices, 261
Sustainable biomordants, 216
Sustainable production, 86
Synergistic Bioactivities and roles, 237–238
Synergistic effects, 238
Synthetic biology, 72, 86, 91–92, 99, 103, 197, 219–220, 357, 367–368, 375
Synthetic dyes, 76, 306–307, 345
    market competition, 353–354
Synthetic pigments, 21, 271, 365
Systems biology, 370

Talaromyces, 275
Tambjamines, 40
Temperature, 69, 174–176
Terpenoid-derived pigments, 231
1,3,6,8-tetrahydroxynaphthalene (THN), 56
Tetrapyrroles, 38, 231
Textile dyeing, microbial pigments for, 205–208
    binding affinity, 209
    biodegradability and impact, 217
    challenges, 218–219
    chemical stability, 208–209
    color intensity, 208
    concentration of dye, 214
    cultivation of pigment, 209
    dye uptake and filtration, 213–214
    energy efficiency, 217–218
    extraction methods and purification process, 209–210
    fabrics, methods for, 211–213
    fastness, techniques for, 216
    innovations and future, 219–220
    light and wash fastness, 209
    techniques for yield and stability, 210–211
    wash, light, and rub fastness, 214, 216
    water pollution and waste production, 216–217
Textile industry, 275, 293, 306–307, 353, 356, 374
Thermal stability, 143, 145–146, 290
Thermophiles, 84
Thermus thermophilus, 84
Thin-layer chromatography (TLC), 99, 141, 288–289, 372
Thymus lotocephalus, 355
Toxicity and biocompatibility, 293–295
Toxicity testing, 290–291
Trace metals, 69
Transcription factor, 70
Transcriptomics, 369
Two-component regulatory system, 70
Tyrosinase, 97, 194

Ultrasound-assisted extraction (UAE), 115–116, 330, 372
Ultraviolet (UV) radiation exposure, 175–176, 190, 251–252
Undecylprodigiosin, 163
UV-Vis spectroscopy, 142, 289

Violacein, 13–14, 40–41, 78, 147–149, 189, 206, 236–239, 255, 285, 304, 309, 323
Vitamin B2 deficiency, 37

Wastewater recycling, 213
Wastewater treatment, 248, 250
Waterless dyeing, 212–213
Water pollution and waste production, 216–217
Water-soluble pigments, 176
Wheat (*Triticum aestivum* L.), 351
Whitening effect, 194–195
Worker safety, 293
World Health Organization (WHO), 90

Xanthomonas species, 253
Xanthophyllomyces dendrorhous, 23, 28
Xanthophylls, 51, 82, 255
X-ray crystallography, 142

Yeast pigments, 23, 83–84
Yield optimization and process efficiency, 350–352
Yogurt, 171

Zero waste approach, 343–344

Printed and bound by CPI Group (UK) Ltd, Croydon, CR0 4YY

16/09/2026

14963127-0001